for my friend Fred Bloor

Bob Simon

INDUSTRIAL AND COMMERCIAL POWER SYSTEMS HANDBOOK

Other Electrical Power Engineering Books of Interest

INDUSTRIAL AND COMMERCIAL POWER SYSTEMS HANDBOOK

F. S. Prabhakara, Ph.D., P.E., S.M. IEEE
Senior Engineer, Power Technologies Inc.
Schenectady, New York

Robert L. Smith, Jr., Fellow IEEE
Consultant (retired), General Electric Co.
President, Volts & Vars Inc., Northville, New York

Ray P. Stratford, P.E., Fellow IEEE
Consultant (retired), General Electric Co.
Senior Consultant (retired), Power Technologies Inc.
Schenectady, New York

McGraw-Hill
New York   San Francisco   Washington, D.C.   Auckland   Bogotá
Caracas   Lisbon   London   Madrid   Mexico City   Milan
Montreal   New Delhi   San Juan   Singapore
Sydney   Tokyo   Toronto

Library of Congress Cataloging-in-Publication Data

Prabhakara, F. S.
    Industrial and commercial power systems handbook / F. S. Prabhakara,
Robert L. Smith, Jr., Ray P. Stratford.
        p.   cm.
    Includes index.
    ISBN 0-07-050624-8 (hardcover)
    1. Factories—Power supply.   2. Commercial buildings—Power supply.
3. Electric power systems. I. Smith, Robert L., Jr. 1925- II. Stratford, Ray P.
III. Title.
TK4035.F3P73   1995
621.319—dc20   95-15990
                                                       CIP

McGraw-Hill

A Division of The McGraw·Hill Companies

*The sponsoring editor for this book was Harold Crawford, the editing
supervisor was Bernard Onken, and the production supervisor was
Suzanne Rapcavage. It was set in Times Roman by Renee Lipton of
McGraw-Hill's Professional Book Group composition unit. Printed
and bound by R. R. Donnelley & Sons Company.*

McGraw-Hill books are available at special quantity discounts to use as
premiums and sales promotions, or for use in corporate training programs. For more information, please write to the Director of Special
Sales, McGraw-Hill, 11 West 19th Street, New York, NY 10011. Or
contact your local bookstore.

This book is printed on recycled, acid-free paper containing
10% postconsumer waste.

CONTENTS

Chapter 3. Electrical System Studies   3.1

Chapter 4. Electrical Power System Detailed Design   4.1

Chapter 5. Voltage Calculations and Control   5.1

Chapter 8. Protective Device Application and Overcurrent Coordination

Chapter 9. System and Equipment Grounding   9.1

Chapter 10. Overvoltages and Surge Voltage Protection   10.1

Chapter 11. Power Quality—Voltage Fluctuations and Harmonics   11.1

Chapter 12. Electric Lighting   **12.1**

Chapter 13. Testing System Components   13.1

Chapter 14. Safety   14.1

Chapter 15. Electrical Maintenance   15.1

Chapter 16. Engineering Economics   16.1

Chapter 17. Energy Conservation and Management   17.1

Chapter 18. Electrical Power System Standards, Publications, and Nameplates

18.1

ACKNOWLEDGMENTS

The authors thank the General Electric Company for permission to reprint the following material.

Figures 7.1, 7.2, 7.3, 7.4, 7.5, and 7A.1 are from the *Industrial Power Systems Data Book.*

Table 7.5 is from GET 7002.

Table 7.6 is from GET 6211A.

Table 7.11 is from GEA 1004E.

The computer output in Section 7.33 is from the IPSEO "DAT" and IPSEO "SC" programs.

Figure 8.1 is from GEZ 7723.

Figure 8.2 is from GES 8134.

Figure 8.3 is from GES 8135.

Figure 8.4 is from GES 6227A.

Figure 8.5 is from GES 6104C.

Figure 8.9 is from GES 9500.

Figure 8.31 is from GES 7015B.

Figure 8.33 is from GES 6137.

Figure 8.34 is from GES 6138.

Figure 8.35 is from GES 7202.

Figure 8.36 is from GES 7200.

Figure 8.39 is from GES 7014A.

Figures 12.1, 12.2, and 12.3 are photos from NELA Park.

Figure 8.13 is reprinted with permission of the International Electrotechnical Commission (IEC) from standard IEC 76-1 (1993-03), copyright IEC.

Figures 7.6, 7A.2, and 7A.3, and Table 7.10 reprinted from IEEE Std C37.010-1979, *IEEE Application Guide for AC High-Voltage Circuit Breakers Rated on a Symmetrical Current Basis (Reaff 1988).* Table 7.2 reprinted from IEEE Std C37.13-1990, *IEEE Standard for Low-Voltage AC Power Circuit Breakers Used in Enclosures.* Tables 7A.1, 7A.2, 7A.3, 7A.4, 7A.8, 7A.9, and 7A.10 reprinted from IEEE Std 241-1990, *IEEE Recommended Practice for Electric Power Systems in Commercial Buildings.* Figure 8.7 reprinted from IEEE Std C57.109, *IEEE Guide for Liquid-Immersed Transformer Through-Fault-Current Duration.* Table 8.4 reprinted from IEEE Std C37.2-1991, *IEEE Standard Electrical Power System Device Function Numbers.*

CHAPTER 1
INTRODUCTION AND HISTORY

1.0 INTRODUCTION AND HISTORY

Electrical industry development worldwide, from the invention of the electric light to the advent of the computer, was very rapid. Events in the United States and Europe paralleled each other. The carbon filament lamp was developed simultaneously in Britain and the United States. France started the first hydroelectric plant on the Valserine River in 1881, with a capacity of 900 horsepower. In 1882, Thomas Edison's first central station went into service in lower Manhattan, and London provided its first supply to the public in 1882. These were all direct-current systems. Because of the limited voltage generated with dc, the generating station had to be close to the load. Ultimately, transmission of electric power over long distances would not be possible until the problem of paralleling ac machines was solved. In 1888, that challenge was met and ac systems came into use.

Many of the early inventors associated with the electrical industry at the end of the 19th century must have felt "necessity's sharp pinch," as did Elihu Thompson when he pointed out the safety advantages of system grounding. There were disappointments also, as when Thomas Edison lost the battle with George Westinghouse over ac versus dc transmission systems. This enabled Stanley's transformers to enhance the growing movement to ac, setting up the bulk power distribution for potential customers in the United States.

In the industrial area, the idea of a generating plant within the industrial facility was common in the late 19th century. Paper mills built on streams utilized water wheels to power machinery through shafts. They soon changed to generate electric power to supply electric motors to run machines. As their loads grew, they connected to utility systems, an early use of cogeneration. Gradually, the small power systems built to serve a community or local area became interconnected, enjoying both the economy of paralleling several units and increasing reliability.

When direct current systems were first developed, conductors carried the power from the generators directly to the load, and the conductor voltage drop limited the transmission distance. When ac systems came into use, the practice of going from the source to the load by the same voltage was still used. Power was delivered to a transformer placed at one location in the plant carrying power to diverse loads with large conductors. In the 1930s and 1940s, metal-enclosed switchgear equipment was developed. This led to the development of the *load center substation* concept, connecting switchgear directly to a transformer, permitting the substation to be placed near the load. This concept eliminated the voltage drop problem associated with long feeders. At the same time, the development of power distribution at high voltages made the system economical. Metal-enclosed low- and medium-voltage switchgear made the distribution of power within a facility economical by use of conductors sized not by voltage drop, but by the current-carrying capacity of the conductor based on temperature rise.

The nature of the load determined the generator frequency. Initially, 25 Hz was a common frequency. There are still systems operating in the world at 25 and 40 Hz, and other nonstandard frequencies. The western hemisphere operates primarily at 60 Hz, while the eastern hemisphere operates primarily at 50 Hz. There are exceptions in both areas.

By the 1920s, the central station concept of generating and transmitting power was well established. Not until the 1940s and the increasing need for power transmission over long distances, did the interconnection of large systems develop. The demand for power in the United States during World War II accelerated this practice.

Today, synchronous and asynchronous (dc) ties interconnect the North American Grid, providing increased reliability. Excess generation from one section is available to other sections of the grid. One prominent systems engineer has stated that it is mathematically impossible to tie the grid together, but the work and study of electrical industry engineers has made it possible.

1.1 INDUSTRIAL AND COMMERCIAL POWER SYSTEMS DOCUMENTATION

Professional Societies helped develop better industrial power system engineering practices. The American Institute of Electrical Engineers (AIEE) published *Electric Power Distribution for Industrial Plants* in 1945. It became known as the "Red Book" because of the color of its cover. There have been several editions since that time; the 1993 edition is the latest.

Through the meetings and technical papers of the AIEE and after 1964, The Institute of Electrical and Electronics Engineers (IEEE), development in the industrial and commercial power distribution systems was shared within the engineering community, leading to today's practices.

Another major publication that codified the practices was D. L. Beeman's *Industrial Power Systems Handbook,* the bible for industrial power system engineers since the mid-1950s, This book is a companion to Beeman's handbook and includes practices developed since that time.

1.2 DEVELOPMENT OF MODERN ELECTRIC POWER SYSTEMS

During the 1930s, personnel safety took priority, and metal-enclosed switching equipment became more prevalent. After World War II, increasing interest in adequate interrupting capabilities for power circuit breakers, sparked development of the load center principle. This allowed for more efficient distribution of power in industrial systems, eliminating long, low-voltage feeders which caused so much difficulty with voltage drops. Increased interest in power factor also led to extensive industrial application of capacitors during the same period.

The 1960s saw increasing industrial demand for selective protection systems using coordinated protection to interrupt faults with the nearest protective device. At the same time, growing interest in reliability initiated system analysis to determine the most reliable installation.

The 1970s brought widespread use of thyristors, which drew power system engineers to the problem of harmonic distortion caused by nonlinear loads and system resonant conditions. Such distortion caused many interesting system phenomena, such as random fuse blowing, relay misoperation, computer crashes, motor and transformer overheating, microprocessor misoperation, and system neutral overloading. Harmonics are the problem of the 1990s.

Cogeneration again came into prominence in the 1980s as industrial systems began to feel necessity's sharp pinch when electric utilities curtailed their installed generation because of environmental factors. Industrial corporations that had relied entirely on utility electric supply were now forced to consider their own generation. (Industrial plants using steam for process heat have used cogeneration for decades.) A stronger interest in power system analysis emerged, and many industrials began to appreciate the importance of early conceptual design studies.

1.3 AC VS. DC ELECTRIC POWER

The advantages of dc (direct current) did not escape many industries. The electrochemical and electrometallurgical industries use large amounts of dc power in manufacturing such products as chlorine and aluminum, which require large rectifiers to convert ac to dc. Dc motors have an advantage over ac motors where a need exists for high torque at low speeds, as in metal rolling mills, cranes, cement kilns, and locomotives. The mercury-arc rectifier saw its glory in the early part of the 20th century and flourished until the advent of the semiconductor rectifier in the 1950s and 1960s.

Silicon technology has brought about wide use of adjustable-speed drives, switched-mode power supplies for computers and other electronic devices, high-efficiency electronic ballasts for fluorescent lamps, Touch-Tone telephones, long distance data transmission, etc., that have all changed the types of loads applied to the power systems. Because of the efficiency and the versatility of the devices based on silicon technology, it has been estimated that by 2020 half of the power generated will pass through some kind of silicon device. All are nonlinear loads. Future power systems need to be designed to accommodate these loads. Chapter 11 discusses the problems of harmonic currents that these loads require, their effect on the power system, and how to minimize their problems. Dc has again come into its own in high-voltage transmission, using thyristor valves for high-voltage dc transmission lines.

By the year 2000, new equipment will be available for power system protection and control. Software logic has replaced relay logic in most processes today. Digital technology will replace analog technology. The trend continues. The basic engineering design criteria described in this book will give a foundation on which to apply this new technology.

1.4 VOCABULARY AND MYTHS

The vocabulary of the distribution engineer promotes a prevalence of false myths. Consider the network protector. This device mainly protects underground urban networks from the loss of voltage. The network protector locks closed when network faults occur, assuming the faults will burn free. In the case of an open copper bus, mounted on porcelain insulators, applied on 208Y/120-V systems, this may be true. But, today's 480-V networks use insulated cables and most faults freely burn instead of burning free.

It takes only 140 V to sustain an arc. Such faults sometimes cause explosions which send manhole covers flying. A network protector does not protect the network from the ravages of short circuits or ground faults. It is used only to prevent backfeed from the network into a fault on the source side of the network protector.

Another myth, dating from prior to the 1940s and probably universally dispelled, is that the maximum short circuit encountered on a low-voltage system is 10,000 A. This may be true for homes served from small distribution transformers but certainly is not true for most industrial systems. Even small industrial systems usually find more available short-circuit current on their premises.

Vocabulary can also be misleading. A three-phase, four-wire system consists of five wires: four current-carrying conductors (three phase wires and one neutral) plus one equipment ground conductor. Likewise, a three-phase, three-wire system has four conductors.

Partial differential relaying is neither a partial nor a differential arrangement. It is a complete sum-of-current design used for backup overcurrent relay protection. When viewing the three-line diagram of such an arrangement, it appears that the designer started to outline a bus differential arrangement but neglected to include feeder current transformers.

Many designers misuse the National Electric Code (NEC) as a design guide, despite the disclaimer in Article 90-1(c) of the Introduction. This code is a guide to the minimum requirements for "the practical safeguarding of persons and property." Article 90-1 (c) states, "This code is not intended as a design specification nor an instruction manual for untrained persons."

Copy engineering, sometimes used by inexperienced engineers, can be dangerous in the rapidly changing power distribution field. What applied on the last job may no longer apply to the current situation.

Reliability calculations can help design a good electrical system, but they cannot assure that this same system will attain the calculated reliability unless the proper protective equipment that is selected is skillfully set, regularly maintained, and frequently examined in light of the new loads or changes imposed.

1.5 PROJECT PROCEDURE

Implementation of any project proceeds in several stages. These stages are:

- Feasibility study
- Conceptual design
- Analytical system analysis
- Detailed design
- Specification of equipment
- Procurement
- Construction and installation
- Commissioning and testing
- Project closing with documentation

These steps form a logical sequence for completion of any project. The complexity, the need to focus on ever-increasing details, and the amount of manpower and time spent increase from one step to the next. In small projects, some of these steps may be implied and may not be obvious.

The feasibility study may be a simple or elaborate exercise. The technical analysis of this stage is cursory. The basic question is whether it makes economic (and risk) sense to study this project further from the idea stage. Sometimes, you hear the phrase *prefeasibility study*. This essentially means that the feasibility study needs to be performed in two steps, with the prefeasibility study being a very simple first step study followed by a more detailed study. The conceptual design is described in Chap. 2. The scope and content of the analytical analysis is described in Chap. 3. Many of the studies and topics addressed in this book provide technical input to specification, bid evaluation, and installation and testing of equipment.

1.6   FACTS ABOUT THIS BOOK

The experience of the authors has included all aspects of industrial and commercial power system design. In many of the examples, the sizes of the power systems may be larger or smaller than one in which the engineer is involved. The numbers of kW, kVA, kVAr, etc., can be scaled to whatever size is needed. The principles involved are the same no matter what the system size. Likewise, most of the discussion is for 60-Hz systems. The same principles apply in 50-Hz systems. The impedances change as a function of frequency. Harmonics are different for the two systems, but, again, it is just a matter of a change in the frequency.

The authors hope this new bible helps define and give guidance for application engineers well into the next century. Good luck on the journey through the maze of industrial and commercial power system distribution.

CHAPTER 2

ELECTRICAL POWER SYSTEM CONCEPTUAL DESIGN AND SYSTEM PLANNING

2.0 INTRODUCTION TO CONCEPTUAL DESIGN

System planning can be divided into two parts, namely:

- Conceptual design
- Detailed design

It is during the system planning stage that studies will be done to prove that the design is the optimum and that it will be practical to build and operate. The details and approach used for these two design steps will be presented in this chapter and in Chap. 4.

2.1 CONCEPTUAL DESIGN

Electrical power system conceptual design is the harmonious, balanced integration of various discrete but competing system design aspects to meet the operational objectives economically. This stage of any project is the most crucial in the chain of engineering accomplishments leading up to the commercial operation of an electrical power system. It determines the success or failure of the system. It is in search of the optimum economic power system that will provide the highest quality power.

Conceptual design requires a well-founded familiarity with discrete design aspects and a keen anticipation of potential conflicts between these aspects. It may require analytical studies to verify the feasibility of the design. It is an engineering-proficiency-dependent task. It requires an engineer who has experience in many types of systems. The engineer should be trained in a research and development environment that has developed unique systems to solve specific problems for unusual loads rather than one that has been trained in an engineering organization that has projects oriented to production.

A conceptual design engineer is familiar with system short circuit, protective device coordination, load flow, and transient stability computer studies. He or she must be familiar with the needs of different types of loads and processes, such as paper making, petroleum and petrochemical, electrochemical, and electrometallurgical. The engineer need not be proficient in each, but must be practiced with whatever type of system he or she is called upon to design. This means knowing the requirements of special loads, such as static power converters, adjustable-speed drives, resistance welding, large motor drives, and arc furnaces. It also means understanding control system theory, which is essential to recognizing the effect of paralleled voltage control systems and their effect on VAR control. Engineers need to understand transient phenomena.

The conceptual design engineer must have knowledge and familiarity with discrete design aspects. He should have a keen anticipation of potential conflicts among different design aspects. Installation of shunt capacitors for power factor improvement can introduce conflicts. For example, switching capacitors can cause switching transient overvoltages; within a harmonic-rich source environment, the capacitors can produce harmonic resonances with the system inductance, and these capacitors can act as harmonic sinks. If such conflicts are unavoidable, then remedy or mitigation and cost of such action should be properly accounted for as early as possible. Resolution of conflicts and/or proving that the proposed design will provide satisfactory performance through analytical studies is worth the time and expense of these studies.

Management must be aware of the importance of the conceptual design study in any project. Management must provide adequate funding. They need to know that there will be a long-term payoff. Payoff considerations need to be planned at the beginning of any project. A good electrical power system will not only take care of the needs of the present project but will be designed with built-in flexibility so that a minimum expenditure will be made on the next expansion. This does not mean that the present project will overexpend, but the system will be arranged so that the additions to it will cause a minimum change to the existing system.

The Construction Industry Institute (CII) has researched costs of projects and come up with the cost influence curve in Fig. 2.1. This curve shows the effect of a conceptual design on the economics of a project. The time and the money spent early in a project on the conceptual design pays off big in any project.

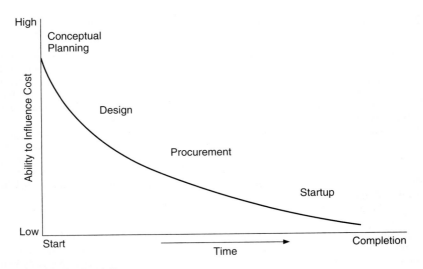

FIGURE 2.1 Cost influence curve.

The conceptual design procedure requirements of equipment capability and system performance should be compatible at a reasonable (and affordable!) cost. The three key aspects of conceptual design are:

- Design requirements
- Design criteria
- Design aspects

2.2 DESIGN REQUIREMENTS

First, establish primary and secondary objectives for the project electrical system design through consultation with process engineers, operators, and maintenance personnel. This would include such objectives as continuity of service, and classifying processes as critical, essential, or general-purpose.

Second, a good understanding of the type of load and application is essential to a good system plan. A complete load survey to determine real and reactive power needs should be undertaken. Depending upon changes made during design and/or construction, some changes to these load estimates may be made. Review loads and processes as to their relative reliability requirements based on economics. That is, what is the cost of lost production or equipment damaged due to power system faults, outages, and voltage sags and surges? Classify loads as critical, essential, or general-purpose loads. This, then, determines the amount of project costs that can be spent to minimize the effects of power system faults, outages, etc. Process and load requirements can be determined by examining existing processes or loads. If none are available, then best estimates of the new loads have to be made based on experience of other similar loads.

Any special loads should be noted. Some of the typical special loads are:

- Starting large motors
- Arc furnace
- Resistance welding
- Static power converters
- Continuous operating loads
- Sensitive electronic loads
- High noise level
- Harmonic current loads
- Coordination of electrical energy with other energy systems

At the conclusion of load survey the following are established:

- Power supply requirements (peak load, maximum demand, diversity factor, load factor, and demand factor)
- Reliability requirements (critical, essential, general purpose)
- Power quality needs (constant voltage supply, no sags or surges)
- Special requirements (harmonic currents, filtering, etc.)
- Physical security needs (locked and isolated substations, motor rooms)

Third, determine the total load, load factors, and demand. This will define the size of the equipment that will be needed to feed the loads. The definitions of some of the terms used are:

Peak load: Maximum load, either instantaneous maximum or maximum average over a time period.

Average load: Load averaged over a period of time, such as one day, week, month, or year.

$$Load\ factor = \frac{average\ load}{peak\ load}$$

Connected load: Sum of electrical ratings of all loads connected.

Demand: Electric load averaged over a time period. This is usually in kW or kVA. Averaging time may be 15 or 30 min or 1 h.

Maximum demand: The greatest of all demands that have occurred during a specific period. The period for billing purposes is usually a month and for design purposes it is the design life of the plant or planning period.

$$Demand\ factor = \frac{maximum\ demand}{connected\ load}$$

$$Diversity\ factor = \frac{sum\ of\ individual\ maximum\ demands}{total\ demand}$$

Coincident demands: Any demand that occurs simultaneously with any other demand.

Fourth, what are the system needs for the next 5, 10, or 20 years? Are processes changing? If there are nonlinear or adjustable-speed drives, the system impedance needs to be small to ensure that the commutating or overlap angle is small. The system should be designed to be expanded to meet these needs. The skeleton of the system should be designed for up to twice the present load. The equipment for the total future system need not be purchased at this time, but the system and equipment should be designed for the future. However, the purchase of the equipment should be as it is needed.

Fifth, consider cogeneration. Is the electric energy to be purchased or generated? If process heat or steam is needed, consideration should be given to generating part or all of the electrical requirements. Local generation will increase the reliability of the electric system for critical loads.

2.3 DESIGN CRITERIA

Uniform criteria or standards are essential for the purpose of selecting equipment and system as well as for comparing performance of different alternatives. Once the load requirements are established, some basic criteria need to be formulated so that a proper distribution system is selected. Comparison of different distribution systems should be performed on a common basis. Some of the basic considerations are:

- Safety
- Reliability
- Simplicity of operation
- Voltage quality

- Maintenance
- Flexibility
- Cost

2.3.1 Safety

Safety is a paramount consideration during the detailed design stage. If sufficient attention is not paid to safety aspects, then personnel may be endangered during operation and maintenance of the system. Catastrophic failure of equipment and consequent loss of power supply may also occur. Thus, all codes and standards requirements should be strictly followed. In addition, some of the key considerations to ensure safety are:

- Verify adequate interrupter rating of switching devices.
- Enclose energized conductors or place them at a higher level.
- Key interlock no-load switches with circuit interrupter.
- Key interlock breaker isolation switches.
- Maintain only deenergized equipment that is tagged and locked out.
- Minimize access to electric equipment rooms and have adequate exits.
- Protect all electric apparatus from mechanical damage; leave areas accessible only for operation and maintenance.
- Consider hazardous areas; use explosion-proof equipment.
- Place warning signs on fences, gates, doors, and conduits.
- Use adequate grounding of both electric power system and equipment.
- Install emergency lights to show exits.
- Provide adequate spares of correct ratings.
- Train operating and maintenance personnel.
- Retrain periodically.

In selecting a distribution system, its design for the safety of personnel is of paramount importance and no compromise of this requirement should be allowed. Second in importance is protecting electric equipment and other nearby equipment and facilities. Rigorous adherence to safety codes and standards is a must.

2.3.2 Reliability

The reliability requirement for the given type of system may be met by many different methods. Some key considerations are:

- *Selection of appropriate voltages for power supply.* Utility distribution systems have lower reliability than subtransmission systems. Distribution systems, 5 kV and 15 kV class, that are used for residential and commercial loads are exposed to many hazards. Higher-voltage systems are designed with more care and have more protection built into them.
- *Redundancy.* Duplication of power paths between the source and the load will provide additional reliability. This also allows maintenance of one path while the other is being used.

- *Adequate protection of system and equipment.* Proper coordination of the protective devices will remove the minimum loads for any fault without affecting the other loads.
- *Monitoring and control.* Proper alarms will inform maintenance of personnel problems as soon as they occur, resulting in the minimum downtime.
- *Selection of appropriate reliable equipment.* Equipment that has conservative ratings will provide service for long periods of time. Recognizing the duty placed on the equipment will allow the engineer to choose the equipment that will be able to meet the duty and operate for its full life.

2.3.3 Simplicity of Operation

Select a simple system that is *simple to operate.* Simply designed systems can provide the service the load requires. This may be obvious, but it is a very important consideration, as long as system requirements are met. Simpler systems are safer and more reliable.

2.3.4 Voltage Quality

The voltage quality of supply systems has become an important consideration in recent years, because of increasing use of sensitive, electronically controlled equipment, including computers. The particular requirements for power supply quality should be established. Typical items to be considered are:

- *Voltage regulation.* This includes transient and dynamic disturbances (sags and surges) that are caused by remote faults.
- *Frequency.* The frequency variation of the power grid in North America is very stable. However, if the local generation is disconnected from the grid, frequency may not be as stable if there are step loads that cause momentary dips in the frequency. These dips can affect timing that is dependent upon sine-wave voltage crossover or zero volts
- *Voltage distortion.* Nonlinear loads have harmonic currents that result in voltage distortion on the power supply as a result of these currents flowing through the impedance of the circuit. Following recommendations in Chap. 11 will minimize voltage distortion caused by these loads. Utility systems have standards that limit the amount of distortion in the voltage they furnish the user.
- *Communication/control interference.* Circuits for control of the power system, close and trip circuits, relay circuits, etc., need special shielding to protect them from false operation due to electromagnetic and electrostatic coupling with power circuits. The use of fiber optics on long runs on these circuits can isolate them from these phenomena.
- *Voltage spikes* from switching transients and lightning need to be minimized by the use of surge protective devices, lightning arresters, and MOVs (metal oxide varistors).

There may be special needs or special types of loads. These should be recognized and documented. Examples of special situations include unusually large loads, motors, frequency controlled drives, and medical services.

2.3.5 Maintenance Requirements

Proper maintenance is a key element in assuring continuity and quality of power supply. Thus, all necessary provisions for effective and efficient maintenance should be incorporated into the detailed design itself. Some of the maintenance-related aspects to be considered during the design stage are:

- Cleanliness
- Moisture control
- Adequate ventilation
- Corrosion reduction
- Thermal inspections
- Visual inspections
- Regular testing
- Record keeping
- Following codes, standards, and manufacturer recommendations

For example, where moisture and/or corrosion problems exist, pressurized electrical rooms with makeup filtered air are employed. During the design stage, provision for using infrared scanning should be made. For maintenance record keeping, sample forms should be included.

2.3.6 Flexibility

The system needs to be designed so that loads can be added and changed. This includes each voltage level starting at the panel board at 120/208 V and up. Spare spaces should be allotted even though circuit breakers are not furnished. At the 480-V level in motor control centers, spare spaces should be allotted for future additions. Transformer capacity should be available for additional loads. The same planning should be done at the medium volt level. Even if spare circuit breakers are not purchased initially, space should be provided in the substation or building for additions to the equipment.

2.3.7 Cost

In all engineering decisions, cost of the system must be balanced against the reliability that the system must provide. The reliability of the system is dictated by the loads requirements. Experienced engineers can make judgments as to what electric equipment and system arrangements are required for different system reliability.

2.3.8 Steady State and Transient System Operation

There are two time frames that need to be considered in any design. The first of these is the operation of the system in the *steady state*. The second is the operation of the system under *transient conditions*.

In the steady state condition, the criteria to be considered include:

1. Keeping the design simple for safety sake
2. Maximizing the equipment utilization
3. Minimizing voltage drop and power losses

Under transient conditions, the criteria to be considered include:

1. Minimizing the magnitude and the effect of motor starting voltage drop by minimizing the impedance between the motor and the power source

2. Minimizing the escalation and effect of ground faults by limiting the ground faults through neutral grounding resistors and using ground sensor instantaneous relaying

3. Minimizing the effect of multiphase faults by instantaneous relaying and coordination with other relays

4. Minimizing the effect of harmonic currents from nonlinear loads such as static power converters by controlling the flow of the harmonic currents with harmonic filters that use power factor improvement capacitors

5. Maximizing induction motor stability by minimizing the impedance between the motor and the generation or power source

6. Maximizing synchronous machine stability by minimizing the impedance between machines

2.4   CONCEPTUAL DESIGN ASPECTS

There are six major aspects that need to be considered in the design of electric power systems. These are:

- Voltage level selection
- Short-circuit level selection
- Ground-fault islanding
- Fast fault clearing
- Selective fault protection
- Avoidance of single-phase operation

2.4.1   Voltage Level Selection

This is important for both the economic and operational aspect of the system. The higher the voltage, the lower the current that is needed for any load. Conductor size and voltage drop in the feeders are directly a result of the current to be carried. However, if the conductors become too small, they will not have the physical strength to be handled in all conditions. Higher voltages can be more expensive because of the increased insulation. Table 2.1 gives a good relationship between voltage, current, and loads.

2.4.2   Short-Circuit Level Selection

Short-circuit level and voltage level are interconnected. A current interrupter has about the same current interrupting capability at any voltage level. That is, an interrupter designed for low voltage has about the same interrupting capability as a medium voltage designed interrupter. As the voltage of a circuit increases, the value of the kVA that is interrupted increases. Thus, larger systems need to operate at higher voltages.

2.4.3   Ground-Fault Islanding

This design aspect allows for the instantaneous relaying of the first line-to-ground fault of the circuit breaker nearest the fault. Since most faults start as line-to-ground faults,

TABLE 2.1 Relationship Between Voltage, Current, kVA, and Load

Voltage (V)	Current (A)	kVA	80% kVA (Load)
480	200	166	133
	400	333	266
	600	500	400
	800	667	532
	1,200	1,000	800
	1,600	1,330	1,065
	2,000	1,663	1,330
	3,000	2,500	2,000
	4,000	3,325	2,660
2400	400	1,663	1,330
	800	3,333	2,660
	1,200	5,000	4,000
	2,000	8,313	6,650
4160	600	4,325	3,460
	1,200	8,650	6,920
	2,000	14,410	11,530
	3,000	21,600	17,280
12,470	600	13,720	10,975
	1,200	25,920	20,735
	2,000	43,200	34,560
	3,000	64,700	51,835
13,200	600	13,720	10,975
	1,200	27,435	21,950
	2,000	45,725	36,580
	3,000	68,590	54,870
13,800	600	14340	11,475
	1,200	28,680	22,950
	2,000	47,800	38,345
	3,000	71,700	57,365

interrupting such faults instantaneously (3 to 5 cycles), results in minimum damage sustained.

Medium-voltage systems, 2.4 to 13.8 kV, should be resistance-grounded to limit the damage of ground faults that may occur in machines connected directly at these voltages. With the ground sensor relaying (50GS), fault currents as low as 15 to 30 A can be detected and relayed instantaneously.

2.4.4 Fast Fault Clearing (8 to 10 cycles for three-phase faults)

Circuit breakers clearing in this time will allow for stability of the system. When line-to-ground faults increase to three-phase faults, there is no limiting impedance for these faults. If there are generators on the system, a bus fault will essentially unload the generator bus so that the generator will tend to overspeed and go out of step with the other machines. If the fault is removed within the 8 to 10 cycles, the machines will not have time to increase their power angle with the system and thus become unstable or out of step.

2.4.5 Selective Fault Protection

This aspect allows the circuit protector nearest the fault to interrupt the fault, thus limiting the disturbance from that circuit with the fault. If that protector does not interrupt the fault, the next protector upstream will operate after a short time delay (0.3 s), removing additional circuits but still minimizing the effect of the fault on the system.

2.4.6 Avoid Single-Phase Operation

A three-phase motor is efficient and operates to deliver the torque needed by the load as long as it has all three phase windings energized. If the circuit is protected from faults with fuses, and one operates, then the motor has only single-phase excitation, the torque is reduced, and the current will increase to the point that the motor will overheat. The use of fuses for circuit protection should be limited, and if they are used, some means of monitoring the fuses must be used so that the circuit will not operate in single phase.

2.5 VOLTAGE, CURRENT, SHORT-CIRCUIT LEVEL VS. LOAD

Table 2.2 lists the relationship among voltage, current, equipment current rating, and load. The currents listed are typical of equipment current ratings. Loads are based on using the equipment at 80 percent of its current-carrying capacity. This is a design criterion as recommended by the National Electric Code. Table 2.2 is a general guide to the voltage levels and short-circuit levels that should be used with various-size loads and motors. It is based on good conceptual design and good system planning. Using this table as a guideline will eliminate many problems that systems have because they are inadequate to start large motors without excessive voltage drop that affect other loads. It also allows induction motor and synchronous motor recovery after system faults, and static power converters to operate with minimum commutation or overlap angles.

TABLE 2.2 Voltage/Short-Circuit Level Selection Based on Load Size

Volts	Transformer Rating, mVA	Transformer Load, mVA	S.C. Interlevel kA	S.C. Interlevel mVA	Transformer full load amperes	S.C.R. load*	Max. motor size†
480	1.0	0.8	22	18.3	1,203	23	200
	1.5	1.2	33	27.5	1,804	23	300
	2.0	1.6	44	36.6	2,406	23	400
	2.5	2.0	55	45.8	3,007	23	500
2.4 kV	7.5	6.0		150	1,804	25	2,000
4.16 kV	15	12		250	2,082	21	3,000
13.8 kV‡	30	24		500	1,255	21	5,000
	40	32		750	1,673	23	7,500
	50	40		1,000	2,092	25	10,000

*Measure of the critical recovery voltage.
†Generally limited by coordination requirements.
‡For loads greater than 24 mVA, it may be preferable to design for multiple 24-mVA load modules.

2.6   ONE-LINE DIAGRAM

One of the tasks of the conceptual design stage is preparation of a preliminary one-line diagram. This preliminary one-line diagram is used during the study stage and includes the major elements of the system. It does not include the details that will be added during the detailed design stage of the project. It includes the transformers, switchgear, cables, power factor improvement capacitors, and the utility system characteristics (short-circuit levels, both minimum and maximum). The detailed one-line diagram will be used in the specification, procurement, installation, and testing of equipment and system. The complete one-line diagram includes:

- Power sources, including voltages and short circuit levels
- Generators, including kVA, voltage, impedances, and grounding method
- Size and type of all conductors, cables, bus, and overhead lines
- Transformer sizes, voltages, impedances, connections, and grounding method
- Protective devices (fuses, relays, circuit breakers)
- Instrument transformer ratios (current and potential)
- Surge arresters and capacitors
- Power factor improvement capacitors
- Identification of all loads, including large motor sizes and impedances
- Type of relays and settings for utility supply
- Future planned additions

The amount of detail on any one-line diagram is determined relative to its use. The conceptual one-line design should not include all of the items in the preceding list because it would clutter the diagram and not easily show what is important at this stage of the project. The diagram should be arranged such that the high-voltage sources are shown at the top of the sheet with the generation. As the diagram proceeds down the page, lower-voltage substations and loads are shown. Figure 2.2 shows such an arrangement. The physical location in the plant should not necessarily be on the preliminary diagram, although, if it is convenient, it is possible to arrange the diagram so that it represents some aspects of the physical location of the equipment.

After commissioning of the plant distribution systems, these one-line diagrams are revised as-built drawings and are used for future reference. A typical preliminary one-line diagram is shown in Fig. 2.4. Chapter 4 shows various one-line diagrams for different types of loads.

Symbols used in one-line diagrams are defined within IEEE Standard 315-1979 (reaffirmed 1989) and IEEE Standard 315A-1986 (supplement to 315). The listing of those symbols that are most used appear at the end of this chapter.

2.7   POWER DISTRIBUTION SYSTEMS

The objective here is to select one or a combination of different types of power supply configurations for supplying the load. Also, the service voltage levels from the utility source point to the load point is established.

Given the same equipment and component quality, the cost of the system increases with increasing reliability. Several different types of systems are available and used in practice, as described in the following sections.

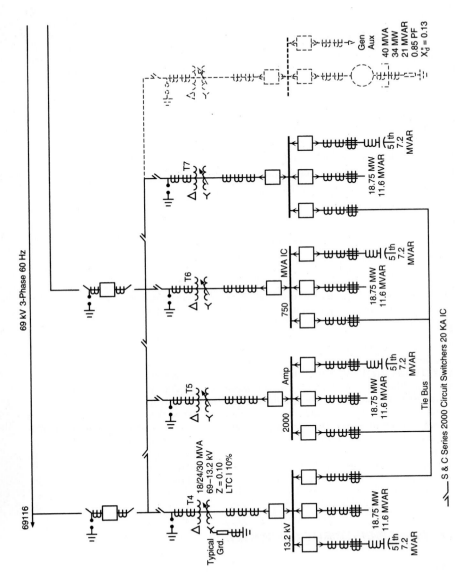

FIGURE 2.2 Typical preliminary one-line diagram.

2.7.1 Simple Radial System

The *simple radial system* (Fig. 2.3) is the simplest system that is used to feed smaller loads. The incoming voltage is stepped down to 480/277 V for distribution to loads. These loads will be lighting at 277 V, heating and air conditioning, and with a stepdown transformer to 120/240 V for receptacle for office equipment and other uses. This can become part of a larger system.

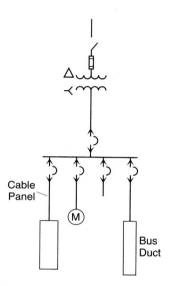

FIGURE 2.3   Simple radial system.

2.7.2 Expanded Radial System

The *expanded radial system* (Fig. 2.4) is made up of two or more radial systems to feed larger loads. Both of these systems make protective device coordination simple because the protective device nearest the load will operate on a fault to remove the faulted circuit without interference with other loads.

2.7.3 Primary Selective System

The *primary selective system* (Fig. 2.5) allows two primary feeds to each substation. This increases the reliability of the system but is more expensive.

2.7.4 Primary Loop System

The *primary loop system* (Fig. 2.6) is the one that is used by electric utilities to feed residential and commercial loads. Each load can be served from either end of the loop from a separate primary feed.

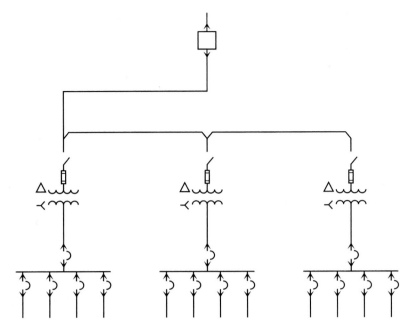

FIGURE 2.4   Expanded radial system.

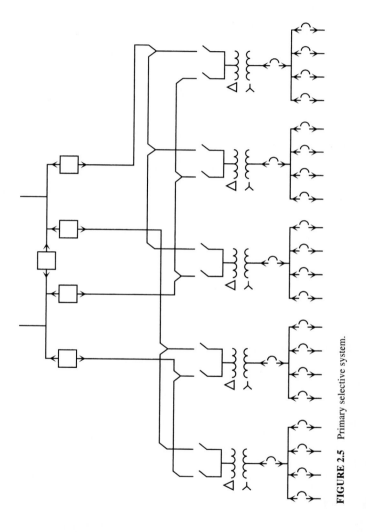

FIGURE 2.5 Primary selective system.

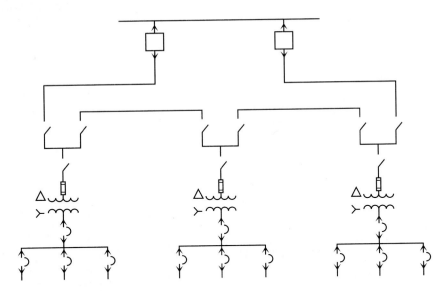

FIGURE 2.6   Primary loop system.

2.7.5   Secondary Selective System

The *secondary selective system* (Fig. 2.7) is the one most used in industrial and commercial systems where two primary feeds are used to feed each bus, which is segregated by a tie breaker. This system can be carried down into the system so that each load has two sources. Figure 2.7 shows the secondary selective system for both the primary distribution as well as the secondary distribution system. Fused switches are shown on the main transformer primaries; however, circuit switchers can be used as they are available in interrupting ratings adequate for most applications, and they have the advantage of three-phase operation.

2.7.6   Utility Secondary Spot Network

The *secondary spot network* (Fig. 2.8) is used in high-density loads such as in metropolitan areas. The secondary voltage can be 120/208 V or 480/277 V. (See Chap. 8.) The theory of operation is that any fault on the secondary bus will burn itself free and service will not be interrupted. For faults in the transformer or primary feeder, the network protector will interrupt the backfeed from the secondary bus into the fault. The network protector operates very fast on reverse current.

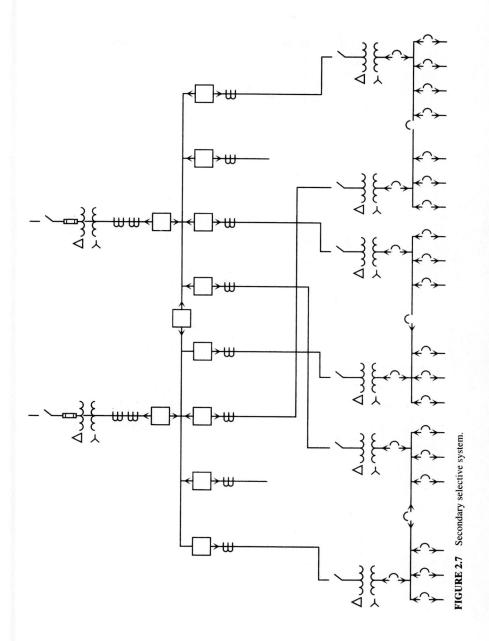

FIGURE 2.7 Secondary selective system.

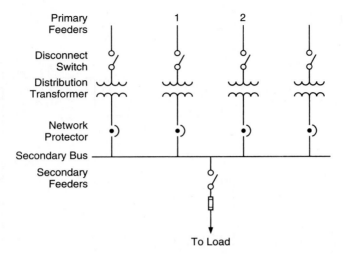

FIGURE 2.8 Secondary spot network.

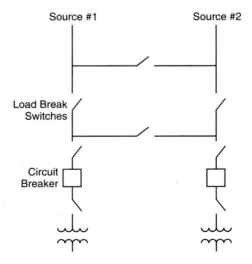

FIGURE 2.9 Ring bus system.

2.7.7 Ring Bus System

The *ring bus* arrangement (Fig. 2.9) is used when the utility feeders are tapped to feed a load. Circuit breakers or switches can be used. Circuit switchers can also be used.

The power supply voltage level from the electric utility can be at any one of these voltages: 138, 69, 34.5, 13.8, 12.5, 4.16 kV levels. In plant distribution, voltage levels usually start at 13.8 kV and step down to the actual load equipment voltage rating, which is usually 480 V. Medium and large motors may be fed at 2.4 to 13.8 kV (see Table 2.2). Thus, the distribution planner has several choices to make for both the type of system and voltage level. A brief comparison of different types of systems is presented in the following:

System type	Advantages/disadvantages
1. Simple radial system	It has the lowest cost.
	It has the simplest possible arrangement.
	Single contingency or maintenance outage interrupts the power supply.
	It is suitable for small installations.
2. Expanded radial system	It has a multiple radial system to supply a larger load.
3. Primary selective system	It has two primary sources.
	The distribution transformer switches either automatically or manually to an alternate source if there is an outage of first source.
	It allows maintenance of cables and primary switchgear.
	There is a brief interruption until switching is completed.
	It costs more than a radial system.
4. Primary loop system	The loop arrangement reduces cost.
	It requires several sectionalizing reclosings on fault, hence it may be dangerous.
	A section may become energized from both ends.
	A primary selective system is preferable.
5. Secondary selective system	A normally open secondary tie circuit breaker allows manual or automatic switching.
	Each unit normally operates as a radial system.
	Sparing transformer arrangement has a lower cost.
	Automatic transfer is not possible for sparing transformer arrangement.
	This system is recommended for reliability.

System type	Advantages/disadvantages
6. Secondary spot network	A common secondary bus is established.
	Utilization equipment is supplied from the secondary bus radially.
	It has parallel distribution transformers with separate sources.
	The cost is high.
	It is used for high-density commercial distribution.
	It may be used to supply low-voltage applications with high load density.
7. Ring bus	The major feature is automatic isolation and restoration for faults.
	There is no interruption of service for single contingencies or maintenance outages.
	The cost is high.

The selection of power supply system voltage is discussed in Sec. 2.5.

2.8 EXAMPLES

The following two examples demonstrate some of the criteria that have been discussed up until now. There are more examples in Chap. 4, which bring out the application of the criteria discussed.

2.8.1 Example 1: Supermarket Power System

A supermarket is to be fed from a utility substation. It will have a load of 1.8 mVA. The load is mainly made up of lighting and refrigeration. The utility has offered service at either 277/480 V or 120/208 V. The market asked that the utility furnish two 1.0-mVA transformers to feed two incoming lines to the market's substation. The refrigeration in the market has motor loads from 3 to 50 hp. These are ideal sizes for three-phase 460-V motors.

The other major load is lighting that is mainly fluorescent and can be fed from 277 V. This leaves only the electronic checkout equipment and the 120-V reciprocals that are used for housekeeping and are distributed throughout the store. Figure 2.10 is the one-line diagram of this system.

2.8.2 Example 2: Large Industrial Plant

The second example is an industrial plant that will have a future load of 50 mVA, which will be installed within five years. The initial load will be half of that, or 25 mVA. From Table 2.2, we see that, with this load, we need to choose 13.8 kV as the voltage level to distribute power throughout the plant. We also see that we need a 30-mVA transformer

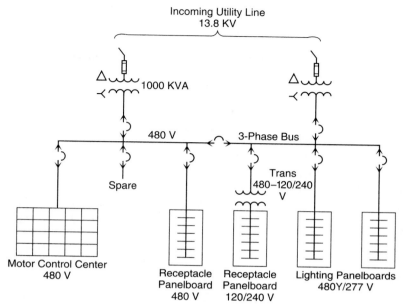

FIGURE 2.10 One-line diagram of supermarket in Example 1.

to feed this load. With the future load, we should plan for two 30-mVA transformers. Figure 2.11 is the one-line diagram of this system.

The initial load has some processes that need high reliability in the power system. With the initial purchase of two 30-mVA transformers, the substation will be in a position to feed the future load and the power system will be in place to accommodate the load as it grows. The initial cost will be higher, but the overall cost will be lower since the main substation will be purchased as a complete unit.

The transformers are protected by a circuit switcher that has the ability to interrupt fault current. These devices have a limited number of operations for interrupting fault current compared to a circuit breaker, but they are much less expensive. Primary protection of transformers is a good application for these devices.

The transformers are connected with a delta primary and a wye secondary. The 13.8-kV winding of the transformer is grounded through a resistor that is rated either 200 or 400 A. The size is determined by how many transformers are in parallel. In this case where there is a normally open tie circuit breaker, the 400-A size will limit any ground fault to a maximum of 400 A.

A secondary selective system is used for the substations in the plant to give the high reliability that is needed for the loads. The process loads are fed from 2000 kVA (2 mVA) double-ended substations. The secondary current for this size substation is 2406 A. This would take a 3000-A, 480-V secondary circuit breaker. This breaker is expensive, and the same protection can be furnished with a primary breaker that has current transformers on the secondary of the transformer with relays that will trip the primary circuit breaker. This saves the cost of the primary fused switch and provides the addi-

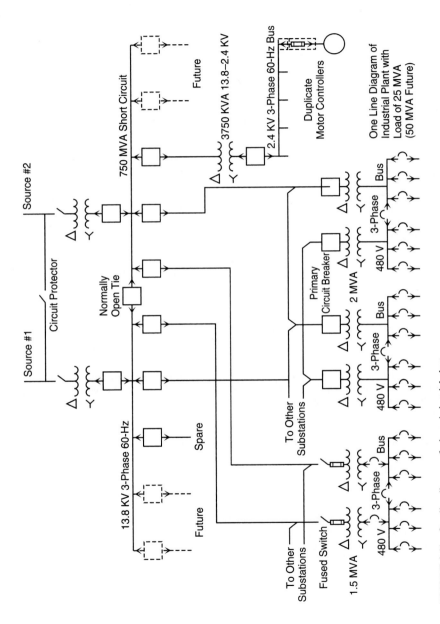

FIGURE 2.11 One-line diagram of a large industrial plant.

Source #2

Source #1

Circuit Protector

13.8 KV 3-Phase 60-Hz

Spare

Future

Normally
Open Tie

750 MVA Short Circuit

Future

3750 KVA 13.8–2.4 KV

2.4 KV 3-Phase 60-Hz Bus

Duplicate
Motor Controllers

One Line Diagram of
Industrial Plant with
Load of 25 MVA
(50 MVA Future)

To Other
Substations

To Other
Substations

Primary
Circuit Breaker

2 MVA

480 V

3-Phase

Bus

480 V

3-Phase

Bus

480 V

3-Phase

Bus

Fused Switch

1.5 MVA

tional protection of a circuit breaker. The 13.8-kV circuit breaker costs less than the 3000-A secondary breaker.

The smaller substation, 1500 kVA (1.5 mVA), has a secondary current of 1804 A. A 2000-A circuit breaker will handle that load and is less expensive than the primary breaker. The fused switch on the primary operates for faults in the transformer. Although there is a possibility of single phasing when one fuse operates, there are not a lot of loads on these substations that are adversely affected by short-duration single phasing. A voltage balance relay on the secondary side of the transformer should be applied to alarm the condition of single phasing.

Loads up to six times the smallest transformer can be fed and protected from the same medium-voltage breaker as long as the substations have a secondary circuit breaker. A fused switch on the primary of the transformer provides protection for that substation and will not have to depend upon the medium-voltage breaker for its protection against faults in the transformer. The load center transformers have a delta primary winding and a wye-connected secondary winding. The secondary winding will be either solidly grounded or high-resistance grounded. (See Chap. 9 for further discussion on grounding.)

There is one area of the plant that has eight drives that are 300 to 800 hp. These will be furnished by a substation that is rated 3.75 mVA at 2.4 kV. This is an ideal voltage for this size motor. The individual motors are fed from motor controllers that consist of a fused contractor. The fuse is to protect against short circuits, and the contractor, which has a make-and-break capability of 10 times its continuous rating, is used for overload protection and starting. The contractor has the ability to operate many thousands of times, where a circuit breaker is designed to operate less frequently (hundreds of times). The motor controllers have all of the necessary protective devices for the motors.

2.9 PROBLEMS

1. Construct a one-line diagram for an industrial plant that has a total load (demand) of 7.5 mVA. It will require substations in eight different load areas, each of which has a load of approximately 1 mVA with the largest motor of 200 hp. The utility has a service voltage of 34.5 kV. It is expected that the future load will grow 50 percent. Only one utility line is available for feeding the plant.

2. What voltage would the following loads be serviced?
   a. Small industrial plant, total load 1.5 mVA.
   b. Shopping mall with individual shops with loads from 2.5 kVA to large department stores with 1.5 mVA. Total load at the mall is 12 mVA.

3. A large industrial plant has a demand of 65 mVA. The future load is expected to grow by another 25 mVA. The largest motor to be started is 5000 hp. Load centers will be 25-2 mVA, except for the office and administration building which will be 1.5 mVA. There is one area that has seven motors ranging from 400 to 1000 hp. Total load of these motors is 3.5 mVA. Construct a one-line diagram of the medium-voltage distribution system, including the load center substations. The utility will serve the plant through two 138-kV subtransmission lines.

TABLE 2.3 One-Line Diagram Symbols

Power Transformers

2 - Winding	
3 - Winding	
Tapped Primary	
Tertiary Winding	
Autotransformer	
Load Tap Changing (LTC)	

Winding	⟩— Wye	△ Delta
Configuration	⊢ Tee	⟩ Zig-Zag
	✳ Diametric	

TABLE 2.3 One-Line Diagram Symbols (*Continued*)

Motors and Generators

Induction Motor and

Generator

Wound-rotor Induction

Motor and Generator

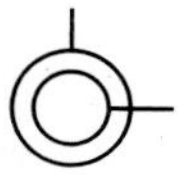

Synchronous Motor,

Generator and Condenser

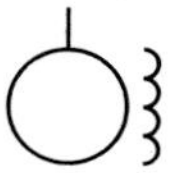

	Shunt Connection	Series Connection

D-C Generator and Motor

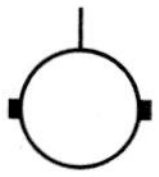

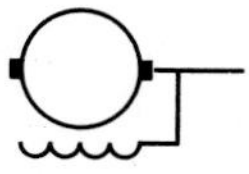

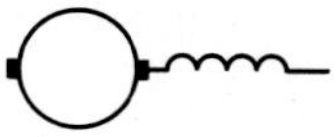

(*Continued*)

TABLE 2.3 One-Line Diagram Symbols (*Continued*)

Static Power Convertors

Diode Rectifier

Thyristor

		Switch with Arcing horns
Switchgear		
Air Break Switch		

Medium or High Voltage	Stationary	Drawout	
Circuit Breaker			

Low Voltage Air	Stationary	Drawout	Fused Circuit Breaker
Circuit Breaker			

Fuse		Disconnect	Drawout

Load Break Fused

Switch

TABLE 2.3   One-Line Diagram Symbols (*Continued*)

Current and Voltage Transformers

Current Transformers

Potential Transformers

Surge Protective Apparatus

Lightning Arrester

Surge Capacitor

Air Gap

Current Limiting Apparatus

Air Core Reactor

Magnetic Core Reactor

Resistor

Power Capacitor

Motor Controllors

Medium Voltage

Low Voltage

CHAPTER 3
ELECTRICAL SYSTEM STUDIES

3.0 INTRODUCTION

This chapter is written to answer three questions: why, what, and how, regarding electric system studies. The reader should be able to get a good understanding of the purpose and results obtained from each of the studies described in this chapter.

The four objectives of electric system studies are:

- To evaluate the performance of the existing or future system
- To determine the effectiveness of alternatives for modifying the existing system or designing a new system
- To select a cost-effective solution or alternative
- To select adequate equipment ratings

One of the major functions of the plant engineer is to keep the existing electric system in good operating condition and provide continuity and quality of power supply to the production and operating units of the company. This is easier said than done. There are usually equipment failures, misoperations, modifications needed for changing operations, requirements to meet laws and regulations, etc.

Another function of the plant engineer is to determine what is needed to modify the existing system or design a new system. The modification may include removing, adding, replacing, and changing the existing system and equipment. It is safe to state that study of modifications may be more difficult and involved, because the performance of the existing system needs to be understood before changes can be made. Any changes made should not adversely affect the existing system. Introducing new equipment and technology may be another important factor in evaluating effective alternatives. For large and complex processes and plants, this type of technical analysis may become quite involved, especially if the performance requirements are very stringent.

Any modification or alternative selected should be cost-effective. As is commonly said, the idea is to keep the cost of per-unit product or service as low as possible. The electric facility costs, both capital and operational, may be a substantial part of the product or plant. Thus, it is not surprising that management would like to keep these costs as low as possible. The economic analysis is discussed further in Chap. 16.

Historically, the electrical system analysis was performed by longhand calculations. Many simplifying assumptions were made so that meaningful answers could be obtained with reasonable commitment of time and manpower. This limited the number of options considered, and the results were not as refined. The systems were built, equipment was installed, and modifications were made as required, and this was expensive. Then analog-type simulators were developed to study the electric system. This gave engineers a quicker method of simulating on a much smaller, scaled-down version of different alternatives, and enabled them to evaluate the performance. This was definitely less expensive, because the study could be much more thorough and the problems and solutions could be identified before implementation. However, the analog-type simulators still limited the scope of studies both in quality and quantity. The introduction of digital computers to make studies made it affordable to evaluate many different options and perform fast evaluations. Thus, making so-called paper studies before implementation became less expensive, there was more confidence in the expected system performance, and a least-cost solution could be identified much more easily.

Recent advances in microcomputer technology has made computing power accessible to every person at a very affordable cost. Electrical power system analysis software packages which are user-friendly and with different degrees of sophistication are available. Further, training in the methods of electrical power system analysis applications, as well as in the use of software, is easily and widely available. It is important for the plant engineers to be knowledgeable in the types of power system studies that are performed.

3.1 GENERAL STUDY PROCEDURE

The major steps in a typical electric system study may be listed as follows:

1. Define purpose and scope
2. State assumptions and methodology
3. Establish performance criteria
4. Perform technical analysis of a baseline system
5. Develop alternatives
6. Conduct screening studies
7. Perform detailed technical analysis of alternatives
8. Perform economic analysis
9. Perform sensitivity analysis
10. Select or recommend cost-effective alternative for implementation

These steps follow somewhat a logical sequence. However, in practice it is not inconceivable to bypass some steps. These steps are also iterative in nature. For example, all the factors in the performance criteria may not have been defined at the beginning of the study. After conducting some technical and economic studies, additions and deletions may be made to the performance criteria. A similar situation holds good for different alternatives. All these steps may be performed formally or informally. However, it is always helpful to document these steps, especially while solving complex system problems. The rigorousness of following this procedure and the study itself is also dependent upon the importance and cost magnitude of the system being studied. The rigorousness of the study is different for a million-dollar project as compared to a thousand-dollar project.

Every study has to start with an initial system. For the majority of planning studies, the existing system or part of the system is used as a convenient starting system, because the system configuration, types of equipment, characteristics, and performance are known and data is usually readily available. For new plants and systems, an initial computer model of the system has to be assembled based on the experience of the engineers and similar plants elsewhere.

Even though exhaustive analysis of the alternatives is desirable, it is necessary to limit the number of alternatives to be studied. There is no point in wasting time and money in studying marginally attractive alternatives. For this purpose, it is beneficial to perform a screening analysis. In the screening stage, only major performance requirements are evaluated and alternatives which do not meet these requirements are discarded from further analysis or deferred for future consideration. Also, widely varying and costly alternatives are discarded. For example, an alternative which has five times the cost of another alternative need not be evaluated in detail or needs to be deferred until the less expensive alternative is found to be unacceptable.

The discussion in this section gives the reader a basic understanding of a planning or study procedure. The types of studies performed for technical analysis are discussed in the following sections.

One of the main tasks of the plant electrical engineer is to evaluate performance of the electrical system and equipment, whether existing or new. For economic reasons, it has become necessary to utilize equipment to its full capability and limit. Also, the new equipment is designed and constructed with a smaller margin from physical loading and performance limits. This is done by performing technical analysis. The calculations required for this type of analysis are fairly complex and are difficult to perform by longhand or calculators, except for simple cases. Thus, digital computer programs are becoming indispensable in technical analysis. This is true both from the point of view of the complexity of the analysis and the time needed for such analysis. Access to and familiarity with digital computers has become common. Hence, we will describe technical studies, with the assumption that, primarily, digital computers will be used for computations.

The engineer who performs technical analysis, irrespective of whether computer programs are used for analysis or not, should be familiar with system modeling issues, methods of analysis, procedure used, interpretation of results, and so forth. The type of studies performed for technical analysis include:

- Load flow
- Short circuit
- Motor starting
- System stability
- Protection and coordination
- Transients and switching
- Reliability
- Harmonics

These studies primarily address the performance of the electrical power system as related to the time period under consideration. This relationship is shown in Table 3.1. For example, lightning is an impulse with time periods of microseconds to milliseconds. A switching transient (for example, by turning on or off a shunt capacitor bank) has a time period of milliseconds. The thermal loading of electric equipment has a time period of minutes to hours. Thus, it is not necessary to consider lightning surges when determining the heating from load carried by a cable. In general, it may be stated that the severity of the applied quantity (voltage or current) is inversely related to the time duration of application. This is an important concept to remember.

Some of the salient points of these studies are summarized in Table 3.2. The engineer needs to acquire a somewhat specialized knowledge in performing these studies. Short courses and training seminars are widely available within the industry to supplement or obtain new training, so this should not be a drawback.

These studies are described in the remaining sections of this chapter. We will assume that the reader has a basic electrical engineering background and familiarity with network analysis concepts such as linear, nonlinear, equivalents, and theorems such as

TABLE 3.1   Electrical Phenomena and Study Tools for Different Time Periods

Time scale	Phenomena	Study tools
Microseconds Milliseconds	Lightning (impulse) Switching surges Transients	EMTP/TNA
Milliseconds Cycles	Relaying Protection Power swing	Short-circuit Relay coordination Stability
Seconds	Motor starting	Power flow; stability
Minutes	Wave distortion	Harmonic analysis
Hours Days Weeks	Operations planning	Power flow Short-circuit Stability Reliability
Years	System planning	Economic analysis

Thevenin, Norton, and Superposition. The direct current network analysis based on Ohm's law and Kirchoff's laws are simpler. Most of the electrical power system analysis deals with 50- or 60-Hz alternating current sinusoidal waveforms. Phasor representation of these ac voltages and currents for three-phase and single-phase systems is also assumed. Knowledge of symmetrical components is essential for unbalanced system analysis. For this reason, symmetrical components are explained in Sec. 3.3. For more sophisticated technical studies (such as calculating control system response and finding harmonics in a distorted ac waveform, understanding the application of Fourier transformation, and Laplace transformation), time domain and frequency domain concepts are necessary. The reader is referred to appropriate introductory texts on network and control system subjects.

3.2   PER-UNIT SYSTEM

While performing calculations, voltage, current, kVA, and impedance are usually expressed as a per-unit of a selected base or reference value of each of these quantities. The per-unit method of calculation is simpler than the use of actual amperes, ohms, and volts.

The per-unit value represents, in a sense, a normalized value of the quantity or the constant. The normalizing numbers are referred to as *base values* in power system terminology. The normalized value of a quantity can thus be expressed as a ratio of its value in physical units to an *appropriate* base value. The normalized values are commonly known as *per-unit* values. The basic formulas are:

$$\text{Per-unit volts} = \frac{\text{volts}}{\text{base volts}}$$

$$\text{Per-unit amps} = \frac{\text{amps}}{\text{base amps}} \tag{3.1}$$

TABLE 3.2   Salient Points of Power System Studies

Type of study	Salient points

Load flow or power flow	Steady state analysis
	Uses network analysis methods
	Used to determine
	    MVA or current rating of lines, cables, transformers
	    Voltage regulation (voltage drop)
	    Load tap changer, range, step size, control
	    Shunt capacitor, location, size (var flow)
	    Losses
	Easy and fast to run on PC
	Can run many cases to determine system performance for
	    different circuit configurations and contingencies
	Output in both graphical and tabular form
	Used as base case for stability and motor starting studies

Short-circuit and fault studies	Steady state analysis with fault conditions
	Uses network analysis techniques
	Used to determine
	    Short-circuit currents for three-phase, line-to-line, and
	      line-to-ground faults
	    Circuit breaker and other equipment duties
	    Relay settings
	    Protective coordination

Motor starting	Quasi–steady state
	Uses steady state analysis
	Used to determine
	    Voltage drop during starting
	    On-load starting characteristics
	    Acceleration time
	    Coordination protection
	    Special starting methods

Dynamics and stability studies	Dynamic analysis with disturbances
	Uses differential equations and network analysis
	Used to determine
	    Synchronous stability
	    Induction motor stability
	    Corrective actions for unstable cases

Protection and coordination studies	Used to determine
	    Relay settings, types, and numbers
	    Time-overcurrent coordination of protective devices
	    Fuse ratings

Transient and switching studies	Transient analysis
	Uses differential equations or analog simulation
	Used to determine
	    Transient overvoltages
	    Breaker interrupting capability
	    Surge arrester ratings
	    Equipment failure investigations

(Continued)

TABLE 3.2   Salient Points of Power System Studies   *(Continued)*

Type of study	Salient points

Reliability	Steady state (long-term) analysis
	Uses deterministic or probabilistic approach
	Used to determine ,
	  Risk (qualitative/quantitative)
	  Effect of outages/failures
	  Redundancy needs

Harmonic analysis	Uses Fourier transformation and network solution techniques
	Used to determine
	  Distortion, shunt filter design
	  Resonance/overvoltage
	  Flicker
	  Telephone interference

$$\text{Per-unit ohms} = \frac{\text{ohms}}{\text{base ohms}}$$

A percent value of a quantity is its per-unit value multiplied by 100. A major advantage of the per-unit (or percent) value is that the various constants of electric equipment of widely different voltage and power ratings lie within reasonably narrow numerical ranges if the rated values are used as base values in computing per-unit values.

Out of the four electrical quantities—namely, voltage, current, kVA, and impedance—selection of base values for any two of them determines the base values of the remaining two. If the base values of current and voltage are selected, then base impedance and base kVA can be determined. In practice, base kVA and base voltage in kV are the quantities selected at the beginning of the calculations. The base kVA in single-phase systems is the product of base voltage in kV and base current in amperes. For single-phase systems, the term *current* refers to line current; the term *voltage* refers to voltage to neutral, and the term kVA refers to kVA per phase. The following formulas relate the various quantities:

$$\text{Base current in amperes} = \frac{\text{base kVA}}{\text{base voltage in kV}} \tag{3.2}$$

$$\text{Base impedance} = \frac{\text{base voltage in volts}}{\text{base current in amperes}} \tag{3.3}$$

$$= \frac{(\text{base voltage in kV})^2}{\text{base kVA} \times 1000}$$

Since the balanced three-phase circuits may be solved as a single-phase system, the bases for quantities in the impedance diagram are kVA per phase and kV from line to neutral. Data are usually given as total three-phase kVA and line-to-line kV. Base impedance and base current can be computed directly from three-phase values of base kV and base kVA. Given, base kVA for the total of the three phases and base voltage for line to line,

$$\text{Base current in amperes} = \frac{\text{base kVA}}{\sqrt{3} \times \text{base voltage in kV}} \tag{3.4}$$

$$\text{Base impedance} = \frac{(\text{base voltage in kV})^2}{\text{base kVA} \times 1000} \tag{3.5}$$

The per-unit value of a given three-phase kVA on a three-phase kVA base is identical to the per-unit value of a given single-phase kVA on a single-phase kVA base.

During calculations, all impedances in any part of the system must be expressed on a common base. Thus, it becomes necessary to convert per-unit impedances from one base to another. The impedance base conversion formulas are:

$$\text{Per-unit ohms} = \frac{\text{actual ohms} \times \text{base kVA} \times 1000}{(\text{base kV})^2} \tag{3.6}$$

$$\text{Per-unit ohms (new base)} = \text{p.u. ohms (old base)} \times \frac{\text{new base kVA}}{\text{old base kVA}} \tag{3.7}$$

$$\text{Per-unit volts (new base)} = \text{p.u. volts (old base)} \times \frac{\text{old base kV}}{\text{new base kV}} \tag{3.8}$$

when both the base kVA and base kV need to be changed; then Eqs. (3.7) and (3.8) may be combined as:

$$\text{Per-unit } Z_{\text{new}} = \text{per-unit } Z_{\text{old}} \left(\frac{\text{base kV}_{\text{old}}}{\text{base kV}_{\text{new}}}\right)^2 \times \left(\frac{\text{base kVA}_{\text{new}}}{\text{base kVA}_{\text{old}}}\right) \tag{3.9}$$

The selection of base values of kVA and kV is made in order to reduce the work required by the calculations as much as possible. The selected base should yield per-unit values of rated voltage and current approximately equal to unity in order to simplify the work of computing as well as get good accuracy in the calculations. When the resistance and reactance of a device are given by the manufacturer in percent or per-unit, the base is understood to be the rated kVA and kV of the equipment (see Table 3.3).

TABLE 3.3　Base Current and Impedances for 10-MVA Base

Line-to-line base voltage (V)	Line-to-ground* base voltage (V)	Base current† (A)	Base impedance‡ (Ω)
480	277.1	12,028.1	0.023
2,300	1,327.9	2,510.2	0.529
4,160	2,401.8	1,387.9	1.73
13,800	7,967.4	418.4	19.0
23,000	13,279.1	251.0	52.9
34,500	19,918.6	167.3	119.0
69,000	39,837.2	83.7	476.1
115,000	66,395.3	50.2	1,322.5
138,000	79,674.3	41.8	1,904.4
230,000	132,790.6	25.1	5,290.0

*$V_{\text{LG}} = V_{\text{LL}}/\sqrt{3}$.
†Assumed to be line current.
‡Line-to-ground (neutral) impedance value.

3.3 SYMMETRICAL COMPONENTS

A three-phase power system or equipment is said to be balanced when:

- Voltage and current magnitudes are equal in all the three phases.
- Voltage and current phase angles are 120° apart in all three phases.
- Impedances in the three phases are the same.

An example of a three-phase system is shown in Fig. 3.1.
This is a simple but general example and is called a balanced system when

1. Generator source voltages are equal in magnitude.

$$E_A = E_B = E_C = E$$

2. The generator voltage phase angles are 120° apart.

$$\theta_B = \theta_A - 120°$$
$$\theta_C = \theta_B - 120°$$
$$\theta_A = \theta_C - 120°$$

3. The generator or source impedances (reactances by neglecting resistance) are equal in three phases.

$$X_A = X_B = X_C = X$$

4. The line impedances are equal.

$$Z_{1a} = Z_{1b} = Z_{1c} = Z$$

5. The load impedances are equal.

$$Z_A = Z_B = Z_C = Z$$

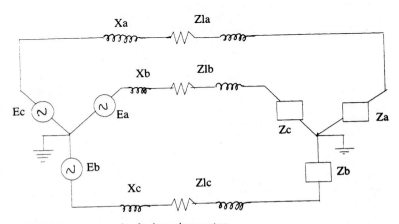

FIGURE 3.1   An example of a three-phase system.

The balanced three-phase system is the type of power system encountered in performing voltage drop and loading calculations in industrial or commercial systems. The calculations are done on a single-phase equivalent basis (see Sec. 2.1). However, it is not uncommon to face the problem of unbalanced power system conditions. In general, the unbalance may be in any one of the quantities in Fig. 3.1. The unbalance may be due to:

- Voltage magnitude and/or angle
- Unbalanced load
- Fault conditions
- Open conductor
- Single-phase supplies
- Special transformer connections
- Load connected line to line (single phase)

For example, calculating line-to-ground fault currents, which is very commonly done, requires solution of unbalanced systems. However, most industrial and commercial power systems are resistance-grounded, which limits the line-to-ground fault currents.

The method used for solution of an unbalanced power system is called *symmetrical components*. The basis of this method is that a given unbalanced three-phase power system condition may be represented by three balanced power system conditions. Once we can transform the unbalanced condition to a balanced condition, we can use the simpler single-phase equivalent methods for solution and recombine the solution quantities to get back to the unbalanced power system quantities. The equations used for transformation of a three-phase system to its components are as follows:

$$V_{A1} = \frac{1}{3}(V_A + aV_B + a^2V_C)$$

$$V_{A2} = \frac{1}{3}(V_A + a^2V_B + aV_C) \qquad (3.10)$$

$$V_{A0} = \frac{1}{3}(V_A + V_B + V_C)$$

or in the matrix form,

$$
\begin{bmatrix} V_{A1} \\ V_{A2} \\ V_{A0} \end{bmatrix} = \frac{1}{3}
\begin{bmatrix} 1 & a & a^2 \\ 1 & a^2 & a \\ 1 & 1 & 1 \end{bmatrix}
\begin{bmatrix} V_A \\ V_B \\ V_C \end{bmatrix} \qquad (3.11)
$$

where V_{A1} = positive sequence voltage
V_{A2} = negative sequence voltage
V_{A0} = zero sequence voltage

$$a = -\frac{1}{2} + j\frac{\sqrt{3}}{2} = 1\ \underline{/+120°}$$

$$a^2 = -\frac{1}{2} - j\frac{\sqrt{3}}{2} = 1\ \underline{/-120°}$$

$$a^3 = 1 + j0 \text{ and } a^4 = a \text{ etc.}$$
$$1 + a + a^2 = 0$$

Phasor operators a, a^2, and a^3 are shown in Fig. 3.2(a).

We have used here the voltage symbol in the equations, but this may be applied to voltage or current. The equations are illustrated by phasor diagrams as shown in Fig. 3.2(*b*) to (*e*.) Examples of three phase and sequence voltage waveforms are shown in Fig. 3.3.

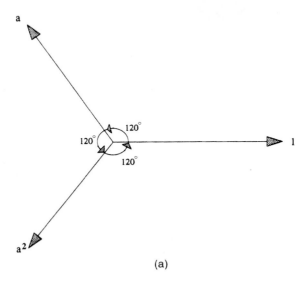

(a)

FIGURE 3.2a Phasor operators.

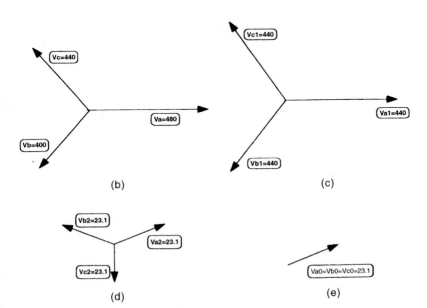

(b) (c)

(d) (e)

FIGURE 3.2b–e Sequence phasor representation.

EXAMPLE 3.1 *Consider a set of three-phase unbalanced (only in magnitude) voltages given as follows:*

$$V_A = 480 + j0 = 480\ \underline{/0°}$$

$$V_B = -200 - j346.4 = 400\ \underline{/-120°}$$

$$V_C = -220 + j0381.1 = 440\ \underline{/+120°}$$

By using Eq. (3.10), we can calculate:

$$V_{A1} = 440 + j0 = 440\ \underline{/0°}$$

$$V_{A2} = +20 - j11.5 = 23.1\ \underline{/-30°}$$

$$V_{A0} = +20 + j11.5 = 23.1\ \underline{/+30°}$$

These are shown in Fig. 3.2. In Figure 3.2(b) the given three-phase unbalanced voltage phasors are shown. The assumed direction of rotation of these phasors is also shown on these diagrams. The positive and negative sequence voltages form a balanced three-phase set, as shown in Fig. 3.2(c) and (d), respectively. Then the power system problem can be solved much more easily because balanced three-phase voltages are applied. Note that the negative sequence phasor rotation direction is in the opposite direction to positive sequence. All three zero sequence voltages are in the same direction and of the same magnitude. Hence, solution of the network with these zero sequence impressed voltages can be accomplished without much difficulty.

Once the three different symmetrical component quantities have been determined, it is necessary to compute the phase quantities. The reverse transformation is given by

$$V_A = V_{A0} + V_{A1} + V_{A2}$$

$$V_B = V_{B0} + V_{B1} + V_{B2} \qquad (3.12)$$

$$V_C = V_{C0} + V_{C1} + V_{C2}$$

However, we know that positive (V_{A1}, V_{B1} and V_{C1}) and negative (V_{A2}, V_{B2}, V_{C2}) sequence quantities form a set, each, of balanced three-phase quantities. The three zero sequence quantities are equal in magnitude and phase. Hence, we can write the following relationships:

$$V_{B1} = a^2 V_{A1}$$

$$V_{C1} = a V_{A1}$$

$$V_{B2} = a V_{A2} \qquad (3.13)$$

$$V_{C2} = a^2 V_{A2}$$

$$V_{A0} = V_{B0} = V_{C0}$$

Here, a^2 represents rotation by $-120°$ and a represents rotation by $+120°$. Substituting Eq. (3.13) in Eq. (3.12) we get

$$V_A = V_{A0} + V_{A1} + V_{A2}$$

$$V_B = V_{A0} + a^2 V_{A1} + a V_{A2} \qquad (3.14)$$

$$V_C = V_{A0} + a V_{A1} + a^2 V_{A2}$$

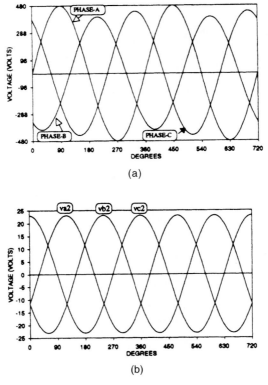

FIGURE 3.3   Three-phase and sequence voltage waveforms.

This represents the reverse transformation to get phase quantities from symmetrical components. This may be written in matrix form as

$$
\begin{bmatrix} V_A \\ V_B \\ V_C \end{bmatrix} = \begin{bmatrix} 1 & 1 & 1 \\ 1 & a^2 & a \\ 1 & a & a^2 \end{bmatrix} \begin{bmatrix} V_{A0} \\ V_{A1} \\ V_{A2} \end{bmatrix}
$$

(3.15)

Note that, in matrix terminology, the 3×3 coefficient matrix in Eq. (3.15) is called the inverse of 3×3 coefficient matrix in Eq. (3.11). There are other types of transformations (for example, see reference 2).

SEQUENCE NETWORKS: Example 3.1 illustrated how an unbalanced set of phase voltages (currents) can be transformed to a set of balanced sequence voltages (currents). On the basis of voltage and current transformations, the unbalanced three-phase imped-

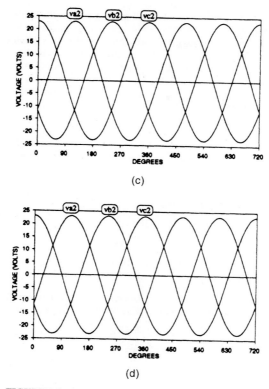

(c)

(d)

FIGURE 3.3 (*Continued*) Three-phase and sequence voltage waveforms.

ances (loads, lines, etc.) may be converted to balanced sequence impedances. These are called positive, negative, and zero sequence networks. Each network corresponding to each sequence can be solved independently.

The positive sequence network will be the same as the power system network. Usually the generator (or source) voltages are considered balanced both in magnitude and phase angle unless some special unbalance cases are being investigated. The negative sequence network is similar to the positive sequence network except there are no source voltages (negative sequence voltage is zero because of only balanced source or generator voltage assumption). There are no source voltages from generators in the zero sequence network. However, the zero sequence network is different from the other types of sequence networks, the main difference being that all neutral connections to ground and the return paths for current from other than the phase conductors should be represented properly. The reason for this is self-evident upon a closer examination. As stated earlier, the positive and negative sequence components are a set of three-phase balanced systems. Hence, at any neutral point, the sum of the positive and negative sequence voltages and currents is zero. Because zero sequence voltages are all equal in the three phases, the net sum is not zero. Hence, neutral and ground are not the same. The treatment of transformer impedance, including the number of windings (two, three, or four windings), type of connection (delta, wye, or auto), and whether it is grounded or ungrounded, are some of the most important considerations in developing a zero sequence network. The positive and zero sequence representation for different types of transformers is available in many references. Reference 1 covers most of the commonly encountered types of transformers. For short-circuit calculations, these three sequence networks are connected together and the fault calculations are performed on this network. The actual short-circuit calculations are discussed in other sections and in Chap. 6.

EXAMPLE 3.2 *Consider a three-phase induction motor with a blown fuse in one of the phases. The current flow (10A) is as shown in Fig. 3.4. Determine the symmetrical components of line currents.*

The currents in the three supply lines are

$$I_A = 100 \underline{/0°}$$

$$I_B = 100 \underline{/180°}$$

$$I_C = 0$$

For convenience, line C is assumed open. Here, we are using phase A as the reference. Using Eq. (3.10), the symmetrical components of the line current are

$$I_{A1} = 50 - j\,28.9 = 57.8 \underline{/-30°}$$

$$I_{A2} = 50 + 28.9 = 57.8 \underline{/+30°}$$

$$I_{A0} = 0$$

Note that the zero sequence component of the current is 0, because there is no current through the neutral conductor or ground.

3.4 LOAD FLOW

One of the basic calculations needed in industrial and commercial power systems analysis is calculation of voltage (or voltage drop) and the current flow in different parts of the system. The current and voltage need to be within a certain range of the rating of the equipment and facilities. The system has to be designed to meet this objective and perform satisfactorily for almost all anticipated operating conditions that may be encountered during the lifetime of the plant or equipment. Hence, the calculations may need to be repeated several times, including consideration of lines, cables, and the like being out of service.

The method of calculating voltage and current in different parts of a system is called *load flow* or *power flow*. The salient points of this type of program and application were summarized in Table 3.1.

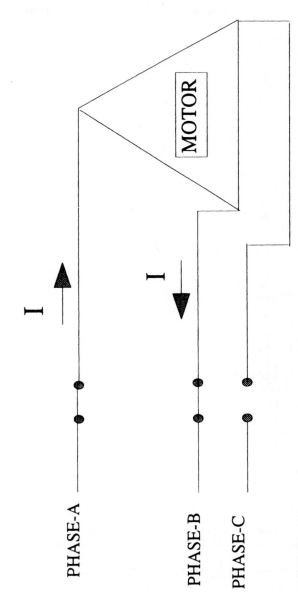

FIGURE 3.4 Motor example.

3.15

EXAMPLE 3.3 *We will illustrate the concept of power flow through a simple direct-current type of network problem (Fig. 3.5). This method of iterative solution is called the Gauss-Seidel method.*

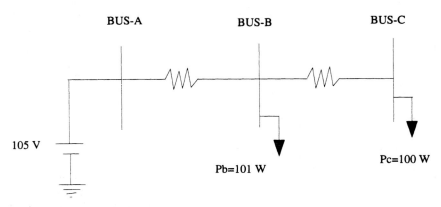

FIGURE 3.5 Sample dc load flow problem.

In this problem, we know

V_A = 105 V
P_B = 101 W
P_C = 100 W

We want to determine

1. Voltages at buses *B* and *C*
2. Current in wires (lines) *A* to *B* and *B* to *C*
3. Power put out by battery source at *A*

Step 1: *We will assume an approximate voltage of 100 V at B and C and compute load currents*

$$I_C = \frac{100}{100} = 1 \text{ A}$$

$$I_B = \frac{101}{100} = 1.01 \text{ A}$$

Step 2: *In a longhand approach, we can sum up the current and compute the bus voltages at B and C. We will do this with a more formal method by equations.*

$$V_B = V_A - Z_{AB}(I_B + I_C) = 105 - 2(1 + 1.01) = 100.98 \text{ V}$$

$$V_C = V_B - Z_{BC} \cdot I_C = V_A - (Z_{AB}) (I_B + I_C) - Z_{BC} \cdot I_C$$

$$= 105 - 2 (1 + 1.01) - 1 \times 1 = 99.98 \text{ V}$$

These voltages are approximate because the power at these buses does not match the given value.

$$P_B = V_B \cdot I_B = (100.98)(1.01) = 101.9898 \text{ W}$$

$$P_C = V_C \cdot I_C = (99.98)(1) = 99.98 \text{ W}$$

The differences in the calculated and given values for power are called mismatches.

$$\Delta P_B = 101.9898 - 101 = 0.9898 \text{ W}$$

$$\Delta P_C = 99.98 - 100 = -0.02 \text{ W}$$

Strictly speaking, mismatch is the algebraic sum (net) of all the real (P) and reactive (Q) power at each node or bus. For correct solution, we want this mismatch to become 0. In practical power system problems, we want this mismatch to become practically 0 (negligible).

Step 3:   Using the voltage calculated in Step 2, we will repeat Steps 1 and 2 iteratively until we get the desired mismatch. The steps and results are summarized in Table 3.4.

Ac power flow: In an ac power flow problem, the equations are:

$$V_i I_i^* = P_i + jQ_i \qquad \text{for } i = 1, 2, ..., n \text{ buses} \qquad (3.16)$$

$$[I] = [Y][V]$$

where V_i = complex voltage at bus i
I_i = complex current at bus i
* = complex conjugate denoting $(a + jb)^* = a - jb$
P_i = real power of load or generator at bus i
Q_i = reactive power of load or generator at bus i

Current vector $[I] = \begin{bmatrix} I_1 \\ I_2 \\ \cdot \\ \cdot \\ \cdot \\ I_n \end{bmatrix}$ Voltage vector $[V] = \begin{bmatrix} V_1 \\ V_2 \\ \cdot \\ \cdot \\ \cdot \\ V_n \end{bmatrix}$

System admittance matrix $[Y] = \begin{bmatrix} Y_{11} & Y_{12} & \cdot & \cdot & \cdot & \cdot & Y_{1n} \\ Y_{21} & Y_{22} & \cdot & \cdot & \cdot & \cdot & Y_{2n} \\ \cdot & \cdot & \cdot & \cdot & \cdot & \cdot & \cdot \\ \cdot & \cdot & \cdot & \cdot & \cdot & \cdot & \cdot \\ \cdot & \cdot & \cdot & \cdot & \cdot & \cdot & \cdot \\ Y_{n1} & Y_{n2} & \cdot & \cdot & \cdot & \cdot & Y_{nn} \end{bmatrix}$

TABLE 3.4   Sample dc Load Flow Problem Results

Iteration no.	V_B^{old}	V_C^{old}	I_B	I_C	V_B^{new}	V_C^{new}	ΔP_B	ΔP_C
1	100	100	1.01	1	100.98	99.98	0.9898	−0.02
2	100.98	99.98	1.000198	1.0002	100.9992	99.999	0.019208	0.019008
3	100.9992	99.999	1.000008	1.00001	101	99.99995	0.000761	0.000951
4	101	99.99995	1	1	101	100	3.41E-05	4.36E-05
5	101	100	1	1	101	100	1.55E-06	1.98E-06
6	101	100	1	1	101	100	7.03E-08	9.01E.08
7	101	100	1	1	101	100	3.19E-09	4.09E.09
8	101	100	1	1	101	100	1.45E-10	1.86E-10
9	101	100	1	1	101	100	6.59E-12	8.46E-12
10	101	100	1	1	101	100	3E-13	3.85E-13
11	101	100	1	1	101	100	2.24E-14	2.22E-14
12	101	100	1	1	101	100	0	0

Just as a reminder, a complex number (voltage, current, impedance, etc., in an ac network) may be expressed as consisting of a real and an imaginary part or as a magnitude and phase angle.

Again, the main idea behind the power flow problem is to solve the two equations (3.16) iteratively until the mismatches are acceptable.

EXAMPLE 3.4   *A solution method for an ac power flow problem will be illustrated through a sample problem, as shown in Fig. 3.6. As discussed in Sec. 3.2 (on the convenience of using per-unit quantities), we will convert the problem data into per-units, as shown in the following:*

		P.U.
Base voltamperes =	1000	
Base voltage =	480 V (line-line)	
Base current =	1.203 A	
Base impedance =	230.400 Ω	
$R12$ =	0.230 Ω	0.001
$X12$ =	2.304 Ω	0.010
$R23$ =	0.461 Ω	0.002
$X23$ =	4.608 Ω	0.020
Real power (at load bus 3) =	1000 W	1.000
Reactive power (at load bus 3) =	250 Vars	0.250
Power factor of load @ bus 3 =	0.970	
Real power (at load bus 2) =	2000 W	2.000
Reactive power (at load bus 2) =	75 Vars	0.750
Power factor of load @ bus 2 =	0.920	
Source voltage (at bus 1) =	540 V (line-line)	1.050

The three steps in Example 3.3 are used iteratively for this ac power flow example. The iterations are shown in Table 3.5. Again, this is based on the Gauss-Seidel method of solution. In order to speed up the solution, certain acceleration factors are used for real and imaginary parts. For example,

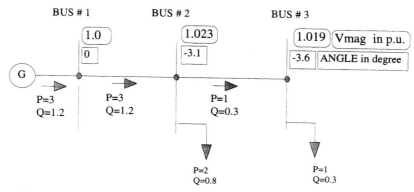

FIGURE 3.6 Checking Gauss-Seidel worksheet by ac power flow solution.

$$V_{B\ \text{old}}^{(2)} = V_{B\ \text{old}}^{(1)} - \text{Accn factor} * (V_{B\ \text{new}}^{(1)} - V_{B\ \text{old}}^{(1)}) \qquad (3.17)$$

where the superscript $^{(\cdot)}$ indicates iteration number.
  Note that the ac power flow calculations are done in complex quantities. The values shown in Table 3.5 are in rectangular coordinates (real and imaginary).
  We wish to make a few points at this time:

- Solving this type of problem by longhand for a system with more than a few buses becomes an impossible task.
- Different iterative solution techniques are available for these types of calculations.
- A huge volume of technical papers and quite a few textbooks have been written on this subject.[2,3,4]

Fortunately, there are power flow software packages available for use on almost any type of computer. These programs may be run interactively, by batch mode, or in a controlled sequence mode. Some are available in the Windows version on PCs. Training and documentation on this software is also available. These programs also have the capability to maintain system data, make changes, and produce selective outputs; hence, the plant engineer can have access to a fairly sophisticated type of power flow at a small cost. We will illustrate a sample power flow problem solved by a computer program.

EXAMPLE 3.5   *An example electrical power system is shown as a one-line diagram in Fig. 3.7. This is a balanced three-phase power system supplied through two 69-kV lines emanating from a utility substation. The industrial supply substation starts from the 69/13.8-kV transformers. There may be many variations to the supply arrangement within the plant. We will use this simpler system for illustration purposes. The electrical data for this system is shown in Table 3.6. The one-line diagram (Fig. 3.7) also shows the voltage magnitudes, angles, and the real and reactive power flows in the system for a given load condition as obtained from a solved load flow case. A typical hardcopy numerical output is shown in Table 3.7. The format of these inputs and outputs may*

TABLE 3.5 Sample ac Load Flow Problem Results

Iteration #	V_2 old real	V_2 old imaginary	V_3 old real	V_3 old imaginary	I_2 real	I_2 imaginary	I_3 real	I_3 imaginary
1	1	0	1	0	2	−0.75	1	−0.25
2	1.0384	−0.0928	1.036	−0.1084	1.9907	−0.9938	1.0152	−0.3524
3	1.0042627	−0.036201	1.0029507	−0.0425203	1.9740112	−0.8372537	0.9952125	−0.2872666
4	1.0319562	−0.0696961	1.0302318	−0.0813683	1.9994374	−0.9354104	1.0145322	−0.3276852
5	1.0107625	−0.0505875	1.0091253	−0.0592924	1.9737813	−0.8754287	0.9981157	−0.3032781
6	1.0263138	−0.0609764	1.0247287	−0.071238	1.9960288	−0.9110225	1.0110697	−0.3175548
7	1.0152745	−0.0557099	1.0135917	−0.0652219	1.978267	−0.8906376	1.0013471	−0.3095285
8	1.0228952	−0.0580812	1.0213122	−0.0679003	1.9916993	−0.9017847	1.0083749	−0.313805
9	1.0177638	−0.0572638	1.016096	−0.0670042	1.9819388	−0.8960804	1.0034478	−0.3117047
10	1.0211394	−0.0573092	1.0195371	−0.0670214	1.9888081	−0.8986955	1.0068121	−0.312594
11	1.0189691	−0.0575982	1.0173193	−0.0673797	1.9841038	−0.897749	1.0045696	−0.3123405
12	1.0203319	−0.0572063	1.018715	−0.0669108	1.9872469	−0.8978578	1.0060303	−0.3122893
13	1.0194977	−0.0575886	1.0178588	−0.0673628	1.9851956	−0.8981196	1.0051005	−0.312463
14	1.0199938	−0.0572625	1.0183691	−0.0669796	1.9865034	−0.8977361	1.0056782	−0.3122609
15	1.0197088	−0.0575203	1.0180754	−0.0672813	1.9856898	−0.8981191	1.0053287	−0.3124476
16	1.0198653	−0.0573266	1.018237	−0.0670554	1.9861826	−0.8977885	1.0055337	−0.3122929
17	1.0197847	−0.0574667	1.0181535	−0.0672183	1.9858932	−0.8980518	1.005418	−0.3124129
18	1.0198221	−0.0573684	1.0181925	−0.0671043	1.9860567	−0.897853	1.00548	−0.312324
19	1.0198082	−0.0574355	1.0181777	−0.067182	1.985969	−0.8979973	1.0054493	−0.3123876
20	1.0198103	−0.0573908	1.0181801	−0.0671303	1.9860125	−0.8978958	1.0054625	−0.3123434
21	1.0198135	−0.0574199	1.0181833	−0.0671639	1.9859937	−0.8979653	1.0054585	−0.3123733
22	1.0198084	−0.0574014	1.0181782	−0.0671426	1.9859995	−0.8979188	1.0054581	−0.3123535
23	1.0198136	−0.0574129	1.0181834	−0.0671558	1.986	−0.8979491	1.0054603	−0.3123663
24	1.0198091	−0.057406	1.0181789	−0.0671478	1.9859969	−0.8979298	1.0054576	−0.3123582
25	1.0198127	−0.05741	1.0181825	−0.0671525	1.9860007	−0.8979418	1.0054602	−0.3123632
26	1.01981	−0.0574077	1.0181797	−0.0671499	1.9859971	−0.8979346	1.005458	−0.3123602
27	1.0198119	−0.057409	1.0181818	−0.0671512	1.9860001	−0.8979388	1.0054597	−0.312362
28	1.0198105	−0.0574084	1.0181803	−0.0671506	1.9859978	−0.8979364	1.0054585	−0.312361
29	1.0198115	−0.0574086	1.0181813	−0.0671509	1.9859995	−0.8979377	1.0054593	−0.3123615
30	1.0198109	−0.0574086	1.0181807	−0.0671508	1.9859983	−0.8979371	1.0054587	−0.3123613
31	1.0198113	−0.0574085	1.0181811	−0.0671508	1.9859991	−0.8979373	1.0054591	−0.3123614
32	1.019811	−0.0574085	1.0181808	−0.0671508	1.9859985	−0.8979373	1.0054589	−0.3123613
33	1.0198112	−0.0574085	1.018181	−0.0671508	1.9859989	−0.8979373	1.005459	−0.3123613
34	1.0198111	−0.0574086	1.0181809	−0.0671508	1.9859987	−0.8979373	1.0054589	−0.3123614
35	1.0198111	−0.0574085	1.018181	−0.0671508	1.9859988	−0.8979372	1.005459	−0.3123613
36	1.0198111	−0.0574086	1.0181809	−0.0671508	1.9859987	−0.8979373	1.005459	−0.3123613
37	1.0198111	−0.0574085	1.0181809	−0.0671508	1.9859988	−0.8979373	1.005459	−0.3123613
38	1.0198111	−0.0574086	1.0181809	−0.0671508	1.9859988	−0.8979373	1.005459	−0.3123613

TABLE 3.5 Sample ac Load Flow Problem Results *(Continued)*

Iteration #	V_2 new real	V_2 new imaginary	V_3 new real	V_3 new imaginary	Delta P_2	Delta Q_2	Delta P_3	Delta Q_3
1	1.024	−0.058	1.0225	−0.06775	−0.0915	0.098	−0.039437	0.062125
2	1.0170642	−0.0574256	1.0153442	−0.0672252	−0.081739	−0.146441	−0.054468	−0.03956
3	1.0215711	−0.0571354	1.0200014	−0.0668003	−0.06443	0.0074718	−0.034308	0.0234681
4	1.0187101	−0.0577532	1.0170402	−0.0675708	−0.09087	−0.087438	−0.053962	−0.014716
5	1.0204821	−0.0570805	1.0188774	−0.0667584	−0.064178	−0.030695	−0.037204	0.0076294
6	1.0194143	−0.0576848	1.0177681	−0.067478	−0.087332	−0.063569	−0.050462	−0.004972
7	1.0200374	−0.0571919	1.018417	−0.0668959	−0.068844	−0.045343	−0.040495	0.0017569
8	1.0196881	−0.0575703	1.0180521	−0.0673402	−0.082828	−0.054876	−0.04771	−0.001566
9	1.0198735	−0.0572922	1.0182467	−0.0670149	−0.072665	−0.050339	−0.042646	−0.000146
10	1.019783	−0.0574898	1.018151	−0.0672454	−0.079819	−0.052138	−0.046107	−0.000564
11	1.0198209	−0.0573533	1.0181916	−0.0670866	−0.074919	−0.051748	−0.043798	−0.000629
12	1.0198105	−0.0574453	1.0181799	−0.0671933	−0.078193	−0.051487	−0.045304	−0.000368
13	1.0198078	−0.0573848	1.0181777	−0.0671233	−0.076056	−0.051989	−0.044345	−0.000677
14	1.0198157	−0.0574236	1.0181855	−0.0671682	−0.077419	−0.051453	−0.044941	−0.00039
15	1.0198066	−0.0573992	1.0181764	−0.0671401	−0.076571	−0.051931	−0.04458	−0.000629
16	1.0198149	−0.0574142	1.0181848	−0.0671572	−0.077084	−0.051543	−0.044792	−0.000443
17	1.0198081	−0.0574053	1.0181778	−0.0671471	−0.076783	−0.05184	−0.044672	−0.000581
18	1.0198134	−0.0574104	1.0181833	−0.0671529	−0.076953	−0.051622	−0.044736	−0.000482
19	1.0198095	−0.0574076	1.0181792	−0.0671497	−0.076862	−0.051776	−0.044704	−0.000551
20	1.0198123	−0.057409	1.0181821	−0.0671513	−0.076907	−0.05167	−0.044718	−0.000504
21	1.0198103	−0.0574084	1.0181801	−0.0671506	−0.076888	−0.051742	−0.044714	−0.000535
22	1.0198116	−0.0574086	1.0181815	−0.0671508	−0.076894	−0.051695	−0.044714	−0.000515
23	1.0198108	−0.0574086	1.0181806	−0.0671508	−0.076894	−0.051725	−0.044716	−0.000528
24	1.0198113	−0.0574085	1.0181812	−0.0671507	−0.076891	−0.051706	−0.044713	−0.00052
25	1.019811	−0.0574086	1.0181808	−0.0671508	−0.076895	−0.051717	−0.044716	−0.000525
26	1.0198112	−0.0574085	1.018181	−0.0671507	−0.076891	−0.051711	−0.044713	−0.000522
27	1.0198111	−0.0574086	1.0181809	−0.0671508	−0.076894	−0.051714	−0.044715	−0.000524
28	1.0198111	−0.0574085	1.018181	−0.0671508	−0.076892	−0.051712	−0.044714	−0.000523
29	1.0198111	−0.0574086	1.0181809	−0.0671508	−0.076894	−0.051713	−0.044715	−0.000523
30	1.0198111	−0.0574085	1.0181809	−0.0671508	−0.076892	−0.051713	−0.044714	−0.000523
31	1.0198111	−0.0574086	1.0181809	−0.0671508	−0.076893	−0.051713	−0.044715	−0.000523
32	1.0198111	−0.0574086	1.0181809	−0.0671508	−0.076893	−0.051713	−0.044714	−0.000523
33	1.0198111	−0.0574086	1.0181809	−0.0671508	−0.076893	−0.051713	−0.044715	−0.000523
34	1.0198111	−0.0574086	1.0181809	−0.0671508	−0.076893	−0.051713	−0.044714	−0.000523
35	1.0198111	−0.0574086	1.0181809	−0.0671508	−0.076893	−0.051713	−0.044714	−0.000523
36	1.0198111	−0.0574086	1.0181809	−0.0671508	−0.076893	−0.051713	−0.044714	−0.000523
37	1.0198111	−0.0574086	1.0181809	−0.0671508	−0.076893	−0.051713	−0.044714	−0.000523
38	1.0198111	−0.0574086	1.0181809	−0.0671508	−0.076893	−0.051713	−0.044714	−0.000523

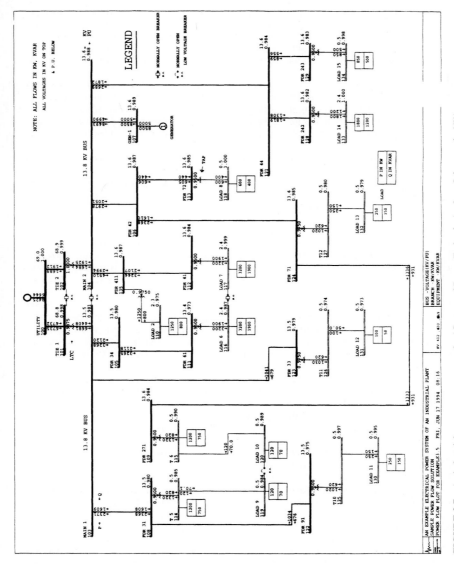

FIGURE 3.7 An industrial power system example. (*Courtesy of Power Technologies Inc., Schenectady, NY.*)

3.22

TABLE 3.6 An Example Electric Power System of an Industrial Plant

Example 3.5 system data for power flow

			Bus data		
Bus no.	Name	Base kV	Voltage (kV)	P_{load} (MW)	Q_{load} (MVAR)
100	Utility	69	69	0	0
101	Tie 1	69	68.832	0	0
102	Tie 2	69	68.942	0	0
103	Main 1	13.8	13.537	0	0
104	Main 2	13.8	13.624	0	0
105	FDR 34	13.8	13.522	0	0
106	FDR 42	13.8	13.613	0	0
107	GEN-1	13.8	13.641	0	0
108	FRD 31	13.8	13.519	0	0
109	FDR 271	13.8	13.574	0	0
110	Load 2	0.48	2.292	1.25	0.8
111	FDR 61	13.8	13.418	0	0
112	FDR 41	13.8	13.575	0	0
113	FDR 72	13.8	13.589	0	0
114	T 5	0.48	0.456	1.2	0.75
115	T 6	0.48	0.458	1.2	0.75
116	Load 6	2.4	2.286	3.2	1.9
117	Load 7	2.4	2.283	3.2	1.9
118	Load 8	0.48	0.457	0.6	0.4
119	Load 9	0.48	0.456	0.12	0.07
120	Load 10	0.48	0.458	0.12	0.07
121	FDR 44	13.8	13.57	0	0
122	FDR 91	13.8	13.457	0	0
123	FDR 33	13.8	13.514	0	0
124	FDR 71	13.8	13.595	0	0
125	T10	0.48	0.459	0.75	0.45
126	T11	0.48	0.458	0.9	0.6
127	T12	0.48	0.458	0.75	0.45
128	FDR 242	13.8	13.55	0	0
129	FDR 243	13.8	13.556	0	0
130	Load 11	0.48	0.458	0.25	0.15
131	Load 12	0.48	0.458	0.1	0.0
132	Load 13	0.48	0.457	0.25	0.15
133	Load 14	2.4	2.285	1.8	1.2
134	Load 15	0.48	0.459	0.85	0.5
135	FDR 411	13.8	13.612	0	0

NOTE: Bus voltages are needed input for generator, utility supply, and synchronous motor buses only. Others can be a best possible guess or 1 p.u.

Generator unit data

			Generation				V_{sched}				Z-Source	
Bus. no.	Name	Base kV	P (MW)	Q (MVar)	Q_{max}	Q_{min}	(p.u.)	P_{max}	P_{min}	M_{base}	R (p.u.)	X (p.u.)
100	Utility	69	8.4	6.7	25.0	−10.0	1.0	100.0	0.0	100.0	0.002	0.030
107	Gen-1	13.8	8.5	5.0	5.0	−2.0	1.0	10.0	1.0	10.0	0.012	0.300

(Continued)

TABLE 3.6 An Example Electric Power System of an Industrial Plant *(Continued)*

Example 3.5 system data for power flow

Branch Data							
From bus	To bus	From name	To name	Line R (p.u.)	Line X (p.u.)	Charging (p.u.)rating	Branch rating (kVAR)
100	101	Utility	Tie 1	0.0015	0.003	0.0051	30,000
100	102	Utility	Tie 2	0.0015	0.003	0.0051	30,000
101	103	Tie 1	Main 1	0.0067	0.0533	0	15,000
102	104	Tie 2	Main 2	0.0067	0.0533	0	15,000
103	105	Main 1	FDR 34	0.0026	0.0011	0	5,000
103	108	Main 1	FRD 31	0.0044	0.0016	0	5,000
103	123	Main 1	FDR 33	0.0146	0.0019	0	2,000
104	106	Main 2	FDR 42	0.0022	0.001	0	5,000
104	107	Main 2	GEN-1	0.0008	0.001	0	10,000
104	121	Main 2	FDR 44	0.01	0.0064	0	5,000
104	135	Main 2	FDR 411	0.0017	0.0005	0	5,000
105	111	FDR 34	FDR 61	0.0189	0.0058	0	5,000
106	113	FDR 42	FDR 72	0.0256	0.0033	0	2,000
106	124	FDR 42	FDR 71	0.0042	0.0015	0	5,000
108	114	FRD 31	T 5	0.0529	0.45	0	1,500
108	122	FRD 31	FDR 91	0.0395	0.0051	0	2,000
109	115	FDR 271	T 6	0.0529	0.45	0	1,500
109	124	FDR 271	FDR 71	0.0102	0.0013	0	2,000
110	135	Load 2	FDR 411	0.0458	0.367	0	1,500
111	116	FDR 61	Load 6	0.0244	0.147	0	3,750
112	117	FDR 41	Load 7	0.0244	0.147	0	3,750
112	135	FDR 41	FDR 411	0.0068	0.0021	0	5,000
113	118	FDR 72	Load 8	0.0958	0.767	0	1,500
114	119	T 5	Load 9	0.0582	0.0243	0	0
115	120	T 6	Load 10	0.0464	0.0239	0	0
121	128	FDR 44	FDR 242	0.0073	0.0009	0	2,000
121	129	FDR 44	FDR 243	0.0104	0.0013	0	2,000
122	125	FDR 91	T10	0.0548	0.383	0	1,500
123	126	FDR 33	T11	0.0548	0.383	0	1,500
124	127	FDR 71	T12	0.0548	0.383	0	1,500
125	130	T10	Load 11	0.0291	0.023	0	0
126	131	T11	Load 12	0.0451	0.013	0	0
127	132	T12	Load 13	0.0291	0.023	0	0
128	133	FDR 242	Load 14	0.0329	0.23	0	3,750
129	134	FDR 243	Load 15	0.0821	0.575	0	1,500

NOTE: 1. Impedance data in p.u. on 10 MVA base.

TABLE 3.6 An Example Electric Power System of an Industrial Plant (*Continued*)
Example 3.5 system data for power flow

From bus	To bus	Tap side	Ratio	Contrld. bus no.	Tap range Max	Tap range Min	V_{max} (kV)	V_{min} (kV)	Tap step
					Additional transformer data (rating & impedance given in the above table)				
101	103	From	0.9875	103	1.5	0.8	14.49	13.11	0.00625
102	104	From	1	—	1.5	0.51	—	—	0.00625
108	114	From	0.95	—	1.05	0.95	0.504	0.456	0.025
109	115	From	0.95	—	1.05	0.95	0.504	0.456	0.025
110	135	To	0.975	—	1.05	0.95	0.504	0.456	0.025
111	116	From	0.95	—	1.05	0.95	2.415	2.185	0.025
112	117	From	0.95	—	1.05	0.95	2.415	2.185	0.025
113	118	From	0.95	—	1.05	0.95	0.504	0.456	0.025
122	125	From	0.95	—	1.05	0.95	0.504	0.456	0.025
123	126	From	0.975	—	1.05	0.95	0.504	0.456	0.025
124	127	From	0.975	—	1.05	0.95	0.504	0.456	0.025
128	133	From	0.95	—	1.05	0.95	2.415	2.185	0.025
129	134	From	0.95	—	1.05	0.95	0.504	0.456	0.025

```
     PTI INTERACTIVE POWER SYSTEM SIMULATOR--PSS/E      FRI, JUN 17 1994  15:00
AN EXAMPLE ELECTRICAL POWER SYSTEM OF AN INDUSTRIAL PLANT          RATING
SAMPLE POWER FLOW SOLUTION                                         SET A

BUS    100 UTILITY 69.0 AREA CKT    KW      KVAR      KVA   %I 1.0000PU   0.00  100
    GENERATION             1     8424.6   6642.6R10728.4   11 69.000KV
    TO   101 TIE 1    69.0  1  1  6735.2   4731.3  8230.9   27
    TO   102 TIE 2    69.0  1  1  1689.4   1911.3  2550.9    9

BUS    101 TIE 1    69.0 AREA CKT    KW      KVAR      KVA   %I 0.9976PU  -0.07  101
                          1                                     68.833KV
    TO   100 UTILITY 69.0  1  1 -6724.9  -4762.4  8240.4   28
    TO   103 MAIN 1  13.8  1  1  6724.9   4762.3  8240.4   55 0.9875RG

BUS    102 TIE 2    69.0 AREA CKT    KW      KVAR      KVA   %I 0.9992PU  -0.01  102
                          1                                     68.943KV
    TO   100 UTILITY 69.0  1  1 -1688.5  -1960.6  2587.4    9
    TO   104 MAIN 2  13.8  1  1  1688.5   1960.5  2587.4   17 1.0000LK

BUS    103 MAIN 1   13.8 AREA CKT    KW      KVAR      KVA   %I 0.9812PU  -1.96  103
                          1                                     13.540KV
    TO   101 TIE 1    69.0  1  1 -6680.5  -4407.7  8003.5   54 0.9875UN
    TO   105 FDR 34  13.8  1  1  3268.5   2119.9  3895.8   85
    TO   108 FDR 31  13.8  1  1  2371.4   1609.2  2865.9   74
    TO   123 FDR 33  13.8  1  1  1040.8    678.7  1242.6   64

BUS    104 MAIN 2   13.8 AREA CKT    KW      KVAR      KVA   %I 0.9876PU  -0.46  104
                          1                                     13.629KV
    TO   102 TIE 2    69.0  1  1 -1684.0  -1924.8  2557.4   17 1.0000UN
    TO   106 FDR 42  13.8  1  1  2978.0   2050.9  3615.9   79
    TO   107 GEN-1   13.8  1  1 -8492.1  -4990.8  9850.1  102
    TO   121 FDR 44  13.8  1  1  2689.2   1871.5  3276.3   56
    TO   135 FDR 411 13.8  1  1  4510.6   2993.7  5413.6  158
    M I S M A T C H              -1.6     -0.5     1.7
```

TABLE 3.7 A Sample ac Power Flow Computer Program Output. (*Courtesy of Power Technologies Inc., Schenectady, NY.*)

vary, but the essential information is almost the same in different programs. Invariably, the solved load flow cases can be plotted onto a predefined one-line diagram and printed output may be obtained. From a practical user's point of view, it should be noted that the amount of time spent on data collection, input, output, and documentation of study results may be as much as 50 percent of the total load flow study. Thus, a user-friendly program should make this task for the user as easy as possible. In addition, it should be possible to use the same input database and model with minor additions and deletions for other studies discussed in this chapter.

Reviewing Fig. 3.7, several operating conditions can be determined. First, at the incoming bus from the utility the voltage is 1.0 per-unit. When the power reaches the plant substation, the voltage at bus 101 has dropped to 0.998 p.u. and on bus 102 only 0.999 p.u. As a result, the VAR flow is higher than the power flow from bus 100 to bus 102. The power flow is greater than the VAR flow from bus 100 to bus 101.

The power and VAR loss in transformers and lines can be calculated from the differences between the incoming power (VAR) and outgoing power (VAR). For example, the power and VAR flow from bus 108 to bus 122 is 1034 kW and 676 kVAR. Through the transformer, the load is 1020 kW and 670 kVAR. Therefore, there is 14 kW and 56 kVAR loss in the transformer.

The power factor of any feeder or total plant load can be calculated. The power factor of the plant at the 69-kV incoming bus is:

$$\text{PF} = \cos\left(\tan^{-1}\left(\frac{1912 + 4732}{1684 + 6735}\right)\right) = 0.785$$

Checking bus 127, the voltage is only 0.980 per-unit and the transformer tap is set at 0.975. By changing the tap to 1.000, the voltage can be raised for better operation of the load to about 1 p.u.

Load flow programs have almost all of the information that you would get from taking readings on the actual power system for the conditions that are set for the load flow.

SINGLE-PHASE OR UNBALANCED SYSTEMS: The sample problem shown in Example 3.5 is a three-phase balanced system. In some industrial and commercial electric distribution systems, there may be single-phase lines, cables, and loads.

The unbalanced system may be modeled as a four-wire (three phase wires plus neutral) arrangement as shown in Fig. 3.8. The neutral wire is not shown in the diagram. The source (stepdown transformer) is a three-phase source. The lines, cables, transformers, motors, loads, and shunt capacitors may be included in these calculations. The system unbalances are taken into account by using the positive and zero sequence impedances (see Sec. 3.3). The voltage drop and other calculations are more difficult as compared to a three-phase balanced system. Load flow software packages (including short-circuit calculations) are available for solving problems in these types of systems. Sometimes this type of solution is called three-phase load flow.

EXAMPLE 3.6   *A sample unbalanced power flow for the system shown in Fig. 3.8 will be illustrated through this example.*

The positive and zero sequence data for this system is given in Table 3.8. The load data is given in Table 3.9. The bus voltages and the line flows are shown in Table 3.10.

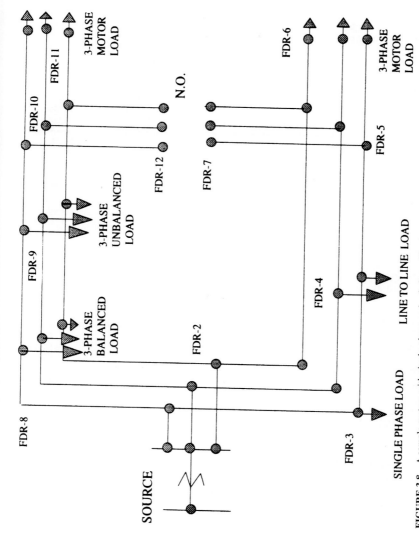

FIGURE 3.8 A sample system with single-phase and unbalanced loads.

TABLE 3.8 Impedance Data for Example 3.6 in Per-Unit on 1000 kVA, 13.8-kV Base

Bus name		Positive sequence impedance		Zero sequence impedance	
From	To	Resistance	Reactance	Resistance	Reactance
SO	FDR1	0.0000	0.0300	0.0000	0.0300
FDR1	FDR2	0.0134	0.0172	0.0283	0.0488
FDR2	FDR3	0.0294	0.0379	0.0624	0.1074
FDR3	FDR4	0.0241	0.0310	0.0510	0.0879
FDR4	FDR5	0.0268	0.0345	0.0567	0.0976
FDR5	FDR6	0.0067	0.0086	0.0142	0.0244
FDR5	FDR7	0.0067	0.0086	0.0142	0.0244
FDR2	FDR8	0.0294	0.0379	0.0624	0.1074
FDR8	FDR9	0.0241	0.0310	0.0510	0.0879
FDR9	FDR10	0.0241	0.0310	0.0510	0.0879
FDR10	FDR11	0.0067	0.0086	0.0142	0.0244
FDR10	FDR12	0.0067	0.0086	0.0142	0.0244

TABLE 3.9 Load Data for Example 3.6

	Phase A		Phase B		Phase C	
Bus	P (kW)	Q (kVAR)	P (kW)	Q (kVAR)	P (kW)	Q (kVAR)
F3	120	60	0	0	0	0
F4	80	40	80	40	0	0
F6	100	Three-phase motor load				
F8	100	50	100	50	100	50
F9	120	60	150	75	80	40
F11	100	Three-phase motor load				

TABLE 3.10 Unbalanced Power Flow Solution for Example 3.6

Section		Load current (A)			Load voltage (L-N in kV)			Line loading		Line losses	
From	To	A	B	C	A	B	C	kW	kVAR	kW	kVAR
	SO	49	57	48	8.501	8.499	8.499				
SO	FDR1	49	57	48	7.798	7.812	7.831	1143.8	612.0	0.00	57.70
FDR1	FDR2	64	57	42	7.737	7.772	7.799	1143.8	554.8	5.17	6.74
FDR2	FDR3	28	16	12	7.675	7.762	7.774	382.2	180.1	1.56	2.12
FDR3	FDR4	10	16	12	7.663	7.739	7.755	260.6	120.2	0.50	0.65
FDR4	FDR5	4	5	5	7.656	7.733	7.746	100.1	41.6	0.08	0.10
FDR5	FDR6	4	5	5	7.654	7.731	7.744	100.0	43.4	0.02	0.02
FDR5	FDR7	0	0	0	7.656	7.733	7.746	0.0	−0.3	0.00	0.00
FDR2	FDR8	36	41	31	7.664	7.696	7.752	756.5	371.1	4.93	6.38
FDR8	FDR9	22	26	17	7.626	7.655	7.735	451.6	216.9	1.50	1.96
FDR9	FDR10	4	5	5	7.620	7.650	7.727	100.1	41.9	0.07	0.09
FDR10	FDR11	4	5	5	7.618	7.648	7.725	100.0	43.6	0.02	0.02
FDR10	FDR12	0	0	0	7.620	7.650	7.727	0.0	−0.3	0.00	0.00

3.5   SHORT-CIRCUIT CALCULATIONS

The entirety of Chap. 7 is devoted to this topic as applied to equipment and protection. In this section, only preliminary background is introduced for system studies.

An electrical power system may experience faults or short circuits. In such an event, the faulted equipment or facility needs to be isolated safely from the rest of the system so that minimal or no damage to faulted and other equipment can occur. The loss of power supply should be limited to the faulted equipment and/or to that part of the system only. There should be no danger to plant personnel. Protective relays, breakers, switches, and fuses are used to detect faulted equipment and isolate it. Proper coordination and application of such devices requires calculation of expected short-circuit current magnitudes for different types of faults and different system operating conditions. Mainly three-phase and line-to-ground (L-G) faults and system conditions, which may produce both maximum and minimum short-circuit currents, are of interest. Thus, the purposes of short-circuit calculations are:

- To determine the duty imposed on interrupting devices such as circuit breakers and fuses
- To apply relays and protective devices
- To coordinate protective devices
- To determine the mechanical and thermal duty on cables, bus ducts, etc.
- To ascertain the minimum short circuit
- To determine operating characteristics of large static converters

All major sources of fault current should be represented in the short-circuit calculation model. The major short-circuit current sources are:

- Utility supply system
- Synchronous machines (generators and motors)
- Induction machines (mostly motors)

The supplying electric utility system is almost always a relatively large system. Hence, the inclusion of current flow to the faults from the supplying utility system is very important. Electric utilities should be requested to furnish the short-circuit levels at their supplying buses without any contribution from the plant electric system. Based on this level, the electric utility system is usually represented by a constant voltage (usually 1 p.u.) behind an equivalent impedance (mostly reactance is used in practice). The p.u. reactance is equal to the reciprocal of p.u. short-circuit level.

After the occurrence of a fault, a synchronous machine continues to be driven by its prime mover and the field windings are supplied with direct current. Thus, an internal voltage is maintained. This internal voltage causes a current to flow from the machine to the fault or short circuit. This current is limited only by the machine and intervening system impedance (reactance). Thus, synchronous machines are represented by a constant voltage behind a machine impedance or reactance. However, the fault current is very large compared to steady state load currents and is established suddenly. But the theorem of constant flux linkages states that the flux in a magnetic circuit cannot be changed instantaneously. The flux changes slowly depending upon the electromagnetic characteristics of a rotating machine. This type of behavior is addressed in the short-circuit calculations by using different machine reactances for different time frames of application. Subtransient reactance (X_d'') is used for obtaining the maximum possible

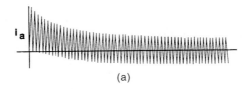

(a)

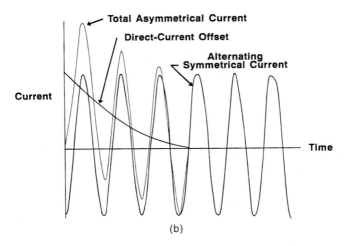

(b)

FIGURE 3.9 Typical system fault current.

short-circuit currents in a time frame of about 6 to 10 cycles (0.1 to 0.15 s on a 60-Hz basis) from the instant of fault. For fault currents beyond this time frame but up to about 2 s, the transient reactance (X_d') is used.

Further, when an ac generator is short-circuited, the resulting current contains both the steady state ac component and the transient component, as shown in Fig. 3.9. The magnitude of the transient component is dependent on the point or instant of fault on the steady state voltage wave. The magnitude is greatest when the fault occurs at a point corresponding to the crest of the steady state voltage; it is zero when the fault occurs at a point corresponding to the zero of the steady state voltage. The transient behavior of the circuit may cause the actual current to rise to a value substantially greater than that predicted by generator subtransient reactance considerations alone.

The asymmetrical short-circuit current is considered to be composed of two components. One is the ac symmetrical component E/Z. The other is a dc component initially of maximum possible magnitude, equal to the peak of the initial ac symmetrical component, assuming that the fault occurs at the point on the voltage wave where it creates this condition. At any instant after the fault occurs, the total current is equal to the sum of the ac and dc components.

Since resistance is always present in an actual system, the dc component decays to zero as the stored energy it represents is expended as I^2R loss. The decay is assumed to be an exponential, and its time constant is assumed to be proportional to the ratio of reactance to resistance (X/R ratio) of the system from source to fault. As the dc component decays, the current gradually changes from asymmetrical to symmetrical.

The symmetrical short-circuit current is calculated by dividing the per-unit voltage by the per-unit equivalent reactance at the point of the fault. This current should be multiplied by the appropriate factor from Table 3 of ANSI/IEEE Standard C37.13-1981 for the corresponding X/R ratio of the total network impedance at the point of fault.

Unlike synchronous machines, induction machines do not have separately excited field windings. However, there is induced electromotive force (emf) inside the rotor. According to the theorem of constant flux, this emf cannot change to zero instantaneously upon the occurrence of fault in the system. However, the terminal voltage of the induction motor will drop; hence, the torque and, consequently, the speed will drop. Thus, the internal emf slowly decays. During this period, the induction machine will contribute current to the fault.

The induction machine is also modeled by a constant voltage behind its subtransient reactance (X_d''). The same type of model is used for both squirrel-cage and wound rotor induction motors. In some wound rotor motors, the external resistance in the rotor winding may cause a very rapid decay of the internal emf. In such cases, these motors need not be included in the calculations.

If motor impedances are unknown, as is the case in a new installation, the contribution from groups of induction motors can be assumed to be 4.0 times the full-load current of the motors. For synchronous motors, the contribution can be assumed to be 4.6 times the full-load current.

In those cases in which the X/R ratio is unknown, or for approximate calculations, a ratio of 20 should be assumed, and the appropriate factor from Table 3 of ANSI/IEEE Standard C37.13-1981 for the type of breaker should be selected.

Transformers, overhead lines, busways, conductors, and cables are represented by their respective impedances. Impedances for components such as bus structures, connections, and current transformers are usually neglected. High-voltage and medium-voltage network resistances are low, compared to the reactance; hence, resistances may be ignored. However, for lower-voltage networks, resistance values may be comparable to reactance values and should not be ignored.

The short-circuit calculation is based on typical data for the equipment and facility. The short-circuit calculation methodology is well known to the practicing engineers. Mostly, readily available computer programs (on PCs these days) are used to make these calculations. However, it is important to have a clear understanding of the methodology and basis for the short-circuit calculations.

The computer programs allow proper representation of all equipment to the necessary detail and fast calculation. Once the model is set up, it is easy to make changes to the model and repeat the calculations for different conditions. Thus, the tediousness of short-circuit calculations has been almost eliminated.

3.5.1 Types of Faults

In industrial systems, three-phase fault currents are higher than line-to-ground fault currents, whereas in utility systems it is usually the other way. Hence, in industrial systems, three-phase fault calculations are more important than the line-to-ground fault calculations. Because three-phase faults need only a balanced system representation, the calculations are also simpler. But with the use of computers, line-to-ground or line-to-line fault levels can also be calculated without much difficulty. The only requirement for these additional types of fault calculations is that the zero sequence impedance data for the system components and equipment need to be input to the computer program. The usual assumptions made for simplified three-phase fault calculations are:

1. The three-phase system is balanced.
2. The balanced three-phase system is represented by p.u. phase impedances (positive sequence) as a single-phase system and line-to-neutral voltage. Note that use of the per-unit system for all calculations is recommended.
3. Impedances of all major components and equipment are included in the calculations.
4. All static loads, shunt capacitors and cable charging, and static power converters are not represented.
5. Faults are considered to be bolted (zero arc resistance) faults.

EXAMPLE 3.7 *The simple three-phase fault level calculation will be illustrated through an example system as shown in Fig. 3.10 with the impedances as shown.*

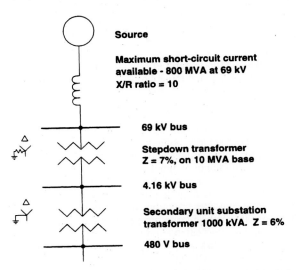

FIGURE 3.10 A simple three-phase short-circuit calculation example.

1. Convert impedances to p.u. value
   a. Select 10 MVA as base MVA.
   b. Compute source impedance or reactance.

$$X_S = \frac{\text{base MVA}}{\text{short-circuit level}} = \frac{10}{800} = 0.0125 \text{ p.u.}$$

Assuming, X_S/R_S to be 10:

$$R_S = 0.00125 \text{ p.u.}$$

c. Compute stepdown transformer impedance. Transformer impedance is 7 percent on 10-MVA base

$$X_{t1} = 0.07 \text{ p.u.}$$

If X/R ratio is known, or losses are known, resistance can also be computed. Resistance value can also be assumed by using data in ANSI/IEEE Std. C37.0-1979.

By assuming an X/R of 16,

$$R_{t1} = 0.0044 \text{ p.u.}$$

d. Compute secondary unit substation transformer impedance. This transformer impedance is given to be 6 percent on 1000-kVA base:

$$X_{t2} = \frac{6}{100} \times \frac{10 \text{ MVA}}{1000 \text{ kVA}} = 0.6 \text{ p.u.}$$

Assuming X/R of 5,

$$R_{t2} = 0.125 \text{ p.u.}$$

2. Form an equivalent circuit (Fig. 3.11).

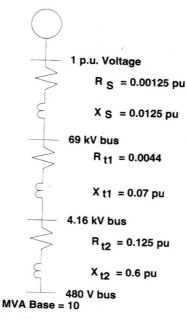

1 p.u. Voltage

$R_S = 0.00125$ pu

$X_S = 0.0125$ pu

69 kV bus

$R_{t1} = 0.0044$

$X_{t1} = 0.07$ pu

4.16 kV bus

$R_{t2} = 0.125$ pu

$X_{t2} = 0.6$ pu

480 V bus

MVA Base = 10

FIGURE 3.11 Equivalent circuit for simple system example.

3. Compute the total equivalent impedance from source to fault location.
4. Compute three-phase fault levels by using E/Z.

5. Three-phase fault current $= \dfrac{\text{calculated fault MVA} \times 1000}{\sqrt{3} \text{ voltage } L{-}L \text{ kV}}$

The calculated fault current values are as shown in Table 3.11.

TABLE 3.11 Calculated Three-Phase Fault Level for Example 3.7

Fault location	Total impedance (pu)	Fault level (MVA)	3-phase short-circuit) current (A)
69-kV bus	$0.00125 + j\,0.0125$	800 (given)	6,693 A
4.16-kV bus	$(R_s + R_{t1}) + j\,(X_x + X_{t1})$ $= 0.00565 + j\,0.0825$	121	16,790 A
480-V bus	$(R_s + R_{t1} + R_{t2}) +$ $j\,(X_s + X_{t1t} + X_{t2})$ $= 0.13065 + j\,0.6825$	14.4	17,320 A

EXAMPLE 3.8 *Calculate line-to-ground fault current for the system in Example 3.7.*

For calculating line-to-ground fault levels, we need to create a zero sequence network similar to the positive sequence network shown in Fig. 3.11. For the sake of simplicity, we will assume that the zero sequence impedance of the source is the same as the positive sequence value. A grounding resistance of 5 Ω (400-A rating) or 0.0036 p.u. on 10-MVA base and 4.16 kV is assumed. The transformer connections are treated differently in zero sequence network,[2] and note that there is no voltage source. The zero sequence equivalent circuit is shown in Figure 3.12.

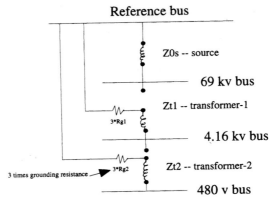

FIGURE 3.12 Zero sequence network for Example 3.8.

Let us say that we want to calculate the L-G fault levels at 69-kV, 4.16-kV, and 480-V buses. The zero sequence network needs to be connected in series with the positive sequence network (symmetrical components), as shown in Fig. 3.13(a), (b), and (c). The negative sequence impedance is assumed to be the same as the positive sequence imped- ance values. Hence, the negative sequence network is not explicitly shown. The L-G fault current is given by

$$I_f = \frac{3\,V_S}{Z_1 + Z_2 + Z_0} \tag{3.18}$$

$$= \frac{3 V_S}{2\,Z_1 + Z_0}$$

where Z_1 = total positive sequence equivalent impedance as seen from the fault point
Z_0 = total zero sequence equivalent impedance as seen from the fault point

The calculated values are shown in Table 3.12.

For calculating L-G fault level, the sequence network system shown in Fig. 3.13(a) applies. Note that in the zero sequence network, the grounded wye connection and the source are in parallel. This net zero sequence impedance is shown in column (3) of Table 3.12. The sum of positive and negative sequence values (two times positive sequence impedance, because of assumption $Z_1 = Z_2$) are shown in column (2) of this table. The total impedance ($2Z_1 + Z_0$) is shown in column (4). The fault level in p.u. is reciprocal of this total impedance. The fault current in p.u. and amps is shown in columns (5) and (6), respectively.

Now consider the L-G fault at a 4.16-kV bus. Because the 69/4.16-kV transformer is wye-connected on the high side, a single-phase current cannot flow into the high side. However, there is a ground path through the grounded wye side of the 69/4.16-kV trans- former. A grounding resistance (R_g) of 5 Ω (rated 400 A) is assumed here. This is shown in the zero sequence network connection in Fig. 3.13(b). The equivalent network imped- ances and the fault levels are as shown in Table 3.9.

EXAMPLE 3.9 *For the sample system shown (example 3.7), the three-phase fault and zero sequence fault currents were computed by using a commercial short-circuit pro- gram. Assume 0.012 p.u. and 0.05 p.u. zero sequence impedance for the utility source on a 100-MVA base and the plant generator is ungrounded. The sequence impedance data and the results are shown in Tables 3.13 and 3.14.*

TABLE 3.12   Calculated L-G Fault Levels for Example 3.8

(1) Fault location	(2) Positive + negative seq. impedance (p.u.)	(3) Zero seq. impedance (p.u.)	(4) Total impedance (p.u.)	(5) L-G fault level (MVA)	(6) L-G fault current (A)
69-kV bus	0.0025 + j 0.025	0.00125 + j 0.0125	0.00375 0.0375	800.1	6,695
4.16-kV bus	0.01130 + j 0.1625	0.0152 + j 0.07	0.0265 + j 0.2325	128.2	17,792
480-V bus	0.2613 + j 1.365	0.125 + j 0.6	0.3863 + j 1.965	14.98	18,019

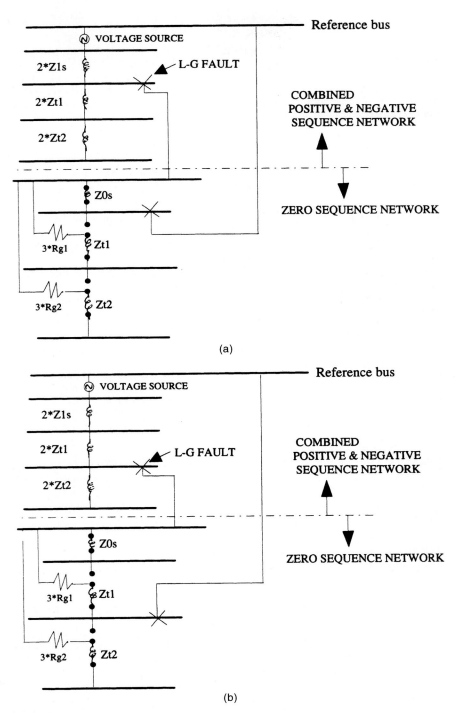

FIGURE 3.13 Sequence network connection for fault calculation of Example 3.7.

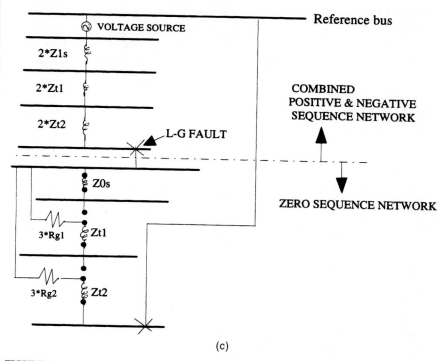

(c)

FIGURE 3.13 (*Continued*) Sequence network connection for fault calculation of Example 3.7.

3.6 MOTOR STARTING

The purposes of studies for motor starting are:

- Calculation of voltage drop during starting
- Determination of motor acceleration capability and ability to come up to speed especially under full-load conditions
- Coordination of overcurrent and undervoltage protection
- Determination of suitable starting methods and any special switching arrangements

The motor starting calculations are needed, especially when large motors (as compared to the power system capacity), special applications, and unique loads are involved.

Starting a motor is an electromechanical problem, because motors are used to drive some type of load, whether fan, pump, compressor, or drives. Many motors have to start on full or partial load. Hence, consideration of both mechanical load and electrical characteristics is important. Because most of the motors used are of the induction motor type, this section will deal only with this type of motor.

An induction motor at no-load runs nearly at synchronous speed and generates torque to overcome friction and windage losses. As the load increases, the induction

TABLE 3.13 Zero Sequence Impedance Data for an Example Electric Power System of an Industrial Plant

Bus		Zero sequence impedance (p.u.)		
From	To	Resistance	Reactance	Susceptance
100	101	0.0045	0.0089	0.0154
100	102	0.0045	0.0089	0.0154
101U	103GR	3.1646	0.0533	0
102U	104GR	3.1646	0.0533	0
103	105	0.0077	0.0034	0
103	108	0.0133	0.0048	0
103	123	0.0438	0.0057	0
104	106	0.0065	0.0029	0
104	107	0.0024	0.0030	0
104	121	0.0300	0.0191	0
104	135	0.0051	0.0015	0
105	111	0.0567	0.0173	0
106	113	0.0768	0.0100	0
106	124	0.0127	0.0046	0
108U	114G	0.0529	0.4500	0
108	122	0.1185	0.0154	0
109U	115G	0.0529	0.4500	0
109	124	0.0306	0.0040	0
110U	135G	0.0458	0.3670	0
111U	116G	0.0244	0.1470	0
112U	117G	0.0244	0.1470	0
112	135	0.0068	0.0021	0
113U	118G	0.2874	2.3010	0
114	119	0.1746	0.0729	0
115	120	0.1392	0.0717	0
121	128	0.0220	0.0029	0
121	129	0.0312	0.0041	0
122U	125G	0.0548	0.3830	0
123U	126G	0.0548	0.3830	0
124U	127G	0.0548	0.3830	0
125	130	0.0873	0.0690	0
126	131	0.1353	0.0390	0
127	132	0.0873	0.0690	0
128U	133G	0.0329	0.2300	0
129U	134G	0.0821	0.5750	0

NOTE: U—Ungrounded side

G—Grounded side

GR—Grounded with 20 Ω resistance

p.u. impedance on 10-MVA base

TABLE 3.14 PTI Interactive Power System Simulator—PSS/E

An example electric power system of an industrial plant fault current, flat conditions (on 10-MVA base)

At bus		Three-phase fault current		Line-ground fault current	
No.	Name	Magnitude (p.u.)	Angle (degrees)	Magnitude (p.u.)	Angle (degrees)
100	[Utility 69.0]	335.39	−86.19	271.11	−81.71
102	[Tie 2 69.0]	164.03	−74.21	110.44	−70.54
101	[Tie 1 69.0]	161.92	−73.94	109.78	−70.45
104	[Main 2 13.8]	20.01	−82.88	0.94	−2.75
107	[GEN−1 13.8]	19.72	−82.42	0.94	−2.82
135	[FDR 411 13.8]	19.72	−81.06	0.94	−2.79
106	[FDR 42 13.8]	19.53	−80.60	0.94	−2.82
124	[FDR 71 13.8]	18.67	−76.40	0.93	−2.94
112	[FDR 41 13.8]	18.46	−74.33	0.93	−2.88
109	[FDR 271 13.8]	17.22	−66.89	0.92	−3.01
121	[FDR 44 13.8]	17.17	−73.86	0.93	−3.27
103	[Main 1 13.8]	16.72	−81.95	0.94	−3.09
105	[FDR 34 13.8]	16.30	−79.72	0.94	−3.18
128	[FDR 242 13.8]	16.25	−67.54	0.92	−3.31
108	[FDR 31 13.8]	16.08	−78.10	0.94	−3.21
129	[FDR 243 13.8]	15.83	−65.10	0.91	−3.33
113	[FDR 72 13.8]	15.71	−57.76	0.90	−3.00
123	[FDR 33 13.8]	15.31	−69.40	0.92	−3.19
111	[FDR 61 13.8]	13.78	−65.71	0.91	−3.59
122	[FDR 91 13.8]	11.88	−51.58	0.88	−3.46
117	[Load 7 2.40]	4.93	−78.91	5.41	−79.36
116	[Load 6 2.40]	4.55	−75.72	5.10	−76.95
133	[Load 14 2.40]	3.42	−78.88	3.67	−79.72
110	[Load 2 2.40]	2.38	−82.67	2.48	−82.73
127	[T12 0.48]	2.27	−81.19	2.37	−81.40
126	[T11 0.48]	2.22	−80.07	2.33	−80.60
125	[T10 0.48]	2.17	−76.58	2.29	−78.14
132	[Load 13 0.48]	2.14	−78.10	2.13	−76.24
131	[Load 12 0.48]	2.11	−74.95	2.13	−71.94
130	[Load 11 0.48]	2.04	−73.90	2.06	−73.46
115	[T 6 0.48]	1.96	−81.45	2.04	−82.02
114	[T 5 0.48]	1.94	−82.67	2.02	−82.86
120	[Load 10 0.48]	1.85	−76.96	1.83	−74.53
119	[Load 9 0.48]	1.82	−76.96	1.80	−73.39
134	[Load 15 0.48]	1.56	−80.25	1.61	−80.75
118	[Load 8 0.48]	1.20	−81.02	0.75	−82.10

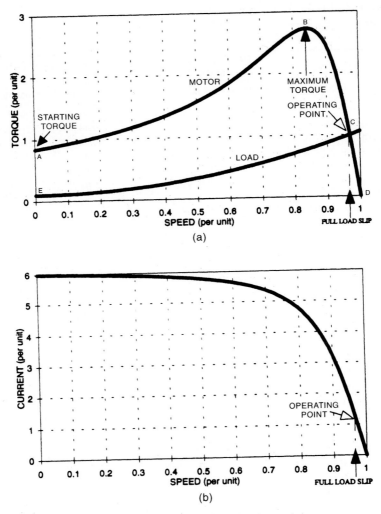

FIGURE 3.14   Induction motor and drive speed–torque characteristic.

motor slows down and has a maximum torque at certain slip (the per-unit deviation from the synchronous speed). The torque-speed characteristic of an induction motor is shown in Fig. 3.14.

The normal operating point is the intersection load and motor characteristic. The motor characteristic may vary depending upon the type, construction, the winding resistances, and reactances. When the motor switch is closed, the starting torque developed is shown by point *A* in Fig. 3.14(a). This torque should be sufficient to overcome the load torque (shown by point *E*) and accelerate the motor along the path *A-B-C* in the figure, ultimately reaching a steady state balance as shown by point *C*. The difference between

·the energy consumed by load and the equivalent mechanical energy put out by the motor is used up as stored kinetic energy (due to inertia of the moving parts). If there is no external mechanical load, the motor reaches point *D*, attaining nearly the synchronous speed.

The torque vs. speed characteristic shown in Fig. 3.14(a) for the motor is based on the assumption of 100 percent voltage at the motor terminals. However, it can be shown that the torque developed by an induction motor is proportional to the square of the voltage. Hence, as voltage dips below 100 percent, the speed-torque characteristic drops down much faster. For example, if the terminal voltage is 70 percent of the rated voltage, then the motor can develop only 49 percent of the full-load torque. It should be further recognized that a motor operating in the *B-C-D* part of the speed-torque characteristic is stable (increasing torque means decreasing speed). Common loads, such as pumps, need power that is proportional to the cube of the speed or torque that is proportional to the square of the speed. Thus, the two mechanical requirements for starting a motor are

1. The speed-torque characteristic should be above the speed-load characteristic.
2. The full-load or normal-load operating point should be in the stable portion of the motor characteristic.

Now, let us turn our attention to the electrical side of the starting problem. The electrical equivalent circuit is as shown in Fig. 3.15. The magnetic core losses and magnetization current have been ignored in this figure.

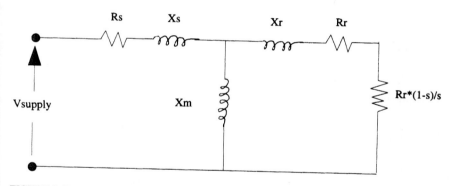

FIGURE 3.15   Simplified equivalent circuit of an induction motor.

At standstill or zero speed, the slip is unity (*A* in Fig. 3.14) and the equivalent load resistance is zero. The current in this situation is limited only by the motor winding impedance, which is mostly inductive. This current can be four to seven times the full-load current. This large current causes a large voltage drop across the system impedance and, consequently, the voltage (V_{supply}) at the motor terminals is reduced. Also, because the motor reactance is several times higher than the resistance, this current drawn is at a low lagging power factor. A 1 per-unit current at unity power factor causes a smaller drop in voltage magnitude as compared to the same magnitude of current at, say, 0.3 power factor. The electrical torque is proportional to the square of the voltage. Thus, any voltage drop will reduce the torque from the motor. This reduced torque during starting may result in several different types of problems, depending upon the type of application:

- The motor will take a longer time to come to full speed.
- The motor does not have sufficient torque to come to full speed or cannot drive the load.
- The voltage drop causes other loads to draw more current, compounding the problem and causing plantwide low-voltage problems.
- Frequent start and stop may also cause a flicker problem.
- Severe drop may cause other loads to drop out due to undervoltage and overcurrent protection actions.

The solutions to avoid or overcome such problems include:

- Recognizing the problem before purchasing and installing the motors
- Selecting proper motors for the load
- Using suitable starters or starting method
- Employing special switching arrangements for large motors or for unusual power system conditions

In the remaining part of this section, we will present a voltage drop calculation method for a motor-starting condition. The calculation methods may be divided into three broad categories:

- Hand calculation
- Use of a load flow type of program
- Stability study for complex analyses

The basic data requirements for these types of calculations are:

- Motor characteristics
- Electrical system data
- Load characteristics

Hand calculation may be used whenever a simple voltage drop type of calculation (Chap. 5) is sufficient. Many common applications require this type of check to determine if there is a problem and whether any further study is needed.

The load flow calculation permits representation of the entire plant or parts of the plant. Most often, a load flow system model is already available. The effect on, as well as the effect of, other loads on motor starting can be evaluated without much difficulty.

In uncommon instances, a more detailed and complex analysis of motor starting may be required. Any hand calculation or load flow for this type of analysis would be either very tedious or not applicable. A stability (see Sec. 3.7) study is used in such cases. Using the stability study, time-dependent values for voltage, torque, current, power, and time to reach steady state may be determined. Also, response of local or isolated generators, etc., may be easily obtained. More involved load representations may also be possible.

We will illustrate some of these methods through some examples:

EXAMPLE 3.10 *Consider a 10,000-hp, 13.8-kV induction motor, which has a full-load power factor of 0.9 (lag). The starting current of this motor is six times the load current and the power factor is 0.3 (lag) at starting.*

Assume a short-circuit level of 100 MVA for the supply system at the motor terminal.

$$\text{Motor load} = 10,000 \text{ hp} = 7.45 \text{ MW}$$

$$\text{Motor P.F.} = 0.9$$

$$\text{Motor load} = 8.28 \text{ MVA}$$

$$\text{Voltage (line-to-line)} = 13,800 \text{ V}$$

$$\text{Full-load current} = \frac{8.28 \times 10^6}{\sqrt{3} \times 13,800} = 346.4 \text{ A}$$

$$\text{System equivalent impedance} = X_{sys}$$

$$= \frac{(\text{line-to-line voltage in kV})^2}{\text{S.C. MVA}}$$

$$= \frac{13.8^2}{100} = 1.9 \ \Omega \text{ (reactance)}$$

$$I_{start} = 6 \times I_{FL} = 2078.4 \text{ A}$$

$$\text{Starting power factor} = 0.3$$

$$I_{start} = 623.5 - j1982.7 \text{ A}$$

The voltage drop in the system due to starting the motor is

$$\Delta V_1 = I_{start} * jX_{sys} = 3767.1 + j1184.6 \text{ V}$$

Note that this ΔV_1 is a line-to-ground value.

$$\text{Motor bus voltage} = \sqrt{3} \left[\frac{13,800}{\sqrt{3}} - (3767.1 + j1184.6) \right]$$

$$= 7559 \ \underline{/-15.8^\circ}$$

The bus voltage magnitude is 55 percent at the instant of starting. As the motor accelerates and picks up speed, the current drawn from the system decreases, ultimately reaching load current.

Hand calculations used here are simple and straightforward. Load flow programs may also be used for these calculations.

The actual steps for using the power flow program for this purpose are dependent on the actual program being used. However, the steps in general are as follows:

1. Solve the load flow of the plant electric system to establish the plant electric system condition before starting the motor. This is sometimes called t^- system condition.

2. Convert the generator and loads to some type of constant characteristic representation. This step is necessary because we are interested in finding the voltage at the instant of the starting motor. For example, the power consumed by a resistance type of load will instantaneously go down with the voltage drop, whereas generator voltage cannot change instantaneously. Other motor loads will not change instantaneously.

3. Connect a shunt impedance corresponding to the motor-starting condition at that bus. The bus voltage before motor starting and the expected starting current are used in calculating this impedance.

4. Solve the load flow for this condition. The resulting bus voltages reflect the condition when starting the motor. However, this will not represent voltage with respect to time, i.e., as the motor accelerates. This is called "t plus p", i.e., time just after closing the motor starter switch.

5. It is possible to get approximate voltage vs. time by repeating step 4 after changing the effective motor impedance. The time relationship and the change of impedance has to be established outside the load flow program. This type of voltage vs. time calculation is best done by using a stability type of program. The same type of stability program is used to study motor reacceleration and out-of-phase reclosing problems after short-duration faults on the utility systems.

EXAMPLE 3.11   *Using the plant electrical system from Example 3.5, compute plant bus voltages when starting a 150-hp motor on bus 130. Assume the starting current is six times the full-load current and the starting power factor is 0.3. Use the power flow computer program for this purpose.*

The results from using load flow for computing voltage for this example are shown in Figs. 3.16 and 3.17.

System condition before starting the motor is shown in Fig. 3.16. At bus 130, there is only a load of 138 kW and 83 kVAR, which is the original load minus the 150-hp motor about to be started. All the bus voltages are above a minimum acceptable level of 0.95 p.u.

By following steps 2 to 4, the system condition after closing the motor breaker will be as shown in Fig. 3.17. The voltage on bus 130 drops approximately 4 percent.

3.7 STABILITY

Most of the electrical power is generated by the electric utilities. However, cogeneration projects and in-plant generation for reasons of reliability and continuity of power supply, as well as cost, have contributed to increasing generation within industrial and commercial establishments. But these in-plant generators supply only a part of the plant's power needs. Also, in some establishments, the excess power is sold to electric utilities. Hence, plant engineers need to deal with the problem of synchronous operation of the in-plant generators, not only within the plant, but also with the electrical power system.

Stability as applied to the electrical power system is an electromechanical phenomenon. The electrical power system operates at a constant nominal frequency (60 or 50 Hz). All the generators are constantly and continuously controlled such that the electrical power (P_{elec}) output from these generators is equal to the total input mechanical power (P_{mech}) from the turbine (or prime mover) which drives the generator. This type of control is performed on an individual generator as well as on a selected area basis. However, there are many situations which may upset this delicate balance. Some of these are:

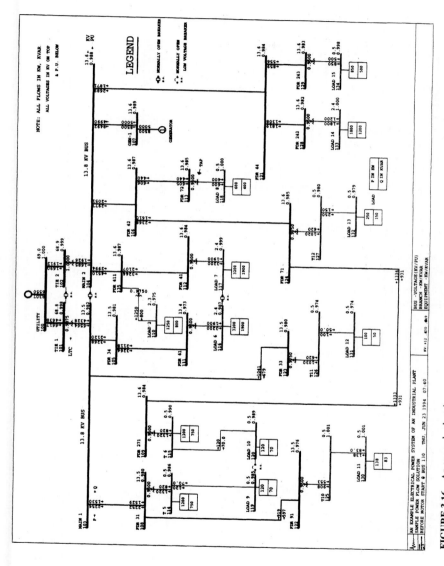

FIGURE 3.16 An example electric power system of an industrial plant sample power flow solution before 125-kW motor start at bus 130. (*Courtesy of Power Technologies Inc, Schenectady, NY.*)

3.45

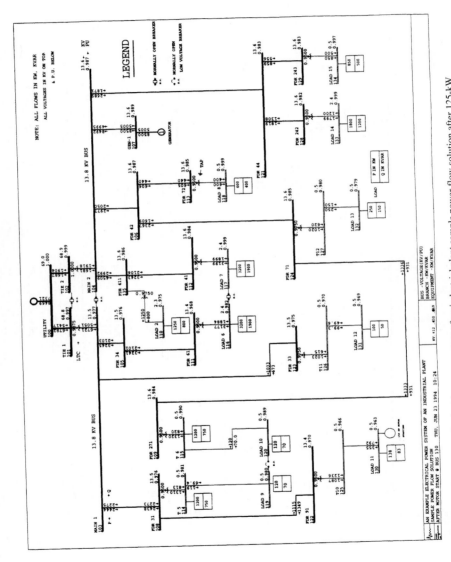

FIGURE 3.17 An example electric power system of an industrial plant sample power flow solution after 125-kW motor start at bus 130. (*Courtesy of Power Technologies Inc, Schenectady, NY.*)

3.46

- Faults (temporary short circuits)
- Loss of tie(s) to the power supply utility
- Loss of in-plant generating units
- Sudden or large step change of load or generation
- Switching of lines, capacitors, etc.

For example, consider a system with two synchronous machines shown in Fig. 3.18. Let us say that a three-phase fault occurs on one of these lines. As soon as the fault is detected by the protective relaying of this system, the faulted line will be disconnected. This may take a few to several cycles (on 60- or 50-Hz basis). During this time, electric power flow into or out of these synchronous machines is reduced or increased. The actual reduction or increase depends upon the type of fault, its location, and the system. However, within this short period it is not possible to decrease or increase the input mechanical power (steam or gas) into the turbine. Thus, the generators either speed up or slow down based on the laws of motion and the connected electrical network. A fault

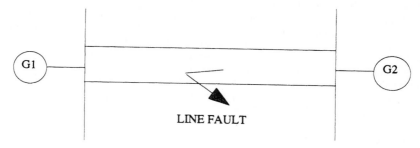

FIGURE 3.18 Two synchronous machines example.

near a generator will reduce the power the generator is furnishing the load, so until the input power to the generator from the prime mover is reduced, the generator will accelerate. The electromechanical motion may be described by the second-order differential equation

$$M \cdot \frac{d^2\delta}{dt^2} = P_{mech} - P_{elec} \tag{3.19}$$

where M = inertia of the generator and turbine rotor assembly
 δ = generator angle
 P_{mech} = input mechanical power
 P_{elec} = output electrical power
 $\dfrac{d\delta}{dt} = w$ = angular frequency

Solving this type of equation for an electrical power system is called *transient stability*. After solving these equations for all the generators within the system, the system is said to be transiently unstable if the difference between angle (δ) of any two generators within the system exceeds 180°. This is also often called an out-of-step condition, slipping of poles, or loss of synchronism. When a generator angle exceeds 180°, the generator is subjected to large current flow and unusually severe torque loading. Hence, appropriate protection is provided to detect such conditions (especially on large generators) and promptly disconnect the generator. The consequences of instability, when it happens, can range from a simple and local problem (such as a small 5-MW generator tripping when it is connected to a large 1000-MW-plus electric power system) or a major systemwide problem including:

- Areawide blackout
- Interruption of loads
- Low-voltage condition
- Damage to equipment

If a blackout or damage to equipment occurs, the restoration to normal operation is time consuming and expensive. We have described the concept of transient stability which pertains only to the performance of the power system for a few seconds to several seconds after the disturbance. There are several other types of stability, such as steady state and dynamic stability. The reader is referred to other books on these topics.

The transient stability studies used to be performed by hand calculations. Then came the analog simulators, followed by digital computers. It is difficult to conceive of any person performing a transient stability calculation other than by using a digital computer program. The state of the art of these programs is such that very sophisticated system equipment, their modeling, controls, and myriad types of switchings can all be simulated. The user can select a simple or complex simulation, depending upon the type of problem being solved. The types of data required include:

- Electrical network data for load flow
- Electrical generator/motor parameters
- Details of control, transfer function, etc.
- Details of faults, fault clearing times, etc.
- Details of other types of switching such as intentional breakup of system, load shedding, etc., to minimize impact of uncontrolled loss of synchronism
- For longer times (1.0 s), the mechanical system parameters such as voltage regulators, prime movers, governors, etc.

The output available from these types of programs includes:

- Change of generator angles [δ in Eq. (3.19)] with respect to time (called *swing curves*)
- Bus voltages (magnitude and angle)
- Line flows (MW, MVAR, MVA)

The input and output of this type of program will be described through the following example.

EXAMPLE 3.12   *Using the plant electrical system from Example 3.5, determine the critical clearing time for a three-phase fault on one of the 69-kV utility tie lines, say from bus 100 to bus 102.*

1. As a first step, before running a stability program, a solved power flow case should be established. This is shown in Fig. 3.7. For this stability example, assume that the tie breaker between buses 103 and 104 is closed and resolve the power flow problem of Fig. 3.7. This will basically make the power flow through the two utility tie lines and the two stepdown transformers equal (see Fig. 3.19).

2. Critical clearing time is the time within which, if the fault is cleared, then the system is stable (i.e., all the generators are in synchronism or stay together). If the fault is cleared after the critical clearing time, then at least one generator loses synchronism or becomes unstable and the generator protection disconnects the unstable generator unit. Unstable means that parts of the system become separated from each other. Some parts of the system may continue to operate, but with loss of load, overload, and low- or high-voltage conditions.

3. Utility generation model: Additional data describing the characteristics of the generators and controls are necessary to run stability computer programs. Typically, the utility generation is very large as compared to the plant generator unit sizes. Hence, utility generation is represented by a constant voltage behind a reactance (based on a three-phase short-circuit level at the utility bus). This method of electrical representation is called the *classical model* for the generator units. Because the total inertia of all the utility generator units is very large compared to the plant generator units, any fault of reasonable duration on the lower-voltage tie lines or within the plant will not cause any change in the frequency of the utility system. Hence, the inertia constant is assumed to be zero. The following values are assumed in this example for the utility generator connected to bus 100.

$$\text{Generator (or source) impedance} = 0.002 + j0.03 \text{ p.u. on 100-MVA base}$$

$$\text{Inertia constant } H = 0$$

4. Plant generator models: Because we are interested in the response of the plant generator units to faults with respect to utility generation and plant load, more detailed representation of generator units in the computer program is desirable. The extent of detail needs to be tempered with the data availability, accuracy of the data, and stability results. This requires familiarity and experience in performing stability studies.

    Almost all generators have an exciter/voltage regulator and a governor/speed control system. It may be desirable to represent these in the stability studies, especially if the system performance or stability problem becomes critical. Whereas in robust systems when the system is stable for a wide range of disturbances, then representing the plant generation by classical models may be sufficient. There are many types of models available in standard software packages for generators, exciters, governors, relays, etc. The data format and so forth are different from one program to another. The plant generator unit model parameters and their values for this example are given in Table 3.15. These values are usually available from the manufacturer of generator units. In this example, we will assume the models shown in Figs. 3.20 and 3.21 for the exciter and governor. The constants for these models are also given in Tables 3.16 and 3.17, respectively.

5. Now, consider a three-phase fault on the 69-kV tie line to the utility (from bus 100 to 102) which is cleared by opening the breakers at both ends of the line in a normal clearing time of 0.15 s (9 cycles based on 60 Hz). This condition was simulated by using Power Technologies, Inc.'s (PTI) PSS/E stability program. Results are shown in a number of plots in Fig. 3.22(a)–(f).

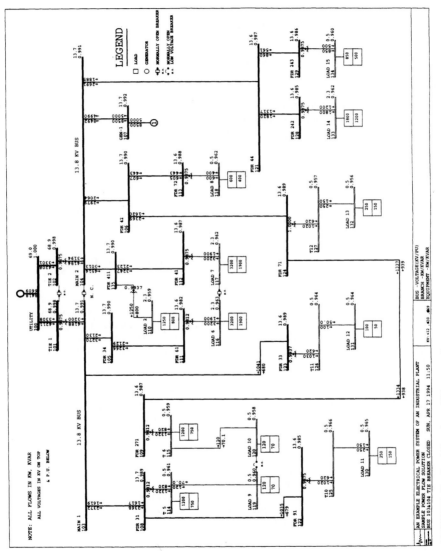

FIGURE 3.19 Load flow for stability example.

TABLE 3.15   Plant Generator Unit Model Constants

All values are in p.u. on generator unit MVA base (10 MVA in this example)

Parameters	Value
T_{do}' (>0) (s)	6.5
T_{do}'' (>0) (s)	0.06
T_{qo}' (>0) (s)	0.2
T_{qo}'' (>0) (s)	0.05
Inertia H	4
Speed damping D	0
X_d	1.8
X_q	1.75
X_d'	0.6
X_q'	0.8
$X_d'' = X_q''$	0.3
X_l	0.15
Saturation points $\quad S(1.0)$	0.09
$S(1.2)$	0.38

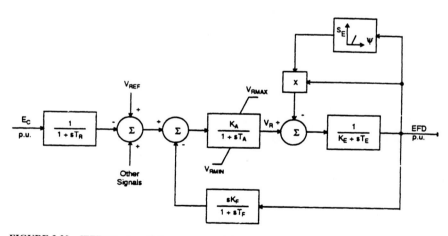

FIGURE 3.20   IEEE type 1 excitation system model.

TABLE 3.16 IEEE Type 1 Excitation
System Parameter Values for Model
Shown in Figure 3.20

Parameter	Value
T_R (s)	0
K_A	40
T_A (s)	0.04
V_{Rmax} or 0	6.615
V_{Rmin}	−6.615
K_E or 0	1
T_E (>0) (s)	0.8
K_F	0.02
T_F (>0) (s)	1
Switch	0
E_1	3.375
$S_E(E_1)$	0.035
E_2	4.5
$S_E(E_2)$	0.47

TABLE 3.17 Steam Turbine Governor
Parameter Values for Model Shown in
Figure 3.21

Parameters	Value
Droop constant − R	0.05
T_1 (>0) (s)	0.05
V_{max}	1.05
V_{min}	0.3
T_2 (s)	0
T_3 (>0) (s)	0.5
D_t	0

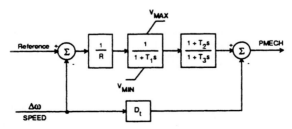

FIGURE 3.21   Steam turbine governor model.

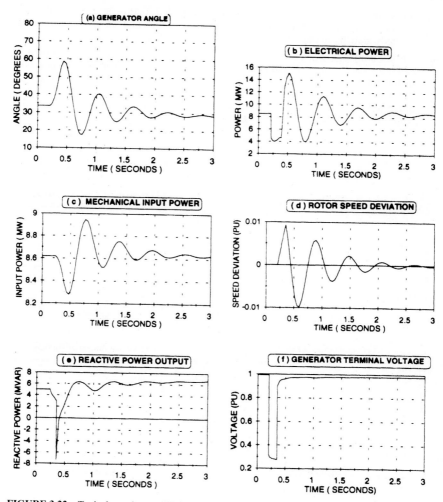

FIGURE 3.22 Typical transient stability program output plots for Example 3.12, stable case.

The angle for the plant generator is shown in Fig. 3.22(*a*). All along we have assumed that the utility system is not affected by this relatively local fault. Hence, the angle plotted in Fig. 3.22(*a*) is relative to the utility generator. In the plot, the fault is applied at 0.2 s. The constant values during this 0–0.2-s period show steady state conditions. When the three-phase fault occurs, the voltage at bus 102 is nearly zero. The mechanical power input to the generator stays nearly constant [Fig. 3.22(*c*)] and, consequently, speeds up the rotor, because the electric output of the generator is drastically reduced [Fig. 3.22(*b*)] due to low or zero voltage at bus 104. The generator rotor angle increases until the fault is cleared. The governor on the turbine tries to reduce the mechanical power [Fig. 3.22(c)]. After clearing the fault, the generator supplies a higher percentage of plant load and slowly reaches a nearly steady state by about 3.0 s. Hence, the system is stable.

6. In this example, the question was to determine the critical clearing time or the lowest fault clearing time when the plant generator becomes unstable. This is accomplished by running several cases with increasing fault clearing times and zeroing-in on the critical clearing time. In this example, the critical clearing time was determined to be 0.41 s. The results shown in Fig. 3.23(a)–(f) for this clearing time exhibit borderline stability.

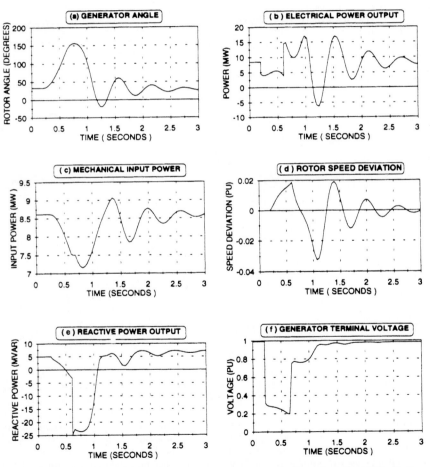

FIGURE 3.23 Typical transient stability program results plots for Example 3.12, stability limit or critical clearing time case.

3.8 PROTECTION AND COORDINATION

Protection and coordination is of extreme importance to a plant engineer. Hence, a separate chapter (Chap. 8) is devoted to the topic in this book. In this section, the discussion is strictly introductory in nature.

Protection and coordination studies go together, because one study without the other is not very useful. However, to keep the concepts of these studies clear, the topic may be defined as two separate studies.

The objective of protection is to provide safety and continuity of service. Here safety means that hazard to personnel, environment, equipment, or operating system is avoided. The continuity of service means that there will be no disconnection of equipment and facilities unless there is a problem (no false trips) and any isolation of equipment is limited to only that part of the system, process, or equipment that is faulty. Sometimes disconnection of a given equipment may be required due to process, safety, or system consideration, even though some other related equipment has developed a problem. Thus, the objective of protection is achieved through suitable application of relays, fuses, circuit breakers, and other protective devices. The relays perform the task of detecting abnormal operating conditions and the interrupting/isolating devices perform the disconnection function. Before starting a protection study, a short-circuit study is required.

The objective of a coordination study is to ensure that the multiple protective relays and devices function in a coordinated manner. Any given equipment and system will have many types of protective devices. It is important to ensure that the application requirements do not conflict with each other. Any ambiguity in the application should be resolved. The starting point in the coordination study is completion of a protection study which may be included within a coordination study itself. Several iterative steps between selection of protective devices and their coordination may be needed before an acceptable solution to the problem is found. Resolving conflicts and arriving at a compromise by trying several options is very important. The knowledge and experience of the application engineer is vital for this purpose.

The protection and coordination studies were done manually using french curve templates, graph paper, and a light table. However, several computer programs are available now. They provide the capability for interactive use, a database of relay, fuse, and device characteristics, and the means for output, including graphical output. The final settings may be stored on hard or floppy disks for later use. Thus, it may be said that these computer programs have removed the tediousness of coordination studies and, at the same time, they offer the design engineer the means to do a much more comprehensive study. The amount of time and money saved by using appropriate computer programs in performing these studies may also be substantial, especially when studies are repeated.

3.9 SWITCHING TRANSIENTS

Switching transients studies may be called the most complex type of study. This type of study deals with the electromagnetic and electrostatic response of the electrical system and equipment from power frequency to 10 kHz. The main reason for this type of study is to determine the extent of exposure of equipment and facilities to overvoltages and high currents. Equipment and facilities are subjected to several types of stresses due to currents and voltages. The currents imposed on a system are full-load, switching, and short-circuit currents. The current causes ohmic losses and creates thermal stress. Fault currents also impose high mechanical stresses. For example, the force between two parallel conductors is a function of I^2. The voltage impressed on equipment includes lightning-induced (impulse), switching surges, and normal operation voltage. These voltages subject the insulation (dielectric material) to stress. The level of stress (voltage magnitude) and time (duration) of exposure of the equipment determine whether the equipment is damaged. For most equipment, the shorter the exposure time, the higher the levels of stress that can be permitted. However, this relationship is not necessarily linear. Also, the thermal time constants are much longer as compared to the dielectric breakdown time constants. The surrounding environment, the type of equipment, and the power system source characteristics have a great influence, both on the level of transients generated and the permissible levels for safe operation (see Chap. 10).

Switching of electric equipment involves applying or removing a voltage or current. Both supplyside as well as the equipment to be switched have resistances (R), inductances (L), and capacitances (C). The actual magnitude of these three parameters and their relative values with respect to each other of the same equipment, as well as with other parts of the system, can vary over a very wide range. Whenever an R, L, and C circuit is switched, a sudden voltage is applied or removed. An inductance stores electromagnetic energy, a capacitor stores electrostatic energy, and the resistor loses energy through heating. Because of the interaction between the electromagnetic energy in the inductor and the electrostatic energy in the capacitor; overvoltages may be generated when sudden changes, such as switching, occur. Any resistance present in the circuit provides damping of these overvoltages through energy or ohmic losses.

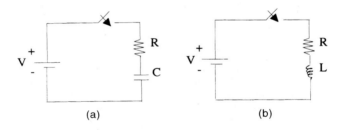

(a) (b)

FIGURE 3.24 Simple switching example.

Consider a simple RC circuit, shown in Fig. 3.24(a). The source is assumed to be an ideal voltage source (constant voltage and zero internal resistance). We can define that the time $t = 0^+$, when the switch is closed. It is a common practice to add superscripts ($-$) to denote time before and (+) for immediately after closing the switch. Assuming that there is no remnant charge (or initial voltage) on the capacitor before the switch is closed, the differential equation governing this switching operation is

$$R \cdot i + \frac{1}{C} \int_0^t i \cdot dt = V \tag{3.20}$$

The solution to this type of equation is given by

$$i = \frac{V}{R} \cdot e^{-t/RC} \tag{3.21}$$

The value of current is highest immediately after the switch is closed and exponentially decreases to zero at steady state ($t = \infty$). If there is any charge left on the capacitor, then the initial capacitor voltage is added (taking into account proper polarity) to the applied voltage V. The constant RC is called the time constant of this RC circuit.

Now consider a simple RL circuit, shown in Fig. 3.24(b). The differential equation for the switching of this circuit is

$$R \cdot i + L \frac{di}{dt} = V \tag{3.22}$$

and the solution to this type of differential equation is

$$i = \frac{V}{R}(1 - e^{-(R/L)t}) \tag{3.23}$$

Because current cannot be established in an inductor instantaneously, the current at $t = 0$ is zero (the exponential term becomes 1) and reaches exponentially to a magnitude of (V/R) at steady state ($t = \infty$, the exponential term becomes 300). The time constant for this type of circuit is given by (L/R).

The combination of R, L, and C in practical problems, of course, makes the solution of switching problems more complex. In switching problems in power systems, the source is an alternating sinusoidal waveform, not a constant dc source. The sinusoidal waveform is nonlinear and varies with time. Thus, analysis of practical power system switching problems tends to be more complex. The reader should consult other books for the physics and mathematics involved in solving such problems. For a simple understanding of the transient voltages in an ac switching problem, typical transients generated while switching R, L, and C circuits are shown in Fig. 3.25(a)–(c). In these figures, the source voltage (V_S) is represented by a sinusoidal voltage (note that cosine and sine differ by 90° in phase angle) waveform with a magnitude of 1 p.u., a frequency of f ($\omega = 2\pi f$), and a source reactance of X_S (or ωL_S). The loads are shown as a resistance, reactance, capacitance, or a combination of these. We will write some basic mathematical equations for three simple cases, so that complexities involved in computing transients in real systems (which may have a number of R, L, and C elements connected in many complicated manners) may be appreciated by the reader.

1. Pure resistive load:

$$V = V_S \cos \omega t$$

$$V_L = R \cdot I_L \tag{3.24}$$

$$V_L + L_S \cdot \frac{dI_L}{dt} = V_S \cos \omega t$$

For example, if the switch is closed when voltage is zero ($\omega t = 90°$), then the current through the resistor is also zero at this instant. Note that, even though the load is purely resistive, due to source reactance, it is an R, L circuit. The final load current magnitude is given by

$$I_L = \frac{V_S}{\sqrt{R^2 + (\omega L_S)^2}} \tag{3.25}$$

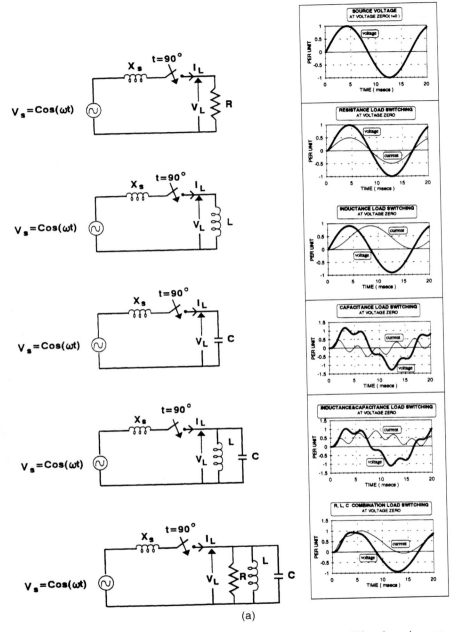

(a)

FIGURE 3.25(a)   Voltage and current waveforms with switch closing at $wt = 90°$ on the cosine waveform (i.e., voltage = 0).

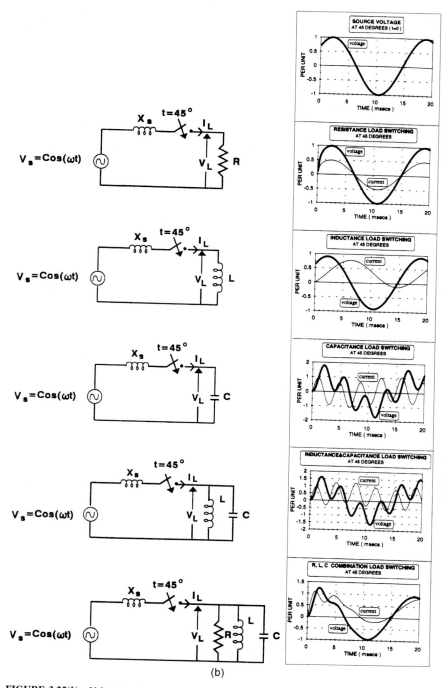

FIGURE 3.25(b) Voltage and current waveforms with switch closing at 45° on the voltage cosine waveform.

3.59

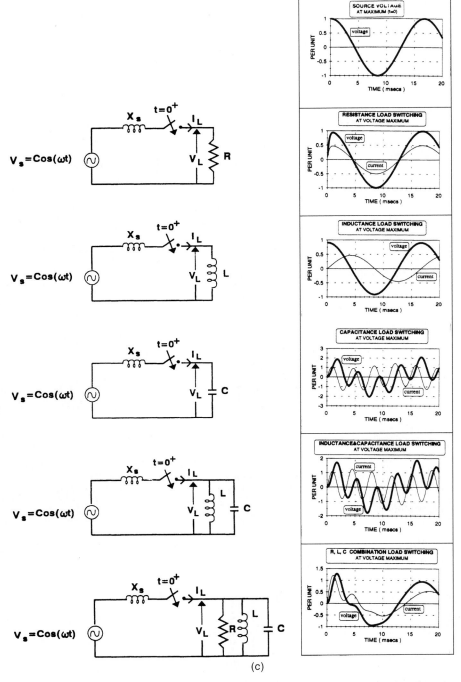

FIGURE 3.25(c) Voltage and current waveforms with switch closing at $wt = 0°$ on the voltage cosine waveform (i.e., voltage = peak).

2. Pure inductive load:

$$V = V_S \cos \omega t$$

$$V_L = L \, dI_L/dt \tag{3.26}$$

$$(L_S + L) \cdot \frac{dI_L}{dt} = V_S \cos \omega t$$

$$I_L = \frac{V_S \sin \omega t}{\omega \, (L_S + L)} \tag{3.27}$$

Note that in Fig. 3.25(c), for the second switching, the voltage (V_L) across the load inductor is the source voltage and load current is zero. In the corresponding waveform there are no transients, because the initial condition of both source and load currents being zero at voltage maximum coincides with the steady state behavior of an inductive circuit, whereas in Fig. 3.25(a), note the unsymmetrical current waveform as well as the higher magnitude.

3. Pure capacitance load:

$$V = V_S \cos \omega t$$

$$V_C = \frac{1}{C} \int I_L \cdot dt \tag{3.28}$$

$$L_S \cdot \frac{dI_L}{dt} + \frac{1}{C} \int I_L \cdot dt = V_S \cos \omega t$$

This is a second-order differential equation. Because of the combination of the source reactance (inductance) and the load capacitance, the mathematical solution is more complicated than the pure inductance case previously mentioned.

Even with these simple examples, it can be seen that switching capacitors, especially at the instant of voltage peak, produces a high-voltage transient. In the third switching in Fig. 3.25(c), the voltage across the capacitor is two times the source voltage. Also, note that switching the capacitor when voltage is zero produces a very low transient [third one in Fig. 3.25(a)]. Note the contrast between switching of inductive and capacitive loads.

The actual overvoltage generated depends on many factors, the main factors being:

- R, L, C values of the equipment being switched
- R, L, C values of the source
- Applied voltage magnitude
- Switching point (angle) on the voltage waveform
- Resonance

The purpose of a switching transient study is to determine the transient voltages (also referred to as *switching surges*) that may be generated due to switching of equipment and facilities. This type of study includes:

- Determining magnitude, duration, and probability (sometimes) of voltage transients
- Determining transient voltages due to switching actions and/or faults
- Evaluating cost-effectiveness of different surge limiting (resistors, reactors) and surge protection (capacitors, lightning, or surge arresters) methods
- Determining the required energy absorption capability of surge arresters
- Establishing special operating procedures
- Providing input to an insulation coordination study

Study of transient overvoltages may be performed either through a transient network analyzer (TNA) or through a digital computer program (electromagnetic transients program, or EMTP). The TNA is a special-purpose analog simulator. The problem to be studied is represented on this simulator and switching actions are performed. The resulting voltages and currents are recorded for analysis. There is a limited number of TNA facilities around the world. Time slots to make these studies need to be reserved, and sometimes the design engineer has to travel to the TNA facility to perform the study. Simple studies may be performed without the physical presence of the design engineer.

In recent years, the EMTP has gained increasing popularity and use. The EMTP is a digital computer program for simulation of transient and steady state operation of multiphase power systems. Problems are solved at discrete time steps, with the simulation proceeding for a user-selected time. Before the simulation begins, a steady state phasor solution is done to establish initial conditions.

The EMTP contains many power system equipment models, including transmission lines, transformers, synchronous generators, motors, sinusoidal sources, nonsinusoidal sources, breakers, diodes, thyristors, surge arresters, and other nonlinear elements. Using these models and other user input, the EMTP calculates a network solution at discrete time steps using trapezoidal integration. In addition to the main simulation program, there are auxiliary programs for calculating transmission line traveling wave modal constants, cable constants, transformer characteristics, and transformer hysteresis, as well as additional supporting programs. Two analog simulation modeling features allow simulation of control systems (generator exciters, SVC thyristor switching, etc), TACS and the newer MODELS.

The waveforms in Fig. 3.25(a)–(c) are based on the computed values from an EMTP program, which simulated the corresponding circuit switching. There are several versions of this type of program available, including one version in the public domain. However, for the uninitiated, the ability to make proper use of this program takes a little bit of effort and time. This statement can be made with respect to any program, but it especially applies to EMTP. There are several consulting organizations and universities that provide introductory and advanced seminars on the use of EMTP. Any design engineer who is planning to get involved in switching transient studies using EMTP is well advised to get the necessary orientation and exposure by attending such a seminar.

The most common switching transient study in an industrial plant or commercial facility is shunt capacitor switching. Shunt capacitors are widely used for power factor improvement, to provide reactive support and voltage control, and to avoid cost penalties for reactive power drawn for electric utility. In addition, these shunt capacitor banks may be used as harmonic filters to minimize harmonic voltage distortion and limit injection of harmonic currents into the electric utility systems. Any time a large shunt capacitor bank is switched, voltage transients are generated. These voltage transients, if excessive, can cause tripping or failure of equipment. Some adjustable-speed drives are especially vulnerable to such voltage transients. Many different mitigation techniques are available to reduce the unacceptable high-voltage transients due to capacitor switching. These include proper sizing of the capacitor banks, including series inductors or preinsertion switching resistors; using zero voltage crossing controlled switches, or locating the capacitors at other electrically suitable buses.

3.10   RELIABILITY

A reliable electrical power supply, whether to a plant or equipment, is a very important consideration in the design, operation, and maintenance of such supply systems.[5] Not only should each piece of equipment or device meet the reliability requirements, but they should be arranged so that the required reliability is realized. If cost is not a factor, a highly reliable system can be designed and installed. However, the total cost of the system has to be as low as possible. Here, the emphasis is on total cost; this concept is illustrated in Fig. 16.2. For example, a low-initial-cost system with low reliability usually has much higher operating and maintenance costs due to frequent and/or long failures with consequent loss of production.[6] On the other hand, a very highly reliable system with high initial cost may not yield a comparable savings in operation and maintenance and increased production. Also, the incremental cost difference for incremental reliability change is nonlinear. Reliability analysis of different alternatives and associated costs is an important aspect of the design process. Even though a smooth curve for cost changes is shown in Fig. 16.2, in practice, the cost changes in steps rather than smoothly.

Reliability analysis may be approached in two different ways—namely, deterministic and probabilistic methods. Deterministic methods, being qualitative rather than quantitative, are not considered to be a strict reliability analysis. However, deterministic methods are much easier to understand and apply. Understanding highly mathematical aspects of analytical models, probability distribution, and validity of data are not essential.

3.10.1   Deterministic Approach

Deterministic methods have evolved from the concept of redundancy. Thus, deterministic methods essentially deal with checking whether there is sufficient redundancy to meet certain outages. This is also called contingency analysis. At the design stage, the deterministic reliability evaluation essentially consists of

- Defining reliability (redundancy) criteria
- Developing alternatives which pass technical requirements
- Performing contingency analysis
- Selecting acceptable alternatives

In deterministic analysis, it is important to define the design reliability at the beginning of the study. A common reliability criterion provides a means to compare systems of comparable reliability. Note that in practice two alternatives very rarely have the same performance. In this sense, the deterministic reliability study basically provides a "go" or "no go" for the alternatives, from the redundancy point of view. Too much redundancy within the system will cost too much as compared to lesser redundant systems. Thus, when alternatives are being compared, higher-cost alternatives are eliminated from further consideration.

A simple example of applying deterministic reliability and contingency analysis is in load flow analysis. The reliability (redundancy) requirement or criterion may be stated as:

With outage of any one of the supply tie lines, stepdown transformers, or one of the feeders serving certain specified buses, the remaining equipment and facilities shall not be loaded beyond their normal continuous rating and all bus voltages shall be within the normal operating range.

Given this type of criterion, the following steps are performed:

1. Establish a base case load flow.
2. Perform contingency analysis by simulating the prespecified equipment or facility being out of service.
3. Analyze the results to verify whether the reliability criterion is being satisfied.
4. Adjust the system, if the reliability criterion is not satisfied, and repeat the steps.

These steps indicate that application of deterministic reliability is not difficult. This analysis is usually done in practice, even though it may not be formal and explicitly recognized as a reliability study. The main drawbacks of deterministic reliability evaluation are:

- Considerable amount of subjectivity is inherent in the method.
- It does not yield a quantitative reliability index.
- It cannot obtain the true cost of incremental reliability improvement.

3.10.2 Probabilistic Approach

The probabilistic approach to reliability evaluation consists of computing the following three types of indices:

- Availability
- Frequency
- Duration

Availability as applied to the electrical supply is defined as the percent of time (usually within a one-year period) the supply is available. This is a long-term average value. Availability, when expressed in per-unit value, is probability of power supply being available. Availability by itself is not a sufficient index, because it does not specify either the frequency of supply interruptions or the duration per interruption. For example, in the computation of availability, 10 interruptions each lasting 1 minute are treated the same as one 10-minute-long interruption.

Frequency of interruption is defined as the long-term average of the number of interruptions in a given period. In an analytical sense, it is an expected (or average) value. The interruption frequency is usually expressed as number of events per year. However, the time period could be different.

Duration is the long-term average of individual interruption durations. Again, this is an expected value. The units are hours, minutes, or seconds per event.

Probability (P), frequency (f), and duration (d) are related to each other by the following formula:

$$P = \frac{f \cdot d}{8760} \tag{3.29}$$

assuming d is expressed in hours and the total period is one year (8760 hours in one 365-day year).

From a practical point of view, these three reliability indices are divided into many different forms. For example, some of the common indices used by a practicing engineer are:

- Load interruption frequency (events/year)
  - Total loss of supply
  - Partial loss of supply to specific buses
  - Loss of supply due to low voltage
- Expected or average duration per interruption (of above types)
- Total average duration of interruptions per year
- System unavailability
- Expected unsupplied energy

Typical steps in a quantitative reliability analysis or studies may consist of:

- System description
- Process or load description
- Reliability criterion
- Reliability analysis
- Indices calculation
- Sensitivity analysis
- Improvements and costs

Straightforward methods of computing reliability indices for several different methods of supplying an industrial load are described in the IEEE Gold Book[7]. There are five examples, starting from the simple radial supply to a secondary selective system. A sixth example illustrates the contribution of spares to the reliability. We urge the reader to study these examples for a basic understanding of practical reliability calculations.

EXAMPLE 3.13 *Two identical motors, each rated at 10 hp, are used in an air-handling system. Each motor is capable of handling the full load, i.e., it is a fully redundant system. Both motors run in parallel. The frequency of failure is 1.825 per year and repair time is 48 h. Compute (1) availability, (2) frequency, and (3) duration of outage of the air-ventilating system.*

Let us designate two motors as M_1 and M_2, with the following reliability data:

Frequency of M_1 or M_2 failure = 1.825 per year
Duration of M_1 or M_2 failure = 48 h/year
Probability of motor failure = 0.01 ($= 48 \times 1.825/8760$)
Motor availability = $1 -$ failure = 0.99

The air-ventilating system has four operating states:

State	Motor M_1	Motor M_2	Probability M_1	M_2	Ventilator
1	On	On	0.99	0.99	0.9801
2	On	Failed	0.99	0.01	0.0099
3	Failed	On	0.01	0.99	0.0099
4	Failed	Failed	0.01	0.01	0.0001
				Total	1.0

Probability of ventilator failure = 0.0001 (State-4)

Availability of ventilator = 0.9999 (States 1, 2, and 3)

f_A = frequency of air-handing system failure

$$= \text{frequency of both motors failing} = P_{M1,M2} \, (f_{M1} + f_{M2}) \qquad (3.30)$$

$$= 0.0001 * (1.825 + 1.825) = 0.000365$$

where $P_{M1, M2}$ = probability of both motors M_1 and M_2 failing

f_{M1} and f_{M2} = frequency of motors M_1 and M_2 failing

$$\text{duration } d_A = \frac{8760 * P_A}{f_A}$$

$$= \frac{8760 * 0.0001}{0.000365} = 24 \text{ h}$$

The average duration of both motors failing is 24 h. This makes sense, because, as a minimum, both motors may not fail at the same time (0 h of overlapping outage period) and, as a maximum, both motors may fail at the same time and any one motor will be restored to service in 48 h; thus, the average is 24 h.

One of the quantitative methods for reliability analysis is called *failure mode and effects analysis.* This type of analysis may be described as finding the minimum amount of equipment, lines, or devices, when outaged as a group or set, that cause interruption of service. This is a more systematic and rigorous method of finding weak spots in the system. Once these failure modes are identified, then the probability, frequency, and duration of system failure may be computed. The type of data needed for this type of analysis includes:

- Forced outage rate of equipment, lines, etc.
- Mean repair time (average time for repair)
- Scheduled maintenance
- Overload ratings
- Undervoltage conditions

Often, it is difficult to find the proper data. The types of failures for power systems are numerous and sometimes difficult to define precisely. For example, an undervoltage condition, if lasting beyond a certain length of time, may trip a motor. The exact voltage and the length of time may vary from motor to motor. Many devices within the power system—for example, breakers—have more than one failure.[3] Computer programs are available for these types of calculations.

Reliability concepts may also be applied to evaluate other types of power system problems, such as:

- Power supply quality
- Welder and other intermittent loads[4]
- Flicker[5]
- Value of service[6]

The reader is referred to other books and references noted above.

3.11 HARMONICS

The invention and development of solid-state electronics technology has introduced large-scale and widespread use of thyristors for rectification, as well as control of adjustable-speed drive (ASD) motors. The use of ASD is widespread in different industries and processes. Thyristors have the capability to precisely chop the ac waveform and supply power to the device with only necessary magnitude and duration. This basic chopping of the current on the ac supply side generates a distorted-voltage waveform. A current waveform resulting from a six-pulse rectifier bridge is shown in Fig. 3.26. Thus, this type of load is called nonlinear.

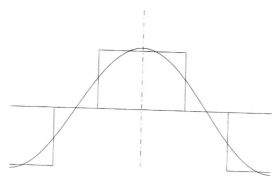

FIGURE 3.26   Idealized supply current for a six-pulse rectifier bridge.

The waveform in Fig. 3.26 may be decomposed into several ac sinusoidal waveforms of different frequencies by applying the Fourier transformation. The components are as follows:

$$I_{ac} = \frac{2\sqrt{3}}{\pi} I_{dc} \left(\cos(\omega t) - \frac{1}{5}\cos(5\omega t) + \frac{1}{7}\cos(7\omega t) \right.$$
$$\left. - \frac{1}{11}\cos(11\,\omega t) + \frac{1}{13}\cos(13\,\omega t) + \dots \right) \tag{3.31}$$

where I_{ac} = supply ac-side current as shown in Fig. 3.26
I_{dc} = dc-side load current
$\omega = 2\pi f$, f being the ac system fundamental frequency
t = time, independent variable
$n\omega t$ = represents, nth harmonic

The first term corresponds to the fundamental frequency ac waveform, followed by the 5th, 7th, 11th, and 13th harmonic components. These harmonic components for a six-pulse rectifier are called characteristic harmonics and are represented by $6p \pm 1$, where $p = 1, 2, 3....$ As can be seen from Eq. (3.31), the higher the harmonic, the lower the magnitude of the harmonic. For example, in Eq. (3.31), the coefficient (1/13) of the 13th harmonic is smaller than the coefficients of the 5th, 7th, and 11th harmonic terms. Other-order harmonics, such as 2nd, 3rd, etc., may be present if there are unbalances, stray capacitances, or the like. In most of the applications, noncharacteristic harmonics may be ignored. However, the engineer should be aware of the possibility of such harmonics.

The major effects of these harmonic currents may be summarized as:

- Interference to communication and control circuits
- Heating of motors, generators, transformers, and capacitors
- Voltage distortion
- Relay and meter misoperation
- Overvoltage due to resonance and damage to equipment

Further discussion of the theory, modeling, criteria, limits, applicable standards, and measurements is presented in Chap. 11. Hence, in this section, only a brief discussion on harmonic studies is presented.

A harmonic study should be performed whenever

- Capacitor and/or filter banks are to be installed.
- A large number of adjustable-speed drives and other nonlinear loads is present.
- System or equipment is experiencing harmonic problems.
- There is a need to limit harmonic current injection to the supplying utility at the point of common coupling (PCC).

The major factors which influence the effect of harmonics on the system and equipment include:

- Electric utility supply system short-circuit level
- Capacitor and filter banks
- Loads
- Balanced/unbalanced conditions

The short-circuit level at the high side of the stepdown transformer represents the strength of the system. The higher the short-circuit level, the lower the system impedance (or reactance), thereby providing a major path for the harmonic current flow as compared to the plant equipment. The capacitor banks offer lower impedance at higher frequencies and can become harmonic sinks, with consequent heating, rupture of tank, fuse blowing, etc. Filter banks are tuned capacitor banks and, as such, act as harmonic sinks. The resistive loads provide some damping to especially harmonic resonance peaks, but motor loads do not provide this type of damping. Unbalanced system conditions can generate noncharacteristic harmonics, and if these coincide with the system resonance frequencies, there could be trouble.

A harmonic study consists of five major steps:

1. Calculating harmonic source currents
2. Modeling of the electric system
3. Solving the electrical problem at each harmonic frequency, one at a time
4. Computing the harmonic performance indices for the system and equipment
5. Evaluating effective mitigation alternatives

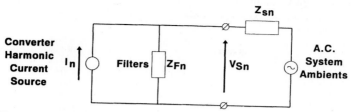

FIGURE 3.27   Conceptual model.

The simplest model, on a conceptual level, used for harmonic analysis of an ac system is shown in Fig. 3.27. Harmonic source current is injected as shown. The capacitor or filter bank acts as a local source admittance (sink). The system-source reactance and the system-source voltage are as shown.

Actual harmonic study is performed by using a computer program. The standard ac load flow techniques have been adapted to perform harmonic studies. Programs for study of both utility transmission systems and industrial-type systems are available. These programs have all the advantages of input data, interactive capability, tabular and graphical outputs, etc. Capability to model multiple harmonic sources is also available. A sample harmonic study system is illustrated in the remaining part of this section.

EXAMPLE 3.14   *For the industrial plant system of Example 3.5, assume that a shunt capacitor/filter bank with a capacitor c = 53.647 μF and an inductor L = 5.263 mh in series is connected to bus 103.*

1. Perform a harmonic frequency scan.
2. Perform a harmonic power flow assuming 5th-harmonic current injection of 0.192 /180° at bus 108.
3. Assume a six-pulse 2.5-MVA converter is connected to bus 130 and compute total harmonic distortion. The six-pulse converter is a harmonic source with the harmonic currents as shown in Table 3.18 (Fig. 3.28) injected at bus 130.
4. Compute distortion based on IEEE Standard 519.

As an exercise, compute the filter bank tuning frequency.

1. The harmonic scan is useful to determine resonant frequencies. A 1-p.u. current of varying frequencies is injected into a given bus. Then the corresponding harmonic voltage is calculated. This voltage represents the *driving-point impedance* of the system as seen from the current injection bus at that particular harmonic frequency, because

$$Z = \frac{V}{I}$$

where *I* was set to 1 p.u. This driving-point impedance is shown in Fig. 3.29.

2. A harmonic power flow shows where the harmonic currents are flowing in the system. Whereas harmonic scan, previously discussed, dealt with harmonic current injection at one bus, of different harmonic frequencies, the power flow may have multiple harmonic sources at different points in the system, but, with harmonic frequency, one at a time. Two sample harmonic power flow outputs (bus voltage and line currents) for the 5th-harmonic current injection (0.192 /180° at bus 108) are shown in Tables 3.19 and 3.20, respectively. As expected, most of the current flows from the source at bus

TABLE 3.18 Spectrum of a Six-Pulse Converter

Harmonic no.	Current injection	
	p.u. mag	Angle
5	0.192	180
7	0.132	0
11	0.073	180
13	0.057	0
17	0.035	180
19	0.027	0
23	0.020	180
25	0.016	0
29	0.014	180
31	0.012	0
35	0.011	180
37	0.010	0
41	0.009	0
43	0.008	0
47	0.008	180
49	0.007	0

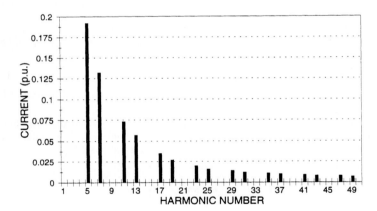

FIGURE 3.28 Harmonic spectrum.

108 to bus 103 and then to the 5th harmonic filter connected to this bus 103. Because of the filter, very little harmonic current flows into the utility system.

3. During the harmonic studies, one of the most important harmonic performance indices computed is called *total harmonic distortion* (see Chap. 11). The computed harmonic distortions are shown in Tables 3.21 and 3.22.

4. In addition to harmonic effects within the plant, the harmonic current injected into the utility supply system and the corresponding harmonic distortion are also important. The connection of the plant electric system to the utility supply system is called the point of common coupling (PCC). IEEE Standard 519 specifies certain limits at PCC (see Chap. 11). These limits may also be computed using a computer program. These computed values for this example are shown in Table 3.23.

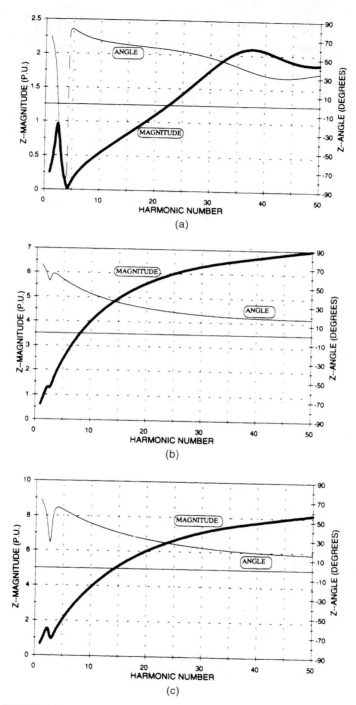

FIGURE 3.29 Harmonic impedance as seen from three different buses.

TABLE 3.19 Sample Harmonic Power Flow Solution—Bus Voltages

Harmonic frequency = 300 Hz

| Bus no. | Harmonic voltage | | Bus kv fundamental |
	Magnitude (p.u.)	Angle (degrees)	
100	0.0168	−98.61	69.00
101	0.0171	−98.40	69.00
102	0.0165	−98.86	69.00
103	0.0228	−96.23	13.80
104	0.0117	−104.27	13.80
105	0.0228	−96.29	13.80
106	0.0117	−104.31	13.80
107	0.0116	−104.17	13.80
108	0.0246	−100.30	13.80
109	0.0116	−104.35	13.80
110	0.0105	−115.73	2.40
111	0.0225	−96.52	13.80
112	0.0116	−104.38	13.80
113	0.0116	−104.30	13.80
114	0.0217	−114.75	0.48
115	0.0103	−118.80	0.48
116	0.0201	−107.92	2.40
117	0.0103	−115.78	2.40
118	0.0105	−115.74	0.48
119	0.0217	−114.82	0.48
120	0.0103	−118.87	0.48
121	0.0116	−104.59	13.80
122	0.0244	−100.30	13.80
123	0.0228	−96.23	13.80
124	0.0117	−104.36	13.80
125	0.0224	−110.09	0.48
126	0.0208	−106.12	0.48
127	0.0106	−114.16	0.48
128	0.0116	−104.59	13.80
129	0.0116	−104.59	13.80
130	0.0223	−110.25	0.48
131	0.0207	−106.15	0.48
132	0.0106	−114.31	0.48
133	0.0105	−114.85	2.40
134	0.0104	−116.58	0.48
135	0.0117	−104.30	13.80

TABLE 3.20 Sample Harmonic Power Flow Solution Line Currents
Harmonic frequency = 300 Hz

From bus no.	To bus no.	Harmonic current	
		Magnitude (p.u.)	Angle (degrees)
100	Utility	0.0034	171.39
100	101	0.0207	15.68
100	102	0.0177	−159.85
101	103	0.0203	16.18
102	104	0.0181	−160.51
103	105	0.0066	−114.69
103	108	0.1894	−1.38
103	123	0.0022	−112.99
103	Filter	0.1675	173.77
104	106	0.0031	−123.71
104	107	0.0116	165.83
104	121	0.0028	−122.72
104	135	0.0047	−122.69
105	111	0.0066	−114.69
106	113	0.0006	−123.34
106	124	0.0025	−123.80
107	Load	0.0116	165.83
108	Harmonic source	0.192	180.00
108	114	0.0029	−121.84
108	122	0.0023	−117.06
109	115	0.0014	−125.88
109	124	0.0014	54.12
110	Load	0.0013	−123.03
110	135	0.0013	56.97
111	116	0.0066	−114.69
112	117	0.0033	−122.55
112	135	0.0033	57.45
113	118	0.0006	−123.34
114	Load	0.0026	−121.87
114	119	0.0003	−121.47
115	Load	0.0012	−125.92
115	120	0.0001	−125.52
116	Load	0.0065	−114.69
117	Load	0.0033	−122.55
118	Load	0.0006	−123.34
119	Load	0.0003	−121.48
120	Load	0.0001	−125.52
121	128	0.0019	−122.45
121	129	0.0009	−123.29
122	125	0.0023	−117.06
123	126	0.0022	−112.99
124	127	0.0011	−121.13
125	Load	0.0017	−117.05
125	130	0.0006	−117.09

(Continued)

TABLE 3.20   Sample Harmonic Power Flow Solution Line Currents
(*Continued*)

Harmonic frequency = 300 Hz

From bus no.	To bus no.	Harmonic current Magnitude (p.u.)	Angle (degrees)
126	Load	0.0019	−113.11
126	131	0.0002	−111.86
127	Load	0.0008	−121.12
127	132	0.0003	−121.15
128	133	0.0019	−122.45
129	134	0.0009	−123.29
130	Load	0.0006	−117.09
131	Load	0.0002	−111.86
132	Load	0.0003	−121.16
133	Load	0.0019	−122.45
134	Load	0.0009	−123.29

TABLE 3.21   Total Harmonic Distortion (THD) for Bus
Voltages

Bus no.	THD	Bus no.	THD
100	0.0212	118	0.0077
101	0.0214	119	0.0168
102	0.0210	120	0.0072
103	0.0266	121	0.0117
104	0.0118	122	0.0272
105	0.0265	123	0.0265
106	0.0118	124	0.0118
107	0.0118	125	0.0185
108	0.0275	126	0.0178
109	0.0117	127	0.0079
110	0.0078	128	0.0116
111	0.0260	129	0.0117
112	0.0117	130	0.0184
113	0.0118	131	0.0177
114	0.0168	132	0.0079
115	0.0072	133	0.0078
116	0.0162	134	0.0074
117	0.0073	135	0.0118

NOTE: These spectrum injections defined:

Bus no.	Size (kVA)	Type
108	2500	6-pulse-Y

Fundamental voltage assumed to be 1 p.u.

TABLE 3.22 Total Harmonic Distortion (THD) for Line Currents

From bus no.	To bus no.	THD	From bus no.	To bus no.	THD
100	101	0.0116	109	124	0.0010
100	102	0.0101	110	135	0.0010
101	103	0.0110	111	116	0.0052
102	104	0.0110	112	117	0.0023
103	105	0.0053	112	135	0.0024
103	108	0.0621	113	118	0.0005
103	123	0.0018	114	119	0.0002
103	Filter	0.0505	115	120	0.0001
104	106	0.0022	121	128	0.0014
104	107	0.0056	121	129	0.0006
104	121	0.0020	122	125	0.0019
104	135	0.0033	123	126	0.0018
105	111	0.0053	124	127	0.0008
106	113	0.0005	125	130	0.0005
106	124	0.0018	126	131	0.0002
108	114	0.0022	127	132	0.0002
108	122	0.0019	128	133	0.0014
109	115	0.0010	129	134	0.0006

NOTE: These spectrum injections defined:

Bus no.	Size (kVA)	Type
108	2500	6-pulse-Y

THD for current based on 0.25 p.u. fundamental.
THD given for the "from" end of these lines.

TABLE 3.23 Harmonic Distortion Computations
(Per IEEE-STD. 519)

Current distortion at PCC in percent of plant loading						
Harmonic number	<11	11–16	17–22	23–34	>34	THD
Odd harmonics						
Calculated values	0.73	0.49	0.24	0.13	0.07	1.16
IEEE limits	12	5.5	5	2	1	15
Even harmonics						
Calculated values	0	0	0	0	0	0
IEEE limits	3	1.38	1.25	0.5	0.25	3.75

Voltage distortion in percent of PCC base voltage		
	Individual (max)	THD
PCC bus	0.77	2.12
IEEE limits	3	5

(a) Point of common coupling (PCC) bus 100 to 101
(b) Plant load kVA = 10000.00 (assumed to be base kVA)
(c) PCC base kV = 69.00
(d) PCC ISC/Iload = 100.00
(e) Harmonic source as in Table 3.21

The reader should note that in Table 3.23, the current THD of 1.16 percent and the voltage THD of 2.12 percent are the same as the first THD value shown in Tables 3.22 and 3.21, respectively. This is as it should be.

3.12 PROBLEMS

PROBLEM 3.1: A single-phase transformer rated at 1000 kVA and 138/13.8 kV has a resistance and reactance of 1.5 and 10 Ω, respectively, as measured from the low-voltage side. Calculate the per-unit resistance and reactance (*a*) from the low voltage side and (*b*) from the high-voltage side. If the measurements were made at the high-voltage side for the same transformer, what would be the actual ohmic values?

PROBLEM 3.2: Given the three symmetrical components of current in phase *A* in Prob. 3.1, determine the corresponding currents for phases *B* and *C*. [*Hint:* Use Eq. (3.13).]

Answer:

$$I_{B1} = 57.8\ \underline{/-30°} * 1\ \underline{/-120°} = 57.8\ \underline{/-150°}$$

$$I_{B2} = 57.8\ \underline{/+30°} * 1\ \underline{/+120°} = 57.8\ \underline{/+150°}$$

$$I_{B0} = 0$$

$$I_{C1} = 57.8\ \underline{/-30°} * 1\ \underline{/+120°} = 57.8\ \underline{/+90°}$$

$$I_{C2} = 57.8\ \underline{/+30°} * 1\ \underline{/-120°} = 57.8\ \underline{/-90°}$$

$$I_{C0} = 0$$

PROBLEM 3.3: Voltages measured on a 480-V, three-phase supply system are as follows:

$$V_A = 470\ V \qquad V_B = 481\ V \qquad V_C = 475\ V$$

Assume the phase angles are 120° apart. Compute three sequence voltages for phases *A*, *B*, and *C*.

PROBLEM 3.4: Current flow measured in a four-wire feeder system is as follows:

$$I_A = 50\ A \qquad I_B = 60\ A \qquad I_C = 55\ A$$

Compute the current in the neutral conductor. (*Hint:* Zero sequence current flows in the neutral conductor.)

PROBLEM 3.5: Redo Example 3.3, assuming 200 W of load power at each of the two load buses and starting at 50 V for step 1. Set up a spreadsheet, as in Table 3.2. Determine how many iterations are required to reach mismatch of 10^{-10}. Can you make the tolerance zero by increasing steps? If not, why?

PROBLEM 3.6: In Example 3.6, you may notice that load unbalance on the three phases is not very large. Solve this as a balanced load flow and see the difference between the two methods of solution.

PROBLEM 3.7:   For Example 3.10, compute the voltage at the motor terminals at full load.

PROBLEM 3.8:   Compute bus voltages, using a load flow program, for starting a 850-kW motor on bus 134 of the electric system in Example 3.5.

PROBLEM 3.9:   For the problem in Example 3.12 (a) determine the critical clearing time by representing the plant generator by a classical model, i.e., ignoring the exciter and governor; (b) provide a detailed generator model representation; (c) if the bus tie breaker between buses 103 and 104 is open, explain the implications for generator unit stability; (d) if one of the tie lines is out of service, and a fault occurs on the second line, explain the implications for plant generator stability and power supply to loads.

PROBLEM 3.10:   For the problem in Example 3.12, determine the critical clearing time if only a line-to-ground fault is considered instead of a three-phase fault. Will the critical clearing time be longer or shorter? Explain why.

PROBLEM 3.11:   If $R = 100 \ \Omega$, $C = 100 \ \mu F$, and $V = 100$ V in the circuit of Fig. 3.24(a), (a) what is the time constant of this circuit? (b) compute the source current and capacitor voltage with respect to time and plot these qualities, and (c) what is the physical reason for steady state current to be zero in this example?

PROBLEM 3.12:   If $R = 10 \ \Omega$, $L = 100$ mH, and $V = 100$ V, (a) compute the time constant, (b) compute the source current and voltage across the inductor with respect to time and plot these two qualities, and (c) why is the steady state current limited by the resistance only?

PROBLEM 3.13:   Consider the sample system shown in Fig. 3.7. (a) Determine the effects (overloading, high voltage, low voltage, and loss of load) of any one branch (line, cable, transformer) or plant generator outage. (b) What is the benefit of closing the tie breaker between buses 103 and 104?

PROBLEM 3.14:   If the failure rate in Example 3.13 is increased by five times due to harsh environmental conditions, determine the number of additional motors needed to limit the average failure duration so as not to exceed 24 h.
Solution:

$$f_{M1} = 5 \times 1.825 = 9.125$$

$$P \ (\text{fail } M_1) = 0.05$$

$$f_A = 0.05 * 9.125 \times 2 = 0.9125$$

$$d_A = 24 \ h$$

Note that the frequency of outage of the air-handling system has increased but the duration is the same.

By increasing the number of motors running in parallel, the frequency and duration may be reduced as shown in the following.

No. of motors	f	d
3	0.0684	16
4	4.56×10^{-4}	12

PROBLEM 3.15:   Repeat Example 3.14 by assuming a 12-pulse converter.

3.13 REFERENCES

1. *Electrical Transmission and Distribution Reference Book,* Westinghouse Electric Corporation, East Pittsburgh, Pennsylvania, Fourth Edition, 1964.

2. G. T. Heydt, *Computer Analysis Methods for Power Systems,* Macmillan, New York, 1988.

3. F. S. Prabhakara, "A Fast Discrete Method for Voltage Drop and Current Calculation for Manufacturing Facilities with Intermittent Loads," *IEEE Transactions on Industrial Applications,* vol. 28, no. 2, March/April 1992, pp. 329–335.

4. F. S. Prabhakara, J. J. Miller, and W. E. Kazibwe, "Frequency and Duration of Voltage Drop in Plant with Intermittent Loads," *presented at IEEE Industrial Applications Society 1991 Annual Technical Conference,* Dearborn, Mich., September 28–October 4, 1991.

5. R. Billinton and R. N. Allan, *Reliability Evaluation of Power Systems,* Plenum Press, New York, 1984.

6. G. Tollefson et al., "Comprehensive Bibliography on Reliability Worth and Electrical Service Consumer Interruption Costs: 1980–1990," *IEEE Transactions,* vol. 6, no. 4, November 1991, pp. 1508–1514.

7. IEEE Standard 493-1990, "IEEE Recommended Practice for the Design of Reliable Industrial and Commercial Power Systems" (IEEE Gold Book).

CHAPTER 4
ELECTRICAL POWER SYSTEM DETAILED DESIGN

4.0 DETAILED DESIGN

The detailed design stage follows the system planning and conceptual design. By this stage, the voltage level, the type of distribution system, and ratings for major components and equipment have been established. At this stage, details of equipment and related facilities design are developed for preparing procurement specification and construction. It is not the purpose of this section to include the construction engineering of the project, but to continue the planning of the more detailed aspects of the project. Some of the key aspects that need to be considered are:

- Utility service
- Monitoring, control, and communications
- Emergency and standby generation
- Uninterruptible power supplies (UPS)
- One-line diagram

4.1 UTILITY SERVICE

Sufficient lead time should be allowed to obtain utility service. The amount of lead time, of course, depends upon the size of demand, location, whether it is a new service or incremental uprating, etc. The lead time may vary anywhere from a few months to a few years. Thus, liaison with the electric utility should be established as soon as possible, especially if the utility has to undertake construction of new lines and substations. The type of data to be furnished to the utility includes the following.

- *Facility layout.*   This allows the utility to determine where the service can best serve the facility.
- *Point of delivery.*   This is a joint decision made by the utility and the facility engineers. Care should be used to make sure that if the process has a corrosive output, the electrical substation is located upwind from any dust or corrosive atmosphere.
- *Facility load.*   This includes the total power (kW) load demand and the types of loads that will be served that will effect the utility source. These include large motors that must be started, nonlinear loads such as static power converters, and adjustable-speed drives.

- *Preferred service voltage.* The reliability of the utility source may determine the voltage that the facility is served. Subtransmission or transmission systems are more reliable than distribution systems, because they have less exposure and usually have better lightning protection. The facility load and negotiations with the utility will determine this voltage. The larger the facility, the more influence the facility will have with the utility.

- *Preferred supply arrangement.* This includes a single feed from the utility or a double feed. The reliability needed by the facility will determine the necessity of the supply arrangement.

- *Start-up schedule.* This is critical for the utility. If there is construction to the supply circuit, enough lead time needs to be given the utility. If cogeneration is involved, the utility needs to know this.

- *Unusual requirements.* These may include large motor loads, processes that cannot be interrupted, nonlinear loads, intermittent loads, etc., which will help the utility plan for the correct service.

- *Power factor.* The utility rates will determine if power factor improvement needs to be made at the plant. Most facilities need to maintain a 0.95 power factor on an integrated monthly basis. That is, the kVAR hours divided by the kW hours (which are metered) will give the tangent of the angle whose cosine is the power factor.

$$PF = \cos \tan^{-1} \frac{\text{kVARh}}{\text{kWh}}$$

- *Load characteristics.* Even if there are no unusual requirements, the utility needs to know if the facility will be operating on a one-, two-, or three-shift basis, and if it will be operating on weekends.

The following data should be requested from the utility:

- *Supply voltages.* Determine availability and reliability of each.

- *Voltage regulation range.* Will a load tap changer be required on the incoming transformer to compensate for voltage spread on the utility system?

- *Point of delivery.* The same problem that is listed above.

- *Billing rates.* The rates may determine whether power will be purchased at the primary voltage or at a secondary voltage and if power factor improvement capacitors (or harmonic filters) are needed.

- *Transformer ownership.* This is determined by the rate schedule (purchase at primary or secondary power), the cost of capital funds, and the ability to maintain the transformer. A contract with the utility may take care of the latter.

- *Space requirements.* If the facility is large and may expand in the future, sufficient space for the substation needs to be allotted. Smaller installations can sometimes be located on an elevated platform.

- *Short-circuit duty (max. and min.).* The maximum short-circuit duty will determine the rating of the switchgear and other power system components. The minimum short-circuit duty determines the operation of static power converters. The impedance of the system determines the time it takes to commutate static power converters. If the facility major loads are static power converters, the time it takes to commutate from one phase to the next can be critical.

- *Metering requirements.* The utility will determine the type of metering, but the facility must provide space for this equipment.

- *Type of grounding.* If the utility supply voltage is the same as the facility distribution voltage, the utility grounding should be compatible with the facility practice. For example, if the facility has generator or motors connected at the distribution voltage, then the utility should provide resistance grounding to limit the ground fault that might occur in the machine to limit the core burning.

- *Protection coordination.* The coordination between the facility relaying and the utility should be such that the utility's relaying will be set long enough to allow the facility's relays to operate. The utility relays should be used as backup to the facility's if they do not operate.

- *Reliability data on supply.* This should be coordinated with the processes in the plant. The higher the service voltage, the more reliable the supply.

- *Backup alternatives.* These include the possibility of spare equipment (normally transformers) that the utility has. If a fault occurs in a major piece of equipment, will the utility have a spare that can be rented until that equipment is repaired or replaced?

Some of these data items are needed to select the type of system and voltage level. Thus, a continued dialog between the facility engineer and the utility is required until the total facility is commissioned.

4.2 MONITORING, CONTROL, AND COMMUNICATIONS

A reliable, uninterrupted, and safe operation of a power system requires monitoring, control, and communication. Because of development and progress in microprocessor technology and the availability of this technology at an affordable cost, monitoring, control, and communications are being increasingly used and relied upon to achieve reliable, safe, and cost-effective operation. Hence, how extensively to incorporate these types of subsystems into the design has to be decided. This will depend upon the needs of the particular type of load. Some of the aspects are:

- Self-contained telephone systems
- Annunciators
- Alarms
- Radio and television monitors
- Fire and smoke alarm circuits
- Security scanning
- Monitoring of critical areas and services

4.3 EMERGENCY AND STANDBY GENERATION

These types of power supply are required whenever the normal electric supply is interrupted. Emergency power supply systems are legally required by municipal, state, federal, or other governmental agencies in some types of facilities. In general, emergency power supply systems are those that are essential for safety to human life. An emergency power supply is required to be available within 10 seconds of normal power supply interruption (Article 701 of NEC 1990).

Standby power systems are required in those facilities where failure can cause physical discomfort, disruption of an industrial process, damage to equipment, or disruption and loss of business activity. Standby power systems are further classified as being legally required or optional. The legally required type is intended to provide power supply for firefighting, rescue operations, control of health hazards, and similar activities. This power supply is required to be available within 60 seconds of normal supply interruption (Article 701 of NEC 1990). The optimal standby power system is intended to provide power supply for continuing industrial process, commercial or business activity without interruption. The reliability requirement for standby generation needs to be established.

The amount of critical load and duration of emergency power supply need to be established. Using an uninterruptible power supply with batteries can handle the critical loads for several minutes (30) or until a standby gas, diesel, or gas turbine generator comes on line. Some of the considerations are:

- *Cogeneration/combined cycle plant.* If the facility is to have its own generation, but not enough to carry the total load, a load-shedding scheme can be used to shed nonessential load if the utility source is lost. This type of facility may generate an excess of power if a fuel source is available and be able to feed energy to the utility, or it may use extraction steam turbines to provide process steam.

- *Standby or emergency generation.* Lengths the time that the facility can be without a utility source (if a UPS is used for continuity of power) will determine the size of the generator.

- *Generator size (kVA, voltage).* The amount of critical load to be served determines the size of the generator. If all of the critical load is on one bus, then the generator should be connected to that bus. The voltage level is determined by the size of the generator and the bus to which it is connected.

- *Generator protection.* This is determined by the size of the generator and how important it is to the system. Consideration should be given to negative sequence current protection if it is to furnish nonlinear loads as static power converters that are used in UPSs. Static power converter harmonics add to the negative sequence loading of a generator. Erroneous operation of a UPS can add additional negative sequence loading by producing even negative sequence harmonic currents. Does the generator have enough fault current to trip the relaying if it is not paralleled with other sources?

- *Metering.* Watt and VAR meters tell how the generator is operating. VAR meters tell more of the operation than power factor meters. Generators have a VAR rating that is dependent upon the amount of energy (watts) it is producing. VAR are also an indication of the field strength of the machine, which is related to the stability of the machine under transient conditions. The stronger the generator field, the more likely it will stay in step with the utility supply.

- *Voltage regulation.* If there is a motor-starting duty on the generator, will the voltage regulator maintain voltage on the system during motor starting?

- *Synchronizing.* If the generator will operate in parallel with other sources, synchronizing equipment must be furnished. This can be automatic; however, potential transformers must be furnished for each voltage source.

- *Grounding.* This must be compatible with the rest of the system.

- *Cost.* Risk cost factors must be evaluated.

- *Maintenance.* Standby and emergency generators should be tested weekly. If the generator equipment is not in good working order, the cost of it is wasted because it will not be available when it is needed.

- *Largest motor(s) to be started.* The generator must furnish the VAR necessary to start motors. The voltage regulator and exciter (besides the generator rating) are the limiting equipment in the amount of VAR the generator can deliver during motor starting. Oversized exciters can force the generator field to produce more VARs needed while starting large motors.

4.4 UNINTERRUPTIBLE POWER SUPPLY (UPS)

There are two types of UPS units: static (solid state) and hybrid (using a static UPS to drive a motor generator set). Static UPS units are more efficient and maintenance-free. The advantage of the hybrid system is the better sine wave output needed for proper operation of some equipment.

4.4.1 Static UPS

A typical functional diagram for a static UPS system is shown in Fig. 4.1. A UPS can be either on-line (make before break) or off-line (break before make) depending on its application.

An *off-line,* or standby, UPS is used for applications such as lighting systems where a time lag between loss of power and switch to a UPS is not critical. But they are not appropriate for protecting microprocessor-based equipment because of its time gap. Any switching action causes a momentary absence of power to the critical load, which often causes a loss of data or damage to the equipment. In addition, standby systems provide no power conditioning for commonly encountered power variations.

An *on-line* UPS acts as a full-time power conditioner by isolating the dependent load from line fluctuations. If power fails, the UPS continues a constant flow of electricity to the critical load to the limits of UPS capacity.

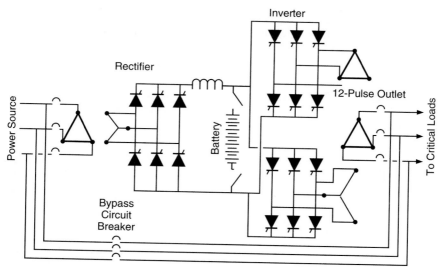

FIGURE 4.1 A typical static UPS functional diagram.

The size of the battery backup on a UPS depends on the applications of the system and what it is protecting. Economic and space considerations usually limit the battery to a size that will sustain the load for 15 to 30 minutes. This time span allows for the supported systems to be turned off safely without causing damage, but does not necessarily include time for a total backup of computer files. Because 50 percent of all blackouts last less than 6 seconds, many users find this is sufficient for their systems. If more than 30 minutes of UPS power is considered necessary, an auxiliary generator used in conjunction with the UPS is the best solution.

Backing up the air-conditioning for computer operation is also a consideration. Most installations can operate for 15 minutes without air-conditioning, but after this amount of time, temperature-sensitive equipment, such as larger computers, may become overheated and damaged or automatically shut down.

Monitoring systems signal dangers, such as, a power overload to help ward off potential problems. A UPS that has self-diagnostics built into the monitoring system can identify and pinpoint malfunctions. The monitoring system can be integrated into existing plant monitoring and security systems. Remote alarms can be installed in two or three alternate locations to provide double or triple backup warning systems. Some UPS equipment allows warnings to be communicated through a standard RS-232 interface.

4.4.2 Hybrid UPS

Hybrid UPS consists of a rectifier, inverter, battery, and synchronous-synchronous motor-generator (MG) set. Some systems use a dc drive motor that can be fed directly from the dc bus without the addition of the inverter. The MG set provides an output for precise voltage control and an almost pure sinusoidal wave. It provides electric isolation from the power system (Fig. 4.2).

In normal operation, power flows from the power supply to the UPS, where the ac power is converted to dc by the rectifier. The dc power is then fed to the battery and the inverter. The inverter changes the dc current to an ac current, which powers the synchronous motor of the MG set. The synchronous motor drives the synchronous generator which has a clean sine-wave output. This design offers electrical isolation between the incoming source and the output to the critical load. The output of this type of UPS is a function of the generator and is independent of the inverter. This type of UPS provides a ± 1 percent voltage regulation during the loss of the ac power source until the return of the power source or until the battery discharges. Since the output waveform is a function of the generator, the output waveform is not changed even with malfunction of the inverter.

When a fault occurs in the power source, the battery supplies dc power to the inverter. This sequence occurs automatically without interruption of power to the MG set because the battery is floating on the dc output of the rectifier. The inverter and the MG set continue to operate.

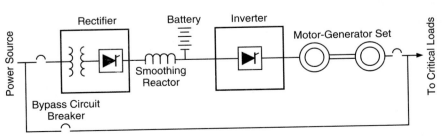

FIGURE 4.2   A typical hybrid UPS.

System bypass switches are used to isolate the UPS for preventive maintenance. Static UPS bypass switches connect the power source directly to the MG set. The switches momentarily parallel the UPS output with the source so that power to the MG set is not interrupted in a make-before-break sequence. The procedure is reversed to bring the UPS back on-line.

4.4.3 Batteries

UPS batteries are normally in a full-float mode, with the UPS rectifier furnishing the requirements of the load plus the power needed to keep the battery at full charge. The battery serves the load only when the power source is lost. Because UPS systems are installed to serve critical loads, it is imperative that batteries be highly reliable.

Given the foregoing, it would appear obvious that considerable attention is given to the specification of the battery for the UPS system. However, this too often is not the case. UPS systems are commonly purchased as a complete system, with batteries selected by the UPS supplier. This results in a least-cost battery, but not necessarily a best-performance battery. The battery specification should be developed as a stand-alone section of the overall UPS specification.

The two basic types of batteries applied in UPS systems are the lead-acid (nominal 2.0 V dc per cell) and the nickel-cadmium (nominal 1.2 V dc per cell) battery. The lead-acid type is predominant, primarily because of cost and availability. The nickel-cadmium is generally favored for smaller self-contained UPS units or in situations where the battery is located close to electronic equipment. Acid-mist or acid-carryover (acid droplets carried out of the cell by exiting gases) from a lead-acid battery can damage sensitive electronic equipment. Lead-acid valve-regulated cells are available to eliminate this problem.

Lead-acid batteries are offered in either lead-calcium or lead-antimony alloy grids. Of these, the lead-calcium is predominant in the UPS applications. Its flat plates have lower resistance than the grids of a lead-antimony cell, which allows them to deliver better short-time performance than an equivalent lead-antimony cell. If the cell is expected to cycle frequently (discharge and charge), lead-antimony might be preferable because of its better cycle-life characteristics.

Several basic factors should be addressed when developing a UPS battery specification:

- Proven technology
- Reliability
- Sizing or capacity
- Duty cycle
- Depth of discharge
- Flame-retardant containers
- Mounting/seismic requirements
- Cable connections
- Warranty
- Maintenance-free potential

The specification should also include information on physical installation details, such as battery-room dimensions and anticipated maximum and minimum battery-room temperature. It is helpful to the supplier if the specification includes a sketch of the battery room. For a replacement battery, it may be a possible to use the existing racks. If so, the racks must be described in the specification.

4.5 CABLING

Although the subject of cabling can get detailed, it is the purpose of this section to define some general principles. The details should be reviewed with the manufacturer who has the data that is needed to specify the cabling in detail. Some of the aspects of a cabling system are:

- *Cables have a voltage and current rating.*   The current-carrying capacity, or ampacity, is based on the ability of the cable to radiate the heat that is produced by the current flowing through the resistance of the cable, the I^2R losses. Cables in a duct bank need to be derated depending upon the number of cables in the duct bank. Current-carrying capacity of cable in cable trays may need to be derated because of the number of cables in the tray.

- *Control and power cables need to be separated.*   The magnetic fields from the power cables can cause erroneous signals in control cables, particularly during transient current surges. Many cases of erroneous operation of control circuits can be avoided by using fiber optics for control circuits. This is particularly advantageous where several signals need to be transmitted over distances. Fiber optics signals can be multiplexed with a single pair carrying several signals.

- *Cables need to be protected from physical damage.*   Cable runs need to be designed so that the cable will not be damaged from some incident associated with an accident. For instance, cable run in overhear trays should be protected from accidents that damage the support system.

- *Control cables routing should be protected.*   For instance, circuit-breaker trip circuits should not be run over switchgear that might have a fault that would result in damage to the trip circuits.

- *In critical processes, it is sometime advisable to use the next higher voltage rating of cable.*   In circuits that have varying voltages from a voltage regulator or transformer taps that raise the voltage above nominal circuit voltage, a higher-voltage-rated cable should be used. Industrial plants that have resistance-grounded power systems, the cable needs to be rated for full line-to-line voltage.

- *Additional ducts should be included in planning for future use or to be used for spare cables in case of a fault in the normal cable.*   Twenty to twenty-five percent additional ducts should be designed into duct banks.

- *Extreme care should be used in the installation of cable.*   Scratches on the insulation can lead to premature failure. Most faults occur at cable terminations and splices. These operations should be made by experienced craftspersons. Voids in the insulation at terminations and splices lead to partial discharge from voltages that break down the insulating materials.

- *A rough estimate of the current-carrying capacity of cables can be determined from Table 4.1.*   The temperature rise on cables is dependent upon the type of insulation. Medium-voltage cable operates at higher temperatures.

- *Conservative practice is to use cables at 80 percent of their current-carrying capacity.*

4.6 SAMPLE SYSTEMS

To learn how an optimum electric power system is designed, several examples will be given with a discussion as to why the different features for the systems were chosen. The

TABLE 4.1 Allowable Current-Carrying Capacity of Insulated Conductors Rated 0–2000 V, 60 to 90°C.

Max. oper. temp.	60°C 140°F	75°C 167°F	85°C 185°F	90°C 194°F	60°C 140°F	75°C 167°F	85°C 185°F	90°C 194°F
	Copper				Aluminum or copper-clad aluminum			
14	20#	20#	25	25#	—	—	—	—
12	25#	25#	30	30#	20#	20#	25	25#
10	30	35#	40	40#	25	30#	30	35#
8	40	50	55	55	30	40	40	45
6	55	65	70	75	40	50	55	60
4	70	85	95	95	55	65	75	75
2	95	115	125	130	65	75	85	85
1	110	130	145	150	75	90	100	100
0	125	150	165	170	100	120	130	135
00	145	175	190	195	115	135	145	150
000	165	200	215	225	130	155	170	175
0000	195	230	250	260	150	180	195	205
250	215	255	275	290	170	205	220	230
300	240	285	310	320	190	230	250	255
350	260	310	340	350	210	250	270	280
400	280	335	365	380	225	270	295	305
500	320	380	415	430	260	310	335	350
600	355	420	460	475	285	340	370	385
700	385	460	500	520	310	375	405	420
750	400	475	515	535	320	385	420	435
800	410	490	535	555	330	395	430	450
900	435	520	565	585	355	425	465	480
1000	455	545	590	615	375	445	485	500

The load current rating and the overcurrent protection for conductor types marked with # shall not exceed 15 A for 14 AWG, 20 A for 12 AWG and 30 A for 10 AWG copper, or 15 A for 12 AWG and 25 A for 10 AWG aluminum and copper-clad aluminum.

systems discussed will grow from a small commercial system through medium-sized systems to a large industrial complex. Some examples are case studies of actual systems.

4.6.1 Typical Small Commercial Installation

Figure 4.3 is a one-line diagram of a small commercial installation that has an estimated load of 400 kVA. Based on Table 2.2, the transformer feeding the load should be 1.25 times the load and the voltage should be 480 V. The service from the utility is 480/277 V, three-phase, four-wire. This voltage is best because the greatest part of the load includes motors for the HVAC (heating, ventilating, and air conditioning), fire protection, elevators, etc., in the range from 5 to 50 hp. It also includes a sizable lighting load. The area lighting load is fluorescent lighting at 277 V. Security lighting around the perimeter of the building and parking will use 277-V ballasts. Distributing power at 480 V to these loads requires smaller and less expensive cables. Other loads are electronic devices such as computers, office machines, accent lighting, etc., at 120 V.

The facility requested that the utility furnish a delta-wye-connected transformer, which will provide a good sine wave for the loads. If a utility-furnished transformer with a grounded wye-grounded wye-connected transformer were used, the voltage wave form on the output side would contain a large third-harmonic content that is the result

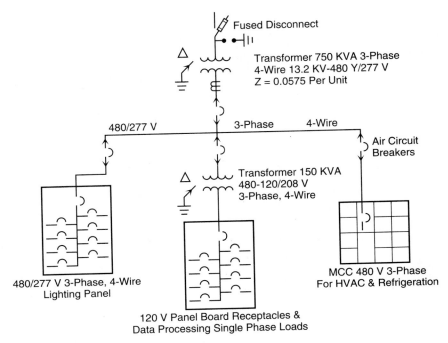

Fused Disconnect

Transformer 750 KVA 3-Phase
4-Wire 13.2 KV-480 Y/277 V
Z = 0.0575 Per Unit

480/277 V   3-Phase   4-Wire

Air Circuit
Breakers

Transformer 150 KVA
480-120/208 V
3-Phase, 4-Wire

480/277 V 3-Phase, 4-Wire
Lighting Panel

MCC 480 V 3-Phase
For HVAC & Refrigeration

120 V Panel Board Receptacles &
Data Processing Single Phase Loads

FIGURE 4.3   One-line diagram of commercial building electrical system.

of the third-harmonic current that is needed to excite the magnetic core. The primary delta winding traps the third-harmonic current in its low third-harmonic impedance. The loads, lighting, and electronic equipment produce third-harmonic currents which will add to the distortion. The delta windings eliminate these currents from the rest of the system. The secondary is solidly grounded to provide good protection for equipment and personnel. It also provides the neutral wire for the 277-V lighting.

The incoming line brings power to the 480-V switchgear, which then distributes power to the mechanical room. A motor control center (MCC) has the motor controllers for the HVAC, fire pumps, etc. Some of the HVAC drives will have adjustable-speed power supplies for the drives. The drive power supplies will be 480 V and will be fed from the MCC. Special care should be given to the fire protection system. A separate circuit coming directly from the transformer secondary and not going through the main switchgear may be required by local codes.

The 277-V fluorescent lighting is fed from a 480/277-V, three-phase, four-wire panelboard. The lighting circuits are arranged so that they are balanced on the three phases. Both electronic and magnetic ballast of the lighting produce third-harmonic currents which will flow back into the 480-V system and, hence, to the utility.

The 208/120-V panelboard will feed the receptacles that will be available for the electronic equipment such as computers and other office machines. If the area to be covered includes large areas, several panelboards can be used. They also can be single-phase units, 480–120/240-V rating. The advantage of using three-phase equipment is that the delta primary windings of the transformers can trap the third-harmonic currents of the switch-mode power supplies of computers and other electronic office equipment.

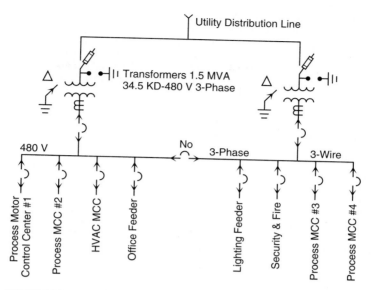

FIGURE 4.4   One-line diagram of small industrial plant electrical system.

4.6.2   Typical Small Industrial Plant

Figure 4.4 is a one-line diagram of a small industrial plant installation. The load is estimated to be approximately 2.2 mVA (2200 kVA). Based on Table 2.2, the total transformer capacity needs to be approximately 3000 kVA. Because some of the processes are critical, two transformers in a secondary selective system are used. With the user purchasing the transformers, the utility rate is based on purchasing power at the primary voltage rate. Although there is only one utility feed, there are two transformers, which will give the plant continuous power even if one transformer has a failure.

A fused disconnect is for protection on the transformer primary. There are also lightning arresters mounted on the transformer for surge voltage protection. The transformers are connected delta-wye to provide a neutral connection for grounding the secondary winding. The grounding can be either solidly grounded or high-resistance grounded. The high-resistance grounding system allows the system to continue to operate with a ground fault (limited to 2 A) on one phase until maintenance personnel can remove it. If a second ground occurs on another phase, then there will be a phase-to-phase fault.

The secondary circuit breaker is used to disconnect the bus from the source if there is a fault on the source. The normally open tie breaker allows for the use of lower-rated switchgear. If the two transformers were paralleled, higher-interrupting-rated switchgear would be needed. It is possible to include an automatic throwover scheme such that if one source of power is lost (e.g., the primary transformer fuse blows), the secondary circuit breaker will open and the tie circuit breaker will close, restoring power to that bus section. A negative sequence voltage relay should be applied on the transformer secondary to indicate single-phase operation if one primary fuse blows.

The switchgear feeder breakers are arranged to feed the various motor control centers. Additional feeders are used for the lighting, offices, etc. The distribution to the panelboard for receptacles and lighting is similar to that discussed under Sec. 4.6.1.

If 277-V lighting is used, then the neutral needs to be carried to the panelboards feeding those loads. If high-resistance grounding is used, the system neutral is connected to the transformer neutral side of the grounding resistor, not to ground.

If the plant covers a large area, the transformers may be separated, but that increases the distance that the power must travel when the tie breaker is used. This would require that a tie breaker be mounted in each switchgear section to protect the cable connecting the two switchgear sections.

4.6.3 Medium-Size Industrial Plant

Figure 4.5 is a one line diagram of a medium-size industrial plant. The same configuration can be used for a distribution voltage of 4.16 kV with a maximum load of 12 mVA, or for a distribution voltage of 13.8 kV with a maximum load of 40 mVA. For larger loads, the utility should be willing to furnish two feeds to increase the reliability of the power system.

The transformers are connected to the utility system with *circuit switchers.* These three-phase devices have the ability of interrupting fault currents up to 20,000 A. Although they do not have the ability of a power circuit breaker, for the application requirements of transformer primary protection, they are an economical solution. The transformers have lightning arresters mounted at the primary bushings for surge protection.

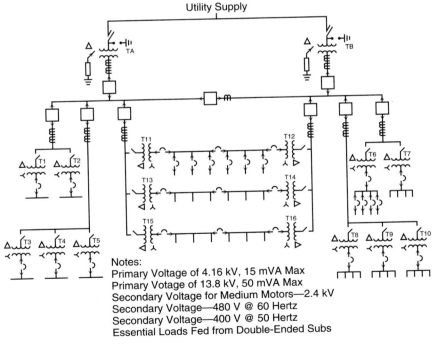

Notes:
Primary Voltage of 4.16 kV, 15 mVA Max
Primary Votage of 13.8 kV, 50 mVA Max
Secondary Voltage for Medium Motors—2.4 kV
Secondary Voltage—480 V @ 60 Hertz
Secondary Voltage—400 V @ 50 Hertz
Essential Loads Fed from Double-Ended Subs

FIGURE 4.5   One-line diagram of a medium-size industrial plant electrical system.

The system is resistance-grounded. The grounding resistors are rated for 200 A. When the two transformers are operated in parallel, an initial ground-fault current of 400 A is produced. With the ground-sensor type ground, or zero sequence relays, faults as low as 10 to 15 A can be detected.

The metal-clad switchgear provides system protection and sectionalizes the power system. Each feeder power circuit breaker feeds a series of load center substations. Substations 11 through 16 have a secondary selective arrangement. These feed the more critical loads. The other substations feed less-critical loads.

Each substation has a primary disconnect switch with secondary air circuit breakers. The medium-voltage power circuit breakers provide protection for the system for faults through the transformers to the source side of the secondary air circuit breakers. Up to six equally rated substations can be fed in this way. If substations of different ratings are used, the total kVA rating of the substations is limited to six times the rating of the smallest substation. If the small substation has a primary fused switch, it provides its own protection. Each of the substation transformers has a delta-wye connection. This provides the neutral bushing for either solid grounding or resistance-grounding of the 480-V secondaries.

With 13.8-kV primary distribution voltage, one of the substations could be for medium-size motors from 250 to 2000 hp where the optimum voltage for these motors would be 2.3 kV to operate from a 2.4-kV substation. The feeders from this substation would be the individual motor controllers, one for each motor.

4.6.4 Case Study of a Large Industrial Plant Expansion

Figure 4.6 is a one-line diagram of a large industrial plant that was in operation and planned for a doubling in size. The original plant was fed by two 18/24/30-mVA transformers. During a previous plant expansion another 18/24/30-mVA transformer was added. This left the switchgear arrangement such that, although the middle section could be paralleled with either end, the ends could not be paralleled without involving the middle section. With need for additional product, a doubling of the plant size was approved. Rather than trying to add to the existing power system, and because the addition was large enough to justify its own power system, a new substation was approved.

Changes to the existing power system were also approved. These included tying the three transformers in parallel so that any two could provide for the load without shutting down. A synchronizing bus scheme was planned. The medium-voltage switchgear was limited to 500 mVA (20,450 A) interrupting capacity. With the three transformers paralleled, the interrupting duty would exceed the switchgear rating. A synchronizing bus ties the three bus sections together with reactors to limit the amount of fault current that the unfaulted buses could feed into a fault. The capacitors were needed to meet the utility's power factor requirement. These were furnished as harmonic filters because the loads included many adjustable-speed drives.

Circuit switchers were added to the primary of the three transformers. The existing two 69-kV power circuit breakers connected the 69-kV bus to two sources. Tie switches allowed the three transformers to be connected to either of the two utility sources. Transformer differential relaying was added for protection.

The new expansion was about the same size as the existing facility. Rather than using the three-transformer arrangement, two 30/40/50-mVA transformers were used. This gave a total capacity about the same as the existing plant but allowed for more flexibility in the switching arrangement. Two power circuit breakers (PCBs) connect to the two utility sources. Lightning arresters were used both at the PCBs and mounted at the primary bushings of the transformers. The transformers were furnished with load tap changers (LTCs) to maintain the voltage on the main 13.2-kV bus to less than ±1 percent.

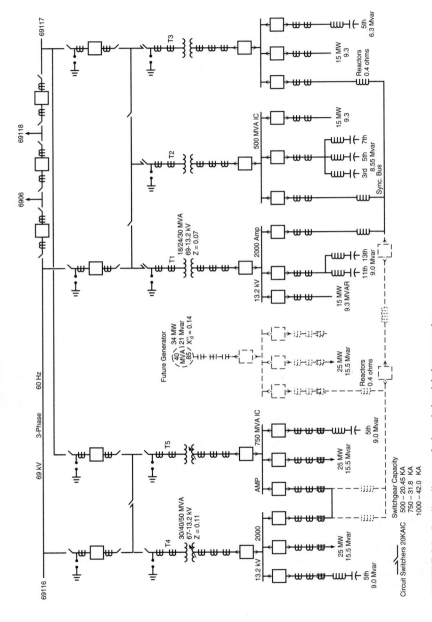

FIGURE 4.6 One-line diagram of a large industrial plant expansion.

4.14

By using 750-mVA (31,800 A) interrupting capacity switchgear, the transformers could be paralleled adding to the reliability of the system. A synchronizing bus was not needed. The one-line diagram shows only one load feeder circuit breaker; however, several feeder circuit breakers were used.

The load shows 25 mW on each bus with a 0.85 power factor. Capacitors were needed for power factor improvement and these were furnished as harmonic filters, because there would be a large number of adjustable-speed drives.

Future plans called for the addition of a 40-mW steam turbine generator. The processes required steam and a study showed some advantage to generating power for additional load and process steam. When this load and generator are added, the synchronizing bus would be extended to include the whole plant. It would normally not operate with all sources paralleled, but if a transformer were to fail, a means of feeding the load connected to its bus would be needed and, thus, a synchronizing bus could be used.

The principles involved in the design of this system included arranging the loads so that each bus has a source (transformer) of the size of the load. Power factor improvement capacitors connected to each bus furnished the VARs needed for the load on the bus. If it becomes necessary to transfer power through the synchronizing bus, only power (watts) would be transferred. This minimizes the voltage drop across the reactors.

The resistance-grounded systems use 200-A resistors. When the two systems are operated independently, the ground faults are limited to 400 or 600 A. If the two are tied together, the ground fault increases to 1000 A and, with the generator, to 1200 A. This shows the advantage of operating the systems separately or as ground-fault islands.

Large industrial installations are similar to this case study in that the power system is broken up into several systems that are limited to approximately 100-mVA maximum. The economical switchgear that is available is limited to 1000-mVA (42,000 A) interrupting capacity. Where this case shows the synchronizing bus tying everything together at 13.2 kV, a better arrangement is to tie a large system together at a higher voltage. The stepdown transformers have impedance that limit the fault current. If generation is present on each bus, this impedance limits the backfeed from the generators. Such a system also has the advantage of *ground-fault islanding*; that is, a ground fault on one system does not affect the other systems.

4.6.5 Case Study of a Power System for Critical Loads

Figure 4.7 is a one-line diagram of a power system that was designed for critical loads for a Federal Aviation Administration facility. It is essential that power be available at all times with no interruptions.

The load characteristics included electronic power supplies rich in harmonics and motor loads for air-conditioning for these critical electronic loads. The normal operation would be from a utility source with standby diesel generators for extended outages from the utility. There is no interruption of power with the loss of the utility source until the standby generators come on-line. The total load of the major facility required 8 mVA of transformers. Smaller facilities required half of the load or two 2-mVA transformers.

Each incoming line from the utility feeds one transformer on each substation. Lightning arresters are located at the primary terminals of each transformer. The solidly grounded neutral allows for 277-V lighting in the facility from the building service feeders. The building service feeder distribution is similar to the facility shown in Fig. 4.3.

Critical loads include all *critical power centers* (CPCs). Essential loads include all *motor control centers* (MCCs) and building service. Each CPC unit consists of an uninterruptible power supply (UPS) at 480 V. The isolation transformers for the UPSs alternate connections from delta-wye to delta-delta so that the power system sees a 12-pulse

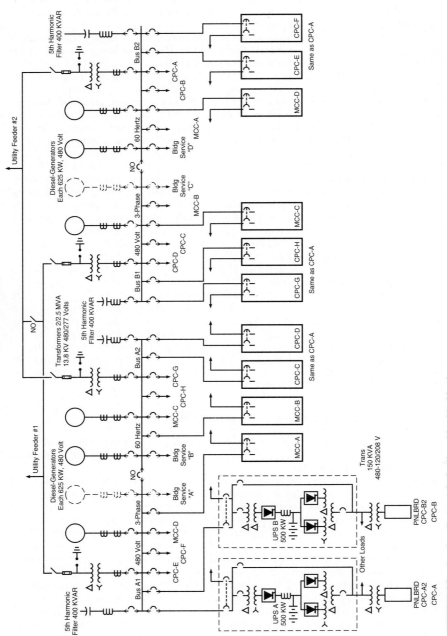

FIGURE 4.7 One-line diagram of a facility with critical loads.

rectifier load. The output from the UPS is a wye-delta/delta three-winding transformer that provides a 12-pulse output. The 480-V output from the UPS is fed to individual three-phase transformers, 480–208/120 V, that feed panelboards where the loads are balanced among the three phases.

Each CPC or MCC can be fed from either bus A or bus B. The facility has two double-ended substations. The loss of one transformer still leaves 6 mVA available on the self-cooled rating. With the engine generators available, full service can be restored to the critical loads with loss of both utility ties. The essential motor loads can also be carried.

The 400-kVAR capacitor bank is to raise the power factor from about 0.84 to 0.95. They are configured into a fifth-harmonic filter to control the flow of harmonic currents and relieve the harmonic current loading on the generators.

With a facility with half the load, one double-ended substation is furnished. The double feed to the CPCs and MCCs can be off each bus section. This allows continuous feed if there is a fault in one of the cable feeds. For other facilities, the rating of the equipment can vary. The operation of the UPSs at 480 V reduces the cost of distributing power as compared to 208/120 V. The individual three-phase delta-wye transformers feeding the panelboards limit the flow of third-harmonic current back to the UPSs.

4.7 SPECIAL CIRCUIT ARRANGEMENTS

There are some equipment combinations that are more economical when size is considered than what is normally used. The cost of a medium-size motor (400 to 1500 hp) applied on a 13.8-kV system may be more than the combination of a stepdown transformer and a lower-voltage motor. Figure 4.8 shows such an arrangement. As shown, the secondary of the transformer is resistance-grounded with a ground sensor instantaneous relay for protection. The other motor protective relays would be included in the power circuit breaker connected to the higher-voltage bus.

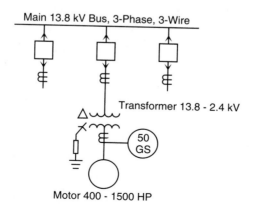

Main 13.8 kV Bus, 3-Phase, 3-Wire

Transformer 13.8 - 2.4 kV

50 GS

Motor 400 - 1500 HP

FIGURE 4.8 One-line diagram of unit motor–transformer arrangement.

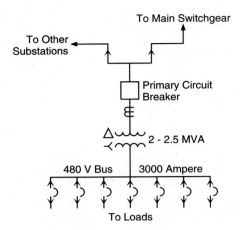

FIGURE 4.9   One-line diagram of large load center substation with primary circuit breaker.

Another arrangement is shown in Fig. 4.9. The cost of a stationary primary circuit breaker on a large (2000 to 2500-kVA) load center substation may be less than a 3000- or 4000-A low-voltage transformer secondary circuit breaker. The protection is actually greater using a primary circuit breaker, in that several substations can be fed from the same feeder and the primary circuit breaker will isolate a faulted transformer as well as a faulted low-voltage bus. Several manufacturers offer this type of arrangement.

CHAPTER 5
VOLTAGE CALCULATIONS AND CONTROL

5.0 INTRODUCTION

As discussed under system planning (Chap. 2), selection of an appropriate voltage level for an electric distribution system is important for a reliable and economical system. Proper voltage regulation during operation is equally important for:

Proper operation of equipment

Avoiding unnecessary failure of equipment and consequent loss of load and equipment

Minimizing system losses

Providing safe operating conditions

Providing economical system operation

Unsatisfactory voltage conditions become evident through overheating, excessive burnout, poor equipment performance, or nuisance tripping conditions in a plant. Both the high- and low-voltage conditions cause problems. For example, the characteristics of gas-filled incandescent-type lamps is shown in Fig. 5.1. The light output is sensitive to the applied voltage; higher voltage gives more lumens out but also reduces the lamp life. The recommended operating voltage range for fluorescent lamps is ±6 percent, as shown in Fig. 5.2. The effect of voltage changes on typical induction motors is shown in Fig. 5.3 and Table 5.1. An indirect benefit of maintaining voltage near 1.0 p.u. during normal load is reduced losses. For example, an induction motor will draw a higher current to drive a given load, if the voltage is lower. The higher current will result in higher I^2R losses. Although the motor will draw less current at higher voltage, the excitation current will increase, resulting in increased core loss which will increase the losses.

Thus, voltage control is essential for proper operation of distribution system facilities. Before proceeding with this discussion, a few commonly used terms are defined.

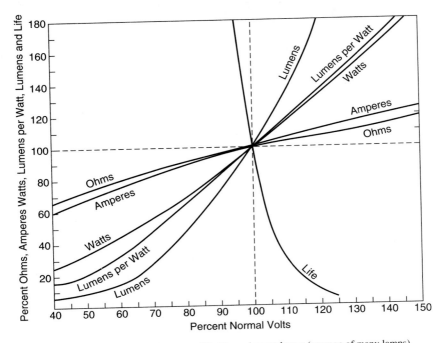

FIGURE 5.1   Characteristics of large gas-filled incandescent lamps (average of many lamps).

5.1 DEFINITIONS

System voltage   The root-mean-square phase-to-phase voltage (fundamental frequency) on an alternating-current electric system. This is the design voltage for the system, e.g., 480, 4160, or 13,800 V.

Nominal system voltage   The root-mean-square phase-to-phase voltage (fundamental frequency) at which the system operates. This is ideally ±5 percent of the system design voltage. The ±5 percent is to allow for voltage variations from no load to full load and variations in the utility supply.

Maximum system voltage   The highest root-mean-square phase-to-phase voltage (fundamental frequency) which occurs on the system under normal operating conditions, and the highest root-mean-square phase-to-phase voltage for which equipment and other system components are designed for satisfactory continuous operation without derating of any kind. Usual conditions allows for ±5 percent. Unusual conditions allow up to 5.8 percent.

Service voltage   The root-mean-square phase-to-phase or phase-to-neutral voltage (fundamental frequency) at the point where the electrical system of the supplier and the user are connected.

Utilization voltage   The root-mean-square phase-to-phase or phase-to-neutral voltage (fundamental frequency) at the line terminals of utilization equipment.

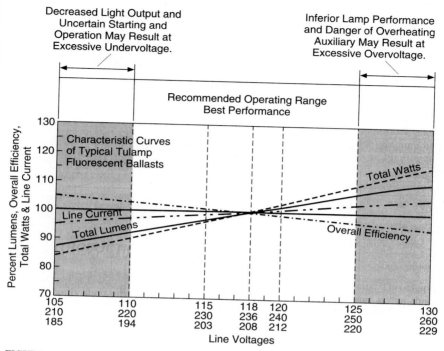

FIGURE 5.2   Characteristics of fluorescent lamps as a function of voltage applied to the ballast.

5.2   *VOLTAGE CLASSES AND NOMENCLATURE*

Low-voltage class (LV)   A class of nominal system voltages 1000 V or less and mostly used to supply utilization equipment.

Medium-voltage class (MV)   A class of nominal system voltages greater than 1000 V and less than 100,000 V and mostly used for primary distribution. The most common voltage for facility distribution within the United States is 480 V for small systems, 4160 V for medium systems and 13,800 V for large systems. (See Chap. 2, Table 2.2). Historically, 12.47 and 13.2 kV have been used, especially where these voltages are used by utilities in the area.

High-voltage class (HV)   A class of nominal system voltages equal to or greater than 100,000 V and equal to or less than 230,000 V. This voltage class is used for subtransmission of electricity to substations.

Extra high voltage class (EHV)   A class of nominal system voltages greater than 230,000 V but less than 800,000 V. These voltages are used for long distance or heavy power transfer transmission lines.

Ultra high voltage class (UHV)   A class of nominal system voltages greater than 800,000 V.

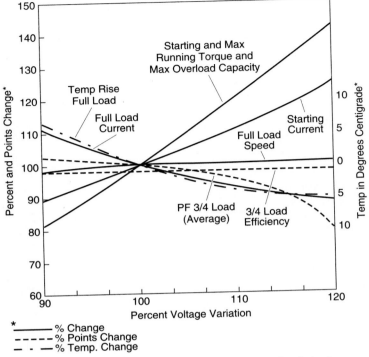

* ——— % Change
* ----- % Points Change
* — · — % Temp. Change

FIGURE 5.3 Characteristics of induction motors as a function of applied voltage.

TABLE 5.1 General Effect of Voltage Variation on Induction Motor Characteristics

	Voltage variation, %	
	90% voltage	110% voltage
Starting or maximum running torque	−19	+21
Percent slip	+23	−17
Full-load speed	−1.5	+1
Efficiency:		
Full load	−2	Small increase
Half load	+2	−2
Power Factor:		
Full load	+1	−3
³⁄₄ load	+3	−4
½ load	+5	−5
Full-load current	+11	−7
Starting current	−10−−12	+11
Temperature rise, full load	+6−+7°C	−1−−2°C
Maximum torque capacity	−19	+21
Magnetic noise, no load in particular	Slightly lower	Slightly higher

+ or − refers to corresponding values at 100% voltage operation

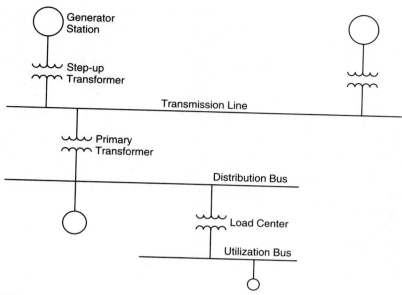

FIGURE 5.4 Major elements of a power system.

The modern electric power system is a complex system with many interconnections forming a large network. However, Fig. 5.4 shows the major elements of power generation, transmission, and distribution in their simplest form. The generating stations are connected to transmission lines or the network through a step-up transformer. The transmission lines at HV and EHV levels transmit power efficiently and economically to substations near major load centers. These substations in one or many stepdown stages bring power to the primary distribution voltage (MV) levels. Then, the secondary distribution system, such as a 13.8- or 4.16-kV electric supply, brings the power to near the utilization equipment where the power is again stepped down to the desired level for the utilization equipment.

Some of the commonly used utilization voltage levels in electrical facilities and distribution systems in the United States and Canada are:

120 V	Single-phase, two-wire supply, used in very small installations.
120/240 V	Single-phase, three-wire (two wires and a neutral) supply, used in residential supply.
240/120 V	Three-phase, four-wire (three phases with one phase midtapped for a neutral) supply, used in older installations. Some single-phase loads are taken off from phase-to-neutral.
120, 240, 480 V; 2.4, 4.16, 13.8, 34.5 kV	Three-phase, three-wire supply, usually delta-connected systems, usually ungrounded. Some low-voltage systems are corner-delta-grounded. Because of NEC requirements on grounding, these systems are not used in new construction and should be considered obsolete.

| 208Y/120, 480Y/277, 600Y/346 V; 4.16Y/2.4, 7.2Y/4.16, 12.47Y/7.2, 13.2Y/7.62, 13.8Y/7.96 kV | Three-phase, three-wire supply with system neutral-grounded. These are common distribution and utilization voltages in industrial plants and commercial buildings. If single-phase loads, phase-to-neutral, are supplied from these systems, then the neutral must be carried through. 277-V (line-to-neutral) lighting is common in commercial and industrial buildings. |

For further discussion of these standard nominal system voltages and voltage ranges, refer to ANSI Standard C84.1.

Historically, system voltages were chosen to fit specific systems. The 7.2Y/4.16-kV systems were common when motor generators were used to convert ac to dc power. The 7.2-kV systems were chosen because that was an ideal voltage for large (2,000 to 15,000 hp) motor design. In residential systems, 110-V lamps and appliances were common in some areas. Standardizing bodies such as IEEE and NEMA (see Chap. 18) have standards that specify the ratings of equipment to be applied to different voltage systems.

Utilization equipment will have a lower voltage rating than the system; e.g., motors applied to 480-V systems will be rated 460 V (older motors were rated 440 V). This allows some voltage drop in the system from the transformer secondary to the motor. Motors applied to a 2.4-kV system will have a 2.3-kV rating. Likewise, a 4.0-kV motor is applied to a 4.16-kV system and a 13.2-kV motor will be applied to a 13.8-kV system.

5.2.1 Voltage Classes and Insulation Levels

IEEE Standard C57.12.00-19XX lists the impulse test levels for liquid-immersed transformers (Table 5.2). IEEE Standard C57.12.01-19XX gives similar values for dry-type transformers (Table 5.3). Rotating machine values are much less and are given in ANSI Standard C50.10-19XX and ANSI Standard C50.13-19XX for synchronous motors and

TABLE 5.2 Liquid-immersed Transformer Voltage Classification and Test Values

Insulation class and nominal bushing rating	Hi-pot tests	BIL full wave (1.2/50)	Switching surge level
kV (rms)	kV (rms)	kV (crest)	kV (crest)
1.2	10	45 (30)	20
2.5	15	60 (45)	35
5.0	19	75 (60)	38
8.7	26	95 (75)	55
15.0	34	110 (95)	75
25.0	50	150	100
34.5	70	200	140
46.0	95	250	190
69.0	140	350	280
92.0	185	450	375
115.0	230	550	460
138	275	650	540
161	325	750	620

Numbers in ( ) are distribution system ratings. Industrial and commercial systems are in this class.

TABLE 5.3 Dry-type Transformer Voltage Classification and Test Values

Nominal winding voltage (volts)		Hi-pot test	BIL (1.2/50)
Delta or ungrounded wye	Grounded wye	kV (rms)	kV(crest)
120–1200		4	10
	1200Y/693	4	10
2520		10	20
	4360Y/2520	10	20
4160–7200		12	30
	8720Y/5040	10	30
8320		19	45
12,000/13,800		31	60
	13,800Y/7960	10	60
18,000		34	95
	22,940Y/13,200	10	95
23,000		37	110
	24,940Y/14,400	10	110
27,600		40	125
	34,500Y/19,920	10	125
34,500		50	150

IEEE Std.141-1993

TABLE 5.4 Rotating Machine Voltage Classifications and Test Values

	Rated motor voltage (V)					
	460	2300	4000	4600	6600	13,200
60-Hz, 1-min hi-pot crest value (kV) per unit of rated	2.71 7.21	7.92 4.22	12.73 3.90	14.43 3.84	20.10 3.73	38.80 3.60
Impulse strength crest value (kV) per unit of rated	3.39 9.01	9.90 5.27	15.91 4.87	18.00 4.80	25.10 4.66	48.50 4.50

NEMA MG-1-19XX for induction motors (Table 5.4). Transformer conductor insulation is not limited by the slot size as rotating machine conduction insulation is, so rotating machines have a lower impulse voltage capability. For this reason, rotating machines need more impulse voltage protection. (See Chap. 10.)

5.2.2. 50-Hz Standard Voltages

The voltages used on 50-Hz systems vary from those listed previously for 60-Hz systems. In magnetizing any iron core, the criterion is to maintain constant volts per hertz to get the same flux density. Therefore, 50-Hz system voltages will be lower (by 17 percent) for the same physical makeup of magnetic cores.

For this reason, 50-Hz system voltages will be: 240 V—same as 60-Hz; 380 V—3.3 kV; 6.6 kV—11.5 kV; etc. In the case of 460-V motors, they will operate at 380 V with the same torque and losses, but with lower horsepower because a four-pole motor, rather than operating at 1800 rpm synchronous speed, will only operate at 1500 rpm because of the reduced system frequency. Transformers rated 13,800–480 V will operate satisfactorily at 11,500–400 V. However, the manufacturer should be consulted when any equipment is to be operated at frequencies that vary from the design frequency.

The principles set forth in this book are applicable to either 50- or 60-Hz systems. Variations have to be made only for the system voltages that are used.

5.3   VOLTAGE DROP CALCULATIONS

A fundamental knowledge of voltage drop calculation is essential for personnel involved with electric facilities engineering. A load has real and reactive power components. For a given kVA load,

$$P = \text{kVA} \times \cos \phi \qquad (5.1)$$

$$Q = \text{kVA} \times \sin \phi \qquad (5.2)$$

where P = the real power in kW
$\quad Q$ = the reactive power in kVAR
$\quad \phi$ = power factor angle
$\quad \text{kVA}$ = kilovolt-ampere of the load
$\quad \cos \phi$ = power factor of the load
$\quad \text{kV}_{\text{load bus}}$ = voltage of the load bus in kV when drawing I_{load} in amperes

Line impedance is:

$$Z = R + jX \qquad (5.3)$$

$$X = 2\pi f L \qquad (5.4)$$

where Z = impedance in ohms
$\quad R$ = resistance in ohms
$\quad X$ = reactance in ohms
$\quad L$ = inductance in henries
$\quad f$ = frequency in hertz
$\quad j$ = operator $\sqrt{-1}$

The approximate relationship between source supply voltage and the load-bus voltage is:

$$V_S = V_L + (IZ \cos \phi + jIZ \sin \phi) \qquad (5.5)$$

The exact relationship is:

$$V_L = V_S - (IR \cos \phi + jIX \sin \phi)\sqrt{V_S^2 - (IX \cos \phi - IR \sin \phi)^2} \qquad (5.6)$$

where j is an operator $= \sqrt{-1}$

The approximate relationship is shown in the phasor diagram (Fig. 5.6) for the simple radial system of Fig. 5.5.

The voltage magnitude is expressed as:

$$|V_S| = \sqrt{(V_L + IR \cos \phi - IX \sin \phi)^2 + (IR \sin \phi + IX \cos \phi)^2} \qquad (5.7)$$

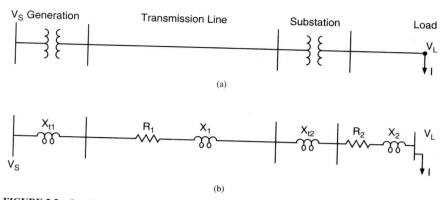

(a)

(b)

FIGURE 5.5   One-line and impedance diagram of a simple radial system.

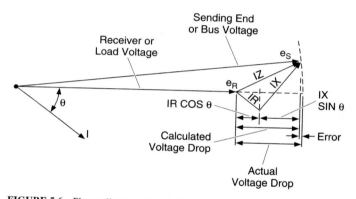

FIGURE 5.6   Phasor diagram of a simple radial power system.

This voltage drop is for a single-phase circuit in one conductor. This value is a line-to-neutral value. For line-to-line values, multiply by 2 for single-phase and $\sqrt{3}$ (1.73) for three-phase systems.

5.3.1 Example of Voltage Drop Calculations

Referring to Fig. 5.5, and using the following impedance and load information, the calculations can be made.

Given: System short circuit is 1500 mVA with an X/R of 8. This includes the system generation, step-up transformer, and transmission line. The primary transformer is 20 mVA with an impedance of 0.09 and an X/R of 22. The cable between the primary transformer, switchgear, and the load-center transformer is 500 MCM and 500 ft in length. The resistance of the cable is 0.0294 Ω/1000 ft and a reactance of 0.0357 Ω/1000 ft. The load center transformer is 1.5 mVA with an impedance of 0.0575 and an X/R of 7. The cable to the motor control center is 250 MCM and 167 ft in length.

The resistance of the cable is 0.0529 Ω/1000 ft and a reactance 0.0285 Ω/1000 ft. The load on the motor control center is 200 kVA or 250 A with a power factor of 0.8.

Using a 10-mVA base, the following calculations can be made.
System impedance:

$$Z = \frac{MVA_{base}}{MVA_{short\ circuit}} = \frac{10}{1500} = 0.0066667 \text{ p.u.} \qquad (5.8)$$

$$R = Z \cos \tan^{-1} \frac{X}{R} = 0.006667 \cos \tan^{-1} 8 = 0.000827 \qquad (5.9)$$

$$X = Z \sin \tan^{-1} \frac{X}{R} = 0.006667 \sin \tan^{-1} 8 = 0.006615 \text{ p.u.} \qquad (5.10)$$

Step-up transformer:

$$Z = Z \frac{MVA_{base}}{MVA_{trans}} = 0.09 \frac{10}{20} = 0.045 \text{ p.u.} \qquad (5.11)$$

$$R = Z \cos \tan^{-1} \frac{X}{R} = 0.002043 \text{ p.u.} \qquad (5.12)$$

$$X = Z \sin \tan^{-1} \frac{X}{R} = 0.04495 \text{ p.u.} \qquad (5.13)$$

Medium-voltage cable:

$$R = R_{ohm} \frac{MVA_{base}}{KV_{base}^2} = \frac{0.0294}{2} \frac{10}{13.8^2} = 0.000772 \qquad (5.14)$$

$$X = X_{ohm} \frac{MVA_{base}}{KV_{base}^2} = \frac{0.0357}{2} \frac{10}{13.8^2} = 0.000937 \text{ p.u.} \qquad (5.15)$$

Load center transformer:

$$Z = Z \frac{MVA_{base}}{MVA_{trans}} = 0.0575 \frac{10}{1.5} = 0.3833 \qquad (5.16)$$

$$R = Z \cos \tan^{-1} \frac{X}{R} = 0.0542 \text{ p.u.} \qquad (5.17)$$

$$X = Z \sin \tan^{-1} \frac{X}{R} = 0.3795 \text{ p.u.} \qquad (5.18)$$

Low-voltage cable:

$$R = R_{ohm} \frac{MVA_{base}}{KV_{base}^2} = \frac{0.0529}{6} \frac{10}{0.48^2} = 0.3827 \text{ p.u.} \qquad (5.19)$$

$$X = X_{\text{ohms}} \frac{\text{MVA}_{\text{base}}}{\text{KV}_{\text{base}}^2} = \frac{0.0285}{6} \frac{10}{0.48^2} = 0.2062 \text{ p.u.} \tag{5.20}$$

Then total $R + jX$ for the circuit is:

$$Z = R + jX = 0.4405 + j0.6382 = \sqrt{R^2 + X^2} = 0.77546 \text{ p.u.} \tag{5.21}$$

$$\frac{X}{R} = 1.449 \tag{5.22}$$

With the total impedance calculated, the voltage drop calculation can now be made. The I_{base} at the 480-V level can be calculated:

$$I_{\text{base}} = \frac{\text{MVA}_{\text{base}}}{\sqrt{3} \text{ KV}} = \frac{10}{\sqrt{3} \, 0.48} = 12,028 \text{ A} \tag{5.23}$$

The per-unit load amperes are then:

$$I_{\text{load}} = \frac{I_{\text{load}}}{I_{\text{base}}} = \frac{250}{12028} = 0.02078 \text{ p.u.} \tag{5.24}$$

The per-unit voltage drop is then:

$$\Delta V = IZ = (0.0207)(0.77546) = 0.01612 \text{ p.u.} \tag{5.25}$$

or 1.61 percent.

Table 5.5 summarizes the impedance values for comparison. Note the X/R ratios of the various equipments and where the major impedances occur.

This does not give the angle between the sending voltage and the load voltage. To calculate this angle, it is best to calculate the IR drop and the IX drop separately.

This load is a small part of the total system. Each section of the system will have additional loads so that the voltage drop in the primary transformer will be greater than with this one small load. Loads also change depending upon the voltage impressed on them. A lower voltage on a constant impedance load will have less current, while a lower voltage on a dynamic load like a motor will have additional current to let the motor have the same output torque that the load requires. For this reason, the voltage drop calculation is a very complicated calculation. The load-flow computer program takes all loads

TABLE 5.5   System Impedance Values for Comparison

	Values on 10 MVA$_{\text{base}}$				
	Impedance Z_{pu}	Resistance R_{pu}	Reactance X_{pu}	X/R ratio	Reference
System	0.006667	0.000827	0.006615	8	1500 MVA$_{\text{sc}}$
Transform	0.045000	0.002043	0.04495	22	20 MVA
Cable		0.000772	0.000937	1.21	9%
LC trans	0.383300	0.054200	0.379500	7	1.5 MVA
Cable		0.382700	0.206200	0.54	
Total	0.77546	0.4405	0.6382	1.45	
Actual Ω	0.01786	0.01015	0.0147	1.45	

into account and makes several calculations as the loads vary with the voltage drop. Several iterations are necessary to get an accurate calculation. It is for this reason that hand calculations do not give the best results.

The X/R ratios in the power system give an indication of the damping effect of the circuit. The X/R ratio on the 13.8-kV bus is almost 22, while the 480-V bus has an X/R of only 7. These numbers are higher than when cable is taken into account. Note that the medium-voltage cable has an X/R of 1.2 and the low-voltage cable is only 0.53. Low-voltage systems do not have the oscillations after a disturbance that medium- and high-voltage systems have because the X/R ratios are much lower. The higher R damps the oscillations. If cable is not taken into account, using just reactance values in the voltage drop calculation will give a reasonable value, especially in medium-voltage systems.

5.3.2 Computer Programs

For large systems consisting of many loads, lines, cables, and stepdown transformers, readily available PC-based computer programs are available. They can model all the components more accurately and with many details. They can perform several types of calculations:

- Load flow
- Voltage drop
- Power factor
- Short circuit
- Relay coordination

A sample load flow output is shown in Fig. 3.2. A simple system is used here for illustration purposes. However, much more complex systems can be analyzed without difficulty. Many different types of reports, including one-line diagrams with outputs (Fig. 3.1), are available. Thus, actual voltage drop calculations need to be made only when very simple calculations are required or when there is no access to a computer program of this type.

If the final cable were ignored in the preceding example, using just the reactance values, the resulting calculation would give a voltage drop that would be within a few percentage points of the actual drop.

5.4 VOLTAGE CONTROL

Any current flow in transformers and lines causes voltage drop due to finite impedance of these elements. In a simple radial system as shown in Fig. 5.5, the voltage drop can be easily calculated. As the load current I changes, the load voltage V_L changes, assuming that the source voltage V_S can be maintained at a constant value.

At no load $\qquad I_L = 0 \qquad V_L = V_S \qquad \Delta V = 0$

At maximum load $\qquad I_L = I_{ma} \qquad \Delta V =$ maximum voltage drop

This maximum voltage drop is also called *voltage spread,* which is defined as the difference between no-load voltage and the voltage at maximum load. This maximum voltage drop should be within allowable limits (-5 percent). At the same time, the maximum voltage at no-load should also be within the permissible limits ($+5$ percent). An example of a voltage drop phasor diagram is shown in Fig. 5.6. Note that if there are shunt capacitors or long cables, the no-load voltage (V_L) may be greater than V_S, the sending end voltage. The method used to maintain voltage within the permissible or desired range is called *voltage control*. Different techniques are used for voltage control, as discussed in the following.

Loads operate more efficiently if the voltage at which they are designed is maintained at the terminals. Several methods of voltage control are available to maintain voltage feeding loads:

- Transformer no-load taps
- Regulating equipment (voltage regulators and load tap changers or LTCs)
- Switched capacitors (capacitors furnish VAR at the load)
- Static VAR compensator (thyristor-switched reactor or capacitor for fast VAR control)
- Higher voltage distribution (less current to cause voltage drop)
- Reduced system impedance (lower transformer's impedance)

5.4.1 Transformer No-Load Taps

Transformers are manufactured with tap points so that different voltage ratios can be obtained. Taps are used to:

- Correct transformer ratio.
- Move the voltage spread.
- Correct for utilization voltage.

No-load taps are fixed taps and can be changed only when the transformer is deenergized. The voltage conversion ratio does not change from no-load to full-load. The no-load taps are on the primary side of the transformer. Two $2\frac{1}{2}$ percent taps above and two $2\frac{1}{2}$ percent taps below nominal voltage are standard. This gives ± 5 percent change to accommodate differences in supply voltage and voltage drops from no-load to full-load in the system. Curves A and B shown in Fig. 5.7 illustrate the primary voltage spread, transformer voltage drop, and feeder voltage drop for two different transformer tap settings. The taps help in increasing or decreasing the secondary voltage but not the range of voltage drop (voltage spread) from no-load to full-load condition. It should be noted that the transformer taps on the primary side should be adjusted in the opposite direction to the secondary voltage change desired; i.e., to raise the secondary voltage, the tap on the primary should be lowered, and vice versa.

5.4.2 Voltage Regulators

A voltage regulator can be incorporated in the low-voltage side of a power transformer. When the regulator is part of a transformer, it is called a *load tap changer* (LTC). The LTC will continuously change taps to maintain a set voltage on the output of the transformer within the range of the taps. The tap changes operate at the rate of 5 to 10 seconds for each tap change. The LTC usually has 32 steps with 5/8 percent voltage for each step. The total range of the LTC is then ± 10 percent. As a rule of thumb, the primary and secondary side ratings should correspond to the corresponding nominal voltage ratings. With LTC applied to the secondary side of the main transformer, the voltage on the plant distribution bus is essential kept at one per-unit voltage for all load conditions.

Separate step voltage regulators operate similarly to the LTC. They consist of separate equipment with a mechanism similar to the LTC normally with the range of ± 10

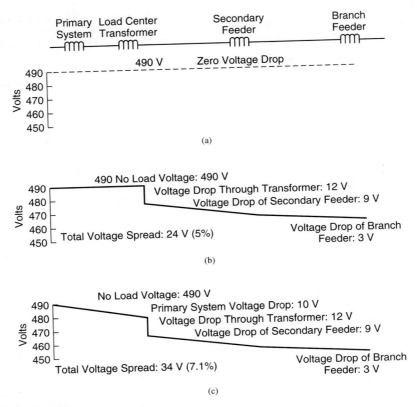

FIGURE 5.7   Voltage spread.

percent. When used with loads that require a large range of voltage, the value between taps can be increased. Some electrochemical processes require larger ranges of voltage so larger steps between taps are needed.

Another voltage regulator is an induction voltage regulator. This is essentially a wound rotor motor that is mounted vertically with a mechanism to rotate the rotor through 360 degrees. The flux on the rotor will add or subtract from the stator voltage to give a continuous variation in voltage output. Because the construction is more like a motor than a transformer, the technology to design and build it are different from either. These apparatus have limited use with the development of semiconductor technology.

5.4.3 Switched Capacitors

The application of fixed and mechanically switched capacitors is the most economical means of correcting low-voltage conditions. They are used extensively in utility systems where they may be switched on during the day at high load periods and off in the evening during low load periods. They are used in subtransmission substations and on distribution systems to compensate for the reactive voltage drop caused by the reactance of the lines. They are not normally used for voltage control in industrial and commercial systems. Some users switch capacitors to provide VARs locally and thus maintain power factor. If, during weekends or light load periods, the capacitors cause too high a voltage on the plant bus, they can be switched off. In general, industrial and commercial power systems do not use switched capacitors for voltage control because of the switching transients that are caused by capacitor switching. The problems associated with capacitor switching are discussed in detail in Chap. 6.

5.4.4 Static VAR Compensation

Thyristor-switched reactors and capacitors are called static VAR compensators (SVC). The term "static" is used to indicate that an SVC does not have inertia or any moving or rotating components. This property of an SVC makes it very fast in responding to changing loads. Unlike mechanical switching devices and rotating machines, the absence of moving components in an SVC minimizes mechanical wear and significantly reduces the need for routine and preventive maintenance. Because an SVC is composed of passive elements, it does not contribute to the short-circuit level at the point of coupling with the network.

An SVC uses standard components which are in common use in other conventional power applications and is, therefore, inherently very reliable. The main components are the thyristor switches, reactors, capacitors, control system, and transformer if needed to match voltage with the thyristor switch. A stepdown transformer is normally required only when an SVC is used for transmission (high-voltage) applications.

The actual design and rating of an SVC is usually determined by the power system planner based on load flow and dynamic studies. These are shunt-connected reactors and capacitors. The capacitors, either fixed or switched, provide a source of reactive power. The thyristor-controlled reactor (TCR) absorbs excess VARs from the capacitors when they are not needed in the circuit to feed the load. The thyristor switch (or valve) controls the current to the reactor and thus the amount of VARs that the reactor will absorb from the circuit. TCRs respond to changing system conditions with a response time of one to two cycles. TCRs are characterized by continuous control of reactive power, no-power transients. Because the reactor current controlled by the thyristor is nonsinusoidal, it contains harmonics. The loads in industrial plants that use SVCs also contain harmonics. The capacitors are used to build harmonic filters to control the flow of these harmonic currents. Figure 5.8 is the schematic of this type of static VARs compensator. The application of this type of SVC is for controlling the VAR and, thus, the voltage on systems with fast fluctuating loads such as electric arc furnaces, resistance welding (large systems), and on systems with other large fluctuating loads that cause excessive voltage drops. The electric arc furnace and resistance welding processes have a weak rectifying action and, thus, produce the harmonics that are associated with rectifiers. Both of these processes are single-phase in their operation (even though the arc furnace is a three-phase device, the loads among the three phases operate independently as single-phase loads). The SVC using TCR gives fast continuous voltage control.

A thyristor-switched capacitor (TSC) consists of a capacitor under the control of a thyristor switch. By suppressing thyristor gate voltage, the reactive power through the capacitor ceases abruptly when the capacitor current reaches a natural current zero. This is at the time when the voltage of the capacitor is at a maximum. The capacitor is left with a trapped charge. The only instant at which the thyristor can be gated on again without causing transients is when the voltage across the thyristor is zero, or when the circuit voltage equals the voltage on the capacitor. TSCs respond to changing system conditions with an average time delay of one cycle. TSCs are characterized by step control rather than continuous control of reactive power. The TSC produces no harmonic

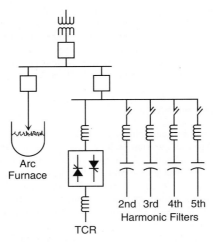

FIGURE 5.8   Static VAR control for arc furnace.

currents, no transients (assuming correct timing of gate voltage), and low losses. This type of SVC is usually used with smaller installations of resistance welders. Since the resistance welders themselves produce harmonic currents, the capacitors are arranged with tuning reactors in fifth-harmonic filters. Figure 5.9 shows such an arrangement. These SVCs are normally applied on low-voltage systems.

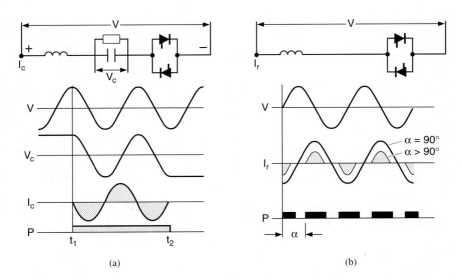

FIGURE 5.9   Voltage and currents in (a) TSC and (b) TCR. V: network voltage. V_c: capacitor voltage. I_c: capacitor current. P: firing pulse. t_1: switch on capacitor. t_2: switch off capacitor. I_r: reactor current. α: delay angle.

5.4.5   Voltage Selection

Tables 2.1 and 2.2 list the relationships among voltage, current, kVA, and load. The minimum voltage drop for a given load will be at the higher voltage. Thus, the higher the voltage, the lower will be the voltage drop. In the first half-century of electric power usage, the voltage drop calculation within a facility was an important factor. With the development of the load center system where the higher voltage was transformed to a lower utilization voltage near the load, the voltage drop problem was minimized.

Primary and secondary distribution voltages are selected based on:

Load density and size

Voltage regulation and control requirements

Utility supply voltages

Overall economics

Future growth

Once primary and secondary voltages have been selected, every effort should be made to locate the stepdown transformers as close to the load as possible. Load center transformers and switchgear can be located on platforms above areas that do not need the full height of the factory, basements, roofs, or any other location that is near the load and does not interfere with production.

5.4.6   Lower Impedance

As can be seen in Table 5.2, the major impedances are the load center transformer and the cables. If lower-impedance primary transformers were to be specified and still not increase the short circuit available on the system, there would be lower voltage drop. Using larger-size cable has always been a method of lowering voltage drop. The voltage drop has to be judged against the short circuit available when comparing impedance ratings of system components.

An additional consideration of system impedance is the operation of static power converters. When a static power converter commutates, it does so by changing the current to the load from one phase to another phase in the power system. The inductance (reactance) of each phase will determine the time it takes to transfer the current from one phase to the next. The lower the inductance, the faster the current will commutate and the higher the voltage on the output of the static power converter. When a large static power converter operates on a weak, high-reactance circuit, the commutating angle can become large and the output of the converter may not have sufficient voltage for the load to operate at its optimum. During commutation, the two phases involved in commutation are shorted and the voltage between them is zero. This may affect the commutation of other static power converters operating on the same system. The short-circuit ratio, S.C.R., should be greater than 15 if the static power converter will not interfere with other loads. See Chap. 11 for more discussion of harmonics and static power converters.

$$\text{SCR} = \frac{\text{MVA}_{\text{short circuit}}}{\text{MW}_{\text{load}}} \tag{5.26}$$

5.5   MOTOR STARTING

A special case of voltage drop calculation is motor starting. Both induction and synchronous motors draw a high reactive current during starting. The speed-torque curves

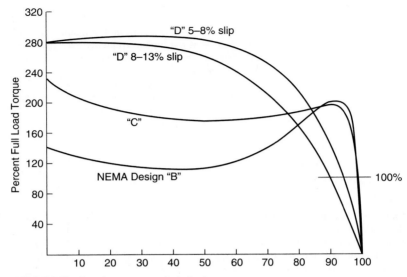

FIGURE 5.10   Speed-torque curve for induction motor.

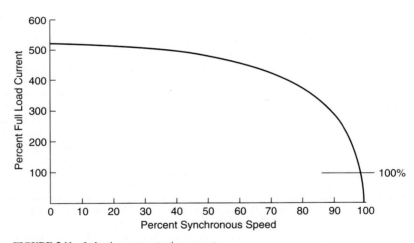

FIGURE 5.11   Induction motor starting current.

of both types of motors are similar (see Fig. 5.10). The starting current characteristics for both types of motors are also similar (see Fig. 5.11). The time that it takes to start any motor depends upon the torque during starting of the motor, the load connected to the motor, and the inertia of the motor and the load. Because of the complex calculations involved, details of such a calculation will not be shown here. Using a load flow computer program that has been altered to calculate voltage drop from motor starting is the usual way of studying a motor starting problem. System dynamics enter into the problem if a large motor (10,000 hp or larger) is to be started. A dynamic study should be made using a transient stability program.

A short, fast calculation of the voltage drop caused by motor starting can be made using the relationship of the starting MVA of the motor compared to the short-circuit MVA of the system.

$$V_{drop} = \frac{MVA_{motor\ starting}}{MVA_{short\ circuit}} \tag{5.27}$$

$$V_{drop} = \frac{kVA_{motor\ starting}}{kVA_{short\ circuit}} \tag{5.28}$$

There are different methods of reducing the voltage drop due to motor starting. The most common way is to reduce the voltage on the motor so that the inrush current is not as great. This can be done by:

• Inserting a series resistor (small motors <200 hp)

• Inserting a series reactor (medium motors, 200–2000 hp)

• Using an autotransformer to reduce voltage (large motors >2000 hp)

• Using a thyristor switch to phase on voltage to give a soft start

There are other methods that require engineering to determine their feasibility. One such method would be to use capacitors that are switched on when the motor is energized so that they may furnish the VARs necessary to start the motor. Since motor-starting current operates at approximately 0.15 to 0.20 power factor, capacitors can be used to furnish the VAR locally and the reactive current does not need to flow through the system impedance. The capacitors are switched on when the motor is being started and switched off just before the motor reaches full speed or when the current has dropped to close to normal value. The voltage drop, then, is only the real portion of the current that represents the energy stored in the rotating inertia of the drive. The reactive portion of the current is furnished by the capacitors. If it is a loaded start, there is additional energy going out the driveshaft; however, this is small compared to the reactive current needed to start.

5.6 VOLTAGE FLICKER

The phenomenon of voltage flicker is experienced whenever there are frequent changes in loads. Intermittent loads, such as resistance welders, electric arc furnaces, or off/on cycling of motors, are examples. Each switching-on of load reduces the voltage and each switching-off of load increases the voltage. The light output of incandescent lamps will show rapid fluctuations. Gaseous discharge lamps (mercury arc or high-pressure sodium) will be affected, though to a lesser extent. These fluctuations, depending upon the intensity and frequency, can be annoying. This is called *flicker.*

The criteria used to evaluate flicker is shown in Fig. 5.12. This figure is the result of testing many individuals by varying the voltage on incandescent lamps at various frequencies. Although this was a subjective test, the curve has been used for several decades. It is dependent upon the thermal inertia of the filament in the lamp. Larger lamps differ from smaller lamps, so the value of the curve has been questioned. The Electric Power Research Institute is reviewing the curve to bring it up to date with today's modern lamps.

In Fig. 5.12, the frequency and magnitude of voltage change is plotted. Consider point A shown in this figure. This point represents a voltage change of 0.5 percent occur-

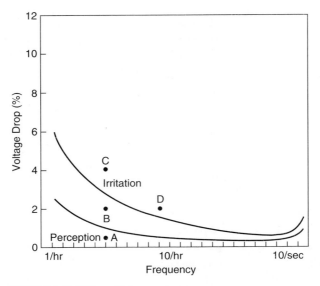

FIGURE 5.12   Voltage flicker curve.

ring six times per hour and is considered to be below perception level; whereas point *B*, with a voltage change of 2 percent, occurs with the same frequency and is perceptible but not irritating. A voltage drop of 4 percent represented by point *C* and a voltage drop of 2 percent represented by point *D* (more frequent) are considered to be irritating. Whenever there are large intermittent type loads, studies should be done to identify flicker problems. Segregating such loads from the system that feeds the lighting is one of the solutions. The same type of load on a weaker system (low short-circuit levels) will show a higher flicker problem. Hence, providing a separate power supply sometimes is a proper solution, if economical and technically feasible. Arc furnaces are the major loads causing flicker. They operate in the range of the minimum value of the curves. Fortunately, a properly applied SVC (TCR/FC) provides control of the VARs to the furnace and, thus, minimizes the voltage fluctuations.

5.7   PROBLEMS

1. What is the voltage drop (estimate) of starting a 15,000-hp motor on a 13.8-kV system that has a short-circuit value of 500 MVA? Is this an acceptable condition? What method would you use to reduce the voltage drop? Would this require a study? Why?

2. In the example voltage drop problem in Sec. 5.3.1, what would be the voltage drop if the primary transformer had an impedance of 12 percent with an *X/R* of 18 and the motor control center had a load of 600 kVA with a power factor of 0.8? The low-voltage cable consists of three 250-MCM cables per phase. What is the angle between the sending end voltage and the load voltage? Draw the phasor diagram.

CHAPTER 6

POWER SYSTEM VAR REQUIREMENTS—POWER FACTOR

6.0 INTRODUCTION

When experiments with electricity first started, direct current and voltages were derived from batteries. At this time, power was defined as the product of voltage times amperes.

$$\text{Power} = \text{volts} \times \text{amperes} \tag{6.1}$$

When alternating current became the means of generating and transmitting electric power, it was discovered that the product of volts and amperes did not give the same result as in direct-current circuits. It developed that the relationship between the alternating current and voltage included a factor. This factor became the cosine of the angle between the voltage and the current.

$$W_{\text{power}} = VI \cos \phi \tag{6.2}$$

Thus, the factor, cosine of ϕ, became known as the *power factor*. From this relationship, the term *reactive power* came into use.

Reactive power, or VARs, represents that current in the circuit that is needed to magnetize iron cores of motors, generators, transformers, etc., and provide the magnetic field around conductors that allows current to flow.

6.1 REACTIVE POWER AND POWER SYSTEM FUNDAMENTALS

In an ac circuit, the current is not necessarily in phase with the corresponding voltage; the current may be lagging, in phase, or leading with respect to the voltage. In Fig. 6.1(*a*), the current is lagging the voltage by 45°. In a pure inductive, no-resistance circuit (in practice there is almost a pure inductance but never a pure inductance unless it is superconducting), the current lags the voltage by 90°. In Fig. 6.1(*b*), the current is in phase with voltage and this is the case for a pure resistive circuit such as electric heaters. In Fig. 6.1(*c*), the current leads the voltage by 45°. In a purely capacitive circuit, the current leads by 90°. These waveforms are usually represented by phasor diagrams as

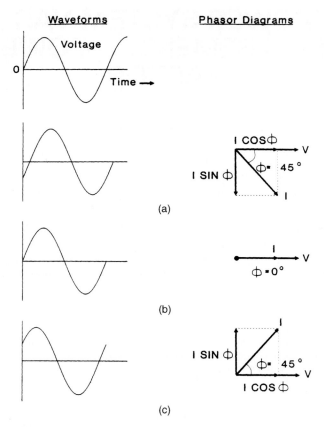

FIGURE 6.1 Current and voltage relationships in ac circuits.

shown in Fig. 6.1. The phasor diagrams represent steady state operating conditions with real quantities represented along the x axis and imaginary (reactive) quantities along the y axis. The current component in phase with the voltage when multiplied by that voltage yields real power (P). The current component in quadrature (along the y axis) with the voltage multiplied by that voltage gives reactive power (Q). In other words,

$$P = VI \cos \phi \qquad (6.3)$$

$$Q = VI \sin \phi \qquad (6.4)$$

where $\cos \phi$ is called the power factor of the circuit. Note that $\cos \phi$ is always positive for both $+$ or $-$ values of ϕ; hence, always specify leading or lagging with power factor.

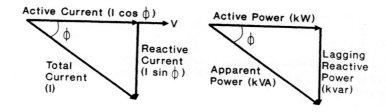

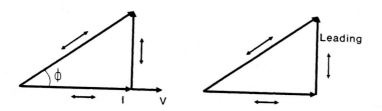

FIGURE 6.2 Angular relationships in ac circuits.

6.1.1 Power Triangle

The real and reactive power relationship is easily shown through triangular representation (Fig. 6.2). The real (kW), reactive (kVAR) and apparent power (kVA) are shown for both leading and lagging power factor conditions. These can be expressed in mathematical form by:

$$\text{Total current } I_t = \sqrt{(\text{active current})^2 + (\text{reactive current})^2} \tag{6.5}$$

$$I_t = \sqrt{(I \cos \phi)^2 + (I \sin \phi)^2} \tag{6.6}$$

$$\text{Apparent power (kVA)} = \sqrt{(\text{active power})^2 + (\text{reactive power})^2}$$

$$VI = \sqrt{(VI \cos \phi)^2 + (VI \sin \phi)^2} \tag{6.7}$$

For a three-phase circuit:

$$P = \sqrt{3} \, VI \cos \phi \tag{6.8}$$

$$Q = \sqrt{3} \, VI \sin \phi \tag{6.9}$$

where V = line-to-line voltage
I = line current
$\cos \phi$ = power factor

If line-to-neutral voltages and line currents are used, then a multiplying factor of 3 (instead of $\sqrt{3}$) is used, assuming that all three phases are identical. Otherwise, individual phase P and Q are added.

The concept of power factor is important. Hence, this will be illustrated through an example. Consider a plant load supplied at 480 V, 1000 A, and 665 kW of power.

$V = 480$ V
$I = 1000$ A
$P = 665$ kW

$$\text{kVA} = \sqrt{3}\, IV = 831.4 \text{ kVA} \tag{6.10}$$

$$\text{Power factor} = \frac{\text{kW}}{\text{kVA}} = 0.8 \text{ (lagging)} \tag{6.11}$$

$$\text{kVAR} = \sqrt{(\text{kVA})^2 - (\text{kW})^2} = 499 \text{ kVAR} \tag{6.12}$$

The typical power factor range for different types of load is shown in Table 6.1. The power factor ranges show that power factor improvement is needed. The technical performance and economic benefits of improving the power factor will be discussed in the following sections.

TABLE 6.1   Typical Industrial Loads and Their Common Power Factors

Load	Typical PF
Incandescent lamps	1.0
Fluorescent lamps	0.95–0.97
Synchronous motors	1.0 to 0.80 leading & lagging
Squirrel-cage motors	
High-speed	0.75–0.90 (at full load)
Low-speed	0.85–0.92 (at full load)
Wound rotor induction motors	0.80–0.90 (at full load)
Induction motors	
Fractional hp	0.55–0.75
1–10 hp	0.75–0.85
Arc Furnace	0.65–0.83

6.1.2   Utility Rate Structure

The electric utilities base their rates on two different types of costs. The first reflects their cost for energy. That is the cost of the fuel that is burned in boilers to generate electric power. This is the kilowatt-hour or *energy charge*. This is reflected in cents per kilowatt-hours.

The second is based on the capital cost that the utility has to furnish the electric energy to the load. This includes the generation equipment, the transformation equipment to step the voltage up to transmission levels, the transmission equipment that is needed to carry the current, and the stepdown equipment and distribution system needed to get the energy to the end user. This capital charge comes in the form of a *demand charge.* It is possible to reduce the demand charge if the amount of equipment the utility must furnish is reduced. The equipment must be sized to carry the current that the user needs. If that current is only the real component—that is, no VARs are furnished by the utility—then the equipment that the utility furnishes needs to furnish only a lower current and, thus, a lower kVA. The reactive power, or VARs, can be furnished locally by using capacitors.

The demand charge, the fixed part of the rate, may use any one of the following methods:

A fixed $/kW demand (no charge for reactive power)

A fixed $/kVA demand (calculated from kW and kVAR)

A fixed $/kW demand with a certain amount of kVAR allowed with an additional charge for kVAR over a fixed amount (e.g., kVAR equal to one-third of kW, power factor approximately 0.95)

A fixed $/kVA (includes both real and reactive power)

A graduated charge for kVAR hour with higher rate for lower power factor

The demand is usually defined as the highest total used in a 15, 30, or 60-min interval during the billing period. In some rate structures, the demand charge is based on the highest demand in the last 12 months (11 month + billing month). This is called a *ratchet clause.*

Except in the case where, specifically, the demand charge is based on $/kW only, either directly or indirectly there is a cost related to the higher equipment rating that is necessary to deliver the energy at a low power factor. Because of this reactive power–related cost in the utility bill, there is sufficient economic incentive to improve low power factors in industrial and commercial systems by furnishing the necessary VARs locally.

6.1.3   Utility Bill Reduction

The demand charge can be reduced by improving the power factor. If a *power factor penalty* is being paid, then, by adding capacitors on the load side of the billing meter, the demand can be reduced. This will be illustrated through the two examples.

A plant with a demand of 1806 kVA, 1350 kW, 1000 kVAR has a contract which stipulates a charge of $3.00 per month for each kVAR of demand in excess of one-third of the kW demand. One-third of the 1350 kW is 450 kVAR. The ratings of the capacitors to be furnished to reduce the kVAR demand to 450 kVAR is:

$$1000 \, kVAR - 450 \, kVAR = 550 \, kVAR \tag{6.13}$$

This excess kVAR cost the user $1650 per month.

$$550 \, kVAR \times \$3 = \$1650 \tag{6.14}$$

If capacitors cost $20 per kVAR installed ($11,000 total),

$$\$20 \times 550 \, kVAR = \$11,000 \tag{6.15}$$

it will take only:

$$\frac{\$11,000}{\$1650} - 6.7 \text{ months} \qquad (6.16)$$

to pay off the cost of installing the capacitors and then there will be a \$1650 per month savings.

A second plant has a demand of 400 kW and 520 kVA. The power contract calls for a demand charge based on a kVA of \$6 per month. The kVA demand can be reduced if the power factor is raised. The present power factor is:

$$\frac{kW}{kVA} = \frac{400 \text{ kW}}{520 \text{ kVA}} = 0.77 \qquad (6.17)$$

If the power factor were to be improved to 0.95, the reduction in kVa would be:

$$kVA_{new} = \frac{kW}{PF} = \frac{400}{0.95} = 420 \text{ kVA} \qquad (6.18)$$

The saving in demand charge is then:

$$(520 \text{ kVA} - 420 \text{ kVA}) \times \$6 = \$600 \text{ per month} \qquad (6.19)$$

The rating of the capacitors to be furnished would be the difference of the kVAR with the original power factor and that of the improved power factor.

$$kVAR = kW \,(\tan \cos^{-1}\phi_1 - \tan \cos^{-1}\phi_2) = 400 \text{ kW}(0.828 - 0.328)$$
$$= 400 \text{ kW} \times 0.5 = 200 \text{ kVAR} \qquad (6.20)$$

Thus, 200 kVAR of capacitors need to be installed. At \$20 per kVAR installed cost, the payoff time is:

$$\$20 \times 200 \text{ kVAR} = \$4000 \qquad (6.21)$$

$$\frac{\$4000}{\$600} = 6.7 \text{ months} \qquad (6.22)$$

These two examples illustrate the power bill reduction through power factor improvement. It also demonstrates the short period of time it takes to pay off the investment of capacitors to start getting the savings from power factor improvement. Note how the kVA is reduced for different power factors in Fig. 6.3. It is rarely economical to correct the power factor to greater than 0.95.

6.1.4 Released Capacity for Existing System

Thermal capacity is proportional to amperes or kVA. Power factor improvement reduces the amperes or kVA, thus, it releases system thermal capacity (Fig. 6.4). Therefore, adding capacitors to an existing system is often the most economical means of obtaining system capacity to serve additional load.

The necessary PF for a certain per-unit (p.u.) release of system kVA is given by:

$$PF_{new} = \frac{PF_{old}}{1 - kVA_{released}} \qquad (6.23)$$

and the p.u. kVA released by correcting to a new power factor:

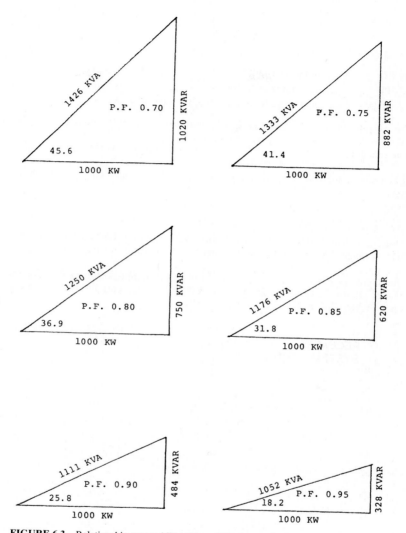

FIGURE 6.3 Relationship among kW, kVA, and kVAR for different power factors.

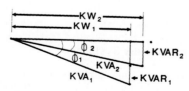

FIGURE 6.4 Illustration of increased power with PF improvement with constant kVA.

$$kVA_{released} = \frac{PF_{new} - PF_{old}}{PF_{new}} \qquad (6.24)$$

where PF_{new} = corrected power factor
  PF_{old} = uncorrected power factor
  kVA = required kVA reduction (in p.u. of existing kVA)

For example, consider a fully loaded system operating at 75 percent power factor. Capacity is needed to serve 20 percent more load.
The preceding calculation is made easier by using Fig. 6.5 as follows:

1. Locate the intersection of original power factor (bold sloped curves) and additional system capacity.
2. Read on the bottom the improved power factor required.
3. New power factor for 20 percent released capacity is 0.94.
4. Determine from Table 6.2 the shunt capacitance required. 1.82 kVAR of shunt compensation per kVAR of released capacity is indicated for this example.

The degree of power factor improvement justified by released capacity depends on the cost of additional system equipment per kW or kVA compared to the cost of capacitors per kVAR. This is discussed in Sec. 6.1.8.

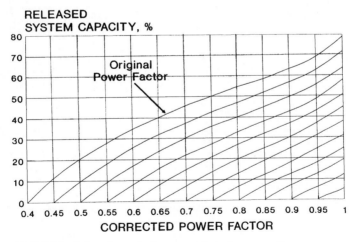

FIGURE 6.5   Released capacity from power factor improvement.

6.1.5 Released Capacity for New System

New or redesigned industrial distribution systems generally are most economical, when power factor improvement permits the use of lower-rated standard components. The most economical power factor can be determined by using the following formula.

TABLE 6.2 kVAR per kVA of Released Capacity

Original PF in %	Final Power Factor of Original Load										
	90	91	92	93	94	95	96	97	98	99	100
95							3.37	3.59	3.88	4.31	5.9
94						3.06	3.21	3.40	3.64	3.82	5.3
93					2.82	2.93	3.11	3.23	3.45	3.78	4.9
92				2.63	2.73	2.83	2.95	3.09	3.29	3.58	4.5
91			2.49	2.55	2.64	2.73	2.84	2.97	3.14	3.40	4.2
90		2.35	2.41	2.48	2.55	2.64	2.74	2.86	3.01	3.24	4.0
89	2.24	2.30	2.33	2.41	2.48	2.56	2.65	2.76	2.90	3.10	3.7
88	2.19	2.24	2.29	2.36	2.41	2.48	2.57	2.67	2.79	2.97	3.5
85	2.05	2.10	2.14	2.19	2.24	2.29	2.35	2.43	2.52	2.66	3.1
80	1.88	1.91	1.94	1.97	2.00	2.04	2.08	2.13	2.20	2.28	2.5
75	1.73	1.75	1.78	1.80	1.82	1.85	1.88	1.91	1.95	2.01	2.2

$$PF = \sqrt{1 - \left[\frac{C}{S}\right]^2} \tag{6.25}$$

where C = cost of capacitors per kVAR
S = cost of system equipment per kVA

For example, a new system which is designed to serve a 1300-kVA load at 70 percent PF requires the application of a 1500-kVA load center substation, and costs $75,000. The cost of this system per kVA is $50 ($S = \$50$). If the distribution system has a 480-V nominal voltage, capacitors (C) cost an estimated $12 per kVAR. By substituting:

$$C = 20 \quad \text{and} \quad S = 60 \text{ in equation (6.25)} \tag{6.26}$$

we obtain the most economical power factor to be 0.94.

The equipment is available in standard sizes. The load power is 910 kW (1300×0.7). Hence, correcting the power factor from 70 to 91 percent reduces the load to 1000 kVA and allows the use of a smaller-rated standard component, a 1000-kVA load center. Additional economies can be realized by installing only those capacitors required to correct the initial load, adding more capacitors at a later date to keep in step with load growth. Depending upon the load growth, the larger transformer may still be the right choice.

6.1.6 Loss Reduction

Losses are proportional to current squared, and since current is reduced in direct proportion to power factor improvement (reduction of kVA). The loss reduction is given by:

$$\text{Loss reduction} = 1 - \left[\frac{PF_{original}}{PF_{new}}\right]^2 \tag{6.27}$$

With most industrial plant power distribution systems, the kW (I^2R) losses vary from 2.5 to 7.5 percent of the load kWh, depending upon hours of full-load and no-load plant operation, conductor size, and length of the main and branch feeder circuits. Consider an example:

1000 kW load, operating for 60 hours per week.

480-V, three-phase service, with 5 percent system losses.

Load power factor = 85 percent.

Energy charge = \$0.04 per kWH.

Improving power factor to 95 percent by adding power capacitors reduces losses by 10 percent.

$$I^2R_{\text{reduction}} = 1 - \left[\frac{0.85}{0.95}\right]^2 = 0.20 \tag{6.28}$$

Then system loss reduction is 1 percent ($0.20 \times .05 = 0.01$).

And $0.01 \times 1000 \times 60 = 600$ kWH per week in energy savings.

This yields $600 \times \$0.04 \times 4.3 = \$103.20$ per month savings

These savings need to be justified by the cost of capacitors. From Sec. 6.1.3, the value of capacitance that is needed is calculated as 291 kVAR. Using 300 kVAR and \$20 per kVAR installed cost of the capacitors, the total cost is \$6000. The time to pay off the investment without taking into account the cost of money, is:

$$\text{Time} = \frac{\$6000}{\$103.2 \text{ per month}} = 58 \text{ months} \tag{6.29}$$

This is a long time for an investment, but if it is added to the savings for power factor penalty, it is an added bonus.

6.1.7 Voltage Improvement

The following approximate expression may be used to calculate the voltage drop in a circuit:

$$E = IR \cos \phi + IX \sin \phi \tag{6.30}$$

The preceding equation for voltage drop may be written:

$$E = (I_{kW}R) + (I_{kVAR}X) \tag{6.31}$$

From this expression it can be seen that kVAR current operates principally on reactance, and since capacitors reduce kVAR current, they reduce the voltage drop by an amount equal to the capacitor current times the reactance.

An estimate of voltage rise from the installation of power capacitors to a factory electrical system can be made:

$$\Delta V = \frac{\text{kVAR}_{\text{capacitor}} \times (\% \text{ impedance of transformer})}{\text{kVA of transformer}} \tag{6.32}$$

More accurate improvement in voltage or voltage regulation can be determined from running a load flow computer program with and without the shunt capacitor. In fact, if many trials have to be made, the load flow computer program should be the preferred method.

6.1.8 Economic Justification for Capacitors

The earlier discussion has shown that the application of shunt capacitors to power systems has several benefits. Among these are:

- Reduced utility billing by improving power factor and reducing kVA or kVAR demand
- Released load capacity for existing or expanding loads
- Lower power system losses
- Improved voltage level at the load

The economic effect of capacitors on distribution systems is most significant in two areas: utility bill reduction and released capacity.

For the examples discussed earlier, the savings are compared to the cost of capacitors and the results are shown in Table 6.3. The payback period for power factor improvement is the most attractive. The capacitor application resulting in investment cost savings shows a benefit-to-cost ratio of 4 to 1, which is very favorable. The released capacity cost of $22/kVA should be compared against other alternative supplies as well as future load and power supply uprating plans. The loss reduction shows a longer payback period, even though it is a reasonable duration.

These benefits are cumulative. Power factor improvement at the very least provides bill reduction, loss reduction, and voltage improvement. Any release in capacity benefits should be taken into consideration only if that capacity is needed or used.

Although capacitors raise voltage levels, it is rarely economical to apply them in industrial plants for voltage improvement alone. Voltage improvements may therefore be regarded as an additional benefit. Section 6.2.5 discusses arrangements that regulate VARs and voltage.

TABLE 6.3 Summary of Economic Justification

Type of application	Text section	Savings $ per month	Capacitor kVAR	Cost	Payback period (months)
Power factor improvement	6.1.3, example 1	1650	550	11,000	6.7
	6.1.3, example 2	600	200	4000	0.7
Released capacity of existing system	6.1.4	$22/kVA			
Released capacity from new system	6.1.5	25,000 (total)	514	6168	4:1 saving
Loss reduction	6.1.6	103.2	300	6000	58

6.2   CAPACITOR UNIT AND BANK RATING

Capacitors are manufactured in individual units that are combined in parallel and series arrangements to give that desired voltage rating and total kVAR needed for the application. Industrial and commercial installations use single units in series for voltage levels 15 kV and below. Low-voltage units are delta-connected; medium-voltage units are wye-connected except 2.4- and 4.16-kV units that are three-phase and delta-connected.

6.2.1   Available kVAR and Voltage Ratings

Capacitors are manufactured in low- and medium-voltage units. Low-voltage units are manufactured by depositing a metal vapor on a polypropylene film. The deposited metal acts as the conductor and the film acts as the dielectric. These have the ability of self-healing if there is a failure in the film. They are packaged in units that are rated up to 2.5 kVAR single-phase. These units are connected in delta and are packaged in groups to give the desired three-phase rating. They have an internal pressure switch that will disconnect the unit if there is an internal fault. They also have an internal resistor that bleeds off the charge within five minutes. Table 6.4 lists the voltage ratings and standard kVAR ratings. The capacitors are fused as a group. A fuse in each phase is recommended. They are available in indoor dustproof and outdoor weatherproof enclosures. The individual units can be combined in any combination to obtain the desired kVAR rating.

Medium-voltage capacitors with ratings of 2.4 and 4.16 kV are packaged in three-phase units and are connected in delta. Table 6.5 lists the kVAR ratings of these units. The individual units can be combined into banks utilizing any of the unit ratings in banks from 25 through 600 kVAR in approximately 25 kVAR steps. These small bank arrangements utilize one-, two-, or three-phase delta-connected units with a terminal enclosure, mounting frame, and current-limiting fuses. This equipment does not require a great deal of space and can be easily and quickly installed. Low voltage three-phase units are available in enclosures up to 300 kVAR in delta-connected units. Standard enclosures are available in indoor dustproof and outdoor weatherproof designs. The two voltage levels are those which are utilized in approximately 90 percent of medium-voltage motor terminal capacitor applications.

In Table 6.6, the kVAR and voltage ratings are for single-phase units which can be obtained either in single-bushing or two-bushing configurations. Two-bushing units are used when series units are used in banks that exceed 34.5 kV. (Depending upon the bank size, series units may be used below 34.5 kV.) The units are connected in a wye configuration and are individually fused with an external fuse. (European practice is to use internal fuses.) The array of voltage ratings makes it possible to connect banks in configurations which will match almost any common industrial voltage level up to 34.5 kV

TABLE 6.4   Low-Voltage Three-Phase Units with Terminal Blocks and Fuses

Voltages	kVAR rating
240	5–25 in 2.5-kVAR steps
25–180	in 10-kVAR steps
480	5–50 in 2.5-kVAR steps
600	50–300 in 25-kVAR steps

TABLE 6.5 2.4- and 4.16-kV Three-Phase Units with Terminal Blocks and Fuses

Voltages	kVAR rating		
	One unit	Two units	Three units
2400	25	175	375
and	50	225	425
4160	75	250	450
(75-kV	100	275	475
BIL)	125	300	500
	150	325	525
	200	400	600

line-to-line. Applications within industrial plants use these units in ungrounded wye banks while electric utility practice uses grounded wye banks. Industrial practice requires that the system be grounded only at the source, either at the incoming transformers or at the generators.

The construction of these units is to place several coils in a stainless steel container and connect them in series and parallel internally to obtain the desired voltage and kVAR rating of the unit. The present construction uses polypropylene film and aluminum foil as the dielectric and conductor. Both materials are thin, the foil is on the order of one mil thick. The voltage gradient at the edge of the coil is high and the failure mechanism is voltage breakdown at the edge of the coil across the dielectric (film). To reduce the voltage gradient, the edges of the foil are folded to give a double thick-

TABLE 6.6 Medium-Voltage Single-Phase Units, One- or Two-Bushing Design

kVAR	Voltages	BIL (kV)*
	2400	75
	2700	75
	4160	75
	4800	75
	6640	95
50	7200	95
100	7620	95
150	7960	95
200	9960	95
300	12,470	95
400	13,280	95
	13,800	95
	14,400	95
	19,920†	125
	21,600†	125

*Basic insulation level only in single-bushing units rated 100 kVAR and above.

†19,920- and 21,600-V units available.

ness, thus reducing the voltage gradient. The voltage is the peak voltage of the circuit that causes this partial discharge or corona.

Individual capacitor units will meet or exceed the criteria described in ANSI C55.1 for shunt power capacitors. The following tolerances in ratings are among those which are considered significant in capacitor applications.

1. Capacitor units have a -0 to $+15$ percent tolerance on rated reactive power at rated voltage and frequency. In actual construction, the average capability above rating will be close to 4 percent because of manufacturing tolerances.
2. Continuous operation at 135 percent of rated reactive power including fundamental frequency and harmonic voltages.
3. Continuous operation at 110 percent of rated terminal voltages, 120 percent voltage including harmonics, but not transients. (Since the capacitor sees peak voltage, the fundamental and the major harmonic present are added arithmetically rather than using the rms value.)
4. Continuous operation at 180 percent of rated rms current including fundamental and harmonics at one per-unit voltage. If capacitors are operating close to this limit, the manufacturer should be consulted regarding fuse selection. (Normal fusing is approximately 140 percent of rated current.)
5. Ambient temperature limits depend upon the mounting arrangements and, hence, the ventilation. The range of 24-hour ambient temperatures is from 35°C in enclosed equipment to 46°C for isolated units in open mountings. The minimum ambient temperature is -40°C.

Other ratings such as transient overcurrent and overvoltage levels are also specified, but these provide the basic parameters for application.

Industrial and commercial systems present many opportunities for the application of low-voltage capacitor units and often they will be the optimum choice from the system operation viewpoint. If larger banks of capacitors are needed to make the desired correction in power factor, then a comparison should be made between the total costs of medium-voltage and low-voltage installations. The cost per kVAR of medium-voltage capacitors is significantly less than that of the low-voltage type (a ratio of 5:1), but this advantage is offset in some instances by the cost of the medium-voltage switching device which is required for the larger bank.

6.2.2 Selection of Capacitor Units for Large Bank Installations

Whenever possible, capacitor units of a standard voltage rating should be used, since they are generally less expensive than those with special voltage ratings.

In general, it is desirable to choose capacitor units with voltage ratings such that each series group contains a large number of units in parallel. With the array of standard voltage ratings available, wye-connected banks without series sections are available up to 34.5 kV. Proper selection of capacitor unit voltage and kVAR rating requires evaluation of these three factors:

• Allowable number of units per group
• Parallel stored energy
• Cost

When more than one series group of capacitors is used, the number of units in parallel per series group is an important factor. The usual requirement is that when one capacitor fuse is open, the other capacitor units should not experience more than 110 percent voltage. The minimum number of parallel units varies depending upon the number of series groups used in each phase and the type of three-phase bank connection (grounded wye, wye, or delta).

The number of capacitors in parallel in each series group determines the stored energy. When one capacitor in a series group fails, the stored energy in the remaining parallel capacitor units of that group may discharge through the fuse to the failed capacitor unit. Excessive energy may cause a fuse to melt or rupture the capacitor unit. If the energy expected to be discharged is higher than permissible limits, then bank grouping has to be reconfigured. The following example is considered a satisfactory wye (ungrounded) three-phase capacitor bank for a 3-MVAR, 13.8-kV application:

Capacitor unit	300 kVAR, 7.96 kV
Number in series per phase	1
Number in parallel per phase	4
Three-phase rating	3.6 MVAR
Voltage with 1 unit out in one phase	109 percent

This increase in voltage is shown in Fig. 6.6.

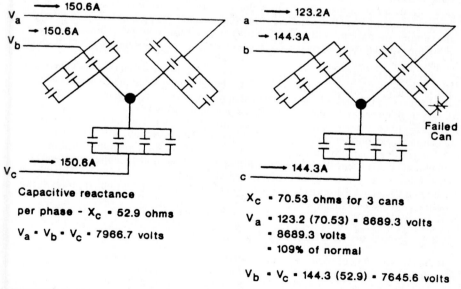

FIGURE 6.6   Overvoltage on capacitor bank with one unit out of service.

6.2.3   Bank Construction and Location

There are three principal types of capacitor bank construction available in the medium-voltage range. The selection will depend on the location in the system where they are to be connected and the kVAR capability of the bank.

The type of bank most often applied is the outdoor rack design. This approach is suitable for large, medium-voltage capacitor banks of any designated kVAR rating and voltage. These designs are utilized by both industrial and utility users. The voltage range starts at 2400 V. The kVAR capacity starts at 300 kVAR and can be designed as large as required. These designs will include racks for mounting and individual unit fuses. A stand and insulated support structure with accessories such as switching devices and controls can be furnished when desired.

The second type of equipment available is the enclosed or house construction. In the medium-voltage range there are two types of construction. One is the pad-mounted design with ratings up to 1800 kVAR at voltages from 2400 V to 34.5 kV. The housing is built to enclose the capacitors, switching device, and the automatic sensing and control equipment. These are used where a relatively small bank of capacitors is needed to improve the voltage profile on a distribution feeder.

Another variety of housed equipment includes larger-rated banks, up to 6000 kVAR. These large installations tend to be more special in their design, meeting specific conditions required by the customer. The capacitor units are individually fused. Vacuum switches or circuit breakers, as well as conventional oil switches, can be utilized and operated by automatic sensing and control. On occasion, the housing may be equipped with forced-air cooling. This type of housed equipment has been utilized by both industrial facilities and utilities.

Since capacitors reduce kVAR flow from the capacitor location to the source, the logical placement is as near the load as possible. This will have the greatest effect on reducing losses. In an industrial system, it means motor terminal installation as well as power utilization buses, either low-voltage or medium-voltage. Concentrations of induction motors are often candidates for shunt capacitor applications, particularly if the motors are only partially loaded. Other possibilities may be induction furnaces, thyristor-type adjustable-speed drive power supplies, and resistance welders, to name a few.

There will be occasions when it will be more advantageous to locate the kVAR supply at the main bus in the plant. Some of the benefits of placing the capacitors right at the load (as in the case of motor terminal applications) are relinquished in favor of a larger bank placed such that the kVAR can be controlled independently from the load. The need to control harmonic voltages and current may indicate that a single capacitor location be established. The economics of purchasing, installing, protecting, and controlling a single large bank can also tilt the decision toward a main bus location. The combination of system needs, system configurations, operational requirements, and cost of purchasing and installing the equipment will all influence the selection of the bank location. Chapter 11 discusses the problem of harmonics and distributed power factor improvement capacitors.

6.2.4   Capacitor Bank Connection

There are a number of ways in which a capacitor bank may be connected. The choice is dependent on the voltage level of the system, the kVAR capacity of the bank, the system grounding, and the desired relay protection. For industrial system voltages up to 34.5 kV, the banks are made up of capacitor units rated at line-to-neutral voltage. Once the individual capacitor units are selected to meet the voltage requirements of the system, then the number of parallel units is selected to meet the bank kVAR requirements.

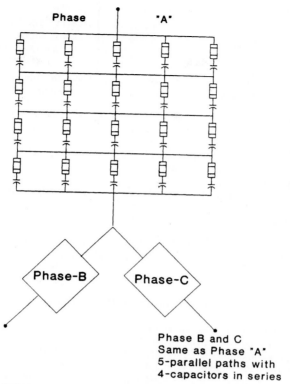

FIGURE 6.7 Series-paralleled connection of a large high-voltage capacitor bank.

Two criteria are applied to determine the minimum allowable number of paralleled capacitor units in each phase:

1. The loss of one capacitor unit in a phase should not produce a voltage across the remaining units in that phase exceeding 110 percent of the rated voltage.
2. In the event of a failure of a unit, sufficient fault current should flow through the individual fuse to ensure clearing in 300 seconds or less. It should be pointed out that the 30 seconds or less is a more desirable time span.

Industrial and commercial capacitor banks are ungrounded wye-connected, with paralleled units to make up the total bank kVAR (Fig. 6.7). Banks at 4.16 kV and lower are usually connected in delta.

The ungrounded wye arrangement has several advantages:

1. For banks installed on ungrounded systems, the rack on which the capacitors are mounted may conveniently and economically be insulated from ground. The rack becomes the neutral connection for the bank and the unit container connected to the rack and are at the neutral potential. When the bank is deenergized, the rack is grounded, thus, all of the capacitor units are also grounded. This usually results in savings in the bank particularly if a circuit voltage is 7.2 kV or higher.

2. Fuses with lower interrupting capacity can generally be used.

3. The ungrounded double-wye arrangement is preferred in large banks, because it offers a better choice of protective relaying schemes. The total bank kVAR should be large enough to permit splitting the bank into two equal-rated wye configurations.

4. In the higher-rated banks, the loss of one fuse may leave enough units in service on that phase so that the 110-percent voltage limit is not violated.

In the case of smaller bank ratings that are used in a utility distribution system, such as 2000 kVAR and less, grounding the neutral of the capacitor bank has advantages. Units are group-fused, which means that several capacitor units are protected by one fuse. The fuse rating is large with respect to the individual capacitor unit, and if one capacitor fails and there is enough fault current to blow the fuse, the neutral of the bank is grounded. There is a requirement that the fuse must clear in 300 seconds or less in the event of a failed capacitor unit. This is easily met when the bank is connected in grounded wye or delta. In these two instances, a unit failure results in a line-to-neutral (wye-connected) or line-to-line fault (delta-connected). In a grounded bank with one series section per phase, when one fuse blows, the voltage across the remaining units stay the same. By contrast, in an ungrounded small bank, the loss of one unit, which is a relatively high percentage of the total phase capacity, and the remaining units are subjected to more than 110 percent voltage.

6.2.5 Types of VAR Supplies Using Static Power Capacitors

The leading current drawn by the capacitors results in a voltage rise through the inductive reactance of the power system, which raises the operating voltage level.

A fixed bank of capacitors will furnish a fixed amount of VARs at a constant voltage to the load. The voltage increases with a decrease in load VAR requirements; the voltage decreases with an increase in load VAR requirements. With an increased voltage, the fixed capacitors will furnish more VARs. VARs from capacitors is proportional to the square of the voltage applied.

Capacitors cannot by themselves control reactive power. Some method of controlling the VARs supplied by the capacitors is necessary. Three methods of controlling VARs using capacitors, in order of increasing complexity, are:

- *Mechanical-switched capacitors (MSC):* Switching capacitors in one or more groups using power circuit breakers, contractors (low-voltage), or vacuum switches.

- *Thyristor-controlled reactor-fixed capacitors (TCR-FC):* Back-to-back phase-controlled thyristor switching of a reactor in parallel with a fixed capacitor bank.

- *Thyristor-switched capacitors (TSC):* Back-to-back thyristor switching of capacitors, which turns the capacitors on or off at current zero.

MSC: Switching power capacitors by circuit breakers; vacuum or sulfur hexifloride (SF_6) switches for controlling reactive power on a continuous basis, requires a switching device that can be operated with high frequency and can interrupt at current zero with a high voltage across the contacts without reignition. Because of these requirements, this method is used only for switching larger banks once or twice a day when a demand changes from normal to light load conditions. The switching device has the special requirements of interrupting a current which leads the voltage by 90 degrees.

This is the least expensive method and can be used for controlling the average VARs over a long period of time. For instance, if during the week a plant operates two shifts a day, this method could be used to disconnect some or all of the capacitors during the period when the plant is at light load or on weekends. This method should not be used for switching the capacitors on and off more than a few times per day because of the transient voltages associated with capacitor switching.

TCR-FC: Back-to-back phase-controlled thyristor switching of a reactor in parallel with capacitors has the advantage of smooth VAR control over the range of the equipment (Fig. 6.8). By controlling the current to the reactor, the problem of switching a leading current is avoided. By modulating the current in the reactor, the VARs from the capacitors can be absorbed more or less, depending upon the system requirements. The thyristor switching of a balanced three-phase load causes fifth, seventh, and other higher harmonic currents. The capacitors may be divided into several sections with tuning reactors to filter these harmonics. The reactor VAR rating is normally equal to the capacitor rating to get full control. More capacitors can be supplied if a bias of VARs are needed on the system. This system is commonly known as the *static VAR compensation* (SVC). This system can control VARs on an individual phase basis (single-phase) and is being used to compensate electric arc-furnace loads, arc welding, and similar loads causing rapid voltage fluctuations.

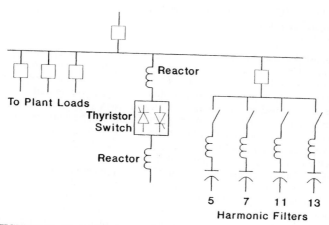

FIGURE 6.8 Static VAR control (TCR-FC).

TSC: Back-to-back thyristor switching of capacitors at zero current leaves the capacitor charged with either a positive or negative full charge on the capacitor (Fig. 6.9). The thyristor's fine control allows switching of the capacitor when the system voltage equals the charged capacitor voltage. This eliminates any transients on the system. The capacitors are switched in finite steps as reactive power is needed. This system can also be used with a fixed bias of capacitors to provide a base VAR with the TSC to be used for variable VARs. This system can be regulated on a per-phase basis. TSC has limited application because of its complications and cost; hence, TSC is used under special situations. Special situations include low-voltage applications for controlling voltages on small and medium systems requiring fast voltage control, such as resistance welding, electric shovels, and draglines.

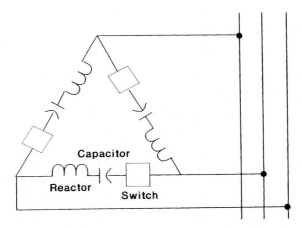

FIGURE 6.9 Three-phase diagram of one bank of thyristor-switched capacitors (TSC).

6.3   PROTECTION OF CAPACITORS AND CAPACITOR BANKS

Complete protection must be provided for a capacitor installation. This will consist of protecting the individual units as well as the total bank. Both fuses and circuit breakers with relays are used, depending upon the rating of the bank, its location in the system, and other factors. Fusing is used to remove a failed medium-voltage capacitor unit from the system. Relays and circuit breakers are applied for overall bank protection and for switching. Normally, the decisions about the unit and group fusing will be made by the manufacturer unless special requirements are spelled out in the procurement specification.

The modern power capacitor unit is extremely reliable, the failure rate being less than 0.1 percent per year. This figure represents the overall average failure rate for all applications for industrial, commercial, and utility applications. In particular, in locations involving frequent switching and/or harmonic duties, the capacitor may not enjoy such high reliability rates and the protection of the units becomes increasingly important. If a failure occurs, then proper protection will limit the extent of the damage. Fuses are the most desirable and economical method of protecting against possible consequences of failure. The major purposes of capacitor fusing are to:

- Remove the failed capacitor unit from the circuit.
- Maintain service continuity of the remaining units.
- Prevent damage to adjacent capacitors and equipment, or injury to personnel.
- Provide visual indication of a failed unit.
- Limit the energy into a faulted unit to prevent case rupture.

6.3.1   Fusing of Individual Capacitor Units

The requirements for proper fuse selection are:

- The rated voltage of the fuse should be at least 110 percent of the capacitor unit rating.

- The maximum interrupting rating of the fuse should be greater than the available short-circuit current which can flow if a capacitor unit is shorted. This includes not only the short-circuit current from the system, but also that current which flows into the faulted unit from adjacent capacitor units. This may require the application of current-limiting fuses in place of expulsion fuses for large bank rating and for banks connected to buses with high short-circuit capacity.
- The fuse should have a time-current clearing characteristic that lies below the time-current case rupture probability characteristic of the capacitor units. This is discussed in Sec. 6.3.4. Industrial and commercial applications should be operated in the safe zone only.
- The minimum fuse link size is determined by its continuous current duty.
- Available fault current and available energy to discharge into a failed unit influence the choice of expulsion or current-limiting fuses and also the connection of the bank.
- A faster fuse clearing characteristic requirement is determined by:
  - Coordination with the capacitor case rupture curve.
  - Minimizing overvoltages on good capacitor units during a unit failure.
  - Coordination with the capacitor bank unbalance detection scheme.

In choosing the best fuse in a given application, it sometimes is not possible to meet all of the requirements. In that case tradeoffs must be made and some risk taken concerning those conditions when the fuses and capacitors may not operate in the most desirable manner. The fuse choice is normally chosen by the manufacturer unless otherwise specified.

6.3.2 Group Fusing

The practice of utilizing one fuse to protect more than one capacitor unit is called *group fusing*. Small capacitor banks on distribution lines or low voltage are often group-fused. On industrial systems with large banks, individual unit fuses are used. Group fusing is considered a special case of individual fusing and the same criteria generally apply, such as the requirements for the fuse voltage rating and interrupting capacity. In the case of the 165 percent of rated current criterion, this now refers to the current of all of the grouped capacitor units protected by the one fuse.

The basis for the 165 percent factor (NEMA Standard CP-1-19xx) is that the actual current flowing through the fuse and capacitor bank usually exceeds the nominal current. This arises because of practical operating conditions, such as

- Operating voltage above the rated capacitor voltage
- The presence of harmonic currents flowing through the bank, actual capacitance values greater than the rated value due to manufacturing tolerances
- Short-duration transient currents resulting from switching and lightning surges

For delta- and floating-wye banks, experience has indicated that it is often satisfactory to reduce this factor from 165 to 150 percent or less if necessary, since there is no path for the flow of third harmonics or multiples of the third.

For the range of fault current available to a failed capacitor unit, the total clearing time-current characteristics of the fuse link must be coordinated with the case rupture curves determined by the manufacturers.

Typical group fusing recommendations are available from manufacturers for ungrounded and grounded wye- and delta-connected capacitor banks. Factors which influence group fusing may be summarized as follows:

- Minimum fuse link size is determined by its continuous current duty.
- Choice of expulsion or current-limiting fuse depends upon the fault-current level.
- T-links can withstand transient currents better because they are the slower melting type.
- K-links minimize overvoltages duty on good units because of their faster melting property.
- K-links provide the best tank rupture margin; hence, it is easy to coordinate protection.

6.3.3 Large Bank Protection

Expulsion fuse protected banks are used until the short-circuit capability exceeds 6000 A for the 200- and 150-kVAR units, 5000 A for the 100-kVAR units, and 4000 A for the 25- and 50-kVAR units.

As the number of parallel capacitor units increases for the higher kVAR rated banks, there will be an increase in the current available from adjacent units in the event of the failure of one capacitor unit. Large banks of capacitors, particularly in an industrial system, will often be associated with a main bus having short-circuit capacity well beyond the values previously given. In these cases, the current-limiting fuses will be applied for individual unit or group protection. Once it is established that the higher short-circuit capability of the current-limiting fuse is needed, the remaining criterion of 165 percent normal current, 300 seconds for three times normal current (if it applies), and coordination with the case rupture characteristics should again be applied.

6.3.4 Time-Current Case Rupture Characteristics

These characteristics have been referred to earlier as part of the capacitor unit protection discussion. As the name implies, these curves display graphically the relationship between fault current and time for different case rupture modes. Figure 6.10 illustrates a typical characteristic.

Within the safe zone, usually no greater damage than slight swelling of the case will occur. It is possible, however, for a case rupture to occur as a result of very low short-circuit currents flowing for extended periods of time. To avoid such case ruptures, the fuse link should be coordinated so that it will clear the fault within 300 seconds. This is a significant consideration generally only for ungrounded wye-connected banks for which fault current is limited to approximately three times normal current.

Some manufacturers have different curves for 10, 50, and 90 percent probability of case rupture. Zone 1 bounded by the 10 and 50 percent curves, is considered suitable for locations where case openings or fluid leakage would present no hazard because violent case ruptures are improbable. Zone 2, the area above the 50 and 90 percent curves, is suitable for locations which have been chosen after careful consideration to possible consequences associated with violent rupture of the case. The area beyond the 90 percent probability curve is in the hazardous zone. The hazardous zone is unsafe for most applications because a failed unit will often rupture with sufficient violence to damage adjacent units. Tests have demonstrated that beyond this value of current, expulsion fuse links will not satisfactorily protect against violent ruptures. Above these fault-current levels, current-limiting fuses should be applied to limit personnel hazard and equipment damage. For the range of fault current available to a failed capacitor unit, the total clearing time of any fuse should be coordinated with the case rupture curves. Industrial and commercial applications should always be below the 10 percent curve.

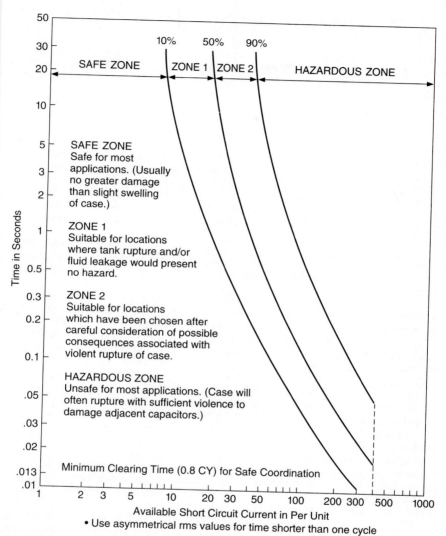

FIGURE 6.10   Typical tank rupture curves.

6.3.5   Relaying Protection

In addition to the individual capacitor unit protection through the use of individual or group fusing, the two basic types of relay protection used with high-voltage capacitor banks are:

1. Overcurrent relaying for major equipment faults
2. Relaying to identify loss of units within a bank

Overcurrent relaying is needed for the removal of the capacitor bank in the event of a fault between the switching device and the bank itself. The relay should be chosen so that the highest magnitude of inrush current associated with the capacitor switching will not trip the circuit breaker immediately as the bank is energized. Also, if a capacitor unit fault occurs, the relay should delay operation until the fuse clears. Relays with an inverse time-current characteristic are usually used and settings are selected to override these two conditions. However, the relay may also have a very inverse or extremely inverse time-current characteristic if coordination with other system protective devices require it. Instantaneous overcurrent relays are not often utilized since they are likely to trip unnecessarily when the bank is energized. Relaying for the detection of the loss of capacitor units and guarding against excessive operating voltages should be used as protective measures supplement to the periodic visual inspection of individual unit fuses.

Except for very short periods of time, the voltage across capacitors should not exceed 110 percent of rated voltage. On large banks, it is recommended that some type of protection be provided which will relay the bank off if the capacitors in a parallel group are subjected to overvoltage because of the loss of a portion of the units in the group. This would also protect the bank when an entire parallel group was shorted by a foreign object.

There are several methods of protecting against overvoltages on a parallel group. Figure 6.12 shows generic relaying methods. For an ungrounded bank, the potential transformer (PT) shown in Fig. 6.11 will detect a change in voltage across the phase if one or more units are removed. The secondary windings of the PTs are connected in broken delta and used in series with the coil of a voltage-sensitive relay. Potential transformers have an advantage over capacitor potential devices in that they furnish a discharge path for the capacitors and will discharge the capacitors in less time than the discharge resistors built into the capacitors. Use of the potential transformers as a discharge path will be effective only if the transformers have the thermal capability to pass the current associated with the stored energy in the bank. The kVAR rating of the bank will determine the energy storage capability and this should be matched against the energy absorption capability of the PT.

The internal resistors built into each individual capacitor unit will also discharge the bank, but this requires about five minutes for high-voltage units and one minute for low-voltage units. For automatically controlled banks, this may be too long a time. If the bank is switched on while still charged from a previous energization, higher-than-normal transient currents may be experienced. The time delays and normal operating sequences of any switching controls should be reviewed to ensure that the bank will have sufficient time to discharge to a safe level before being reenergized.

The schemes shown in Fig. 6.11(b) and (c) are known as the double-wye schemes, and they have considerable merit. They are less expensive since they require only a single PT or current transformer (CT). For smaller banks, it may be expensive to split the banks into two wye groups, or it may be impossible to do so and still keep the voltage on the remaining units below 110 percent of rated in the event of the loss of an individual unit. On larger banks, it may be easier to connect the bank into two wye configurations, particularly on industrial systems, where there is no advantage to grounding the neutral connection of the bank.

When conditions are such that grounded single-wye banks must be used, the scheme of Fig. 6.11(d) and (e) can be used for overvoltage protection. This provides a low-cost protective arrangement, but it requires that the neutral point be grounded. As noted earlier, the grounded arrangement is not used in industrial systems for several reasons, one of them being possible interference with ground-fault relaying. Use of a third-harmonic filter would permit setting the relay at fairly low values of pickup.

Relay arrangement, illustrated in Fig. 6.11(f), is applicable to a wye-wye bank where the neutral of the bank can be solidly grounded. This method is not sensitive to residual

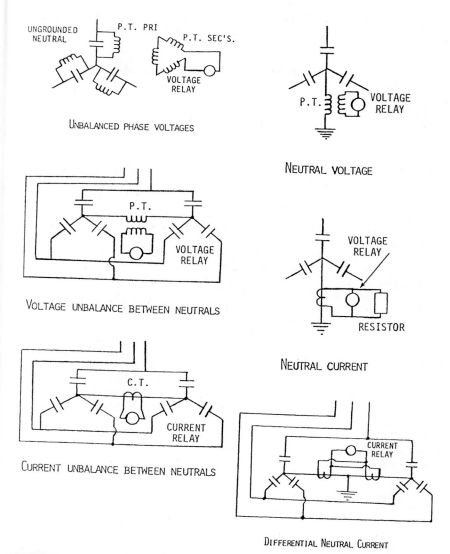

FIGURE 6.11   Typical capacitor bank protection schemes.

currents resulting from harmonic voltages to zero sequence currents as a result of system faults or load unbalance.

The characteristics of these relay arrangements are summarized in Table 6.7.

No overvoltage protection scheme will provide positive protection against overvoltage in all cases. The basic theory of all the schemes is the detection of current or voltage unbalance. There is some inherent unbalance in capacitor banks, and the relaying must be set above the maximum inherent unbalance which can occur if false trip-outs

TABLE 6.7 Characteristics of Protective Relaying Methods for Capacitor Banks

Protective relaying type	Sensitive to			Number required	
	Switching transients	Harmonics	Voltage unbalance	CTs	PTs
Unbalance phase voltage	Yes	No	No	—	3
Unbalance voltage between neutrals	Yes	No	No	—	1
Unbalance current between neutrals	Yes	No	No	1	—
Neutral voltage	Yes	Yes	Yes	—	1
Neutral current	Yes	Yes	Yes	1	—
Differential neutral current	Yes	No	No	2	—

are to be avoided. The inherent unbalance is due mainly to harmonic currents and voltages, and varying capacitance of the capacitors.

The tolerance of capacitance of a power capacitor is -0 to $+15$ percent with the average being about $+4$ percent. When individual units are placed in racks and stacked for high-voltage banks, the capacitance of the racks in each phase leg may vary. This causes normal condition unbalance and makes it more difficult to detect small unbalance due to removal of faulted units in a parallel group. If it is considered necessary to balance the phases closely, it is possible by special arrangement at the factory. The problem of phase unbalance tends to diminish as the bank rating increases.

In cases where a large number of capacitor units, say twenty or more, are used in a parallel group, it may not be possible to detect the loss of one unit, since the relays cannot be set too sensitive because of natural unbalances. This is not too important, since the loss of one unit in such a large group does not raise the voltage on the remaining units above the allowable 10 percent overvoltage. In most cases, the relaying can be set to operate before dangerous conditions exist without serious danger of false operation. On a large installation, it is good practice to use two relays. One will sound an alarm when one or more units have failed but dangerous voltages are not yet present. The second relay will trip if allowable overvoltage is exceeded. Such a procedure has the advantage of keeping the bank in service when possible while indicating that capacitors have failed, yet still protect the capacitors from serious overvoltage.

6.3.6 Lightning Protection

The choice between surge arresters rated for grounded neutral service or ungrounded neutral service should be in accordance with established industry practices. The rating of the surge arresters (line-to-ground) is determined by the system grounding practice and is not related to the grounding, or lack thereof, of the capacitors themselves. See Chap. 10 for a discussion on surge arresters.

Capacitors wye-connected with grounded neutral have some ability to slope off the steep front of an incoming wave and to reduce its crest value. However, this ability is limited by the size of the capacitor bank and the amount of energy in a given surge or lightning stroke. It is difficult to predict the energy content of switching or lightning surges. There are generally switches or circuit breakers associated with the capacitor installation which require surge arresters. Their need for protection is greatest when in the open position, with the capacitor bank deenergized. Hence, the surge arresters should be connected on the line side of any circuit-interrupting or isolating device at the capacitor installation, but as close to the capacitor bank as possible.

In the usual case for an industrial system with an ungrounded capacitor bank, properly applied surge arresters will help protect the capacitors and other system components by shunting the surge current to ground. The capacitor bank will not discharge through the arrester, avoiding possible damage to the arrester as would be a possibility if the bank were grounded.

6.3.7 Grounding of Capacitor Cases

Single-phase capacitor units are built with either one or two bushings. In single-bushing units, one of the terminals is a stud which is electrically connected to the case. The case is grounded when connected to a circuit having one conductor grounded, such as a three-phase, four-wire multigrounded neutral system. When the terminal common to the case cannot be connected to ground, the capacitor case must be insulated from the ground. Under this condition, the case and the rack will be at some potential above ground and are a hazard. Therefore, the rack is mounted on a stand and insulated from ground. Other safety precautions, like fences or other suitable enclosures, are usually provided. When ground connections are used, they must be adequate to dissipate fault current with no hazardous rise of voltage. Economics determines whether the cases should be insulated from ground or whether this insulation should be built into the capacitor unit. When capacitor units are installed on ungrounded systems of 12 kV or more, and on all systems where units are connected in series, it is usually more economical to insulate the cases from ground.

The case of the two-bushing unit may be grounded, provided it has sufficient terminal-to-case insulation. On lower-voltage circuits, it is more economical to standardize on two bushing units having the insulation built into the capacitor, since low-voltage units are normally delta-connected and they have a more universal application.

6.4 CAPACITOR SWITCHING

Power circuit breakers are designed primarily to interrupt inductive short-circuit currents. The abilities of circuit breakers to interrupt short-circuit currents are not necessarily the same. High-frequency voltage and current oscillations occur during capacitor switching, which, if uncontrolled, may result in damage to the apparatus or system outages. The following paragraphs outline the most common conditions encountered in switching and provide information on available switching devices.

6.4.1 Inrush Currents When Energizing Capacitors

Since an uncharged capacitor offers practically zero impedance to the flow of current at the exact instant voltage is applied, it is possible for large high-frequency transient currents to flow during the energizing period. It is good to have an understanding of the relative magnitudes of these currents. For a single bank, the inrush current is always less than the short-circuit value at the bank location. For parallel banks, the inrush current is always much greater than for a single bank. The actual magnitude of the inrush current depends upon circuit and capacitor characteristics and may exceed the short-circuit value at the bank location.

The calculated inrush current, although of a high natural frequency, should not exceed the published 60-Hz momentary current rating of the switching device unless specifically noted otherwise. Although there may be little correlation between the effects of high-frequency currents and 60-Hz currents, the fact remains that the 60-Hz rating is the only one recognized in the industry standards for circuit-breaker performance. The user, therefore, has the alternative of possibly utilizing a larger breaker than would otherwise be necessary or inserting additional reactance to limit the inrush current.

Energizing Single Bank.   The energizing of a single capacitor bank may be represented as shown in Fig. 6.12(*a*). The maximum rms value of inrush current for the single bank may be calculated using the following formula:

$$I_{\text{max rms}} = \frac{E_{l-n}}{X_c - X_L}\left(1 + \sqrt{\frac{X_c}{X_L}}\right) \tag{6.33}$$

where E_{l-n} = line-to-neutral rms voltage
$\quad X_c$ = capacitor reactance in ohms of one phase-to-neutral of the capacitor bank at fundamental frequency
$\quad X_L$ = inductive reactance in ohms per phase of the source at fundamental frequency (based on short-circuit level)

The preceding formula applies to delta-connected capacitor banks if X_c is determined as the reactance of the equivalent line-to-neutral capacitor kVAR. The line-to-neutral reactance of any three-phase capacitor bank, whether it be wye- or delta-connected, can be calculated by:

$$X_c = \frac{(kV_{1-1})^2}{\text{MVAR}} \tag{6.34}$$

where kV_{1-1} = line-to-line kV

In the derivation of this equation, the effect of resistance in the circuit was neglected, thus giving a simple and conservative expression. It was also assumed that there was no residual charge on the capacitors. Inspection of the formula indicates that the peak inrush current to a single capacitor bank is always less than the short-circuit current at that point.

EXAMPLE: A 5400-kVAR capacitor bank is connected to a 13.8-kV system which has a short-circuit level of 250 MVA. This means

$$X_L = \frac{(13.8)^2}{250} = 0.76 \ \Omega \tag{6.35}$$

and

$$X_C = \frac{(13.8)^2}{5.4} = 35.3 \ \Omega \tag{6.36}$$

Substituting these values in Eq. (6.33), we set the maximum inrush current:

$$I_{\text{max rms}} = \frac{7960}{35.3 - 0.76}\left[1 + \sqrt{\frac{35.3}{0.76}}\right] = 1801 \ A \tag{6.37}$$

The symmetrical short-circuit current is:

$$I_{\text{SC}} = \frac{250,000}{\sqrt{3} \times 13.8} = 10,459 \ A \ \text{rms} \tag{6.38}$$

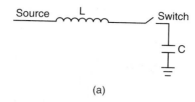

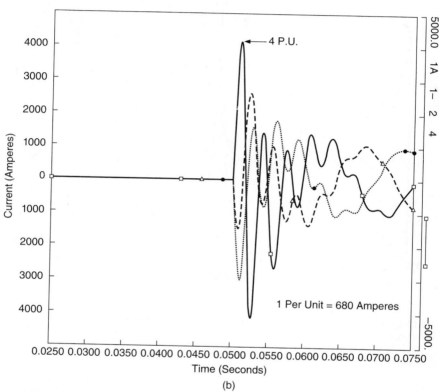

FIGURE 6.12 Inrush current and voltage when energizing a single 16.2-MVAR capacitor bank: (a) Circuit; (b) Current.

Thus, the momentary rating of a switching device for control of a single bank should be based on the short-circuit current which is higher than on the maximum inrush current to the bank.

Energizing Parallel Banks (Multistep Banks).   When energizing a parallel bank of a capacitor, one or more steps in a capacitor bank may already be energized. In such a case, the maximum peak current that flows into the next capacitor group to be energized is determined predominantly by the momentary discharge from those units already in service. Since the impedance between the charged and uncharged group is very small,

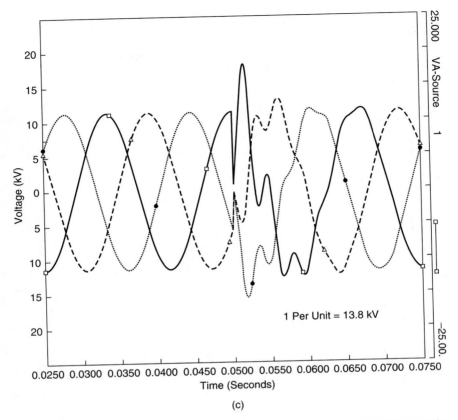

FIGURE 6.12 (*Continued*) Inrush current and voltage when energizing a single 16.2-MVAR capacitor bank: (*c*) Voltage.

high peak inrush currents can be expected (see Fig. 6.13). The maximum inrush to any switch occurs when all other switches have been closed previously (i.e., when the last bank is being energized).

    The energizing of the last step in a three-step capacitor bank is shown in Fig. 6.14. If no charge is on the step being energized, the maximum peak inrush current may be approximately determined by the following formula:

$$I_{max,peak} = \sqrt{2}\, E_{l-n}\, \sqrt{\frac{C}{L}} \tag{6.39}$$

where E_{l-n} = line-to-neutral rms voltage
         C = equivalent capacitance of circuit in μF
         L = inductance in μH between the energized steps and the step being switched on

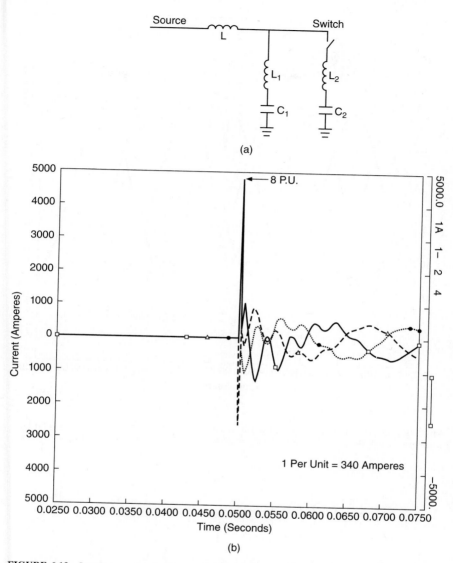

FIGURE 6.13 Inrush current and voltage when energizing an 8.1-MVAR capacitor bank when one 8.1-MVAR capacitor bank is already energized (back-to-back switching): (*a*) circuit; (*b*) current.

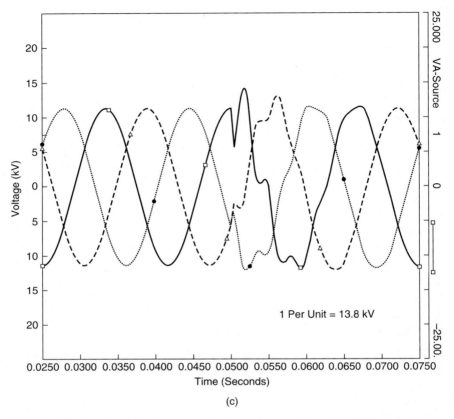

FIGURE 6.13 (*Continued*) Inrush current and voltage when energizing an 8.1-MVAR capacitor bank when one 8.1-MVAR capacitor bank is already energized (back-to-back switching): (*c*) voltage.

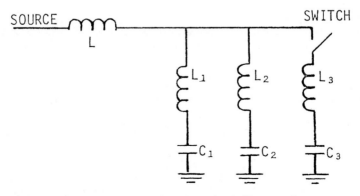

FIGURE 6.14   One-line diagram of back-to-back switching example.

The frequency of the inrush current may be calculated by the formula:

$$f = \frac{10^6}{2\pi \sqrt{LC}} \text{ Hz} \qquad (6.40)$$

where L and C are in the same units as previously specified.

EXAMPLE: Three capacitor banks, 5400 kVAR each, are connected to a 13.8-kV system with a short-circuit level of 250 MVA. Assume that two 5400-kVAR banks are already energized, and the third is to be energized. Assume that the distance from the terminals of each capacitor bank through its circuit breaker to a common point on the bus is 20 feet. Using 0.5 μH/ft of conductor run, each of the inductances L_1, L_2, and L_3 would be 10 μH. C_1, C_2, and C_3 are each equal to 75 μF.

$$L = \frac{(L_1 \times L_2)}{(L_1 + L_2)} + L_3 \qquad (6.41)$$

$$C = \frac{(C_1 + C_2)\, C_3}{(C_1 + C_2 + C_3)} \qquad (6.42)$$

$$I_{\text{max,peak}} = \sqrt{2}\,(13\,800/\sqrt{3})\, \sqrt{\frac{50}{15}} = 20{,}600 \text{ A} \qquad (6.43)$$

The frequency of the inrush current would be:

$$f = \frac{10^6}{2\pi \sqrt{(50)\,(15)}} = 5812 \text{ Hz} \qquad (6.44)$$

Because of the interrupting requirement for this particular application, a 500 MVA breaker would be chosen that has a short-circuit interrupting capacity of 19,500 A and a momentary current rating of 37,000 total rms A. Thus, the short-circuit current would dictate the momentary rating of the breaker for controlling the third step of the capacitor bank. There may be unusual cases, however, where the reverse is true and the formula provides a quick and convenient means of arriving at the momentary requirements of the breaker. For typical applications, formulas for calculating inrush current and frequency may be found in ANSI Standard C37.073-19xx.

Reenergizing Capacitors.   The formula for inrush current previously discussed applies only when energizing uncharged capacitors. If the capacitors are fully charged, as in the case where a bank is being deenergized, and a restrike occurs in the switch, the inrush current may be twice that given by the formula.

Since there is the possibility of doubling the inrush currents upon energizing a charged capacitor bank, it is best to leave the bank deenergized for a period of time long enough to permit discharge to the low voltage before reenergizing. The internal discharge resistor will reduce the potential to 50 V in five minutes. This is not an excessive delay under most conditions, and it will avoid the high inrush currents which otherwise may be experienced.

In some applications, there may be transformers connected in parallel with the capacitor bank on a deenergized circuit. Then, transformers will provide a path for a more rapid discharge of the bank. The residual voltage caused by the trapped charge is dc and the capacitors will discharge quite rapidly through the connected transformers, generally within one cycle. Even with high-speed reclosing (on the order of 20 cycles), the capacitors will be completely discharged before the circuit is closed in again. The precaution is that the transformer or transformers which serve as the discharge path must have the ability to dissipate the stored energy in the form of heat without distress.

Because of the high inrush current associated with back-to-back switching individual capacitor fuses can become fatigued. The silver and copper of the fuse link has an S-N (stress vs. number of cycles) similar to that of steel, except the final value is zero.

6.4.2 Transient Voltage Attending the Switching of Capacitors

The maximum voltage across a capacitor being energized does not exceed twice the steady state voltage, assuming the capacitor switch closes cleanly, i.e., there is no pre-strike, bounce, etc. Contact bounce is the intermittent and undesired opening of contacts during the closure of open contacts or opening of closed contacts. An irregular wavefront during transition is implied. Prestrike is arcing between contacts before they are closed. Restrike is a resumption of current between the contacts of a switching device during an opening operation after an interval of zero current at one-quarter cycle at normal frequency or longer. When a capacitor is deenergized, assuming no restriking in the switch, the maximum voltage across the capacitor is equal to the crest source voltage. These two statements are true for either the single or parallel banks. Although a voltage in excess of two times rated might be tolerated, it is highly desirable to limit the voltage to two times line-to-neutral for normal switching conditions.

Voltage Across Switch Contacts with No Restriking.   When leading kVA (capacitive current) is interrupted, certain phenomena result. For example, consider a grounded neutral capacitor bank connected to a grounded system through a circuit breaker (Fig. 6.14). When the switch or breaker is opened, the initial voltage across the breaker is practically zero, since the capacitor maintains the same crest voltage that existed on the bus side of the breaker at and immediately following current zero. This permits any circuit breaker to interrupt quite easily at the first current zero. However, a half-cycle later, the source voltage has changed from plus to minus, while the capacitor voltage has remained at a plus value nearly equal to crest source voltage. Thus, approximately double voltage is impressed on the open breaker. If the dielectric (insulation) recovery of the breaker has been sufficient so that restriking does not occur, then the circuit can be considered cleared.

Voltage Across Switch Contacts with Restriking.   Assume that the breaker in the preceding case did not gain sufficient dielectric between its contacts during the one-half-cycle interval following the current interruption so that a restrike occurred when the source voltage was maximum in the negative direction. The associated current reverses the potential on the capacitor. The frequency of this current is determined by the capacitance and inductance of the circuit (the natural frequency of the circuit) and in the majority of cases will be between 2000 and 6000 Hz.

The high-frequency voltage associated with the high-frequency current has the same frequency as the current. If the high-frequency current is interrupted at first current zero (see Fig. 6.16), the high-frequency voltage (which is at a maximum due to 90 percent phase lead) is trapped on the capacitor. This voltage may be as high as three times normal crest depending upon the instant at which restrike occurs. Every successive restrike, theoretically adds the maximum voltage to the existing voltage. However, extensive field tests have seldom discovered overvoltages due to restriking that

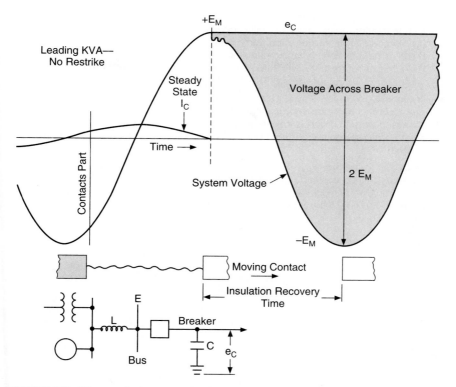

FIGURE 6.15 Diagram of voltage across breaker contacts when interrupting capacitor current without restriking.

exceeded three times normal value. If the restriking is not limited, the voltage stresses may cause failure of the capacitors and other pieces of apparatus, such as lightning arresters and transformers.

Capacitor Switching Criteria. The criteria used for selection of capacitor switching devices are as follows:

1. *Voltage rating:* The voltage rating of the switching device must be equal to or greater than the maximum operating voltage of the circuit to which the equipment is to be connected. Both the 60-Hz high-potential and impulse voltage ratings of the switch should be consistent with those of other apparatus in the same voltage class.

2. *Continuous current rating:* The continuous current rating of the switching device should be at least 135 percent of rated capacitor bank current in accordance with ANSI C37.06-19xx Standard for Circuit Breakers Switching Capacitors. The actual currents in capacitor banks may exceed rated or nameplate values because of the manufacturing tolerance of individual units and possibility of overvoltage and harmonic current flow. The 35 percent margin takes into account these conditions. For ungrounded banks, this value is 125 percent.

3. *Interrupting rating:*   If the switching device is used as a means of short-circuit protection in addition to its duty as a capacitor switch, it must have an interrupting rating adequate to handle short circuits occurring on the capacitor side of the switch. If the switching device has no short-circuit interrupting rating, protection against short circuits in the capacitor banks must be provided by other means, such as fuse cutouts or other devices having adequate interrupting rating.

4. *Momentary rating:*   The switching device must have sufficient momentary current rating to adequately withstand both system short-circuit currents for faults at its terminals and inrush currents associated with energizing. This consideration must be met even though the device is not used as a means of short-circuit protection.

5. *Frequency of operation:*   The mechanical and electrical design should be such that the switch will withstand repetitive switching operation.

Tables 6.8 and 6.9 list capacitor switching characteristic of switching devices.

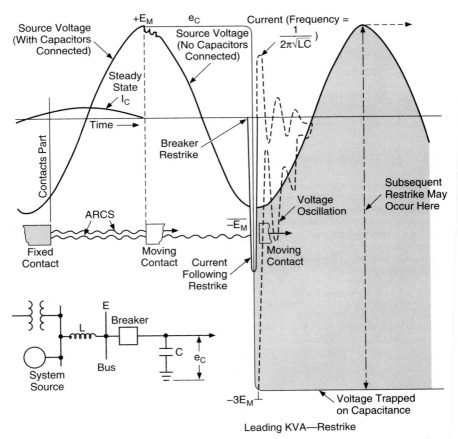

FIGURE 6.16   Diagram of voltage across circuit-breaker contacts when interrupting capacitor current with restriking.

TABLE 6.8 Typical Capacitor Switching Capabilities of Different Types of Switches

Type of switch	No. of phases	Voltage (kV)	Voltage (BIL)	Rated current (A)	Rated capacitor current (A)	Making asym current (A)	Rated switching peak current (kA)	Capacitor current frequency (Hz)
Small
Oil	1	14.4	95	200	150	9000	12	*
Oil	3	14.4	95	400	300	20000	20	*
Vacuum	3	15	110	600	444	20000	*	*
Large
Oil	3	34.5	150	1200	500	6100	*	*
Air	3	4.16	60	1200	890	58000	15	2000
Air	3	4.16	60	2000	1480	58000	15	1270
Air	3	13.8	95	1200	890	37000	15	2000
Air	3	13.8	95	2000	1480	37000	15	1270

*Contact manufacturer for detailed capability.

TABLE 6.9 Capacitor Switching Capability of Special Vacuum Circuit Breakers

Capacitor voltage	Circuit-breaker continuous current	Maximum nameplate capacitor bank rating—mVAR	
		Ungrounded bank	Grounded bank
2400	1200	4	3
	2000	7	6
	3000	10	9
4160	1200	7	6
	2000	11	10
	3000	17	16
7200	1200	12	11
	2000	20	18
12,470	1200	20	19
	2000	34	32
	3000	45	36
13,800	1200	23	21
	2000	30	23
	3000	30	23

NOTES:

1. These circuit breakers have a definite purpose rating per Table 2A, ANSI C37.06-19XX, except these values are higher, based on actual test of the circuit breakers.

2. Bank inrush currents are limited to 15 kA at 2000 Hz or to 30 kA-Hz if the inrush current is less than 15 kA.

3. Check with manufacturer for actual circuit-breaker capability.

6.5 MOTOR TERMINAL APPLICATION OF CAPACITORS

Power factor improvement of induction motor loads by means of shunt capacitors at the motor terminals has been widespread. It permits switching the capacitors and the motors as a unit so that the capacitors are on the system only when required. The kVAR requirements in an induction motor are fairly steady across its load range. Therefore, power factor correction matched to the motor results in a higher power factor at all values of motor loading. With the addition of adjustable-speed ac drives with their static power converter power supplies, harmonic currents are introduced into the power system. With the multiplicity of capacitive and inductive reactance, these harmonic currents are increased through the parallel resonances of the circuit. To minimize the effect of these harmonic currents, and to minimize the number of resonant circuits, motor terminal application of capacitors is no longer recommended.

The two factors which limit the value of capacitors to be switched with the motor are:

1. Overvoltage due to self-excitation.

2. Inrush currents due to out-of-phase reclosing. These limitations apply when the capacitor is connected to the load side of the motor starter and the motor and capacitor are switched as a unit.

6.5.1 Overvoltage from Self-Excitation

A capacitor located at the terminals of a motor can supply part or all of motor magnetizing requirements. Thus, when the motor line switch is opened and the motor disconnected from its source, the capacitor furnishes the motor magnetizing current and the motor will self-excite; that is, it will act as an induction generator. The magnitude of the generated voltage will depend upon the value of the capacitor current and the motor speed. When the capacitor kVAR is higher than the motor magnetizing kVAR, the motor will self-excite. The magnitude of overvoltage depends upon the speed and type of motor, including its magnetization (saturation) curve.

In the usual motor application, the motor slows down rapidly after the switch is opened, so the voltage rapidly decreases. However, the rate of slowdown is dependent on the inertia of the motor and load combined and on the amount of load torque applied at the motor shaft. In the case of a high-inertia drive, the motor may continue to run near rated speed for several seconds. If more capacitor kVAR are applied to the motor than is required to meet the magnetizing requirements, then the windings of the motor will be subjected to overvoltages until the speed declines and the energy in the load capacitor circuit is dissipated. Capacitors are not used with two-winding motors.

6.5.2 Inrush Current Due to Out-of-Phase Reclosing

As discussed in the previous paragraphs, application of motor terminal capacitors extends the time that a substantial voltage exists in the motor after a switch opening. This increases the opportunity for a reclosure with high inrush currents.

The system side is assumed to maintain its voltage E_s while the motor voltage E_m will decay at some rate depending upon factors such as machine time constants and load inertia.

If the recloser-controlled circuit breaker is closed without regard for phase relationship or the two driving voltages, it is possible for the voltages E_s and E_m to be 180 degrees out of phase, producing a net single driving voltage equal to the arithmetical sum of their magnitudes. The maximum symmetrical current between the systems would then be:

$$I = \frac{E_s + E_m}{X_s + X_d''} \tag{6.45}$$

where X_s is the system (service) short-circuit reactance

X_d'' is the locked rotor inductance for induction motor and direct axis subtransient reactance for synchronous motors

Assuming the voltage magnitudes have the following relation

$$E_s = E_m = E \tag{6.46}$$

then:

$$I = \frac{2E}{X_d'' + X_s} \tag{6.47}$$

If some estimate of the rate of decline of machine voltage and reclosure time of the switching device is known, the expression can be modified to consider the phasor difference (ΔE) between the two voltages at the time of reclosure. In such cases,

$$I = \frac{\Delta E}{X_d'' + X_s} \tag{6.48}$$

To obtain asymmetrical current, a suitable dc offset factor must be applied. The application of this technique is much like that of determination of short-circuit currents, except the net driving voltage may be higher.

The maximum symmetrical rms current for which rotating machine windings are normally braced is

$$I_M = \frac{\text{rated voltage}}{X_d''} \tag{6.49}$$

where all values are expressed in per-unit on the rated machine base. When the value of current I_M representing motor withstand capability is less than I representing the inrush current, then the motor is in danger of being damaged. When the application of motor terminal capacitors significantly lengthens the time that voltage is maintained after a switch opening and maintains that voltage closer to rated conditions, the opportunity for high currents is greatly increased. The increased voltages represent torque transients on the output shafts and can be large. Torque is proportional to voltage squared. Motor voltage up to 25 percent is considered sufficiently low to avoid excessive currents in the windings and torques on motor shafts.

6.6 CONTROLS FOR SWITCHED CAPACITORS

Shunt capacitor banks, if left on the system during light load conditions, may result in overvoltages. Thus, the capacitors need to be switched off during light load periods and switched back at predetermined conditions.

There is a variety of ways by which circuit conditions can be sensed and switching actions initiated. Time, voltage and/or current, kilovars, and some combinations of two inputs can be utilized in order to add or remove capacitors from the system to meet varying conditions. To utilize the capacitors most effectively and to select the most suitable control will require that the daily and weekly variations in circuit conditions be known. This includes the time of day when change can be expected and the magnitude of change in terms of current, voltage, and/or kilovars. Some of the most commonly encountered controls and some of the factors in their selection and application are discussed in the following paragraph.

6.6.1 General

Switched capacitors are not used for fine voltage control. For best economy, the voltage changes should be as large as system conditions allow. Only two or four switching operations should occur per day in most applications. A time delay is always used to prevent unnecessary switching due to momentary disturbances. With some types of voltage-regulating relays, a separate time delay relay is used. If an induction disk–type voltage-regulating relay is used, the inverse-time characteristic of the relay will usually provide sufficient time delay. Where separate timers are used, a common delay setting is one minute.

Coordination with other voltage-regulating equipment is required when using voltage control for switching capacitors, so that operation of one device (switched capacitor or regulator) will not cause an operation of another device, resulting in excessive operations and, possibly, hunting.

6.6.2 Time-Switch Control

Time-switch or time-clock control is one of the most common types of control used with switched capacitor banks. The control simply switches the capacitor bank on at a certain time of the day and takes it off at a later time. Its greatest application is with small single-step banks where the daily load cycle is known and consistent. Utilities use this method on distribution and subtransmission systems.

A carryover device is required for each time clock to keep the clock running during temporary power outages. Most carryover devices are of the mechanical spring type and can keep the clock running for up to 36 hours. The spring is continually kept in a wound position by the small electric motor which runs the clock. During a power outage, the spring begins to unwind. If power is restored before the carryover period has passed, the motor restores the spring to its wound position. If a carryover device is not used, it will be necessary for each capacitor location that is affected to be manually reset after a power outage.

An omitting device is also required for each time clock to omit switching the capacitors on or off on days where the known load cycle will change, such as Sundays and holidays. On some feeders there may be a definite reduction in feeder loading on these days, and if the capacitors were switched on, overvoltage could result.

One of the greatest advantages of time-switch control is its low cost. A disadvantage of time-switch control is its inflexibility. The sequence of operation is the same throughout any unusual load conditions. The switching cycle is fixed and it receives no intelligence which would enable it to respond to any unusual conditions which might prevail.

6.6.3 Voltage Control

Voltage alone can be used as a source of intelligence only when the switched capacitors are applied at a point where the circuit voltage decreases as circuit load increases. Generally where they are applied, the voltage should decrease four to five percent with increasing load.

Voltage is the most common type of intelligence used in substation applications. It has the advantage of initiating a switching operation only when the circuit voltage conditions request an operation, and it is independent of the load cycle.

The bandwidth setting of the voltage regulating relay is wider than that used with step or induction regulators and normally ranges from about 4 to 10 percent voltage. The bandwidth setting will depend upon the rating of the capacitor bank, the number of steps, and whether other voltage regulating equipment is also applied on the same circuit. The setting must be larger than the voltage change due to the switching operation of one step. Since the voltage change due to one step will depend upon the system characteristics towards the source, it is also necessary to check the voltage change for possible abnormal conditions.

6.6.4 Current Control

Current control alone is used only where voltage is not a satisfactory signal. Such applications would be at locations where the voltage reduction as load increases is not enough for effective relaying. For effective current control, there must be a load change such that the ratio of maximum demand to minimum demand is three or more.

The greatest applications of current control are with single-step capacitor banks applied on feeders or in substations where large intermittent loads are either on or off. The loads can be fluctuating loads if the capacitor "off" control setting is below the maximum dip.

Current control relays are similar to voltage control relays and can be either of the solenoid type or the induction-disk type. The solenoid type is most often used with large substation capacitor banks, while the induction-disk type is used with smaller, single-step capacitor banks. The current transformer should always be connected on the load side of the capacitor bank in order to measure load current and not load current plus capacitor current.

When using current control, no recognition is given to the voltage conditions of the circuit. Therefore, circuit voltage conditions throughout the load cycle must be known in order to determine when the capacitors should be switched on and off. Abnormal circuit conditions which affect the circuit loading and the voltage rise should be checked to be sure the capacitors will not switch on if overvoltage conditions occur.

6.6.5 Voltage Sensitive with Time Bias

This control scheme is available for use where the voltage profile at the bank location remains relatively flat over 24-hour periods, thus preventing proper switching by voltage-only controls. One type of timing device is a phototimer, and compensation is in the form of a step bias. This step-biasing by time permits a day-night operating shift which generally removes the capacitors at night and applies them during the day on the basis of a relatively narrow on-off band.

The phototimer combines a photoelectric control with a synchronous motor and cams. The photosensitive unit starts the unit at sunrise and cam adjustment is used to obtain switching at the proper time. The primary advantage of this control is that it will resynchronize itself after a power outage of any duration.

6.6.6 Kilovar Controls

Kilovar controls are utilized at locations where the voltage level is closely regulated and not available as a control variable. This can occur on an industrial bus which is served by an LTC-equipped transformer or a generator system with automatic voltage regulators. In these cases, the capacitors can be switched to respond to decreasing power factor as a result of change in system loading. This type of control can also be used to avoid a utility power factor penalty clause by adding capacitor as the system power factor begins to lag.

Since the kilovar control requires two inputs, both current and voltage, it will have a higher cost than a single-input control. Thus, if a single step of switching is all that is required, the needs may be met satisfactorily by a current-sensitive control. If the power factor needs to be more accurately controlled, particularly if several steps are involved, then the kilovar control will be used.

6.7   FIELD TESTING OF POWER CAPACITORS

Power capacitors have been in use on power systems since the 1920s. They are considered to be one of the most reliable electrical apparatus. By far, the most common failure is a "puncture" of the solid dielectric or a breakdown at the edge of the coil (overvoltage), usually resulting in a short circuit. Other failures are due to such things as failure of the case insulation, loose connections, and mechanical failures of the case or bushing.

There are three common steps to check capacitors in the field in order to determine if they have failed or not. These are:

1. Visual inspection
2. Capacitance measurement
3. Low-voltage energization tests

6.7.1   Visual Inspection

When examining new units before they are energized, inspect for obvious external damage such as broken bushings, large dents in the case, broken mounting brackets, and fluid leaks. If any of these are found, the unit should be returned to the factory.

Where there is reason to suspect a unit has been damaged in service, the first step is to perform a visual examination, mentioned previously. In addition, look for case sides which are bulged more than $\frac{1}{2}$ in beyond the corners of the case. Capacitors built in the last several years can bulge about $\frac{1}{4}$ in at normal room temperature. If the narrow (mounting bracket) side of the case is concave or the broad side has a pronounced crease extending from the corner, the unit should be considered suspect. If possible, compare the suspect unit to several others of the same type. Take a large coin and tap the uppermost part of the capacitor. A normal unit is full of fluid and has a solid sound. A unit that is bulged due to gas generated by an internal fault will sound hollow due to the gas bubble. Again, compare several units. Remember that not all failures cause gassing and that most of the low loss "all-film" units with "paper-free" dielectric systems, will not gas noticeably.

6.7.2   Capacitance Measurement

A capacitance meter or a low-voltage bridge can be utilized to measure the capacitance of all-film capacitors to determine whether or not they are partially failed. The voltage used to test the capacitance with this type of equipment is only a few volts. A failure in an all-film capacitor has a very low impedance even to these low voltages and, as a result, partial failures in an all-film capacitor can be easily identified. A 10 percent increase in capacitance generally means a partially failed capacitor.

The same method can also be used to identify partially failed paper or film dielectric capacitors with reasonably good results. Occasionally, at these low test voltages, the fault in a paper or film capacitor will appear to be an open circuit resulting in a normal capacitance reading for a partially failed capacitor. Severely swollen capacitors may indicate a normal reading even though the capacitor has failed. This is because the inter-pack connections are burned open. All capacitors which are excessively bulged should always be considered as failed units. Do not energize any unit failing this test, as the chances of violent rupture are greatly increased. Because the number of series sections within the capacitor is greater for higher voltage units, the failure of a single element may produce less than a 15 percent increase in capacitance in units rated above 7200 V. This depends on the number of series sections in the unit, and this has varied between manufacturers over the years and even between different designs by the same manufacturer. The only certain way to determine this is to give the catalog and serial number to the factory. On units manufactured since 1975, it is safe to assume that a normal unit will not exceed 110 percent of the rated capacitance. And, finally, compare several units of the same type and use the results of the visual inspection when trying to decide about a questionable unit.

6.7.3 Low-Voltage Energization Test

An alternative method to capacitance meter or a bridge determining the capacitance of a unit is to utilize a test circuit as shown in Fig. 6.17. When the capacitor test current and test voltage are known, the capacitance of the unit can be calculated using the following formula:

$$C = \frac{i}{\omega v}$$

(6.50)

where C = Capacitance of unit under test
$\omega = 2\pi f$
f = frequency of test voltage
i = test current
v = test voltage

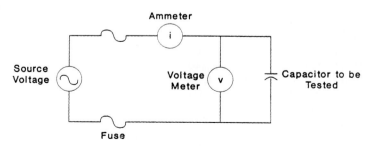

FIGURE 6.17 Test circuit for capacitance measurement.

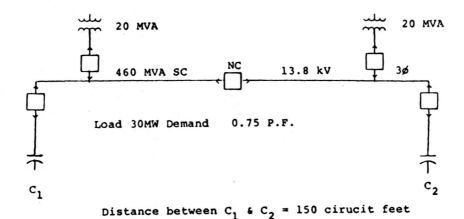

FIGURE 6.18   One-line diagram for example.

6.8   *EXAMPLE OF CAPACITOR APPLICATION*

An industrial plant, as illustrated in Fig. 6.18, has a demand of 30 mW at 0.75 power factor. With a demand charge of $8 per kVA and capacitors costing $12 per kVAR, installed

1. What capacitor bank size would be chosen and what would be the payback period for correcting the power factor to (*a*) 0.90, (*b*) 0.95, and (*c*) 0.98?
2. What would the inrush current be for a bank of 15.2 MVAR?
3. What would the voltage rise be if the load on weekends dropped to 20 percent or 6 mW?

With a load of 30 mW at 0.75 PF, the following table can be set up.

PF	kVA	Rate	Total $	Savings $	Cap kVA	Rate	Cost $	Payback (month)
0.75	40,000	8	320,000	—	—	—	—	—
0.90	33,333	8	266,667	53,333	12,000	12	144,000	2.70
0.95	31,579	8	252,632	67,368	17,100	12	205,200	3.05
0.98	30,612	8	244,896	75,104	20,700	12	248,400	3.31

To correct to: kVA = kW/PF   kVAR = kWx tan φ

0.90	30,000 kW×0.398* = 11.94 MVAR	Use 12 MVAR	20-200 kVAR/phase
0.95	30,000 kW×0.553* = 16.59 MVAR	Use 17.1 MVAR	19-300 kVAR/phase
0.98	30,000 kW×0.673* = 20.19 MVAR	Use 20.7 MVAR	23-300 kVAR/phase

*This factor is the difference between the tangents of the two power factors (see Eq. 6.20).

If the power factor were to be corrected to almost 0.95, then 16.2 MVAR of capacitors would be used with a total of 18-300 kVAR units per phase used. This number is easily divided by 2 to get two equal banks.

The inrush current to the bank of 16.2 kVAR would be:

$$I_{max\ rms} = \frac{E_{l-n}}{X_c - X_L}\left[1 + \sqrt{\frac{X_c}{X_L}}\right] = \frac{7.96}{11.76 - 0.414}\left[1 + \sqrt{\frac{11.}{0.41}}\right] = 2307\ amperes$$

(6.51)

where

$$X_C = \frac{kV^2}{MVAR} = \frac{13.8^2}{16.2} = 11.76\ \Omega$$

$$X_L = \frac{kV^2}{MVA_{sc}} = \frac{13.8^2}{460} = 0.414\ \Omega$$

6.9 HOMEWORK PROBLEM

If the bank of #2 in the preceding example were to be purchased in two units, what would the current inrush to bank 2 be if bank 1 were energized? What would the frequency of current oscillation between the banks be? Would this cause any problems?

CHAPTER 7

SHORT-CIRCUIT CALCULATIONS AND EQUIPMENT RATINGS

7.0 INTRODUCTION

Electric power system faults include both short circuits and open circuits. This chapter concerns only short circuits. The IEEE dictionary[1] defines a short circuit as "an abnormal connection (including an arc) of relatively low impedance, whether made accidentally or intentionally, between two points of different potential." This definition corresponds to that given in ANSI/IEEE Standard C37.100.[2]

Electric power system short circuits fall into the following categories:

- Three-phase
- Phase-to-phase
- Single-phase-to-ground
- Three-phase-to-ground
- Phase-to-phase-to-ground

Short-circuit calculations are performed to determine short-circuit magnitudes for comparing power distribution equipment ratings with calculated duties. Results of these calculations are also used to determine the characteristics, settings, and/or ratings of protective devices in an electric power system.

The most severe type of short circuit occurring in a power distribution system is usually a three-phase fault. See Table 7.1 for a comparison of the magnitudes of different types of faults.

The calculated duty for a three-phase fault is compared to power distribution equipment ratings or capabilities. Since equipment ratings and testing procedures used by manufacturers in the United States are specified by ANSI/IEEE standards, calculations to assure equipment adequacy should follow ANSI/IEEE calculation standards.

TABLE 7.1   Magnitude Comparison of Short-Circuit Currents

Fault type	Magnitude
3-phase (most severe)	$(E/Z) \times$ multiplier
Line-to-line	About 0.87 × 3-phase fault
Line-to-ground (usually least severe)	Depends on system grounding

Ground-fault currents	
System grounding	Fault magnitude
Solidly grounded transformer	1–1.15 × transformer 3-phase fault
Solidly grounded generator	>generator 3-phase fault
Low-resistance-grounded system	100–2000 A (usually 400 A)
Feeding trailing cable to portable equipment	Either 25 or 50 A
High resistance	2–20 A
Reactance (for generators serving a line-to-ground load at generator voltage)	About 0.25 × 3-phase fault current

7.1   AC SHORT-CIRCUIT CURRENT FUNDAMENTALS

In ac circuits, voltages and currents essentially assume a sine-wave shape. The presence of both resistance and reactance in a circuit displaces the voltage and current waves. The cosine of the angle of this displacement is the power factor of the circuit. When a short circuit occurs, the power factor of the circuit (not the load), determines the amount of displacement of the short-circuit current and source voltage waves. Most medium- and high-voltage circuits consist of components with a relatively high ratio of reactance to resistance (*X/R* ratio), resulting in a circuit power factor of close to zero for the circuit from the source to the fault location. Low-voltage circuits usually have lower *X/R* ratios (higher power factor) than medium- and high-voltage circuits.

7.1.1   Asymmetry

Figure 7.1 shows the resulting symmetrical current waveshape in a zero power factor circuit for a fault occurring when the voltage is at its maximum. Figure 7.2 shows the resulting asymmetrical current waveshape for a fault occurring when the voltage is at zero on a purely reactive circuit. This waveshape results because, at the instant of short circuit, the current will start at zero but cannot follow a sine wave symmetrically about the zero axis because such a current would be in phase with the voltage. The waveshape must be the same as the voltage but approximately 90 degrees behind. That can occur only if the current waveshape is displaced from the zero axis as shown.

7.1.2   Dc Component and dc Decrement

Short-circuit currents are nearly always asymmetrical during the first few cycles after a short circuit occurs. This asymmetry is maximum at the instant a short circuit occurs and gradually becomes symmetrical a few cycles after the initiation of the short circuit because of the circuit resistance. Figure 7.3 shows a total asymmetrical current broken down into two components: an ac component and a dc component. Resolving the total

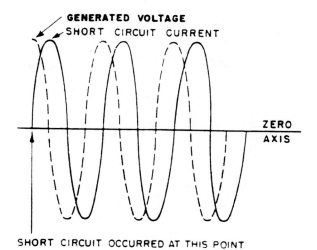

FIGURE 7.1 Symmetrical current and voltage in a zero power factor circuit. (*Reprinted from "Industrial Power Systems Handbook", Donald L. Beeman, Ed., McGraw-Hill, 1955.*)

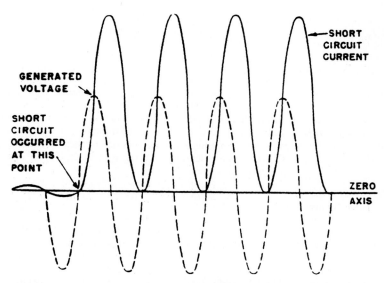

FIGURE 7.2 Asymmetrical current and voltage in a zero power factor circuit. (*Reprinted from "Industrial Power Systems Handbook", Donald L. Beeman, Ed., McGraw-Hill, 1955.*)

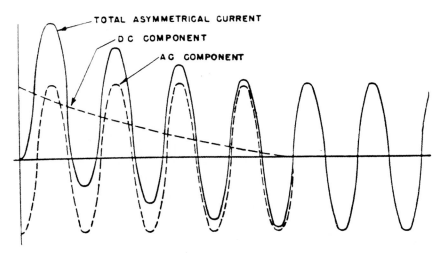

FIGURE 7.3 Oscillogram showing the decay of dc component and effect of current asymmetry. (*Reprinted from "Industrial Power Systems Handbook", Donald L. Beeman, Ed., McGraw-Hill, 1955.*)

short-circuit current in this manner makes it easier to visualize for discussion and easier to handle in calculations.

The presence of resistance in every electric power distribution circuit causes a decay in the dc component of the total short-circuit current. The rate of decay depends upon the amount of circuit resistance compared to reactance: the circuit X/R ratio.

7.1.3 Ac Decrement

Current is delivered to a short circuit through cables and transformers from rotating machine sources: motors and generators. Depending on the electrical distance these sources are from the fault point, there will be a decrease in the short-circuit current due to impedance and time, resulting from both the presence of resistance in the system and the reduction in excitation to those rotating machines that derive their excitation from the system voltage. The system voltage decreases during a fault depending upon the electrical distance of the source from the fault location.

7.1.4 Total Short-Circuit Current

The sum of the local motor contribution, local generation source, and remote generation (usually a utility) contribution, modified to account for the ac and dc decrements constitutes the total short-circuit current. Figure 7.4 shows the different symmetrical components in the total short-circuit current and Fig. 7.5 shows a few cycles of the resulting wave with the symmetrical and dc components. See App. 7B for simulated oscillograms of total short-circuit currents plus contributions from machines.*

---

*Oscillograms adapted from R. P. Stratford, J. R. Dunki-Jacobs, B. P. Lam, "A Comparison of ANSI-Based and Dynamically Rigorous Short-Circuit Current Calculation Procedures," *IEEE/IAS Transactions,* vol. 24, no. 6, Nov./Dec. 1988, pp. 1180–1194.

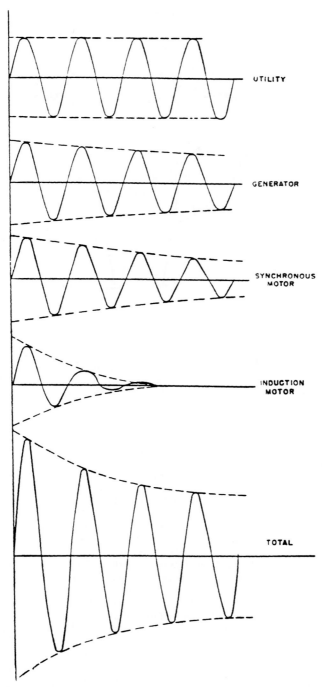

FIGURE 7.4 Symmetrical short-circuit current from three sources (utility, motors, and generator) combined into a total. (*Reprinted from "Industrial Power Systems Handbook", Donald L. Beeman, Ed., McGraw-Hill, 1955.*)

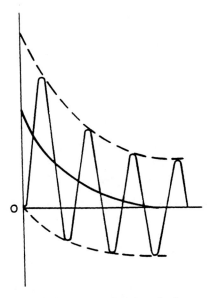

FIGURE 7.5   Asymmetrical short-circuit current from all sources plus dc component. (*Reprinted from "Industrial Power Systems Handbook", Donald L. Beeman, Ed., McGraw-Hill, 1955.*)

7.2   CALCULATED FAULT DUTY AND EQUIPMENT RATINGS

All short-circuit calculations are based on the formula $E = IZ$ where Z is the complex number $R + jX$. E is the voltage behind the impedance Z. R and X are the circuit resistance and reactance. In per-unit short-circuit calculations, $E = 1.0$ since the initial bus voltage is approximately unity and the load current is neglected. I, the per-unit fault current, is the inverse of the total of the circuit impedance.

The internal voltage of rotating generators and motors changes with time as the fluxes within the iron cores decrease during the time a short-circuit persists. Mathematically, the same effect is accomplished by modeling the rotating equipment with increasing impedances. The complex part of the short-circuit calculation concerns the multiplying factors assigned to the circuit reactances during the time the fault persists until it is finally interrupted. Rules for short-circuit calculations in the United States are outlined in ANSI/IEEE standards for different equipments. Internationally, the IEC standards govern. Presently, the IEC and IEEE are attempting a correlation of standards. Equipment manufactured in the United States is rated in accordance with ANSI/IEEE standards. Most of the equipment manufactured in other countries and sold in the United States meets ANSI/IEEE standards. The following sections discuss calculation procedures and capabilities or ratings for different equipment to which calculation results are compared.

TABLE 7.2 Short-Circuit Calculation Multiplying Factor Selection for Low-Voltage Power Circuit Breakers at Different Short-Circuit Power Factors

System power factor %	System X/R ratio	Multiplying factor for calculated short-circuit current	
		Unfused circuit breakers	Fused circuit breakers
20	4.9	1.00	1.00
15	6.6	1.00	1.07
12	8.27	1.04	1.12
10	9.85	1.07	1.15
8.5	11.72	1.09	1.18
7	14.25	1.11	1.21
5	20.0	1.14	1.26

(From IEEE Std C37.13-1990, *IEEE Standard for Low-Voltage AC Power Circuit Breakers Used in Enclosures.*)

7.2.1 Low-Voltage (120–600-V) Equipment

DEFINITIONS:

Low-voltage Power circuit breaker A mechanical switching device (rated 1000 V ac or below, or 3000 V dc and below, but not including molded-case circuit breakers), capable of making, carrying, and breaking current under normal conditions and also, making, carrying for a specified time, and breaking currents under specified abnormal circuit conditions such as those of short circuit. (ANSI/IEEE Standard C37.100-1981)

Molded-case circuit breaker One that is assembled as an integral unit in a supporting and enclosing housing of molded insulating material. (ANSI/IEEE Standard C37.100-1981)

Insulated-case circuit breaker A rugged, glass-reinforced case circuit breaker similar to a molded-case circuit breaker with higher continuous current and interrupting ratings.

Low-Voltage Power-Circuit-Breaker Fault Calculations. For three-phase circuits, the calculation procedure outlined in ANSI/IEEE Standard C37.13[3] requires the calculation of the maximum available rms symmetrical current of the three phases started at a time a half-cycle after the short-circuit occurs. Circuit power factor requirements may require the application of a multiplying factor to the short-circuit current calculated. Table 7.2 tabulates the appropriate multiplying factors.

The system X/R ratio is calculated using separate X and separate R. The complex R + jX or recombined separate R and X components are used to calculate the short-circuit current. X_d'' for a motor is approximately the inverse of the locked rotor current in per-unit of the full load current on the motor kVA base.

Low-Voltage Power-Circuit-Breaker Interrupting Ratings. Table 7.3 tabulates the preferred ratings for low-voltage power circuit breakers with instantaneous direct-acting phase trip elements. For low-voltage power circuit breakers without a direct-acting instantaneous trip element but using a short-time-delay element or remote relay, Table 7.4 applies. The data in these tables are derived from ANSI/IEEE Standard C37.16-1988.[4]

Insulated-Case and Molded-Case Circuit-Breaker Fault Calculations. The interrupting duty for molded-case and insulating-case circuit breakers is calculated per IEEE Standard C37.13,[3] which includes contributions from all sources including all ratings of

TABLE 7.3[13] Preferred Ratings for Low-Voltage Power Circuit Breakers with Instantaneous Direct-Acting Trip Elements

System nominal voltage (V)	Rated maximum voltage (V)	Insulation (dielectric) withstand (V)	3-phase short-circuit current rating (symmetrical amperes)	Frame size (A)	Trip device current rating range (A)
600	635	2,200	14,000	225	40–225
			22,000	600	40–600
			22,000	800	100–800
			42,000	1600	200–1600
			42,000	2000	200–2000
			65,000	3000	2000–3000
			85,000	4000	4000
480	508	2,200	22,000	225	40–225
			30,000	600	100–600
			30,000	800	100–800
			50,000	1600	400–1600
			50,000	2000	400–2000
			65,000	3000	2000–3000
			85,000	4000	4000
240	254	2,200	25,000	225	40–225
			42,000	600	150–600
			42,000	800	150–800
			65,000	1600	600–1600
			65,000	2000	600–2000
			85,000	3000	2000–3000
			130,000	4000	4000

NOTE: 5000-A frame also available with 85,000-A short-circuit current rating at all voltages without instantaneous direct-acting trip elements (130,000 A at 240 V with instantaneous direct-acting trip elements).

motors. Multiplying factors (derating factors) are applied to the interrupting rating of molded-case circuit breakers within three ranges: up to 10,000 A, above 10,000 up to 20,000 A, and above 20,000 A, according to the X/R ratio of the circuit. Table 7.5 lists the applicable derating factors,* power factors, and X/R ratios for each of the three ranges of molded-case circuit-breaker ratings. Table 7.6 gives the multiplying factors for insulated-case circuit breakers.

Molded-case circuit breakers come in many varieties, some with UL labels and some without. Special ratings apply for frequencies other than 50/60 Hz, single-pole applications, 120/240 V, and many other special applications. Table 7.7 gives only a few of the molded-case circuit-breaker ratings available. Table 7.8 gives the insulated-case ratings available. Both molded-case and insulated-case circuit breakers are available without overload trip devices to be used as switches. Consult with the manufacturer to determine the limitations of this application and obtain the withstand ratings of these switches.

---

*Often a multiplying factor is applied to the calculated symmetrical short-circuit current. The short-circuit calculation multiplying factor = 1 + (1−the interrupting rating derating factor).

TABLE 7.4[13]   Preferred Ratings for Low-Voltage Power Circuit Breakers Without Instantaneous Direct-Acting Trip Elements (Short-Time-Delay Element or Remote Relays)

System nominal voltage (V)	Rated maximum voltage (V)	Insulation (dielectric) withstand (V)	3-phase short-circuit current rating (symmetrical amperes)	Frame size (A)	Short-time-delay trip element setting range (A)		
					Minimum time band	Intermediate time band	Maximum time band
600	635	2,200	14,000	225	100–225	125–225	150–225
			22,000	600	175–600	200–600	250–600
			22,000	800	175–800	200–800	250–800
			42,000	1600	350–1600	400–1600	500–1600
			42,000	2000	350–2000	400–2000	500–2000
			65,000	3000	2000–3000	2000–3000	2000–3000
			65,000	3200	2000–3200	2000–3200	2000–3200
			85,000	4000	4000	4000	4000
480	508	2,200	14,000	225	100–225	125–225	150–225
			22,000	600	175–600	200–600	250–600
			22,000	800	175–800	200–800	250–800
			50,000	1600	350–1600	400–1600	500–1600
			42,000	2000	350–2000	400–2000	500–2000
			65,000	3000	2000–3000	2000–3000	2000–3000
			65,000	3200	2000–3200	2000–3200	2000–3200
			85,000	4000	4000	4000	4000
240	254	2,200	14,000	225	100–225	125–225	150–225
			22,000	600	175–600	200–600	150–600
			22,000	800	175–800	200–800	150–800
			42,000	1600	350–1600	400–1600	600–1600
			50,000	2000	350–2000	400–2000	600–2000
			65,000	3000	2000–3000	2000–3000	2000–3000
			65,000	3200	2000–3200	2000–3200	2000–3200
			85,000	4000	4000	4000	4000

Low-Voltage Fuse Fault Calculations.   Low-voltage fuse interrupting ratings are specified in rms symmetrical amperes. Short-circuit calculations to compare with these ratings normally are made in accordance with ANSI/IEEE Standard C37.13.[3] Section 8.81 discusses the different classes of low-voltage fuses and their characteristics. Table 7.9 summarizes the interrupting ratings of the different classes of fuses.

7.2.2   Medium-Voltage (2400–69000-V) Equipment

Medium-Voltage Power Circuit-Breaker Fault Calculations.   ANSI/IEEE Standard C37.010[5] describes the procedure for calculating medium- and high-voltage short circuits. Per-unit calculations are the simplest calculation procedure to use for these systems. Table 7.10 gives the multiplying factors applied to rotating machine impedances for the first cycle and interrupting calculations required.

TABLE 7.5 Derating Factors for Interrupting Rating of Molded-Case Circuit Breakers

Power factor (%)	X/R ratio	Rated maximum interrupting rating		
		1 to 10 kA	11 to 20 kA	21 kA and higher
		Multipliers		
4	24.980	0.61	0.72	0.81
5	19.974	0.62	0.74	0.82
6	16.637	0.63	0.75	0.83
7	14.251	0.64	0.76	0.84
8	12.460	0.65	0.77	0.85
9	11.066	0.66	0.78	0.87
10	9.950	0.67	0.79	0.88
11	9.036	0.68	0.80	0.89
12	8.273	0.69	0.81	0.90
13	7.627	0.69	0.82	0.91
14	7.072	0.70	0.83	0.93
15	6.591	0.71	0.84	0.94
16	6.169	0.72	0.85	0.95
17	5.797	0.73	0.86	0.96
18	5.465	0.74	0.87	0.97
19	5.167	0.75	0.88	0.98
20	4.899	0.76	0.89	1.00
21	4.656	0.77	0.90	1.00
22	4.434	0.77	0.91	1.00
23	4.231	0.78	0.92	1.00
24	4.045	0.79	0.94	1.00
25	3.873	0.80	0.95	1.00
26	3.714	0.81	0.96	1.00
27	3.566	0.82	0.97	1.00
28	3.429	0.83	0.98	1.00
29	3.300	0.83	0.99	1.00
30	3.180	0.84	1.00	1.00
31	3.067	0.85	1.00	1.00
32	2.961	0.86	1.00	1.00
33	2.861	0.87	1.00	1.00
34	2.766	0.88	1.00	1.00
35	2.676	0.88	1.00	1.00
36	2.592	0.89	1.00	1.00
37	2.511	0.90	1.00	1.00
38	2.434	0.91	1.00	1.00
39	2.361	0.91	1.00	1.00
40	2.291	0.92	1.00	1.00
41	2.225	0.93	1.00	1.00
42	2.161	0.94	1.00	1.00
43	2.100	0.95	1.00	1.00
44	2.041	0.95	1.00	1.00
45	1.984	0.96	1.00	1.00
46	1.930	0.97	1.00	1.00
47	1.878	0.97	1.00	1.00
48	1.828	0.98	1.00	1.00
49	1.779	0.99	1.00	1.00
50	1.732	1.00	1.00	1.00

TABLE 7.6   Derating Factors for Interrupting Rating of Insulated-Case Circuit Breakers

Power factor (%)	Multiplier
7	0.85
8	0.87
9	0.88
10	0.89
11	0.90
12	0.91
13	0.92
14	0.93
15	0.95
16	0.96
17	0.97
18	0.98
19	0.99
20	1.00

The multiplying factor applied to the symmetrical current for determining the medium- and high-voltage circuit-breaker interrupting duty depends on the speed of opening of the circuit breaker, the X/R of the circuit, and the remoteness of the source of generation of the short-circuit current. If the X/R ratio of the circuit is 15 or less, the source of generation is remote, and the speed of the circuit-breaker opening is about five cycles, a 1.0 multiplier usually applies to most medium-voltage system calculations. Figure 7.6 shows the curves by which the multiplying factors are determined. As in the low-voltage calculation, the system X/R ratio is calculated using separate X and separate R networks. The resulting complex $R + jX$ is used to calculate the short-circuit current.

Note that medium-voltage fault calculations in accordance with the ANSI/IEEE standards ignore all motors below 50 hp and all single-phase motors whereas low-voltage fault calculations ignore no source. Some engineers include all motors in the first-cycle calculation because they can use one first-cycle calculation for both the low- and high-voltage breaker evaluation and their inclusion probably affects the medium-voltage results minimally.

Tables 7.11 and 7.14 show equipment ratings applicable to medium-voltage and high-voltage power circuit breakers.

Medium-Voltage Fuse and Fused Equipment Fault Calculations.   Medium-voltage power fuses are available for individual mounting and mounting in load break switch equipments, motor controllers, and similar equipment. Current-limiting and non-current-limiting fuses are available. "Smart" fuses used in combination with an electronic module to assure that all three phases interrupt a circuit and/or to send a signal to open a remote or local three-phase device are either available or soon to be available. Power fuse interrupting ratings are specified in rms symmetrical amperes. An asymmetrical multiplier between 1.56 and 1.6 applies to these ratings. Current-limiting fuses may have either an E or an R rating. These ratings define the time-current characteristics of the fuses. R-rated fuses find applications primarily in motor controller equipment. These fuses have characteristics that allow them to withstand repeated motor-starting

TABLE 7.7 Representative ac 50/60 Hz, Three-Pole, Molded-Case Circuit-Breaker Interrupting Ratings

Frame (A)	Trip unit range (A)	Voltage		
		240 (A)	480 (A)	600 (A)
100	15–100	10,000	—	—
		22,000	—	—
150	15–150	18,000	14,000	14,000
		42,000	25,000	18,000
		100,000	65,000	25,000
225	70–225	10,000	—	—
		25,000	22,000	18,000
		65,000	25,000	18,000
400	125–400	42,000	30,000	22,000
600	250–600	42,000	30,000	22,000
		65,000	35,000	25,000
800	300–800	42,000	30,000	22,000
		65,000	50,000	25,000
1200	600–1200	42,000	30,000	22,000
		65,000	35,000	25,000
Current-limiting circuit breakers				
150	15–150	100,000	65,000	25,000
		200,000	100,000	25,000
		200,000	200,000	50,000
250	70–250	100,000	65,000	25,000
		200,000	100,000	25,000
	125–225	200,000	200,000	50,000
400	250–400	200,000	200,000	50,000
600	125–600	100,000	65,000	65,000
		200,000	100,000	65,000
Internally fused circuit breakers				
150	15–100	200,000	200,000	200,000
400	125–400	200,000	200,000	200,000
600	300–600	200,000	200,000	200,000
800	600–800	200,000	200,000	200,000

inrush currents. Both full-range and backup fuses are available. Backup fuses have limited interrupting capabilities at low fault currents and are used in conjunction with an overload detector, such as an overcurrent or thermal relay.

The first-cycle calculation in accordance with IEEE Standard C37.010[5] is used to determine the available short-circuit current a power fuse is required to interrupt. ANSI/IEEE Standard C37.46-1981[7] details the capabilities of the fuse to interrupt asymmetrical and peak currents. Table 7.12 shows some of the ratings which apply.

TABLE 7.8 Insulated-Case Circuit-Breaker Interrupting Ratings

Frame (A)	Trip rating (A)	Interrupting ratings in A (rms), symmetrical		
		240 V	480 V	600 V
800	100–800	65,000	65,000	42,000
		100,000	100,000	65,000
1600	400–1600	85,000	65,000	50,000
		125,000	100,000	65,000
2000	800–2000	85,000	65,000	50,000
		125,000	100,000	65,000
3000	400–3000	100,000	100,000	85,000
		200,000	150,000	100,000
4000	1600–4000	100,000	100,000	85,000
		200,000	150,000	100,000

TABLE 7.9 Representative ac Fuse Interrupting Ratings

UL class	Volts	Amperes	Interrupting rating (rms A)	Current limiting
H	250	0.125–600	10,000	No
	600	1–600	10,000	No
G	300	1–60	100,000	Yes
J	600	1–600	200,000	Yes
L	600	800–6000	200,000	Yes
CC	600	0.25–10	200,000	Yes
K-5	250	1–60	50,000	Yes
	600	1–60	200,000	Yes
RK-1	250	1–600	200,000	Yes
	600	1–600	200,000	Yes
RK-5	250	0.10–600	200,000	Yes
	600	0.10–600	200,000	Yes
T	300	1–1200	200,000	Yes
	600	1–800	200,000	Yes
Type				
Limiters	250	Matched to cable	100,000	Yes
	600	Matched to cable	200,000	Yes
Plug fuses	125	0.30–30	10,000	No

Circuit-Switcher Fault Calculations. Circuit switchers appeared on the scene in the 1970s and '80s as interrupting devices with high momentary ratings and limited interrupting ratings. Certain relaying arrangements or fuse applications allow these devices to interrupt low-level faults while shifting the higher duties to backup circuit breakers or fuses. No IEEE standards presently exist specifying the method for calculating circuit-switcher short-circuit duties. The accepted method is the first-cycle duty calculation for power circuit breakers per ANSI/IEEE Standard C37.010.[5] Manufacturer's rating data specifies all relevant circuit-switcher characteristics in rms symmetrical amperes. Table 7.13 presents representative circuit-switcher characteristics from one manufacturer.

TABLE 7.10 Rotating Machine Positive Sequence Reactances for Short-Circuit Calculations per ANSI/IEEE Std. C37.010[5]

Type of machine	Interrupting duty (per-unit)	Closing and latching duty (per-unit)
All turbogenerators	$1.0\,X_d''$	$1.0\,X_d''$
All hydrogenerators with amortisseur windings		
All synchronous condensers		
Hydrogenerators without amortisseur windings	$0.75\,X_d'$	$0.75\,X_d'$
All synchronous motors	$1.5\,X_d''$	$1.0\,X_d''$
Induction motors		
Above 1000 hp at 1800 rpm or less	$1.5\,X_d''$	$1.0\,X_d''$
Above 250 hp at 3600 rpm		
Induction motors		
50–1000 hp at 1800 rpm or less	$3.0\,X_d''$	$1.2\,X_d''$
50–250 hp at 3600 rpm		
Neglect		
All 3-phase induction motors below 50 hp		
All single-phase motors		

X_d'' of synchronous rotating machines = rated-voltage (saturated) direct-axis subtransient reactance.
X_d' of synchronous rotating machines = rated-voltage (saturated) direct-axis transient reactance.
X_d'' of induction motors = 1.00 divided by per-unit locked rotor current at rated voltage.

(From IEEE Std C37.010-1979, *IEEE Application Guide for AC High-Voltage Circuit Breakers Rated on a Symmetrical Current Basis* [Reaff 1988.])

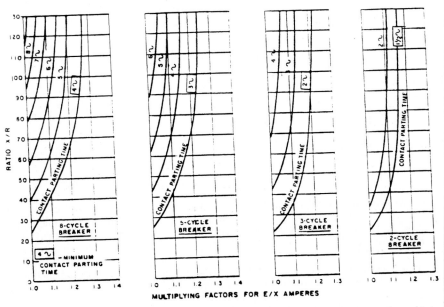

FIGURE 7.6 Three-phase medium-voltage multiplying factors which include effects of ac and dc decrements. (*From IEEE Std C37.010-1979, "IEEE Application Guide for AC High-Voltage Circuit Breakers Rated on a Symmetrical Current Basis" [Reaff 1988.]*)

TABLE 7.11[14] Medium-Voltage Power-Circuit-Breaker Characteristics (ANSI/IEEE Std. C37.06[6])

Identification		Rated values								Related required capabilities			
		Voltage		Insulation level Rated withstand test voltage		Current				Current values			
Nominal rms voltage class (kV)	Nominal 3-phase class (MVA)	Rated maximum rms voltage (kV)[1]	Rated voltage range factor K[2]	Low-frequency rms voltage (kV)	Crest impulse voltage (kV)	Continuous rms current rating at 60 Hz (A)	Short-circuit rms current rating (at max kV)[3,4] (kA)	Rated I interrupting time (cycles)	Rated permissable tripping delay, Y (seconds)	Rated maximum rms voltage divided by K (kV)	Maximum symmetrical interrupting capability[5] / K times related short-circuit rms current (kA)	3-s short-time current-carrying capability (kA)	Closing and latching capability rms (kA)
4.16	250	4.76	1.24	19	60	1200	29	5	2	3.85	36	36	58
4.16	250	4.76	1.24	19	60	2000	29	5	2	3.85	36	36	58
4.16	250	4.76	1.24	19	60	3000	29	5	2	3.85	36	36	58
4.16	350	4.76	1.19	19	60	1200	41	5	2	4.0	49	49	78
4.16	350	4.76	1.19	19	60	2000	41	5	2	4.0	49	49	78
4.16	350	4.76	1.19	19	60	3000	41	5	2	4.0	49	49	78
7.2	500	8.25	1.25	36	95	1200	33	5	2	6.6	41	41	66
7.2	500	8.25	1.25	36	95	2000	33	5	2	6.6	41	41	66
7.2	500	8.25	1.25	36	95	3000	33	5	2	6.6	41	41	66
13.8	500	15	1.30	36	95	1200	18	5	2	11.5	23	23	37
13.8	500	15	1.30	36	95	2000	18	5	2	11.5	23	23	37
13.8	500	15	1.30	36	95	3000	18	5	2	11.5	23	23	37
13.8	750	15	1.30	36	95	1200	28	5	2	11.5	36	36	58
13.8	750	15	1.30	36	95	2000	28	5	2	11.5	36	36	58
13.8	750	15	1.30	36	95	3000	28	5	2	11.5	36	36	58
13.8	1000	15	1.30	36	95	1200	37	5	2	11.5	48	48	77
13.8	1000	15	1.30	36	95	2000	37	5	2	11.5	48	48	77
13.8	1000	15	1.30	36	95	3000	37	5	2	11.5	48	48	77

7.15

TABLE 7.11 Medium-Voltage Power-Circuit-Breaker Characteristics (ANSI/IEEE Std. C37.06[6]) (*Continued*)

Identification		Rated values								Related required capabilities			
		Voltage		Insulation level		Current				Current values			
				Rated withstand test voltage								3-s short-time current-carrying capability	Closing and latching capability
Nominal rms voltage class (kV)	Nominal 3-phase class (MVA)	Rated maximum rms voltage (kV)[1]	Rated voltage range factor K[2]	Low-frequency rms voltage (kV)	Crest impulse voltage (kV)	Continuous rms current rating at 60 Hz (A)	Short-circuit rms current rating (at max kV) (kA)[3,4]	Rated interrupting time (cycles)	Rated permissable tripping delay, Y (seconds)	Rated maximum rms voltage divided by K (kV)	Maximum symmetrical interrupting capability[5] K times related short-circuit rms current (kA)	K times related short-circuit rms current (kA)	rms (kA)
4.16	250	4.76	1.24	19	60	1200 2000 3000	29	5	2	3.85	36	36	78
7.2	500	8.25	1.25	36	95	1200 2000 3000	33	5	2	6.6	41	41	78
13.8	500	15	1.30	36	95	1200 2000 3000	18	5	2	11.5	23	23	58
13.8	750	15	1.30	36	95	1200 2000 3000	28	5	2	11.5	36	36	77

High close and latch capability circuit breakers (these ratings exceed ANSI Std. C37.06)

High interrupting capability circuit breakers (these ratings exceed ANSI Std. C37.06)

| 13.8 | 1500 | 15 | 1.0 | 36 | 95 | 1200 2000 3000 | 63 | 5 | 2 | 15.0 | 63 | 63 | * |

*170 kA crest $\quad I_{crest} = I_{symmetrical} \times \sqrt{3} \times \sqrt{2} \times S = 63 \times \sqrt{3} \times \sqrt{2} \times 1.1$

1. Maximum voltage for which the circuit breaker is designed and the upper limit for operation.

2. K is the ratio of rated maximum voltage to the lower limit of the range of operating voltage in which the required symmetrical and asymmetrical interrupting capabilities vary in inverse proportion to the operating voltage.

3. To obtain the required symmetrical interrupting capability of a circuit breaker at an operating voltage between $1/K$ times rated maximum voltage and rated maximum voltage, the following formula shall be used:

Required symmetrical interrupting capability = rated short-circuit current $\times \dfrac{(\text{rated max.voltage})}{(\text{operating voltage})}$

For operating voltages below $1/K$ times rated maximum voltage, the required symmetrical interrupting capability of the circuit breaker shall be equal to K times rated short-circuit current.

4. With the limitation stated in 5.10 of ANSI/IEEE Std. C37.04-1979, all values apply for polyphase and line-to-line faults. For single-phase-to-ground faults, the specific conditions stated in 5.10.2.3 of ANSI/IEEE Std. C37.04-1979 apply.

5. Current values in this column are not to be exceeded even for operating voltages below $1/K$ times rated maximum voltage. For voltages between rated maximum voltage and $1/K$ times rated maximum voltage, follow (3) above.

In accordance with ANSI/IEEE Std. C37.06, users should confer with the manufacturer on the status of various circuit-breaker ratings.

TABLE 7.12 Medium-Voltage Power Fuse and Fused Equipment Rated Symmetrical Interrupting Currents

Maximum voltage (kV)							
2.8	5.5	8.3	15.5	25.8	38.0	48.3	72.5
Rated interrupting current (rms symmetrical kA)							
							2.5
						3.15	
		4.0	4.0	4.0		4.0	4.0
		5.0	5.0	5.0	5.0	5.0	5.0
		6.3	6.3	6.3	6.3	6.3	6.3
		8.0	8.0	8.0	8.0	8.0	8.0
		10.0	10.0	10.0	10.0	10.0	10.0
		12.5	12.5	12.5	12.5	12.5	12.5
		16.0	16.0	16.0	16.0	16.0	16.0
		20.0	20.0	20.0	20.0	20.0	20.0
		25.0	25.0	25.0	25.0	25.0	25.0
31.5	31.5	31.5	31.5	31.5	31.5	31.5	
40.0	40.0	40.0	40.0	40.0	40.0		
50.0	50.0	50.0	50.0	50.0			
63.0			63.0				
			80.0				

TABLE 7.13 Circuit-Switcher Characteristics

kV	34.5, 46		69, 115, 138, 161		115, 138, 230	
Current values in rms kA symmetrical						
Momentary	61	61	64	64	70	80
3 s	40	40	—	—	43.75	50
1 s	—	—	40	40	—	—
Fault closing (1 time duty cycle)	—	—	40	40	—	—
Fault closing (2 times duty cycle)	30	30	—	—	40	40
Load switching (A)						
Continuous current	1200	1200	1200	1200	1600	2000
Load dropping	1200	1200	1200	1200	1600	2000
Line charging, cap. switching, cable charging*	400	400	400	400	400	400
Shunt reactor switching	600	600	600	600	600	600
Fault interrupting in rms kA symmetrical						
Primary faults	8†	8†	20‡	20‡	8†	8†
Transformer secondary faults	4	4	4	4	4	4

*Values shown are minimum. Consult manufacturer for specific values for any of these applications.
†Duty cycle is O + 0s + CO + 0s + CO.
‡Duty cycle is O or CO.

TABLE 7.14[14]   Preferred Ratings for Outdoor Circuit Breakers, 15.5 to 242 kV, including Circuit Breakers Applied in Gas-Insulated Substations

(Derived from ANSI/IEEE Std. C37.06-1987[6])

Rated maximum voltage (1) kV, rms	Rated voltage range factor K (2)	Rated continuous current at 60 Hz A, rms	Rated short-circuit current at rated max. kV (3) (4) kA, rms	Rated inter-rupting time cycles	Rated maximum voltage divided by K kV, rms	Maximum symmetrical interrupting capability and rated short-time current (5) kA, rms	Closing and latching capability 2.7K times rated short-circuit current kA, crest
15.5	1.0	600, 1200	12.5	5	15.5	12.5	34
15.5	1.0	1200, 2000	20.0	5	15.5	20.0	54
15.5	1.0	1200, 2000	25.0	5	15.5	25.0	68
15.5	1.0	1200, 2000, 3000	40.0	5	15.5	40.0	108
25.8	1.0	1200, 2000	12.5	5	25.8	12.5	34
25.8	1.0	1200, 2000	25.0	5	25.8	25.0	68
38.0	1.0	1200, 2000	16.0	5	38.0	16.0	43
38.0	1.0	1200, 2000	20.0	5	38.0	20.0	54
38.0	1.0	1200, 2000	25.0	5	38.0	25.0	68
38.0	1.0	1200, 2000	31.5	5	38.0	31.5	85
38.0	1.0	1200, 2000, 3000	40.0	5	38.0	40.0	108
48.3	1.0	1200, 2000	20.0	5	48.3	20.0	54
48.3	1.0	1200, 2000	31.5	5	48.3	31.5	85
48.3	1.0	1200, 2000, 3000	40.0	5	48.3	40.0	108
72.5	1.0	1200, 2000	20.0	5	72.5	20.0	54
72.5	1.0	1200, 2000	31.5	5	72.5	31.5	85
72.5	1.0	1200, 2000, 3000	40.0	5	72.5	40.0	108
121	1.0	1200	20.0	3	121	20.0	54
121	1.0	1600, 2000, 3000	40.0	3	121	40.0	108
121	1.0	2000, 3000	63.0	3	121	63.0	170
145	1.0	1200	20.0	3	145	20.0	54
145	1.0	1600, 2000, 3000	40.0	3	145	40.0	108
145	1.0	2000, 3000	63.0	3	145	63.0	170
145	1.0	2000, 3000	80.0	3	145	80.0	216
169	1.0	1200	16.0	3	169	16.0	43
169	1.0	1600	31.5	3	169	31.5	85
169	1.0	2000	40.0	3	169	40.0	108
169	1.0	2000	50.0	3	169	50.0	135
169	1.0	2000	63.0	3	169	63.0	170
242	1.0	1600, 2000, 3000	31.5	3	242	31.5	85
242	1.0	2000, 3000	40.0	3	242	40.0	108
242	1.0	2000	50.0	3	242	50.0	135
242	1.0	2000, 3000	63.0	3	242	63.0	170

See notes 1–5 to Table 7.11, page 7.17.

7.3   THREE-PHASE SHORT-CIRCUIT CALCULATIONS

7.3.1   Per-Unit System

The sample calculation in this section uses the per-unit system to convert all circuit element impedances from their individual bases to a common base. The formulas involved are:

Circuit Element Data   *Formula for Per-Unit*

Available short circuit:

$$Z_{pu} = \frac{\text{per-unit MVA base}}{\text{utility MVA}}$$

MVA plus *X/R*

$$R_{pu} = Z_{pu} \times \cos\left(\tan^{-1}\frac{X}{R}\right)$$

$$X_{pu} = Z_{pu} \times \sin\left(\tan^{-1}\frac{X}{R}\right)$$

Transformer:

$$Z_{pu} = \frac{\%Z}{100} \times \frac{\text{pu MVA base}}{\text{transformer MVA}}$$

kVA, %Z and estimated *X/R*

(transformer self-cooled rating)

$$R_{pu} = Z_{pu} \times \cos\left(\tan^{-1}\frac{X}{R}\right)$$

$$X_{pu} = Z_{pu} \times \sin\left(\tan^{-1}\frac{X}{R}\right)$$

Cables and busway:

$$R_{pu} = R_{ohms} \times \frac{\text{MVA}_{base}}{\text{kV}_{base}^2}$$

R & *X* in Ω/ft

$$X_{pu} = X_{ohms} \times \frac{\text{MVA}_{base}}{\text{kV}_{base}^2}$$

Motors:

$$\text{kVA}_{motor} = \frac{\text{hp} \times 0.746}{\text{efficiency} \times \text{pf}}$$

HP, rpm, estimated *X/R* and
  estimated kVA or

$$X_{pu} = \frac{\text{FLA}}{\text{LRA}} \times \frac{\text{MVA}_{base}}{\text{MVA}_{motor}} \times \frac{\text{kV}_{motor}^2}{\text{kV}_{base}^2}$$

Full-load and locked rotor amperes
  plus estimated *X/R*

$$R_{pu} = \frac{X_{pu}}{(X/R)}$$

7.3.2 Sample Calculation

First Step—Draw a System One-Line Diagram (Fig. 7.7)

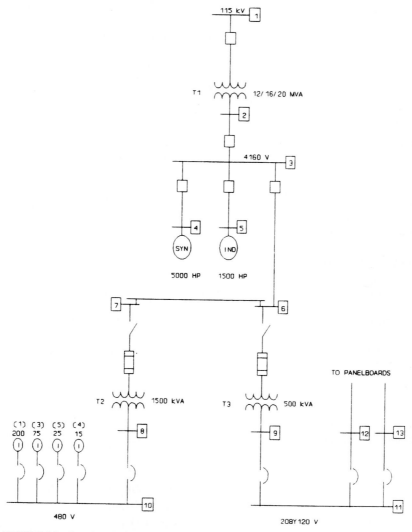

FIGURE 7.7 Sample calculation one-line diagram.

Second Step—Collect System Data

Utility available short circuit	115 KV, 5000 MVA available, $X/R = 7$
Transformers:	
Main transformer	12 MVA (self-cooled rating)
	115 kV $-4160Y/2400$ V, $Z = 8\%$
Motor bus transformer	1500 kVA, $4160-480Y/277$ V
	$Z = 5.75\%$
Lighting & power transformer	500 kVA, $4160-208Y/120$ V
	$Z = 5\%$
Busway:	
4160 V	50 ft—3000 A, copper
480 V	15 ft—2000 A, copper
$208Y/120$ V	30 ft—2000 A, copper
Cables:	
4160 synchronous motor	75 ft: 6 - 1/C - 500 MCM,
4160 induction motor	nonmagnetic conduit
4160 V to lighting transformer	100 ft: 3 - 1/C - 500 MCM, nonmagnetic conduit
4160 V to motor substation	25 ft: 3 - 1/C - 500 MCM, nonmagnetic conduit
$208Y/120$ V to 1st lighting panelboard	100 ft: 3 - 1/C - 500 MCM, nonmagnetic conduit
$208/Y120$ V to 1st power panel	20 ft: 3 - 1/C - 500 MCM, nonmagnetic conduit
	50 ft: 3 - 1/C - 500 MCM, nonmagnetic conduit
Motors:	
4-kV synchronous motor	
4-kV induction motor	5000 hp, 0.8 pf, 600 rpm
460-V induction motors	1500 hp, 1180 rpm
	1—200 hp, 1180 rpm
	3—75 hp, 1180 rpm
	5—25 hp, 1180 rpm
	4—15 hp, 1180 rpm

Third Step—Reduce System Data to a Common Per-Unit Base (Table 7.15)

TABLE 7.15   Sample Per-Unit Calculations on a 10-MVA Base

Element	Calculation	Z_{pu}	R_{pu}	X_{pu}
Source 5000 MVA available	$Z_{pu} = \dfrac{\text{base MVA}}{\text{available MVA}}$ $R_{pu} = Z_{pu} \cos\left(\tan^{-1}\dfrac{X}{R}\right)$ $X_{pu} = Z_{pu} \sin\left(\tan^{-1}\dfrac{X}{R}\right)$ $Z_{pu} = \dfrac{10}{5000} = 0.0020$	0.0020	0.00028	0.00198
Transformer $T1$ 12000 kVA $Z = 8\%$ $X/R = 17.9$ (Table 7A.6)	$Z_{pu} = Z \times \dfrac{\text{study base}}{\text{transformer base}}$ $Z_{pu} = 0.08 \times \dfrac{10}{12}$	0.06667	0.00372	0.06656
Transformer $T2$ 1500 kVA $Z = 5.75\%$ $X/R = 5.9$ (Table 7A.6)	$Z_{pu} = 0.0575 \times \dfrac{10}{1.5}$	0.38333	0.06405	0.37794
Transformer $T3$ 500 kVA $Z = 5\%$ $X/R = 4.3$ (Table 7A.6)	$Z_{pu} = 0.05 \times \dfrac{10}{0.5}$	1.000	0.22651	0.9740
Synchronous motor 4160-V cables $Z_{pu} = Z_{ohms} \times MVA_{base}/kV_{base}^2$	$R = \frac{1}{2}\left(0.220 \times \dfrac{75}{1000}\right) \times \dfrac{(10)}{4.16^2}$ $X = \frac{1}{2}\left(0.369 \times \dfrac{75}{1000}\right) \times \dfrac{(10)}{4.16^2}$		0.0004767	0.0007996
Induction motor 4160-V cables $Z_{pu} = Z_{ohms} \times MVA_{base}/kV_{base}^2$	$R = \left(0.220 \times \dfrac{100}{1000}\right) \times \dfrac{(10)}{4.16^2}$ $X = \left(0.369 \times \dfrac{100}{1000}\right) \times \dfrac{(10)}{4.16^2}$		0.001271	0.002132

(Continued)

TABLE 7.15 Sample Per-Unit Calculations on a 10-MVA Base *(Continued)*

Element	Calculation	R_{pu}	X_{pu}
Transformer $T3$ feed 4160-V cables $Z_{pu} = Z_{ohms} \times MVA_{base}/kV_{base}{}^2$	$R = \left(0.220 \times \dfrac{25}{1000}\right) \times \dfrac{(10)}{4.16^2}$ $X = \left(0.369 \times \dfrac{25}{1000}\right) \times \dfrac{(10)}{4.16^2}$	0.0003178	0.0005331
Transformer $T2$ feed 4160-V cables $Z_{pu} = Z_{ohms} \times MVA_{base}/kV_{base}{}^2$	$R = \left(0.220 \times \dfrac{100}{1000}\right) \times \dfrac{(10)}{4.16^2}$ $X = \left(0.369 \times \dfrac{100}{1000}\right) \times \dfrac{(10)}{4.16^2}$	0.001271	0.002132
Lighting panelboard 208-V feeder cables $Z_{pu} = Z_{ohms} \times MVA_{base}/kV_{base}{}^2$	$R = \left(0.220 \times \dfrac{20}{1000}\right) \times \dfrac{(10)}{.208^2}$ $X = \left(0.369 \times \dfrac{20}{1000}\right) \times \dfrac{(10)}{.208^2}$	0.1017	0.1401
Power panelboard 208-V feeder cables $Z_{pu} = Z_{ohms} \times MVA_{base}/kV_{base}{}^2$	$R = \left(0.220 \times \dfrac{50}{1000}\right) \times \dfrac{(10)}{.208^2}$ $X = \left(0.369 \times \dfrac{50}{1000}\right) \times \dfrac{(10)}{.208^2}$	0.2543	0.3502

Element	X_{pu} Calculation	R_{pu} Calculation
$T1$ secondary busway 4.16 kV, 3000 A, copper (Table 7A.10) $R = 0.000245 \ \Omega/100$ ft $X = 0.00450 \ \Omega/100$ ft	$X = \left(\dfrac{0.00450}{100} \times 50\right) \times \dfrac{(10)}{4.16^2} = 0.0013$	$R = \left(\dfrac{0.000245}{100} \times 50\right) \times \dfrac{(10)}{4.16^2} = 0.0007$
$T2$ secondary busway 480 V, 2000 A, copper (Table 7A.9) $R = 0.00066 \ \Omega/100$ ft $X = 0.00051 \ \Omega/100$ ft	$X = \left(\dfrac{0.00051}{100} \times 15\right) \times \dfrac{(10)}{0.48^2} = 0.0033$	$R = \left(\dfrac{0.00066}{100} \times 15\right) \times \dfrac{(10)}{0.48^2} = 0.0043$
$T3$ secondary busway 208 V, 2000 A, copper (Table 7A.9) $R = 0.00066 \ \Omega/100$ ft $X = 0.00051 \ \Omega/100$ ft	$X = \left(\dfrac{0.00051}{100 \times 30}\right) \times \dfrac{(10)}{0.208^2} = 0.0354$	$R = \left(\dfrac{0.00066}{100 \times 30}\right) \times \dfrac{(10)}{0.208^2} = 0.00458$

TABLE 7.15 Sample Per-Unit Calculations on a 10-MVA Base (*Continued*)

Element	X_{pu} Calculation	R_{pu} Calculation
5000-hp synchronous motor		
4 kV, 0.8 pf, 600 rpm $X_d'' = 20\%$ kVA = hp = 5000 $X/R = 30.1$	$X_{pu} = \dfrac{\%X}{100} \times \dfrac{\text{Equipment } V^2}{\text{Study } V^2} \times \dfrac{\text{MVA}_{base}}{\text{MVA}_{motor}}$ $X_{pu} = \dfrac{20}{100} \times \dfrac{4^2}{4.16^2} \times \dfrac{10}{5} = 0.3698$	$R_{pu} = \dfrac{X_{pu}}{(X/R)} = \dfrac{0.3698}{30.1} = 0.01229$
1500-hp induction motor		
4 kV, 1180 rpm $X_d'' = 16.7\%$ kVA = 0.8×hp = 1200 $X/R = 28.5$	$X_{pu} = \dfrac{16.7}{100} \times \dfrac{4^2}{4.16^2} \times \dfrac{10}{1.2} = 1.2867$	$R_{pu} = \dfrac{X_{pu}}{(X/R)} = \dfrac{1.2867}{28.5} = 0.04515$
200-hp induction motor		
460 V, 1180 rpm $X_d'' = 16.7\%$ kVA = 0.8×hp = 160 $X/R = 11.6$	$X_{pu} = \dfrac{16.7}{100} \times \dfrac{460^2}{480^2} \times \dfrac{10}{0.16} = 9.5858$	$R_{pu} = \dfrac{X_{pu}}{(X/R)} = \dfrac{9.5858}{11.6} = 0.8264$
3–75-hp induction motors		
460 V, 1180 rpm $X_d'' = 16.7\%$ kVA = 0.8×hp = 60 $X/R = 7.2$	$X_{pu} = \dfrac{16.7}{100} \times \dfrac{460^2}{480^2} \times \dfrac{10}{0.06} = 25.5622$ (per motor)	$R_{pu} = \dfrac{X_{pu}}{(X/R)} = \dfrac{25.5622}{7.2} = 3.5504$ (per motor)
5–25-hp induction motors		
460 V, 1180 rpm $X_d'' = 16.7\%$ kVA = hp = 25 $X/R = 3.8$	$X_{pu} = \dfrac{16.7}{100} \times \dfrac{460^2}{480^2} \times \dfrac{10}{0.025} = 61.349$ (per motor)	$R_{pu} = \dfrac{X_{pu}}{(X/R)} = \dfrac{61.349}{3.8} = 16.145$ (per motor)
4–15-hp induction motors		
460 V, 1180 rpm $X_d'' = 16.7\%$ kVA = hp = 15 $X/R = 3.0$	$X_{pu} = \dfrac{16.7}{100} \times \dfrac{460^2}{480^2} \times \dfrac{10}{0.015} = 102.249$ (per motor)	$R_{pu} = \dfrac{X_{pu}}{(X/R)} = \dfrac{102.249}{3.0} = 34.083$ (per motor)

Fourth Step—Draw a System Impedance Diagram (Fig. 7.8)

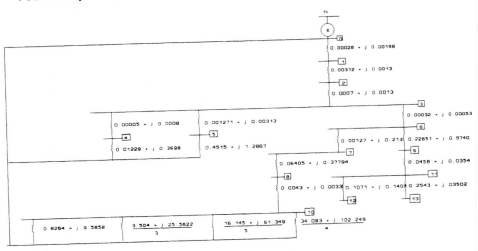

FIGURE 7.8   Sample calculation impedance diagram.

Fifth Step—Reduce System Impedances to a Single Impedance

Table 7.16 outlines the method of reducing all the system impedances shown in Fig. 7.8 corresponding to Fig. 7.7 equipment to one impedance for a fault on bus 3. The formulas for handling mathematically complex impedance quantities follow the table. After reducing the impedances to one impedance, calculate the MVA and symmetrical available short-circuit current as shown in step 6. The result is the first-cycle symmetrical available short-circuit current at bus 3. Multiply this current by 1.6 to obtain the asymmetrical available short-circuit current. The calculation shown is for a fault on one particular bus. Computer programs discussed in Sec. 7.3.3 perform this calculation for all buses in the system and eliminate much of the tedious calculation procedures necessary to do this manually, as illustrated by Table 7.16.

Parallel impedance formulas used in Table 7.16:

$$G = \frac{R}{R^2 + X^2} \qquad B = \frac{X}{R^2 + X^2} \qquad R = \frac{G}{G^2 + B^2} \qquad X = \frac{B}{G^2 + B^2}$$

Impedance equivalents
    Series impedances:

$$Z_T = Z_1 + Z_2 + Z_3 \ldots + Z_N$$

    Parallel impedances:

$$\frac{1}{Z_T} = \frac{1}{Z_1} + \frac{1}{Z_2} + \frac{1}{Z_3} \ldots \frac{1}{Z_N}$$

7.3.3 Computer Programs

Several commercial computer programs are available for calculating short-circuit currents for large or complex systems. These programs make calculations in accordance

TABLE 7.16   Sample Short-Circuit Calculation

Parallel source impedances bus 0 to bus 10					
Motor	R	X	$R^2 + X^2$	G	B
200 hp	0.8264	9.5858	92.57050	0.0089273	0.1035514
3–75 hp	1.1835	8.5207	74.00300	0.0159926	0.1151399
5–25 hp	3.2290	12.2698	160.97443	0.0200591	0.0762221
4–15 hp	8.5208	25.5623	726.03521	0.0117361	0.0352081
				0.0567151	0.3301215

$G^2 + B^2$	R	X		
0.1121968	0.50550	2.94234	←Combined motor R and X, bus 0 to bus 10	

Series source impedances bus 0 to bus 3					
Bus to bus	R	X	Bus to bus	R	X
0–10	0.5055	2.94234	0–1	0.00028	0.00198
10–8	0.0043	0.0033	1–2	0.00372	0.06656
8–7	0.6405	0.37794	2–3	0.0007	0.0013
7–6	0.00127	0.00213	0–1–2–3	0.00470	0.06984
6–3	0.00032	0.00053			
0–10–8–7–6–3	1.15189	3.32624	0–4	0.01229	0.3698
			4–3	0.00005	0.0008
0–5	0.04515	1.2767	0–3–4	0.01234	0.3706
5–3	0.00127	0.0021			
0–5–3	0.04642	1.2888			

Parallel source impedances bus 0 to bus 3					
Bus	R	X	$R^2 + X^2$	G	B
0–1–2–3	0.00470	0.06984	0.0048997	0.9592394	14.253889
0–4–3	0.01234	0.3706	0.1374966	0.0897477	2.6953387
0–5–3	0.04642	1.2888	1.6631603	0.0279107	0.774903
0–10–8–7–6–3	1.15189	3.32624	12.390723	0.0929639	0.268446
				1.1698617	17.992577

Combining parallel source impedances bus 0 to bus 3					
G	B	$G^2 + B^2$	R	X	Z
1.1698617	17.992517	325.1014	0.0035985	0.055345	0.055462

Sixth Step—Calculate Short Circuit Current

Bus 3 results

$$MVA_{SC} = \frac{MVA_{base}}{Z} = \frac{10}{0.055462} = 180.3$$

$$I_{sym} = \frac{MVA_{SC}}{kV\sqrt{3}} = 25{,}063 \text{ A}$$

with the applicable standards and are great for larger systems or systems for which a manual calculation shows that the calculated fault currents are close to the rated capabilities of the interrupting devices.

An easy way to check the outputs of these programs is to do a simplified calculation including the most prominent (lowest contributing source) impedance elements of the system, assuming the impedance is all reactance and neglecting small fault contributions. Such a calculation for bus 3 appears as follows.

12-MVA transformer	$12/.08$	= 150 MVA
5000-hp motor	$5/.20$	= 25 MVA
1500-hp motor	$5/.167$	= 9 MVA
Total fault MVA		184 MVA

This value is close to both the manual calculation (180.8 MVA) and the computer calculation which follows (183.85 MVA).

The pages which follow illustrate one computer program examining the system shown in Sec. 7.32 with the Fig. 7.7 one-line diagram. Both a data reduction[8] and a short-circuit program[9] are used. The short-circuit program prints out the duties for different types of circuit breakers at all voltages plus the medium-voltage fuse duties. Most good commercial programs print out data in a format similar to the one shown.

```
DATA REDUCTION FOR SAMPLE SHORT-CIRCUIT CALCULATION PAGE - 1
GE INDUSTRIAL POWER SYSTEMS ENGINEERING - SCHENECTADY, NY
DATA REDUCTION PROGRAM - VERSION 1.43
(c) 1973-1988, General Electric Company
DATA REDUCTION PROGRAM
10 MVA BASE - 60 HERTZ


SAMPLE CALCULATION
INDUSTRIAL PLANT

UTILITY SOURCE IMPEDANCE ON A 10 MVA BASE
BUS        MVA         X/R     P.U.R   P.U.X
----------------------------------------------
115 KV UTILITY TIE
1        5000        7.0    0.00028 0.00198

TRANSFORMER IMPEDANCE ON A 10 MVA BASE
IDENT     KVA        %Z       X/R     P.U.R      P.U.X      BUS   BUS
-------------------------------------------------------------------
MAIN TRANSFORMER
T1       12000      8.00     17.9    0.00372    0.06656     1     2
480 V SUBSTATION
T2        1500      5.75      5.9    0.06391    0.37797     7     8
208Y120 V SUBSTATION
T3         500      5.00      4.3    0.22821    0.97361     6     9

CABLE IMPEDANCE ON A 10 MVA BASE AT 60 HZ-RES. AT 25 C
CABLE      CONDUCTOR    CONDUIT LEN.  VOLTS  P.U.R    P.U.X BUS BUS
-------------------------------------------------------------------
4160 V MOTOR CABLES
SYN     6-1C-500MCM CU  N'MAG   75FT  4160 0.00051 0.00078  3   4
IND     3-1C-500MCM CU  N'MAG  100FT  4160 0.00135 0.00208  3   5
4160 V SUBSTATION CABLES
```

```
T3      3-1C-500MCM CU  N'MAG   25FT   4160  0.00034 0.00052   3    6
T2      3-1C-500MCM CU  N'MAG  100FT   4160  0.00135 0.00208   6    7
PANELBOARD CABLES
LIGHTS  3-1C-500MCM CU  N'MAG   20FT    208  0.10668 0.14856  11   12
POWER   3-1C-500MCM CU  N'MAG   50FT    208  0.26670 0.37139  11   13
DATA REDUCTION FOR SAMPLE SHORT-CIRCUIT CALCULATION PAGE - 2
-----------------------------------------------------------------
BUSWAY IMPEDANCE ON A 10 MVA BASE AT 60 HERTZ
IDENT   AMPS    LENGTH    VOLTS    P.U.R     P.U.X    BUS    BUS
-----------------------------------------------------------------
MAIN    3000    50FT CU    4160   0.00013   0.00009    2      3
T2      2000    15FT CU     480   0.00460   0.00283    8     10
T3      2000    30FT CU     208   0.04903   0.03014    9     11

P.U. MOTOR X''D OR LOCKED ROTOR IMPEDANCES ON 10 MVA BASE
BUS    HP   MOTOR KVA  RPM   PF   QUAN  %X   X/R  P.U.R  P.U.X  CODE
-----------------------------------------------------------------
THE FOLLOWING HAVE AN EQUIP. VOLTAGE TO SYSTEM VOLTAGE RATIO OF .96
 4 5000.0 SYN  5000   600  0.8  1.0  20.00 30.1  .01223  .36864  3
 5 1500.0 IND  1350  1180   -   1.0  16.70 28.5  .03994 1.1400   4
10  200.0 IND   190  1180   -   1.0  16.70 11.6  .69701 8.1003   5
10   75.0 IND    75  1180   -   3.0  16.70  7.2  .95503 6.8403   5
10   25.0 IND    25  1180   -   5.0  16.70  3.8 3.2833  12.312   6
10   15.0 IND    15  1180   -   4.0  16.70  3.0 8.4102  25.651   6


CASE: 1-FCY PAGE - 1 -
GE INDUSTRIAL POWER SYSTEMS ENGINEERING - SCHENECTADY, NY
THREE PHASE SHORT CIRCUIT PROGRAM - VERSION 1.44
FIRST CYCLE CALC. FOR BREAKER DUTIES PER ANSI C37.13-1981
TOTL. CURRENT & FLOWS FROM COMPLEX NETWORK, X/R FROM SEPARATE R & X
(c) 1973-1988, General Electric Company
       09-17-1993    10 MVA BASE    60 HERTZ

SAMPLE CALCULATION
INDUSTRIAL PLANT

CASE: 1-FCY
SAMPLE SHORT-CIRCUIT CALCULATION FOR
RADIAL SYSTEM WITH 115 kV SOURCE

INPUT DATA
BUS   TO   BUS    R P.U.     X P.U.     CODE
 0          1    0.00028    0.00198      1
 1          2    0.00372    0.06656      0
 7          8    0.06391    0.37797      0
 6          9    0.22821    0.97361      0
 3          4    0.00051    0.00078      0
 3          5    0.00135    0.00208      0
 3          6    0.00034    0.00052      0
 6          7    0.00135    0.00208      0
11         12    0.10668    0.14856      0
11         13    0.26670    0.37139      0
 2          3    0.00013    0.00009      0
```

8	10	0.00460	0.00283	0
9	11	0.04903	0.03014	0
0	4	0.01223	0.36864	3
0	5	0.03994	1.14005	4
0	10	0.48353	4.45033	5
0	10	3.94360	13.89325	6

*BUS 1   E/Z = 25.259 KA (5031.19MVA) AT −81.98°,X/R= 7.18, 115 KV
          Z = 0.000277 + j 0.001968
          1.6*ISYM = 40.41 IASYM BASED ON X/R = 34.20

          MAX. HIGH VOLTAGE POWER FUSE DUTY = 25.26 SYM

CONTRIBUTIONS IN KA

BUS TO	BUS	MAG	ANG	BUS TO	BUS	MAG	ANG
REMOTE	1	25.106	−81.950	2	1	0.153	−87.377

CASE: 1-FCY PAGE - 2 -
*BUS 2   E/Z = 25.543 KA (184.04MVA) AT −86.84°, X/R = 19.16, 4.160 KV
          Z = 0.002995 + j 0.054252
          1.6*ISYM = 40.87 IASYM BASED ON X/R = 39.91

          MAX. HIGH VOLTAGE POWER FUSE DUTY = 26.08 SYM

CONTRIBUTIONS IN KA

BUS TO	BUS	MAG	ANG	BUS TO	BUS	MAG	ANG
1	2	20.215	−86.660	3	2	5.329	−87.523

*BUS 3   E/Z = 25.516 KA (183.85MVA) AT −86.76°, X/R = 18.83, 4.160 KV
          Z = 0.003070 + j 0.054305
          1.6*ISYM = 40.83 IASYM BASED ON X/R = 39.80

          MAX. HIGH VOLTAGE POWER FUSE DUTY = 26.01 SYM

CONTRIBUTIONS IN KA

BUS TO	BUS	MAG	ANG	BUS TO	BUS	MAG	ANG
4	3	3.755	−88.024	5	3	1.214	−87.930
6	3	0.364	−81.338	2	3	20.186	−86.557

*BUS 4   E/Z = 25.250 KA (181.93MVA) AT −86.43°, X/R = 17.43, 4.160 KV
          Z = 0.003423 + j 0.054858
          1.6*ISYM = 40.40 IASYM BASED ON X/R = 39.07

          MAX. HIGH VOLTAGE POWER FUSE DUTY = 25.59 SYM

CONTRIBUTIONS IN KA

BUS TO	BUS	MAG	ANG	BUS TO	BUS	MAG	ANG
3	4	21.489	−86.137	SYNMOT	4	3.763	−88.098

*BUS 5   E/Z = 24.629 KA (177.46MVA) AT −85.64°, X/R = 14.03, 4.160 KV
          Z = 0.004283 + j 0.056188
          1.6*ISYM = 39.41 IASYM BASED ON X/R = 37.17
          MAX. HIGH VOLTAGE POWER FUSE DUTY = 24.63 SYM

CONTRIBUTIONS IN KA

BUS TO	BUS	MAG	ANG	BUS TO	BUS	MAG	ANG
3	5	23.413	−85.519	INDMOT	5	1.217	−87.992

CASE: 1-FCY PAGE - 3 -
*BUS 6   E/Z = 25.273 KA (182.10MVA) AT −86.45°, X/R = 17.02, 4.160 KV
          Z = 0.003402 + j 0.054810

1.6*ISYM = 40.44 IASYM BASED ON X/R = 39.01

MAX. HIGH VOLTAGE POWER FUSE DUTY = 25.56 SYM

CONTRIBUTIONS IN KA

BUS TO	BUS	MAG	ANG	BUS TO	BUS	MAG	ANG
9	6	0.000	−3.033	3	6	24.911	−86.523
7	6	0.364	−81.324				

*BUS 7   E/Z = 24.339 KA (175.37MVA) AT −85.25°, X/R = 12.49, 4.160 KV
Z = 0.004717 + j 0.056826

1.6*ISYM = 38.94 IASYM BASED ON X/R = 36.18

MAX. HIGH VOLTAGE POWER FUSE DUTY = 24.34 SYM

CONTRIBUTIONS IN KA

BUS TO	BUS	MAG	ANG	BUS TO	BUS	MAG	ANG
8	7	0.364	−81.349	6	7	23.976	−85.314

*BUS 8   E/Z = 30.780 KA (25.59MVA) AT −81.09°, X/R = 6.52, 0.480 KV
Z = 0.060504 + j 0.386070

MAX. LOW VOLTAGE FUSE DUTY = 32.61 SYM
MAX. LOW VOLTAGE POWER CIRCUIT BREAKER DUTY = 30.78
MAX. LV MCCB OR ICCB (RATED >20KA INT.) DUTY = 32.61
MAX. LV MCCB OR ICCB (RATED 10-20KA INT.)DUTY = 36.28

CONTRIBUTIONS IN KA

BUS TO	BUS	MAG	ANG	BUS TO	BUS	MAG	ANG
7	8	27.272	−81.047	10	8	3.507	−81.455

*BUS 10   E/Z = 30.566 KA (25.41MVA) AT −80.63°, X/R = 6.21, 0.480 KV
Z = 0.064045 + j 0.388264

MAX. LOW VOLTAGE FUSE DUTY = 32.09 SYM
MAX. LOW VOLTAGE POWER CIRCUIT BREAKER DUTY = 30.57
MAX. LV MCCB OR ICCB (RATED >20KA INT.) DUTY = 32.09
MAX. LV MCCB OR ICCB (RATED 10-20KA INT.)DUTY = 35.70

CONTRIBUTIONS IN KA

BUS TO	BUS	MAG	ANG	BUS TO	BUS	MAG	ANG
8	10	27.056	−80.518	INDMOT	10	2.687	−83.798
INDMOT 10		0.833	−74.152				

CASE: 1-FCY PAGE - 4 -

*BUS 9   E/Z = 26.331 KA (9.49MVA) AT −77.31°, X/R = 4.44, 0.208 KV
Z = 0.231613 + j 1.028419

MAX. LOW VOLTAGE FUSE DUTY = 26.33 SYM
MAX. LOW VOLTAGE POWER CIRCUIT BREAKER DUTY = 26.33
MAX. LV MCCB OR ICCB (RATED >20KA INT.) DUTY = 26.33
MAX. LV MCCB OR ICCB (RATED 10-20KA INT.)DUTY = 28.65

CONTRIBUTIONS IN KA

BUS TO	BUS	MAG	ANG	BUS TO	BUS	MAG	ANG
6	9	26.331	−77.308	11	9	−0.000	−108.888

*BUS 11   E/Z′ = 25.346 KA (9.13MVA) AT −75.15°, X/R = 3.77, 0.208 KV
Z = 0.280643 + j 1.058559

MAX. LOW VOLTAGE FUSE DUTY = 25.35 SYM
MAX. LOW VOLTAGE POWER CIRCUIT BREAKER DUTY = 25.35

```
                    MAX. LV MCCB OR ICCB (RATED >20KA INT.) DUTY = 25.35
                    MAX. LV MCCB OR ICCB (RATED 10-20KA INT.)DUTY = 26.50
```

CONTRIBUTIONS IN KA

BUS TO	BUS	MAG	ANG	BUS TO	BUS	MAG	ANG
12	11	0.000	−75.151	13	11	0.000	−25.433
9	11	25.346	−75.152				

*BUS 12   E/Z = 21.895 KA (7.89MVA) AT −72.21°, X/R = 3.12, 0.208 KV
          Z = 0.387323 + j 1.207119

```
                    MAX. LOW VOLTAGE FUSE DUTY = 21.90 SYM
                    MAX. LOW VOLTAGE POWER CIRCUIT BREAKER DUTY = 21.90
                    MAX. LV MCCB OR ICCB (RATED >20KA INT.) DUTY = 21.90
                    MAX. LV MCCB OR ICCB (RATED 10-20KA INT.)DUTY = 21.90
```

CONTRIBUTIONS IN KA

BUS TO	BUS	MAG	ANG	BUS TO	BUS	MAG	ANG
11	12	21.895	−72.210				

*BUS 13   E/Z = 18.129 KA (6.53MVA) AT −69.05°, X/R = 2.61, 0.208 KV
          Z = 0.547343 + j 1.429949

```
                    MAX. LOW VOLTAGE FUSE DUTY = 18.13 SYM
                    MAX. LOW VOLTAGE POWER CIRCUIT BREAKER DUTY = 18.13
                    MAX. LV MCCB OR ICCB (RATED >20KA INT.) DUTY = 18.13
                    MAX. LV MCCB OR ICCB (RATED 10-20KA INT.)DUTY = 18.13
```

CONTRIBUTIONS IN KA

BUS TO	BUS	MAG	ANG	BUS TO	BUS	MAG	ANG
11	13	18.129	−69.055				

```
CASE: 1-INT PAGE - 1 -
GE INDUSTRIAL POWER SYSTEMS ENGINEERING - SCHENECTADY, NY
THREE PHASE SHORT CIRCUIT PROGRAM - VERSION 1.44
INTERRUPTING CALC. FOR BKR DUTIES PER ANSI C37.010-1979,C37.5-1979
TOT. CURRENT & FLOWS FROM COMPLEX NETWORK, X/R FROM SEPARATE R & X
(c) 1973-1988, General Electric Company
        09-17-1993    10 MVA BASE    60 HERTZ


SAMPLE CALCULATION
INDUSTRIAL PLANT


CASE: 1-INT
SAMPLE SHORT-CIRCUIT CALCULATION FOR
RADIAL SYSTEM WITH 115 kV SOURCE
INPUT DATA
```

BUS	TO	BUS	R P.U.	X P.U.	CODE
0		1	0.00028	0.00198	1
1		2	0.00372	0.06656	0
7		8	0.06391	0.37797	0
6		9	0.22821	0.97361	0
3		4	0.00051	0.00078	0
3		5	0.00135	0.00208	0
3		6	0.00034	0.00052	0
6		7	0.00135	0.00208	0
11		12	0.10668	0.14856	0
11		13	0.26670	0.37139	0

2	3	0.00013	0.00009	0
8	10	0.00460	0.00283	0
9	11	0.04903	0.03014	0
0	4	0.01835	0.55296	3
0	5	0.05991	1.71008	4
0	10	1.20882	11.12583	5

*BUS 1   E/Z = 25.212 KA (5021.88MVA) AT −81.98°, X/R = 7.15, 115 KV
          Z = 0.000278 + j 0.001972

CIRCUIT BREAKER TYPE	8TOT,SYM	5SYM	5TOT	3SYM
MAX DUTY	25.21	25.21	25.34	25.21
MULT. FACTOR	1.000	1.000	1.005	1.000

CONTRIBUTIONS IN KA

BUS TO	BUS	MAG	ANG	BUS TO	BUS	MAG	ANG
REMOTE	1	25.106	−81.950	2	1	0.107	−87.710

SOURCE BUS	TYPE SOURCE	CONTRIBUTIONS AT FAULT BUS LOCAL	REMOTE	TOTAL	P.U. GEN VOLTS
1	REMOTE	0.00	25.11	25.11	0.000

REMOTE/TOTAL = 1.000 SUM 0.00   25.11   25.11

CASE: 1-INT PAGE - 2 -

*BUS 2   E/Z =23.648 KA (170.39MVA) AT −86.83°, X/R =18.72, 4.160 KV
          Z = 0.003241 + j 0.058599

CIRCUIT BREAKER TYPE	8TOT,SYM	5SYM	5TOT	3SYM
MAX DUTY	25.21	24.20	26.62	24.32
MULT. FACTOR	1.066	1.023	1.126	1.028

CONTRIBUTIONS IN KA

BUS TO	BUS	MAG	ANG	BUS TO	BUS	MAG	ANG
1	2	20.215	−86.660	3	2	3.434	−87.859

SOURCE BUS	TYPE SOURCE	CONTRIBUTIONS AT FAULT BUS LOCAL	REMOTE	TOTAL	P.U. GEN VOLTS
1	REMOTE	0.00	20.21	20.21	0.971

REMOTE/TOTAL = 1.000 SUM 0.00   20.21   20.21

*BUS 3   E/Z =23.620 KA (170.19MVA) AT −86.75°, X/R = 18.32, 4.160 KV
          Z = 0.003333 + j 0.058664

CIRCUIT BREAKER TYPE	8TOT,SYM	5SYM	5TOT	3SYM
MAX DUTY	25.09	24.06	26.46	24.16
MULT. FACTOR	1.062	1.019	1.120	1.023

CONTRIBUTIONS IN KA

BUS TO	BUS	MAG	ANG	BUS TO	BUS	MAG	ANG
4	3	2.505	−88.049	5	3	0.810	−87.952
6	3	0.120	−83.591	2	3	20.186	−86.557

SOURCE BUS	TYPE SOURCE	CONTRIBUTIONS AT FAULT BUS LOCAL	REMOTE	TOTAL	P.U. GEN VOLTS
1	REMOTE	0.00	20.19	20.19	0.971

REMOTE/TOTAL = 1.000 SUM 0.00   20.19   20.19

*BUS 4   E/Z =23.366 KA (168.36MVA) AT −86.40°, X/R= 16.80, 4.160 KV
          Z = 0.003730 + j 0.059280

```
CIRCUIT BREAKER TYPE    8TOT,SYM    5SYM    5TOT    3SYM
MAX DUTY                  24.51     23.39   25.73   23.43
MULT. FACTOR              1.049     1.001   1.101   1.003

CONTRIBUTIONS IN KA
BUS TO    BUS    MAG      ANG     BUS TO   BUS     MAG       ANG
 3         4   20.858   -86.196   SYNMOT    4    2.508    -88.098

SOURCE       TYPE     CONTRIBUTIONS AT FAULT BUS    P.U. GEN
BUS          SOURCE      LOCAL   REMOTE   TOTAL       VOLTS
 1           REMOTE       0.00    19.94   19.94       0.972
REMOTE/TOTAL = 1.000 SUM  0.00    19.94   19.94
```

```
CASE: 1-INT PAGE - 3 -

*BUS 5    E/Z = 22.836 KA(164.54MVA) AT -85.67°, X/R = 13.82, 4.160 KV
          Z = 0.004584 + j 0.060601

CIRCUIT BREAKER TYPE    8TOT,SYM    5SYM    5TOT    3SYM
MAX DUTY                  22.84     22.84   24.28   22.84
MULT. FACTOR              1.000     1.000   1.063   1.000

CONTRIBUTIONS IN KA
BUS TO    BUS    MAG      ANG     BUS TO   BUS     MAG       ANG
 3         5   22.026   -85.589   INDMOT    5    0.811    -87.992

SOURCE       TYPE     CONTRIBUTIONS AT FAULT BUS    P.U. GEN
BUS          SOURCE      LOCAL   REMOTE   TOTAL       VOLTS
 1           REMOTE       0.00    19.49   19.49       0.972
REMOTE/TOTAL = 1.000 SUM  0.00    19.49   19.49

*BUS 6    E/Z = 23.407 KA(168.66MVA) AT -86.45°, X/R = 16.71, 4.160 KV
          Z = 0.003670 + j 0.059178

CIRCUIT BREAKER TYPE    8TOT,SYM    5SYM    5TOT    3SYM
MAX DUTY                  24.54     23.41   25.74   23.44
MULT. FACTOR              1.048     1.000   1.100   1.002

CONTRIBUTIONS IN KA
BUS TO    BUS    MAG      ANG     BUS TO   BUS     MAG       ANG
 9         6   0.000    -0.840     3        6    23.288   -86.466
 7         6   0.120   -83.638

SOURCE       TYPE     CONTRIBUTIONS AT FAULT BUS    P.U. GEN
BUS          SOURCE      LOCAL   REMOTE   TOTAL       VOLTS
 1           REMOTE       0.00    20.00   20.00       0.971
REMOTE/TOTAL = 1.000 SUM  0.00    20.00   20.00

*BUS 7    E/Z = 22.589 KA(162.76MVA) AT -85.33°ᵛX/R= 12.54, 4.160 KV
          Z = 0.005006 + j 0.061236

CIRCUIT BREAKER TYPE    8TOT,SYM    5SYM    5TOT    3SYM
MAX DUTY                  22.59     22.59   23.68   22.59
MULT. FACTOR              1.000     1.000   1.048   1.000

CONTRIBUTIONS IN KA
BUS TO    BUS    MAG      ANG     BUS TO   BUS     MAG       ANG
 8         7   0.120   -83.660     6        7    22.469   -85.335

SOURCE       TYPE     CONTRIBUTIONS AT FAULT BUS    P.U. GEN
BUS          SOURCE      LOCAL   REMOTE   TOTAL       VOLTS
 1           REMOTE       0.00    19.30   19.30       0.972
REMOTE/TOTAL = 1.000 SUM  0.00    19.30   19.30
```

7.4   GROUND-FAULT CALCULATIONS

Grounded systems separated from each other by delta-delta and delta-wye transformer banks are considered separate systems and the ground-fault calculations recognize this. In the design of systems, the best systems make ground-fault current sources the same as phase-fault sources. (Positive and zero sequence sources should be the same.) This simplifies the relaying and makes coordination of overcurrent relaying easier. (Sometimes ground-fault coordination is impossible if this is not done.) Locate zigzag, Scott tee, wye-delta grounding banks as close to the supply transformer as possible.

Ground-Fault Current Sources

1. Grounded supply or service transformer
2. Four-wire utility network
3. Grounded generators
4. System grounding transformers

7.4.1   Low-Resistance-Grounded Systems

Most single-source medium-voltage industrial power systems are resistance-grounded using a grounding resistor in the neutral of the supply transformer. This resistor is usually rated to deliver 400 A to a ground fault. Since iron burning in motors starts at about 2000 A, the maximum ground-fault current on multisource resistance-grounded systems should be limited to this value. The use of ground sensor protection on industrial cable-fed circuits permits using 400-A grounding, allowing sensitive backup ground-fault relaying.

Most ground-fault calculations for resistance-grounded systems do not use a computer program for the calculation. If it is really essential to produce a computer output, an easy way to obtain a reasonably accurate ouput is to insert the value of the resistance $(3R_0)$ in series with each ground current source, using a three-phase short-circuit program. The approximate resistance of the ground return path (if known) could be added to this resistance. The result will be approximate but close to the same result obtained by assuming the maximum value the grounding resistor will produce.

Figure 7.9 shows the flow of ground-fault currents in a single-source three-wire resistance-grounded system. The load impedance factor k usually is so low that the component of ground-fault current flowing through the connected load is almost negligible. The one impedance usually not considered in most ground-fault calculations is the

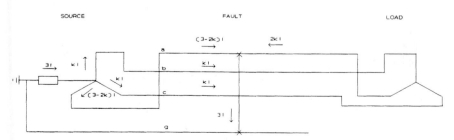

FIGURE 7.9   Ground-fault current flow in a single-source, three-wire resistance-grounded system.

impedance of the ground-fault return path. If this impedance is quite high compared to the resistance of the grounding resistor, it could decrease the magnitude of available ground-fault current significantly. If the system is installed in accordance with the NEC,[10] the impedance of the ground-fault return path is quite small, decreasing the available ground-fault current insignificantly.

Multigrounded-source systems require special consideration, as discussed in Chap. 8. Figures 8.38 and 8.40 show (on one-line diagrams) the approximate ground-current flow for different ground-fault locations on a multigrounded system.

7.4.2 Solidly Grounded Systems

The available ground-fault current for a single-source solidly grounded transformer-fed system is approximately equal to the available three-phase fault current at the transformer secondary terminals. For systems fed from a solidly grounded generator, the available ground-fault current at the generator terminals is greater than the three-phase fault current because zero sequence reactance of generators is significantly lower than the positive or negative sequence reactances. For this reason, generators are grounded through an impedance, usually a resistor. Long low-voltage circuits with ground return conductors smaller than the phase conductors significantly reduce the available ground-fault current the further away from the source the fault is located. Such circuits deserve special consideration, including the ground-fault return impedance in the short-circuit calculation. This calculation may be made manually or with a computer program with input data arrangements. Usually, this calculation is not performed because most industrial circuits are short enough not to decrease the available ground-fault current below the values at which most protective devices are set.

7.4.3 High-Resistance-Grounded Systems

High-resistance-grounded systems limit the ground-fault current to somewhat more than twice the system charging current as determined by the system capacitance. Table 7.17 gives data from which the system capacitance can be calculated. Three major applications for high-resistance grounding are:

1. Unit-connected generators
2. Single adjustable-speed drives supplied by a single isolating transformer
3. Important circuits, the loss of which would create havoc

The usual method of ground detection in the first two cases is a voltage relay connected across the grounding resistor. For medium-voltage systems, the grounding resistor usually is connected to the neutral of the generator on the secondary of a distribution transformer. For ungrounded three-wire systems, the resistor is connected to the open delta three-phase auxiliary transformer bank secondary, the primary of which is connected in grounded wye. The ground connection of the primary wye sometimes requires a resistor to prevent ferro-resonance. Figure 7.10 shows these three methods of high resistance grounding.

TABLE 7.17 Estimated System Capacitance Data

Voltage	System element	Charging current $3I_{c0}$ (A)
13,800	Surge capacitors	2.25 A/ set
	3/c shielded cable in conduit	
	1000 MCM	1.15 A/ 1000 ft
	750 MCM	0.93 A/ 1000 ft
	350 MCM	0.71 A/ 1000 ft
	4/0	0.65 A/ 1000 ft
	2/0	0.55 A/ 1000 ft
	Transformers	Negligible
	Motors	0.15 A/ 1000 hp
4,160	Surge capacitors	1.360 A/ set
	3/c shielded cable in conduit	0.23 A/ 1000 ft
	3/c nonshielded cable in conduit	0.10 A/ 1000 ft
	Motors	0.05 A/ 1000 hp
2,400	Surge capacitors	0.783 A/ set
	3/c nonshielded cable in conduit	0.05 A/ 1000 ft
	Motors	0.03 A/ 1000 hp
600	Surge capacitors	0.424 A/ set
	Cable in conduit	
	350–500 MCM	0.10 A/ 1000 ft
	2/0–3/0	0.05 A/ 1000 ft
	Transformers	Negligible
	Motors	0.01 A/ 1000 hp

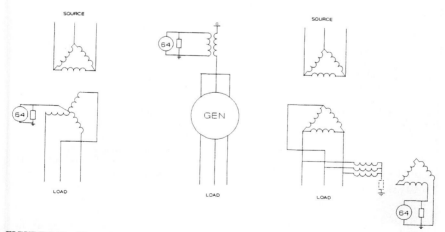

FIGURE 7.10 Three methods of high-resistance-grounding.

7.5 DC SYSTEMS

Industrial power systems include at least five types of dc systems:

- Dc battery systems
- UPS systems
- Rotating machinery buses
- Large rectifier systems
- Individual dc drives

This section discusses only control battery systems.

Dc arcs are more difficult to interrupt than many ac arcs because no current zero is present. The interrupter has the task of breaking a rapidly increasing current. The rate of current increase depends on the time constants of the circuit: the resistance, inductance, and capacitance of the circuit. These values are all difficult to estimate for any circuit; hence, uncertainties plague any attempt to calculate both the magnitude and the rate of rise of current during a short circuit.

Dc interrupters are special. Some ac fuses and circuit breakers carry dc ratings that significantly differ from the ac ratings of the same device. Special dc fuses and circuit breakers exist. Consult the manufacturer for special ratings that are available.

Figure 7.11 shows the calculation method for determining the theoretical rate of rise of battery short-circuit current for a fault at the battery terminals. These data are from R. G. Nailen, "Battery Protection—Where Do We Stand?"[11] Nailen states that battery

---

To calculate the rate of rise of the dc current, use the following:

$$i = \frac{V}{R} - A\epsilon^{-t/T}$$

V = battery voltage
R = circuit resistance
T = circuit time constant = L/R
A = max. current after $t = 0$
L = circuit inductance
i = instantaneous peak current
ϵ = 2.71828

t (s)	i
$0.01726(1 - \epsilon^{-1}) = 726 \times 0.6321 = 459$	
$0.02726(1 - \epsilon^{-2}) = 726 \times 0.8647 = 627$	
$0.03726(1 - \epsilon^{-3}) = 726 \times 0.9502 = 690$	
$0.04726(1 - \epsilon^{-4}) = 726 \times 0.9817 = 718$	

For dc faults at the terminals of a battery capable of delivering 726-A short-circuit current:

$$\frac{V}{R} = 726 \text{ A}$$

$$V = 120 \text{ V}$$

$$R = \frac{120}{726}$$

If $T = 10 \times 10^{-3}$ s = L/R, then:

$$L = 0.16529 \times 10 \times 10^{-3}$$

If fuses respond to rms currents, then the currents shown on the time-current curves for the rate of rise of battery short-circuit current should be calculated by:

$$I_{rms} = \int_{t_1}^{t_2} i^2 \, dt$$

In this equation, t_1 is the beginning of any time-constant period and t_2 is the end.

FIGURE 7.11 Calculation method for rate of rise of battery short-circuit current. (*From Robert L. Smith, Jr., "DC Control Battery System Protection & Coordination", IEEE Conference Record 93CH3255-7, p. 185, ©1993 IEEE.*)

time constants range from 10 to 80 ms. Previously published data from the *Industrial Power Systems Data Book*[12] indicates that such time constants might be as low as 0.32 ms. Figure 7.12 shows the rate of rise graphically, but such calculations ignore the drop in battery voltage during a short circuit. Figure 7.13 shows a more practical concept of what happens to the current and voltage during a short circuit.

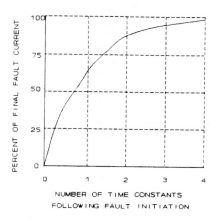

FIGURE 7.12 Theoretical rate of rise of battery current during short-circuit conditions per equations in Fig. 7.11. (*From Robert L. Smith, Jr., "DC Control Battery System Protection & Coordination", IEEE Conference Record 93CH3255-7, p. 185, ©1993 IEEE.*)

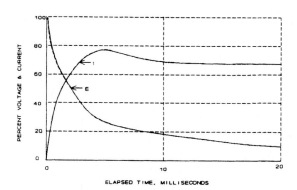

FIGURE 7.13 Practical rate of rise of battery current during short-circuit conditions. (*From Robert L. Smith, Jr., "DC Control Battery System Protection & Coordination", IEEE Conference Record 93CH3255-7, p. 185, ©1993 IEEE.*)

7.6 REFERENCES

1. ANSI/IEEE Standard 100-1988, *Standard Dictionary of Electrical and Electronics Terms,* Institute of Electrical and Electronics Engineers.

2. ANSI/IEEE Standard C37.100-1981 (reaffirmed 1989), *IEEE Standard Definitions for Power Switchgear,* Institute of Electrical and Electronics Engineers.

3. ANSI/IEEE Standard C37.13-1990, *IEEE Standard for Low-Voltage AC Power Circuit Breakers Used in Enclosures,* Institute of Electrical and Electronics Engineers.

4. ANSI/IEEE Standard C37.16-1988, *Standard for Low-Voltage Power Circuit Breakers and AC Power Circuit Protectors—Preferred Ratings, Related Requirements, and Application Recommendations*, Institute of Electrical and Electronics Engineers.

5. IEEE Standard C37.010-1979 (reaffirmed 1988), *Standard Application Guide for AC High-Voltage Circuit Breakers Rated on a Symmetrical Current Basis,* Institute of Electrical and Electronics Engineers.

6. ANSI/IEEE C37.06-1987, *Standard for AC High-Voltage Circuit Breakers Rated on a Symmetrical Current Basis—Preferred Ratings and Related Capabilities,* Institute of Electrical and Electronics Engineers.

7. ANSI/IEEE Standard C37.46-1981 (reaffirmed 1987), *Standard Specifications for Power Fuses and Fuse Disconnecting Switches,* Institute of Electrical and Electronics Engineers.

8. DAT program, General Electric Company, Industrial Power Systems Engineering Operation, Schenectady, N.Y.

9. SC program, General Electric Company, Industrial Power Systems Engineering Operation, Schenectady, N.Y.

10. ANSI/NFPA 70 National Electrical Code, National Fire Protection Association, Quincy, MA, 1993.

11. R. G. Nailen, "Battery Protection—Where Do We Stand?" *Trans. IEEE, Ind. Applications,* vol. 27, no. 4, July/Aug. 1991, pp. 658–667.

12. *Industrial Power Systems Data Book,* Section .173, General Electric Company, 1956, pp. 1–2.

13. This material is reproduced with permission from American National Standard ANSI C37.16-1988 copyright 1988 by the American National Standards Institute. Copies of this standard may be purchased from American National Standards Institute, 11 West 42nd Street, New York, NY 10036.

14. This material is reproduced with permission from American National Standard ANSI C37.06-1987 copyright 1987 by the American National Standards Institute. Copies of this standard may be purchased from American National Standards Institute, 11 West 42nd Street, New York, NY 10036.

APPENDIX A TO CHAPTER 7
IMPEDANCE TABLES

NOTES FOR TABLES AND FIGURES

TRANSFORMERS: Tables 7A.1 through 7A.8.
Impedances tabulated in these tables are "standard" impedances. The actual impedance of a "standard" impedance transformer may be $\pm 7\frac{1}{2}\%$ of the listed value because of manufacturing tolerances allowed and still meet the specified impedance value. For existing transformers, the exact impedance is stamped on the transformer nameplate. For new transformers, the exact impedance is listed on transformer test reports.

TABLE 7A.1   Typical Per-Unit R and X Values for Indoor, Open Dry-Type 150°C Rise Three-Phase Transformers

(Delta primaries, wye or delta secondaries)

kVA	HV (kV)	LV (V)	%Z	X/R	R	X
15			3.00	0.5	0.0268	0.0134
30			5.00	1.0	0.0354	0.0354
45			5.00	1.0	0.0354	0.0354
75			5.50	2.0	0.0246	0.0492
112.5			4.50	1.5	0.0250	0.0374
150			4.50	2.0	0.0201	0.0402
225	2.5–15	208Y–600	5.00	2.5	0.0186	0.0464
300			5.00	2.8	0.0168	0.0471
500			5.00	4.0	0.0121	0.0485
750			5.75	2.0	0.0257	0.0514
1000			5.75	2.5	0.0214	0.0534
1000	2.5–15	480Y	8.00	3.8	0.0214	0.0773
1500			5.75	3.3	0.0166	0.0550
2000	2.5–15	208Y–600	5.75	4.0	0.0139	0.0556
2500			5.75	4.3	0.0130	0.0560

(From IEEE Std 241-1990, "IEEE Recommended Practice for Electric Power Systems in Commercial Buildings".)

TABLE 7A.2   Typical Per-Unit R and X Values for Indoor, Open Dry-Type 150°C Rise Single-Phase Transformers

(120/240 wye or delta secondaries)

kVA	HV (kV)	LV (V)	%Z	X/R	R	X
25	5–15	120/240	4	2	0.0178	0.0358
500	5–15	120/240	6	4	0.0146	0.0582

(From IEEE Std 241-1990, "IEEE Recommended Practice for Electric Power Systems in Commercial Buildings".)

TABLE 7A.3 Typical Range of Per-Unit R and X Values for Indoor, Open Dry-Type 150°C Rise Three-Phase Transformers

(480-V delta primary, 208-V wye secondary)

kVA	HV (kV)	LV (V)	%Z	X/R	R	X
15	480	208Y120	4.5	0.41	0.0416	0.0171
500	480	208Y120	5.9	2.09	0.0255	0.0532

(From IEEE Std 241-1990, "IEEE Recommended Practice for Electric Power Systems in Commercial Buildings".)

TABLE 7A.4 Typical Range of Per-Unit R and X Values for Indoor, Open Dry-Type 150°C Rise Single-Phase Transformers

(240 × 480-V, 480-V, 600-V primaries, 120/240-V secondaries)

kVA	HV (kV)	LV (V)	%Z	X/R	R	X
5	240 × 480	120/240	3	0.6	0.0257	0.0154
167	600	120/240	6	2.0	0.0268	0.0537

(From IEEE Std 241-1990, "IEEE Recommended Practice for Electric Power Systems in Commercial Buildings".)

TABLE 7A.5 Typical Per-Unit R and X Values for Liquid-Filled 55/65 or 65°C Rise Three-Phase Power Transformers

(Delta or wye primaries, wye or delta secondaries)

kVA	HV (kV)	HV BIL (kV)	LV (kV)	LV BIL (kV)	%Z	X/R	R	X
			Secondary substation transformers					
112.5					2.0	1.9	0.0095	0.0176
150.0					2.0	2.2	0.0083	0.0182
225					2.0	2.6	0.0071	0.0187
300					4.5	2.9	0.0145	0.0425
500					4.5	4.3	0.0103	0.0438
750					5.75	4.9	0.0116	0.0563
1000	2.4–13.8	≤36	.120–.600	10	5.75	5.3	0.0165	0.0565
1000		≥110			8.0	5.3	0.0148	0.0786
1500					5.75	5.9	0.0096	0.0567
2000					5.75	6.3	0.0090	0.0568
2500					5.75	6.7	0.0085	0.0569
3000					5.75	7.0	0.0082	0.0569
3750					5.75	10.8	0.0053	0.0573

TABLE 7A.6 Typical *X/R* Values for Liquid-Filled 55/65 or 65°C Rise Three-Phase Power Transformers

(Delta or wye primaries, wye or delta secondaries)

Primary substation transformer X/R ratios			
kVA	X/R	kVA	X/R
750	4.9	12000	17.9
1000	5.3	15000	19.6
1500	5.9	20000	21.9
2000	6.3	25000	23.8
2500	6.7	30000	25.4
3000	7.0	50000	30.2
3750	10.8	75000	34.2
5000	12.0	100000	37.2
7500	14.1	200000	44.6
10000	15.9		

TABLE 7A.7 Typical Impedance Values for Liquid-Filled 55/65 or 65°C Rise Three-Phase Power Transformers

(Delta or wye primaries, wye or delta secondaries)

HV (kV)	HV BIL (kV)	LV (kV)	LV BIL (kV)	%Z	R_{pu}	X_{pu}
		Primary substation transformer impedances [add ½% for LTC]				
13.8	110	2.4	60	5.5	*	†
		4.16	75	5.5	*	†
		7.2	95	5.5	*	†
22.9	150	2.4	60	5.5	*	†
		4.16	75	5.5	*	†
		7.2	95	5.5	*	†
		13.8	110	5.5	*	†
34.5	200	2.4	60	6.0	*	†
		4.16	75	6.0	*	†
		7.2	95	6.0	*	†
		13.8	110	6.0	*	†
		22.9	150	6.5	*	†
43.8	250	2.4	60	6.5	*	†
		4.16	75	6.5	*	†
		7.2	95	6.5	*	†
		13.8	110	6.5	*	†
		22.9	150	6.5	*	†
		34.5	200	7.0	*	†
67.0	350	2.4	60	7.0	*	†
		4.16	75	7.0	*	†
		7.2	95	7.0	*	†
		13.8	110	7.0	*	†
		22.9	150	7.0	*	†
		34.5	200	7.0	*	†
		43.8	250	7.5	*	†

(Continued)

TABLE 7A.7 Typical Impedance Values for Liquid-Filled 55/65 or 65°C Rise Three-Phase Power Transformers (*Continued*)

(Delta or wye primaries, wye or delta secondaries)

Primary substation transformer impedances [add ½% for LTC]

HV (kV)	HV BIL (kV)	LV (kV)	LV BIL (kV)	%Z†	R_{pu}	X_{pu}
115	550	13.8	110	8.0	*	†
		22.9	150	8.0	*	†
		34.5	200	8.0	*	†
		43.8	250	9.0	*	†
		67.0	350	9.0	*	†
138	650	13.8	110	8.5	*	†
		22.9	150	8.5	*	†
		34.5	200	8.5	*	†
		43.8	250	9.5	*	†
		67.0	350	9.5	*	†
		115	550	10.5	*	†
161	750	13.8	110	9.0	*	†
		22.9	150	9.0	*	†
		34.5	200	9.0	*	†
		43.8	250	9.0	*	†
		67.0	350	10.0	*	†
		115	550	11.0	*	†
		138	650	11.0	*	†

*$R_{pu} = (\%Z/100) \times \cos(\tan^{-1}[X/R])$

†$X_{pu} = (\%Z/100) \times \sin(\tan^{-1}[X/R])$

TABLE 7A.8 Approximate Impedance Data—Insulated Conductors in Conduit, 60 Hz (Ω/1000 ft per conductor)

| Size AWG or MCM | Resistance (25°C) | | | | Reactance (600 V—THHN) | | | |
| | Copper | | Aluminum | | Several 1/C | | 1 Multiconductor | |
	Metallic conduit	Nonmetallic conduit	Metallic conduit	Nonmetallic conduit	Metallic conduit	Nonmetallic conduit	Metallic conduit	Nonmetallic conduit
14	2.5700	2.5700	4.2200	4.2200	0.0493	0.0394	0.0351	0.0305
12	1.6200	1.6200	2.6600	2.6600	0.0468	0.0374	0.0333	0.0290
10	1.0180	1.0180	1.6700	1.6700	0.0463	0.0371	0.0337	0.0293
8	0.6404	0.6404	1.0500	1.0500	0.0475	0.0380	0.0351	0.0305
6	0.4100	0.4100	0.6740	0.6740	0.0437	0.0349	0.0324	0.0282
4	0.2590	0.2590	0.4240	0.4240	0.0441	0.0353	0.0328	0.0285
2	0.1640	0.1620	0.2660	0.2660	0.0420	0.0336	0.0313	0.0273
1	0.1303	0.1290	0.2110	0.2110	0.0427	0.0342	0.0319	0.0277
1/0	.1040	0.1020	0.1680	0.1680	0.0417	0.0334	0.0312	0.0272
2/0	0.0835	0.0812	0.1330	0.1330	0.0409	0.0327	0.0306	0.0266
3/0	0.0668	0.0643	0.1060	0.1050	0.0400	0.0320	0.0300	0.0261
4/0	0.0534	0.0511	0.0844	0.0838	0.0393	0.0314	0.0295	0.0257
250	0.0457	0.0433	0.0722	0.0709	0.0399	0.0319	0.0299	0.0261
300	0.0385	0.0362	0.0602	0.0592	0.0393	0.0314	0.0295	0.0257
350	0.0333	0.0311	0.0520	0.0507	0.0383	0.0311	0.0388	0.0311
400	0.0297	0.0273	0.0460	0.0444	0.0385	0.0308	0.0286	0.0252
500	0.0244	0.0220	0.0375	0.0356	0.0379	0.0303	0.0279	0.0250
600	0.0209	0.0185	0.0319	0.0298	0.0382	0.0305	0.0278	0.0249
750	0.0174	0.0150	0.0264	0.0301	0.0376	0.0301	0.0271	0.0247
1000	0.0140	0.0115	0.0211	0.0182	0.0370	0.0296	0.0260	0.0243

(Continued)

TABLE 7A.8 Approximate Impedance Data—Insulated Conductors in Conduit, 60 Hz (Continued)

(Ω/1000 ft per conductor)

Size AWG or MCM	Reactance (5 kV)				Reactance (15 kV)			
	Several 1/C		1 Multiconductor		Several 1/C		1 Multiconductor	
	Metallic conduit	Nonmetallic conduit	Metallic conduit	Nonmetallic conduit	Metallic conduit	Nonmetallic conduit	Metallic conduit	Nonmetallic conduit
8	0.0733	0.0586	0.0479	0.0417	0.0842	0.0674	0.0584	0.0508
6	0.0681	0.0545	0.0447	0.0389	0.0783	0.0626	0.0543	0.0472
4	0.0633	0.0507	0.0418	0.0364	0.0727	0.0582	0.0505	0.0439
2	0.0591	0.0472	0.0393	0.0364	0.0701	0.0561	0.0487	0.0424
1	0.0571	0.0457	0.0382	0.0332	0.0701	0.0561	0.0487	0.0424
1/0	0.0537	0.0430	0.0360	0.0313	0.0661	0.0529	0.0458	0.0399
2/0	0.0539	0.0431	0.0350	0.0305	0.0614	0.0491	0.0427	0.0372
3/0	0.0521	0.0417	0.0341	0.0297	0.0592	0.0474	0.0413	0.0359
4/0	0.0505	0.0404	0.0333	0.0290	0.0573	0.0458	0.0400	0.0348
250	0.0490	0.0392	0.0324	0.0282	0.0557	0.0446	0.0387	0.0339
300	0.0478	0.0383	0.0317	0.0277	0.0544	0.0436	0.0379	0.0332
350	0.0469	0.0375	0.0312	0.0274	0.0534	0.0427	0.0371	0.0326
400	0.0461	0.0369	0.0308	0.0270	0.0517	0.0414	0.0357	0.0317
500	0.0461	0.0369	0.0308	0.0270	0.0516	0.0413	0.0343	0.0309
600	0.0439	0.0351	0.0290	0.0261	0.0500	0.0400	0.0328	0.0301
750	0.0434	0.0347	0.0284	0.0260	0.0482	0.0385	0.0311	0.0291
1000	0.0421	0.0337	0.0272	0.0255				

(From IEEE Std 241-1990, "IEEE Recommended Practice for Electric Power Systems in Commercial Buildings".)

For Table 7A.7, read the *X/R* ratio from Table 7A.6 and calculate the *X* and *R* values using the formulas following Table 7A.7.

Data in these tables was compiled from manufacturer's data, ANSI/IEEE C57 product standards,[1] IEEE Red,[2] Gray,[3] and Buff[4] books.

CABLES: Table 7A.8.
Resistance values are given at 25°C. This provides the lowest impedance for initially energizing a circuit, the circumstance most likely to produce a fault.

Data for this table compiled from the *Industrial Power Systems Data Book,*[5] The IEEE Gray Book,[2] and manufacturer's data.

BUSWAY: Tables 7A.9 and 7A.10.
Data for these tables compiled from manufacturer's data and the IEEE Gray Book.[3] The 5- and 15-kV data is for the configuration noted. If any different configuration exists, check with the manufacturer.

OVERHEAD LINES: Table 7A.8 and Fig. 7A.1.
Data for Table 7A.8 compiled from tables in *Circuit Analysis of AC Power Systems.*[6] Figure 7A.1 is from *Industrial Power Systems Data Book.*[5]

For Fig. 7A.1, the equivalent spacing of unsymmetrical arrangements of conductors can be derived from the relation:

$$S = \sqrt[3]{(S_1)(S_2)(S_3)}$$

S = center-to-center conductor separation in ft

MOTORS AND GENERATORS: Table 7A.12, Figs. 7A.2, and 7A.3.
For large motors and generators, data is best obtained from motor and generator data sheets provided by the manufacturer. The data given in Table 7A.12 should be accurate enough for low-voltage induction motors with normal efficiency. High-efficiency motors have lower impedance than normal efficiency motors by about 2 percent. Small generators are approximately the same as the motor values given in Table 7A.12 but a check with the manufacturer is advised. *X/R* ratios can be estimated from Figs. 7A.2 and 7A.3.

Data for Table 7A.12 compiled from manufacturer's data. Figures 7A.2 and 7A.3 from ANSI/IEEE Standard C37.010-1979.[7]

TABLE 7A.9 600-V Busway Impedance Data

		Ω/100 ft line-to-neutral, 60 Hz		
Busway type	Ampere rating	Resistance (R)	Reactance (X)	Impedance (Z)
Feeder with	600	0.00331	0.00228	0.00402
aluminum	800	0.00210	0.00081	0.00226
bus bars	1000	0.00163	0.00079	0.00181
	1350	0.00143	0.00052	0.00153
	1600	0.00108	0.00051	0.00119
	2000	0.00081	0.00037	0.00089
	2500	0.00064	0.00030	0.00071
	3000	0.00054	0.00024	0.00059
	4000	0.00041	0.00018	0.00045
	5000	0.00032	0.00013	0.00035
Feeder with	800	0.00200	0.00228	0.00304
copper	1000	0.00132	0.00081	0.00156
bus bars	1350	0.00099	0.00079	0.00126
	1600	0.00088	0.00052	0.00102
	2000	0.00066	0.00051	0.00083
	2500	0.00059	0.00037	0.00062
	3000	0.00040	0.00030	0.00050
	4000	0.00034	0.00024	0.00042
	5000	0.00025	0.00018	0.00031
Plug-in with	800	0.00210	0.00114	0.00238
aluminum	1000	0.00163	0.00110	0.00197
bus bars	1350	0.00143	0.00069	0.00159
	1600	0.00108	0.00066	0.00127
	2000	0.00081	0.00044	0.00092
	2500	0.00064	0.00035	0.00073
	3000	0.00054	0.00028	0.00061
	4000	0.00041	0.00021	0.00046
	5000	0.00032	0.00016	0.00036
Plug-in with	800	0.00200	0.00460	0.00500
copper	1000	0.00132	0.00114	0.00174
bus bars	1350	0.00099	0.00110	0.00148
	1600	0.00088	0.00069	0.00112
	2000	0.00066	0.00066	0.00093
	2500	0.00050	0.00044	0.00067
	3000	0.00040	0.00035	0.00053
	4000	0.00034	0.00028	0.00044
	5000	0.00025	0.00021	0.00036
Current-limiting	1000	0.00220	0.0069	0.0072
with aluminum	1350	0.00200	0.0064	0.0067
bus bars	1600	0.00148	0.0064	0.0066
	2000	0.00112	0.0058	0.0059
	2500	0.00090	0.0054	0.0055
	3000	0.00077	0.0050	0.0051
	4000	0.00059	0.0042	0.0042
Current-limiting	1000	0.00177	0.0069	0.0071
with copper	1350	0.00134	0.0069	0.0070
bus bars	1600	0.00121	0.0064	0.0065
	2000	0.00090	0.0064	0.0065
	2500	0.00070	0.0058	0.0058
	3000	0.00058	0.0054	0.0054
	4000	0.00041	0.0046	0.0046

(From IEEE Std 241-1990, "IEEE Recommended Practice for Electric Power Systems in Commercial Buildings.")

TABLE 7A.10   5- and 15-kV Nonsegregated-Phase Metal-clad Busway Impedance Data

Ω/100 ft line-to-neutral, 60 Hz				
Current rating (A)	Conductor material	Conductor size (in)	Resistance at 50°C (R)	Reactance (X)
1200	Copper	$1\frac{1}{4} \times 4$	0.00102	0.00049
2000		$1\frac{3}{8} \times 6$	0.00049	0.0042
3000		$2\frac{3}{8} \times 6$	0.00025	0.0045
4000		6-in square tube	0.00014	0.0029
1200	Aluminum	$1\frac{3}{8} \times 4$	0.00118	0.0045
2000		$1\frac{5}{8} \times 6$	0.00059	0.0037
3000		$2\frac{5}{8} \times 6$	0.00025	0.0041
4000		6-in round tube	0.00019	0.0036

(*From IEEE Std 241-1990, "IEEE Recommended Practice for Electric Power Systems in Commercial Buildings".*)

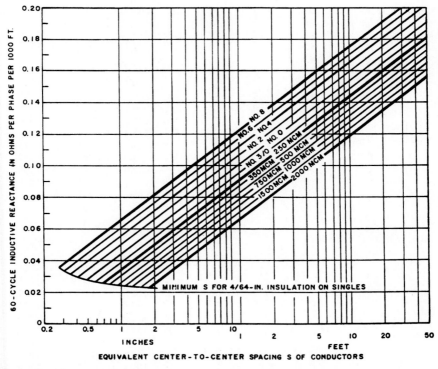

FIGURE 7A.1   Calculated positive- or negative-sequence reactances. (*Reprinted from "Industrial Power Systems Handbook", Donald L. Beeman, Ed., McGraw-Hill, 1955.*)

TABLE 7A.11 Overhead Line Conductor Resistances for Solid Copper, Stranded Copper, and ACSR

60-Hz resistance in Ω/mi at 25°C							
Solid copper		Stranded copper			ACSR		
AWG	Resistance	AWG or MCM	Strands	Resistance	AWG or MCM	Strands Al/steel	Resistance
4/0	0.271	1000	61	0.064	954	54/7	0.0982
3/0	0.342	900	61	0.070	900	54/7	0.104
2/0	0.431	800	61	0.077	874	54/7	0.108
1/0	0.543	750	61	0.082	795	54/7	0.119
#1	0.686	700	61	0.087	795	26/7	0.117
#2	0.864	650	61	0.093	795	30/19	0.117
#3	1.090	600	37	0.101	715.5	54/7	0.132
#4	1.373	550	37	0.109	715.5	26/7	0.131
#5	1.733	500	37	0.120	715.5	30/19	0.131
#6	2.185	450	37	0.133	666.6	54/7	0.141
#7	2.75	400	19	0.149	636	54/7	0.148
#8	3.47	350	19	0.169	636	26/7	0.147
#9	4.38	300	19	0.197	636	30/19	0.147
#10	5.52	250	19	0.235	605	54/7	0.155
#11	6.96	4/0	19	0.276	605	26/7	0.154
#12	8.78	4/0	7		556.5	26/7	0.168
#13	11.08	3/0	7		556.5	30/7	0.168
#14	13.97	2/0	7		500	30/7	0.187
#15	17.62	1/0	7		477	26/7	0.196
#16	22.2	#1	7		477	30/7	0.196
		#2	7		397.5	26/7	0.235
		#3	7		397.5	30/7	0.235
		#4	7		336.4	26/7	0.278
		#5	7		336.4	30/7	0.278
		#6	7		300	26/7	0.311
		#1	3		300	30/7	0.311
		#2	3		226.8	26/7	0.350
		#3	3		226.8	6/7	0.351
		#4	3		4/0	6/1	0.445
		#5	3		3/0	6/1	0.560
		#6	3		2/0	6/1	0.706
					1/0	6/1	0.888
					#1	6/1	1.12
					#2	6/1	1.41
					#3	6/1	1.78
					#4	6/1	2.24
					#5	6/1	2.82
					#6	6/1	3.56
					203	8/7	0.461
					203.2	16/19	0.461
					211.3	12/7	0.443
					190.8	12/7	0.491
					176.9	12/7	0.529
					159	12/7	0.588
					134.6	12/7	0.695
					110.8	12/7	0.845
					101.8	12.7	0.918
					80	8/1	1.170

TABLE 7A.12 Approximate Motor Impedance Data to be Used in the Absence of Better Data

Rating	Speed	%X	X/R	kVA
Induction motors				
≤100 hp, >100 hp, ≤1000 hp, >1000 hp	≤1800 rpm	16.7	See Fig. 7A.2	kVA = hp, kVA = hp/0.95, kVA = hp/0.9
Synchronous motors				
0.8 pf, 1.0 pf	<1200 rpm	15.0	See Fig. 7A.3	kVA = hp, kVA = hp/0.8
0.8 pf, 1.0 pf	≥1200 rpm	20.0		kVA = hp, kVA = hp/0.8

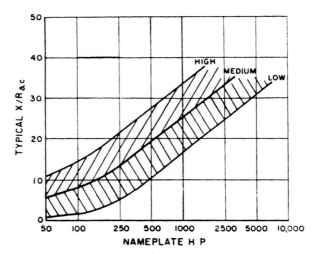

FIGURE 7A.2 *X/R* range for three-phase induction motors. (*From IEEE Std C37.010-1979, "IEEE Application Guide for AC High-Voltage Circuit Breakers Rated on a Symmetrical Current Basis"* [*Reaff 1988*].)

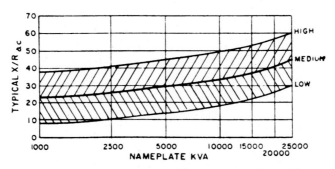

FIGURE 7A.3 *X/R* range for small solid rotor and salient pole generators and synchronous motors. (*From IEEE Std C37.010-1979, "IEEE Application Guide for AC High-Voltage Circuit Breakers Rated on a Symmetrical Current Basis"* [*Reaff 1988*].)

REFERENCES

1. *Distribution, Power, and Regulating Transformers Standards Collection, 1992 Edition (C57)*, Institute of Electronic and Electrical Engineers.

2. IEEE Standard 141-1986 (reaffirmed 1992), *IEEE Recommended Practice for Electric Power Distribution for Industrial Plants* (IEEE Red Book), Institute of Electronic and Electrical Engineers.

3. IEEE Standard 241-1990 *IEEE Recommended Practice for Electric Power Systems in Commercial Buildings* (IEEE Gray Book), Institute of Electronic and Electrical Engineers.

4. IEEE Standard 242-1986, *IEEE Recommended Practice for Protection and Coordination of Industrial and Commercial Power Systems* (IEEE Buff Book), Institute of Electronic and Electrical Engineers.

5. D. L. Beeman, *Industrial Power Systems Data Book,* General Electric Company, Schenectady, N.Y., 1956.

6. E. Clarke, *Circuit Analysis of AC Power Systems, Volume I,* John Wiley & Sons, New York, 1943.

7. ANSI/IEEE C37.010-1979, *IEEE Application Guide for AC High-Voltage Circuit Breakers Rated on a Symmetrical Current Basis* (reaffirmed 1988), Institute of Electronic and Electrical Engineers.

APPENDIX B TO CHAPTER 7
SIMULATED OSCILLOGRAMS OF MACHINE SHORT-CIRCUIT CONTRIBUTIONS

7B.0 NOTES ON FIGURES

The simulated oscillograms of the figures in this appendix were calculated using detailed machine data and a program called MNT/E. The machines were modeled using the d and q axes by the differential equations of fluxes and inertia. Park's equations for d, q, and o voltages were calculated using the rate of change of fluxes. The d, q, and o inverse transforms were used to obtain the phase voltages. The network R, L, and C elements with differential equations were used for the nonlinearities to obtain the phase currents. Again the d, q, and u transforms gave the current in the d, q, and o axes for the feedback into the machine model. The resulting oscillograms model the exact current contributions to a short circuit.

The ANSI Short Circuit Calculation Standard is conservative in its modeling of the machines. However, as can be seen in the induction motor contribution to the short circuit, it is not conservative for large multipole induction machines. The more poles (slower speed) an induction motor has, the more flux in the air gap and the longer that flux takes to decay.

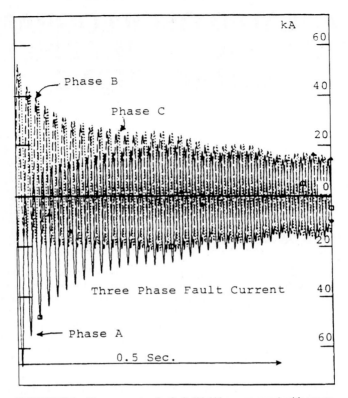

FIGURE 7B.1 Phase currents of a fault (11,111 amperes rms) with generation and motor contributions. Note the relationship of phase with maximum offset (A) and the other two phases (B and C). The oscillations in the current values are a result of the motor contributions adding and subtracting from the total current as the machines cone into and go out-of-step with the system.

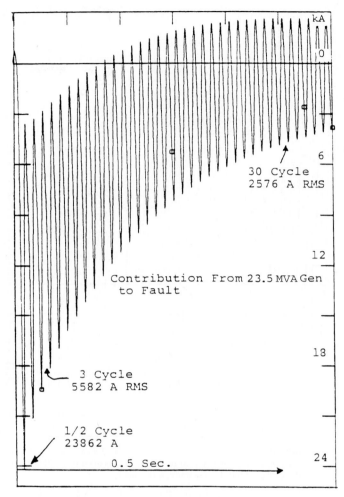

FIGURE 7B.2 Fault-current contribution from a Generator, 23.529 MVA, 0.85 PF, 13.8 kV, $X''_{dv} = 0.102$, $X'_{dv} = 0.154$. The high X/R ration of 73 accounts for the large dc offset.

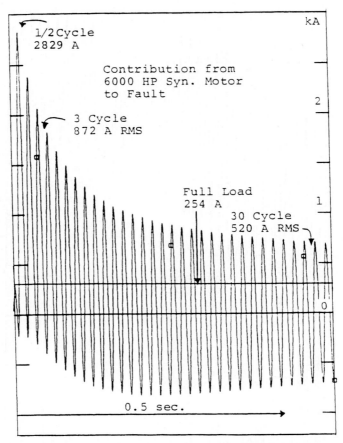

FIGURE 7B.3 Fault-current contribution from a 13.8 kV, 6000 HP, 6-pole, synchronous motor. Note that there are only 27 cycles of current within the 0.5 seconds because of the slowing down of the motor.

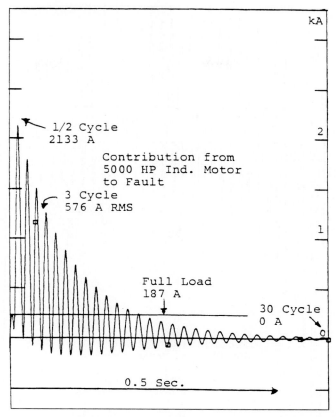

FIGURE 7B.4   Fault-current contribution from a 13.8 kV, 5000 HP, 4-pole induction motor. Note the lengthening of the cycle time as the motor slows down. The 12 to 13 cycle decay to full-load current is a result of the large flux in the air gap.

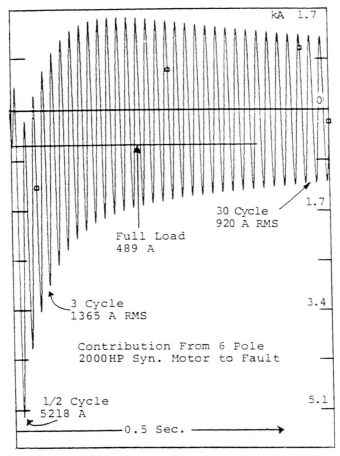

FIGURE 7B.5 Fault-current contribution from a 2.4 kV, 2000 HP, 6-pole synchronous motor. Note: there are on 27 cycles in 0.5 seconds. (Same as Figure 7B.3)

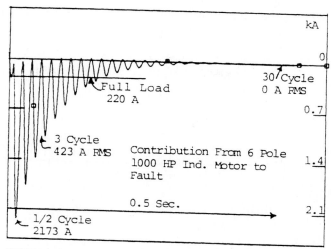

FIGURE 7B.6 Fault-current contribution from a 2.4 kV, 1000 HP, 6-pole induction motor. Note the slowdown of the motor by the lengthening of the cycle time. The nine cycles decay to full-load current is the result of the large flux in the air gap.

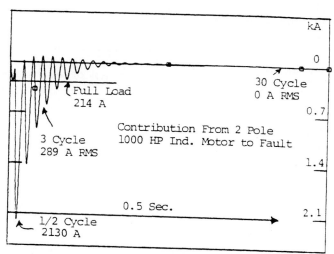

FIGURE 7B.7 Fault-current contribution from a 2.4 kV, 1000 HP, 2-pole induction motor. The five cycles decay to full-load current is faster on the 2-pole machine because of less flux in the air gap.

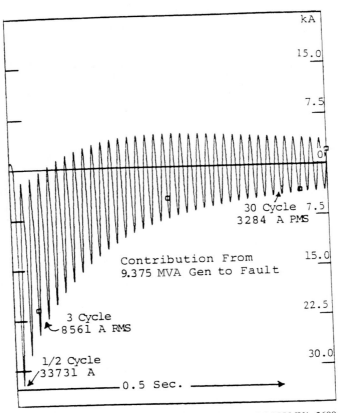

FIGURE 7B.8   Fault-current contribution from a 4.16 kV, 9.375 MVA, 3600 rpm generator with $X''_{dv} = 0.096$ and $X'_{dv} = 0.137$ and an X/R ratio of 40. The offset is less than for the generator of Figure 7B.2.

CHAPTER 8
PROTECTIVE DEVICE APPLICATION AND OVERCURRENT COORDINATION

8.0 INTRODUCTION

PROTECTION GOALS: Three major functions of electric power system protection are: (1) personnel injury prevention, (2) equipment damage control, and (3) system coordination and selectivity. This chapter concerns mainly equipment protection and system coordination of protective devices. Personnel protection, including guarding, equipment accessibility, and safety procedures, is covered by the National Electrical Code (NEC)[1] and the National Electrical Safety Code (NESC).[2]

NATIONAL ELECTRICAL CODE: The purpose of the NEC is "the practical safeguarding of persons and property from hazards arising from the use of electricity" [Article 90-1 (a)]. It is *not* "a design specification nor an instruction manual for untrained persons" [Article 90-1 (c)]. The United States incorporated the NEC in the legislation which created the Occupational Safety and Health Administration (OSHA) and many municipalities adopted the NEC in local codes. The NEC applies to public and private buildings, including mobile homes, recreational vehicles, floating buildings, industrial substations, conductors and equipment that connect to the supply of electricity, other outside conductors and equipment, and optical fiber cable installations. It does not apply to ships, watercraft other than floating buildings, railway rolling stock, underground mines and mobile surface mining machinery and trailing cables, or installations under the exclusive control of public utilities including communication utilities [Article 90-2].
The main thrust of the NEC is overload protection to prevent excessive heat from causing a fire. The NEC contains many provisions for accident prevention concerning guards, clearances, good housekeeping, etc. This chapter concerns only electric relays and trip devices used for equipment protection. It does not cover personnel protection in detail, particularly requirements for enclosures and guards.

NATIONAL ELECTRICAL SAFETY CODE: The purpose of the NESC is "the practical safeguarding of persons during the installation, operation and maintenance of electrical supply and communication lines and associated equipment" [Article 1.010].

8.1 PROTECTION OBJECTIVES

Prevent deterioration by disconnecting from the system circuits and equipments subjected to:

Overload

Overtemperature

Under- and overvoltage

Unbalanced current

Under- and overfrequency

Loss of field

Loss of synchronism

Antimotoring

Reverse power

Reverse current

Phase reversal and open phase

Limit damage by disconnecting from the system circuits and equipment subjected to:

Short circuits

Ground faults

8.2 PRIMARY VS. BACKUP PROTECTION

Primary protection is the first line of defense, the most important protective device for a circuit or equipment, usually instantaneous in operation. Two examples of this type of protection are high-speed differential relaying and instantaneous overcurrent relaying. *Backup protection* operates if the primary protection does not remove a fault from the system. It usually operates with a time delay. Most time overcurrent relaying and trip device operation is backup operation. Some backup protection duplicates the primary protection and may be instantaneous in operation.

Table 8.1 shows some examples of typical primary protection for medium-voltage systems. Table 8.2 shows some examples of typical backup protection for medium-voltage systems.

8.3 ZONES OF PROTECTION

The zone of protection of a relay or circuit-breaker trip device is that segment of the power system in which the occurrence of assigned abnormal conditions should cause the protective device to operate.

A simplified one-line diagram of a system may be sketched with various zones of protection outlined to assure that no portion of the system lacks primary protection of some sort. Making this sketch facilitates the overall evaluation of system protection.

TABLE 8.1 Typical Medium-Voltage Primary Protection Systems

Protected equipment	Protection (with device number)	Sensors	Sensitivity
Transformer	High-speed phase differential (#87T)	6 CTs	1/3 maximum transformer rating
	Ground differential (#87TG)	1 neutral CT, residual connection of 3 main breaker CTs plus aux. transformer	5% available ground-fault current
Motor or generator	High-speed phase differential (#87M or G)	6 CTs or 3 window-type CTs	5% machine full load or 15 primary amperes
Bus	High-speed differential (#87B)	3 CTs in each bus-connected circuit	0.5% of the available short-circuit current
Lines	High-speed pilot wire (#87L)	6 CTs plus pilot wire	0.5% of the available short-circuit current
Feeders	Instantaneous phase overcurrent (#50)	CT in each phase	CT rating dependent
	Instantaneous ground overcurrent (#50N)	Residual connection of 3-phase CTs	CT rating dependent
	Instantaneous ground sensor overcurrent (#50GS)	Zero sequence CT	15 primary amperes

8.4 PROTECTIVE DEVICES AND CHARACTERISTICS

8.4.1 Medium- and High-Voltage Systems

Protective relays can be either of electromagnetic or solid-state construction. Many multifunction relays are of solid-state construction and require external dc power supplies. Single-function overcurrent relays usually do not require external power supplies and derive the required power from the current in the protected circuit. Figure 8.1 shows a comparison of the time-current characteristics of the different curve shapes for overcurrent relays. Figure 8.31 shows the complete typical characteristic curves for a very inverse time overcurrent relay of one manufacturer. While the characteristic shape of the curves of all manufacturers are similar, some slight differences exist when comparing manufacturers. These characteristic curves show the operating time of the relays and do not include the circuit-breaker operating time.

TABLE 8.2   Examples of Backup Protection for Medium-Voltage Systems

Backup protection (with device number)	Backed up protection (with device number)	Sensors	Sensitivity
Phase time over-current (#51)	Downstream differential (#87M, T, B, or L) Local differential (#87 M, T, B, or L) Downstream phase overcurrent (#50/51, 50, or 51)	1 CT in each phase	CT dependent
Ground time over-current (#51N)	Downstream ground over-current (#50GS, 51GS, 50N, or 51N)	Residual connection of 3-phase CTs	
Partial differential phase time overcurrent (#51PD)	Local differential (#87M, T, B, or L) Downstream phase overcurrent (#50/51, 50, or 51)	3-Phase CTs in main, tie, and other source circuits	CT dependent
Partial differential ground time over-current (#51N/PD)	Local ground overcurrent (#50GS, 51GS, 50N, or 51N) Downstream ground over-current (#50GS, 51GS, 50N, or 51N) Local differential (#87M, T, B, or L)	Residual connection of 3-phase CTs in main, tie, and other source circuits	Must be selective with downstream overcurrent protection

Power fuses can be of the expulsion type or self-contained. Self-contained fuses can be of the current-limiting type. Figures 8.2 and 8.3 show minimum melting and total clearing times for one class of fuses of a particular manufacturer. Fuses are rated to carry their rating and operate at some current value above their rating. IEEE/ANSI Standard C37.41-1988[3] specifies the times and currents at which power fuses operate. In the time ranges for which time-current coordination curves are drawn, no operation can be expected in less than 1000 seconds for currents of almost twice the fuse rating for current-limiting power fuses.

8.4.2   Low-Voltage Systems

Low-voltage power circuit breakers (fused or unfused) and *insulated-case circuit breakers* are equipped with trip devices which operate at their settings. Figure 8.4 shows the time-current characteristics of one manufacturer's low-voltage power-circuit-breaker trip device.

Molded-case circuit breakers (fused or unfused) and *low-voltage fuses* (current-limiting or non-current-limiting) are rated to carry their rating and operate at some value of current above their rating. Figure 8.5 shows the characteristic of one manufacturer's molded-case circuit breaker. This curve shows the difficulties encountered when trying to coordinate molded-case circuit breakers of the same rating with low-voltage power circuit breakers.

OVERCURRENT RELAYS

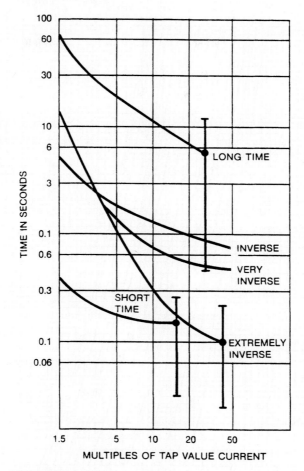

FIGURE 8.1   Comparison of typical curve shapes for overcurrent relays.

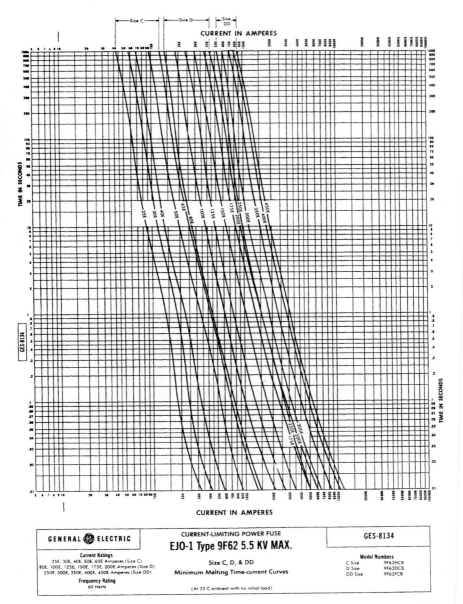

FIGURE 8.2 Typical minimum melting time-current curves for medium-voltage current-limiting power fuses.

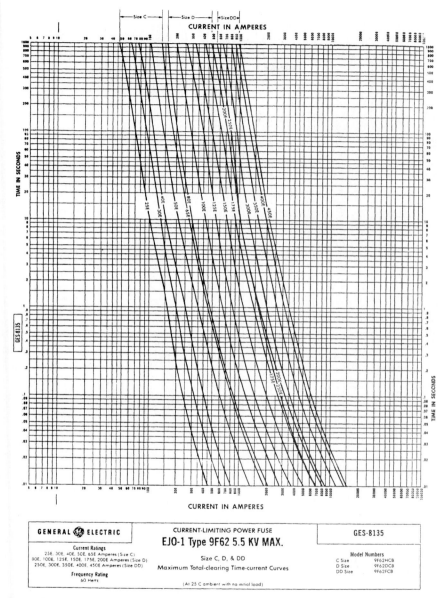

FIGURE 8.3 Typical maximum clearing time-current curves for medium-voltage current-limiting power fuses.

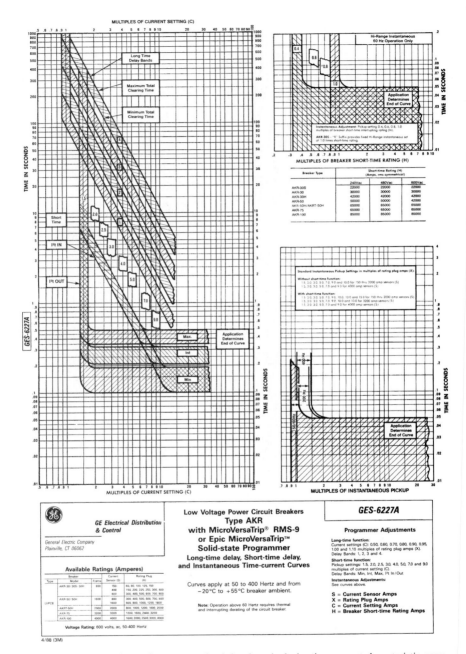

FIGURE 8.4 Typical low-voltage power-circuit-breaker trip device time-current characteristic curve.

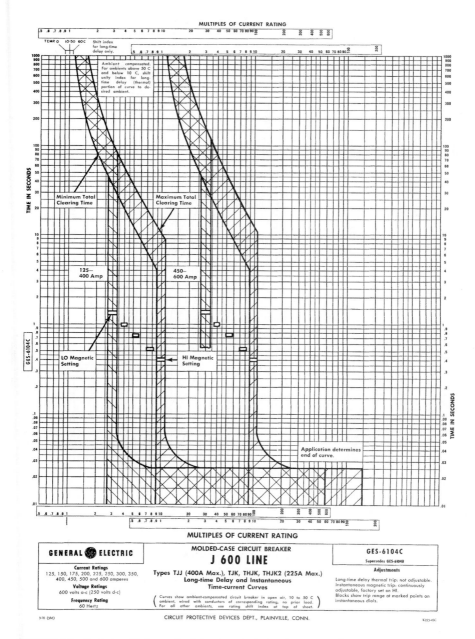

FIGURE 8.5   Typical molded-case circuit-breaker trip device time-current characteristic curve.

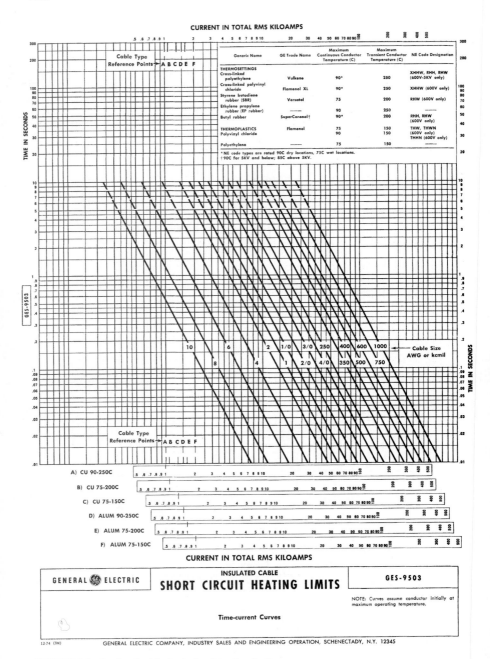

FIGURE 8.6 Insulated cable short-circuit heating limits.

Category II Transformers

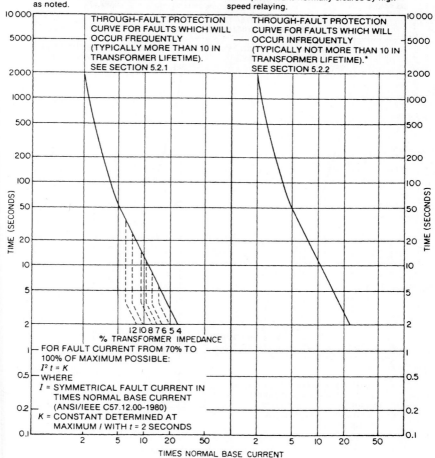

FIGURE 8.7 Transformer through-fault protection curve. (*From IEEE Std C57.109-1993, "IEEE Guide for Liquid-Immersed Transformer Through-Fault-Current Duration".*)

8.5 PROTECTED EQUIPMENT CHARACTERISTICS

Cable   Short-circuit heating limit. (See Fig. 8.6 or obtain curve from cable manufacturer.)

Transformers Transformer through-fault-current duration. (See Fig. 8.7 derived from ANSI/IEEE Standard C57.109-1985[4] and Table 8.3 [how to construct these curves].)

Category III Transformers

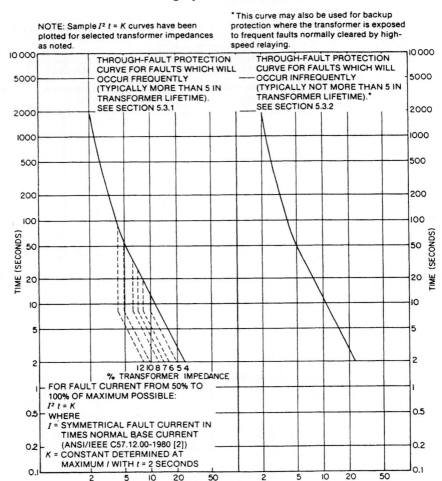

NOTE: Sample $I^2 t = K$ curves have been plotted for selected transformer impedances as noted.

*This curve may also be used for backup protection where the transformer is exposed to frequent faults normally cleared by high-speed relaying.

THROUGH-FAULT PROTECTION CURVE FOR FAULTS WHICH WILL OCCUR FREQUENTLY (TYPICALLY MORE THAN 5 IN TRANSFORMER LIFETIME). SEE SECTION 5.3.1

THROUGH-FAULT PROTECTION CURVE FOR FAULTS WHICH WILL OCCUR INFREQUENTLY (TYPICALLY NOT MORE THAN 5 IN TRANSFORMER LIFETIME).* SEE SECTION 5.3.2

% TRANSFORMER IMPEDANCE

FOR FAULT CURRENT FROM 50% TO 100% OF MAXIMUM POSSIBLE:

$I^2 t = K$

WHERE

I = SYMMETRICAL FAULT CURRENT IN TIMES NORMAL BASE CURRENT (ANSI/IEEE C57.12.00-1980 [2])

K = CONSTANT DETERMINED AT MAXIMUM I WITH t = 2 SECONDS

TIME (SECONDS)

TIMES NORMAL BASE CURRENT

FIGURE 8.7   (Continued) Transformer through-fault protection curve.

Motors	Thermal limit characteristic. (See Fig. 8.8 for a typical curve. Obtain exact curve from the motor manufacturer.)
Capacitors	Case-rupture characteristic. (See Fig. 8.9 for typical characteristic. Obtain exact curve from capacitor manufacturer.)

TABLE 8.3 How to Construct Transformer Through-Fault Withstand Curves

Example is for two delta-wye-connected transformers (ANSI/IEEE Std. C57.109-1985)

Rating	Mechanical			Thermal			Inrush
For transformers with supplemental cooling, use the self-cooled rating.	t (s)	$I_1 = \dfrac{0.58 \times I_{FL}}{Z}$ (A)	$I_1^2 t$ (× 10^6)	$0.5I_1$ (10 MVA) / $0.7I_1$ (2.5 MVA)	$I_0 = \dfrac{0.58 I_{FL}}{0.04}$ (A)	$I_0^2 t$ (× 10^6)	$12 I_{FL}$ @0.1 s
10 MVA 12.47 kV $Z = 7.25\%$ $I_{FL} = 463$	*2.0	*3704	27.4↓		‡6714→	90.14↓	5556
	†8.0		←27.4	†1852↓			
	§26.2			§1852		←90.14	
2.5 MVA 12.47 kV $Z = 5.75\%$ $I_{FL} = 116$	*2.0	*1170	2.738↓		‡1682→	5.658↓	1344
	†4.08		←2.738	†819			
	§8.44			§819		←5.658	

*First point for mechanical curve
†Transition point from mechanical to thermal curve
‡Thermal curve reference point (origin at 2 s)
§Thermal curve in effect (continuation of $I^2 t$ curve bending toward overload limit)
I_1 = positive sequence three-phaser through fault current in symmetrical rms amperes.
$I^2 t = I_1^2$ times time (in seconds).

Procedure:

1. Calculate the 2-s mechanical limit current (I_1) at 2 s using the formula given in the above table.
2. Calculate the current for the transition from the mechanical to the thermal curve which takes place at $0.5I_1$ for Category III transformers ($0.7I_1$ for Category II transformers).
3. Calculate the origin of the thermal limit current curve at 2 s and 25 times full-load current.
4. The thermal curve is an $I^2 t$ curve from the transition point which starts to bend back toward the overload limit of the transformer at about 3 to 4 times full-load current. The origin of this curve is at I_0 and 2 s.

Category	3-phase kVA	Single-phase	comment
I	15–500	5–500	Thermal curve applies
II	501–5000	501–1667	Transition at $0.7I_o$
III	5001–30,000	1668–10,000	Transition at $0.5I_o$
IV	Above 30,000	Above 10,000	Transition at $0.5I_o$

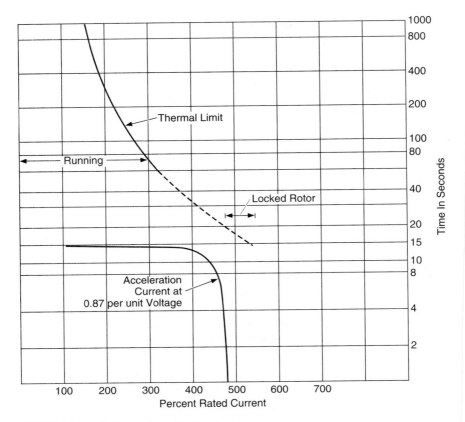

FIGURE 8.8   Typical motor thermal limit protection curve.

8.6 DEVICE FUNCTION NUMBERS

To conserve space on a one-line diagram, device function numbers are used instead of printing the words which describe the relay function. This standard contains the following definition:

> A device function number, with appropriate prefix and suffix where necessary, is used to identify the function of each device in all types of partial automatic switchgear and automatic switchgear and in many types of manual switchgear. These numbers are to be used on drawings, on elementary and connection diagrams, in instruction books, in publications, and in specifications. In addition, for automatic switchgear, the number may be placed on, or adjacent to, each device on the assembled equipment so that the device may be readily identified.

A list of device function numbers compiled from ANSI/IEEE Standard 37.2-1991[5] appears in Table 8.4.

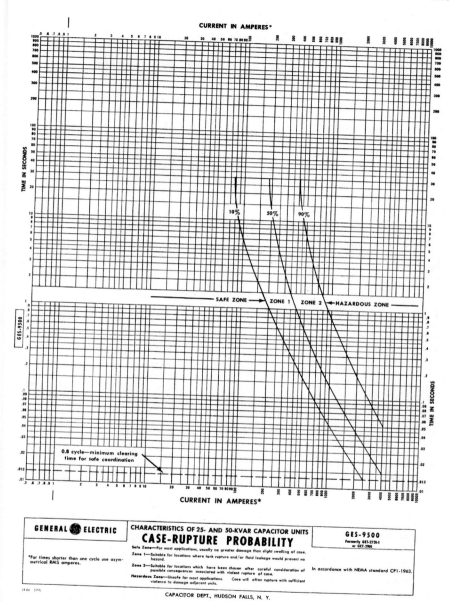

FIGURE 8.9   Typical capacitor case-rupture probability curve.

TABLE 8.4 Device Function Numbers per IEEE Standard C-37.2-1991 (Draft)

No.	Function	No.	Function
1	Master element	50	Instantaneous overcurrent relay
2	Time-delay starting or closing relay	51	ac time overcurrent relay
3	Checking or interlocking relay	52	ac circuit breaker
4	Master contactor	53	Exciter or dc generator relay
5	Stopping device	54	Turning gear engaging device
6	Starting circuit breaker	55	Power factor relay
7	Rate-of-rise relay	56	Field application relay
8	Control power disconnecting device	57	Short-circuiting or grounding device
9	Reversing device	58	Rectification failure relay
10	Unit sequence switch	59	Overvoltage relay
11	Multifunction device	60	Voltage or current balance relay
12	Overspeed device	61	Density switch or sensor
13	Synchronous-speed device	62	Time-delay stopping or opening relay
14	Underspeed device	63	Pressure switch
15	Speed or frequency matching device	64	Ground detector relay
16	Reserved for future application	65	Governor
17	Shunting or discharge switch	66	Notching or jogging device
18	Accelerating or decelerating device	67	ac directional overcurrent relay
19	Starting-to-running transition contactor	68	Blocking relay
20	Electrically operated valve	69	Permissive control device
21	Distance relay	70	Rheostat
22	Equalizer circuit breaker	71	Level switch
23	Temperature control device	72	dc circuit breaker
24	Volts-per-hz relay	73	Load-resistor contactor
25	Synchronizing or synchronism-check device	74	Alarm relay
26	Apparatus thermal device	75	Position-changing mechanism
27	Undervoltage relay	76	dc overcurrent relay
28	Flame detector	77	Telemetering device
29	Isolating contactor	78	Phase-angle measuring or out-of-step prot. relay
30	Annunciator relay	79	ac reclosing relay
31	Separate excitation device	80	Flow switch
32	Directional power relay	81	Frequency relay
33	Position switch	82	dc load-measuring reclosing relay
34	Master sequence device	83	Automatic selective control or transfer relay
35	Brush-operating or slip-ring short-circuiting device	84	Operating mechanism
36	Polarity or polarizing voltage device	85	Carrier or pilot-wire receiver relay
37	Undercurrent or underpower relay	86	Locking-out relay
38	Bearing protective device	87	Differential protective relay
39	Mechanical condition monitor	88	Auxiliary motor or motor generator
40	Field relay	89	Line switch
41	Field circuit breaker	90	Regulating device
42	Running circuit breaker	91	Voltage directional relay
43	Manual transfer or selector device	92	Voltage and power directional relay
44	Unit sequence starting relay	93	Field-changing contactor
45	Atmospheric condition monitor	94	Tripping or trip-free relay
46	Reverse-phase or phase-balance current relay	95	95-99 used only for specific
47	Phase-sequence or phase-balance voltage relay	96	installations if none of the
48	Incomplete sequence relay	97	functions assigned to numbers
49	Machine or transformer thermal relay	98	between 1 through 94 is
		99	suitable.

(From IEEE Std C37.2-199, *IEEE Standard Electrical Power System Device Function Numbers.*)

8.7 MEDIUM-VOLTAGE EQUIPMENT PROTECTION

Table 8.5 describes the minimum protection for moderate-size, radial, medium-voltage systems. Tables 8.6 through 8.9 describe the optimum protection for transformer, feeder, motor, and generator circuits.

8.7.1 Overcurrent Protection

Medium-voltage overcurrent protection includes relays and fuses in circuits with circuit breakers or contactors.

Relay types include solid-state and electromagnetic in both the protective relay and industrial control categories. Most industrial control relays are electromagnetic except where the overcurrent function is included in a multifunction module for motor or circuit protection. Figure 8.1 shows the standard characteristics for overcurrent relays. Thermal overcurrent relays include a variety of heater characteristics curves which are available from manufacturers. Two samples of some of one manufacturer's curves are shown in the coordination example later in this chapter. Protective relays operate to initiate an action, such as tripping the circuit, at their settings.

TABLE 8.5 Minimum Protection for Moderate-Size, Radial, Medium-Voltage Power Distribution Systems

System component	Protection	Fault detector	Interrupter
Main transformer	Short-circuit	Primary power fuses, circuit breaker or circuit switcher	
	Overload Single-phasing	Phase overcurrent relays (#51) Timed voltage balance relay (#60V)	Main secondary power circuit breaker
Main distribution bus	Short-circuit	Phase & ground overcurrent relays (#51, 51N, 51G)	Main secondary power circuit breaker
	Overload	Phase overcurrent relays (#51)	
Medium-voltage feeders	Short-circuit	Phase & ground overcurrent relays (#50/51, 51N, 51G, 50GS)	Feeder power circuit breaker
	Overload	Phase overcurrent relays (#50/51)	
Medium-voltage motor controllers	Short-circuit Overload Undervoltage	Power fuses Thermal relay Contactor holding coil	Power fuses Contactor
Secondary unit substation transformer	Short-circuit	Phase & ground overcurrent relays (#50/51, 50G, 51G)	Feeder power circuit breaker
		OR transformer primary fuses	
	Overload	Main secondary circuit breaker trip device	Main secondary circuit breaker

TABLE 8.6   Optimum Protection for Main Medium-Voltage Transformers

System component	Protection	Fault detector	Circuit interrupter
Complete transformer plus primary and secondary conductors	Short-circuit	High-speed differential relays (#87T) Backup phase & ground overcurrent relays (#50/51)	Primary and secondary circuit breakers Primary power circuit breaker
Transformer windings and other internal components	Short-circuit Incipient fault Overload	Fault-pressure relay (#63FP) Gas detector relay (#63GDR) Top oil or winding temperature detecting relay (#49)	Primary power circuit breaker Primary or secondary power circuit breaker
Transformer secondary windings and conductors	Ground-fault	Sensitive ground differential relay with or without through-fault restraints (#87TG)	Primary power circuit breaker

TABLE 8.7   Optimum Medium-Voltage Bus and Line Protection

System component	Protection	Fault detector	Interrupter
Main or distribution subdistribution bus	Short-circuit and ground-fault	High-speed differential relay (#87B)	Incoming line, feeder, and tie circuit breakers
Tie lines to subdistribution bus		Pilot wire (line differential) (#87L)	Tie line circuit breakers

Power fuses include current-limiting and non-current-limiting types. Many non-current-limiting distribution power fuses expel the fuse element from the casing during operation. Current-limiting fuse types include both full-range and backup fuses. Backup fuses, used in medium-voltage motor controllers, operate at currents more than a certain value set by the manufacturer and require some other device to remove faults below this current value. All fuses operate at values of current above their rating and carry their rated current with no damage.

For medium-voltage feeders with only phase and ground overcurrent protection, the instantaneous element of time overcurrent relays acts as primary protection for the circuit conductors and is set to detect faults in these conductors and the primary windings of any connected transformers. The time overcurrent element is selective with and acts as backup for any primary protection located at the connected transformers. If transformers do not have individual protection or if the circuit is dedicated to one transformer, the time overcurrent element provides transformer through-fault protection and is set lower than the transformer through-fault withstand limit. Ground-fault protection

TABLE 8.8 Optimum Medium-Voltage Motor Circuit Protection

System component	Protection	Fault detector	Interrupter
Motor & motor feeder conductors	Overload & locked rotor	Multifunction motor protecting relay with thermal curve biased by RTD	Motor controller contactor or power circuit breaker
	Ground-fault	Ground sensor relay with zero sequence CT (#50GS)	
Motor	Short-circuit	High-speed differential relay (#87M)	
Motor and (stability of) power system	Out-of-phase reenergization	Undervoltage relay (#27) Sensitive power directional relay (#32) Frequency relay (#81)	Main transformer secondary breaker
	Single-phasing or phase unbalance	Current or voltage negative sequence relays (#46) or (#60V)	Motor controller or power circuit breaker

may be provided by either residually connected or zero sequence CT-connected instantaneous or time-delay ground overcurrent relays, as Fig. 8.10 shows. The NEC requires three relays for each three-phase circuit. This requirement can be met with two phase and one ground relay, although most users prefer three-phase relays plus a ground relay.

For medium-voltage buses, the main breaker overcurrent protection acts as backup to any bus differential relaying and to the feeder overcurrent protection with which it is

TABLE 8.9 Optimum Medium-Voltage Generator Circuit Protection

System component	Protection	Fault detector	Interrupter
Generator stator windings	Short-circuit	High-speed differential relay (#87G)	Generator power circuit breaker
	Unbalanced load	Negative sequence current (#46)	
	Overtemperature	Temperature relay—RTD signal (#49)	
Field windings	Ground-fault	Ground detector (#64G)	Generator power circuit breaker
	Field loss	Loss of field relay (#40)	
	Overexcitation	Overexcitation relay (#59)	
System backup (overcurrent)	Short-circuit	Overcurrent—voltage restrained or controlled (#51V)	Generator power circuit breaker
	Ground-fault	Overcurrent relay (51G)	
Turbine blades	Antimotoring	Sensitive power directional relay (32)	Generator power circuit breaker
Turbine speed	Overfrequency Underfrequency	Frequency relay (81)	Generator power circuit breaker

Phase Overcurrent

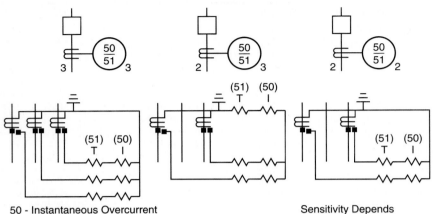

50 - Instantaneous Overcurrent
51 - Time Overcurrent

Sensitivity Depends
on CT Ratio

Ground Overcurrent

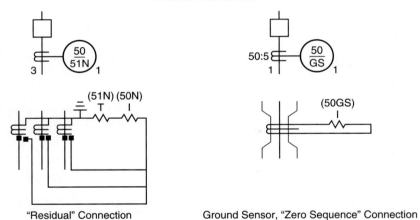

"Residual" Connection

Ground Sensor, "Zero Sequence" Connection

50 - Instantaneous Ground Overcurrent
51 - Time Delay

50GS - Instantaneous Ground Sensor
Overcurrent

Sensitivity Depends on CT Ratio

Sensitivity: 15 Primary Amperes

FIGURE 8.10   Phase and ground overcurrent relaying.

set to be selective. If bus differential relaying is not provided, the overcurrent relaying is primary protection for the bus. Usual practice dictates that primary bus phase overcurrent protection should be set to trip in less than 2 s for the maximum short-circuit current feeding the bus. This limit is usually too high to limit extensive damage; hence, most settings are considerably less than that, maintaining about a 0.3-s margin of operation at the instantaneous setting of the largest feeder breaker overcurrent relay. Ground overcurrent relays are set to be selective with downstream ground overcurrent relays on the largest feeder. Instantaneous elements are either omitted or blocked on incoming lines to distribution buses to allow selectivity of the main and feeder breakers. Sum-of-current main breaker backup time overcurrent relaying enhances bus protection for double-ended substations by eliminating the need for two steps of backup relaying, which would be required if overcurrent relays are provided for the tie breaker. In this arrangement, a fault on one bus would receive current from two sources which are totalized by the CT connections for that bus. The fault current from the main source of the unfaulted bus would be canceled by the CT connection and only the main and tie breakers would be tripped. Figure 8.11 shows this connection, which sometimes is called *partial differential*. This vernacular nickname is a misnomer since there is nothing partial about the circuit and it is not a differential arrangement.

8.7.2   Transformer Protection

High-speed transformer differential relays include harmonic filters to prevent false tripping on initial energization. Sensitivity of these relays is about one-third of their tap setting. For delta-wye-connected transformers supplying resistance-grounded systems rated about 10,000 kVA and above, phase differential relays should be supplemented with secondary ground differential relays as shown in Fig. 8.12. Note the CT connections in the figure. These connections are necessary to compensate for the phase shift

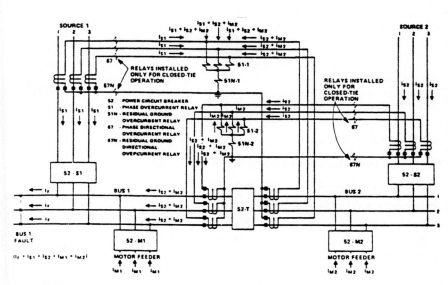

FIGURE 8.11   Sum-of-current overcurrent relaying connections.

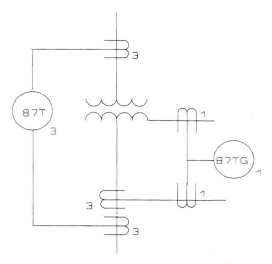

FIGURE 8.12   Transformer phase and ground differential protection (relay and CT connections).

between the primary and secondary connections. For transformers made in the United States, the transformer connection for delta-wye transformers carries an International Electrotechnical Commission (IEC) designation of Dy1. Figure 8.13 shows some of the commonly used winding designations specified in IEC Publication 76. Transformer nameplates show the winding connection and vector diagram for all transformers manufactured domestically or abroad. Figure 8.14 shows the current flows in a Dy1-connected transformer and relay circuit.

For single-phase transformers made in the United States, the polarity is subtractive. Figure 8.15 describes the difference between additive and subtractive polarity, giving a test for polarity. Polarity is important when connecting single-phase transformers in a three-phase bank to assure a standard three-phase connection. High-speed transformer differential relaying is recommended for all transformers rated 10,000 kVA and above and for most transformers where both windings are rated more than 600 V.

Fault-pressure, gas detection, and winding temperature relays provide additional protection for transformers larger than 10 MVA.

8.7.3   Motor Protection

Recommendations for motors 1500 hp and larger include:

High-speed differential	#87M
Overcurrent (RTD bias)	#49
Ground sensor	#50GS
Negative sequence	#46 or #60V
Undervoltage	#27
Sensitive power directional	#32
Frequency	#81

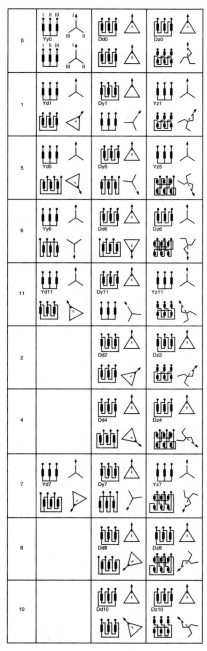

FIGURE 8.13 IEC connection designations for some commonly used separate winding three-phase transformers.

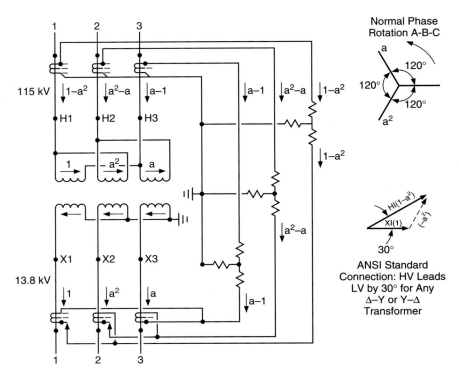

FIGURE 8.14   Current flows in a Dy1-connected transformer and differential relay circuit.

Motor high-speed differential protection (#87M) is recommended for all medium-voltage motors but is rarely applied for smaller motors. Most motor protection protective relays do not include differential protection but do include the ability to communicate the operation of differential relays to a master control system. Figures 8.16 and 8.17 show the connections for two types of motor differential protection. The three-CT method of motor differential is the most commonly applied method since it is less expensive than the six-CT method and may be more sensitive for larger motors. Differential protection requires that all six motor leads be available.

Thermal overcurrent (RTD biased) and instantaneous overcurrent protection (#49/50), negative sequence (unbalanced current or voltage protection (#46 or #60V), ground-fault protection (#50GS or #50N)) may be provided by individual relays or by a single solid-state multifunction motor protective relay. These relays usually incorporate instrumentation and metering functions as well as provide communication capability to a central control system. Synchronous motors and large induction motors (1500 hp and larger) may require protection from sources subject to instantaneous reclosing to prevent motor damage as well as helping to maintain system stability under fault conditions. This type of protection can be obtained by using a frequency relay (#81) set to operate about half a hertz below the nominal system frequency. For systems with wide frequency swings, a sensitive reverse power relay (#32) with its contacts in series with the frequency relay will prevent operation except for system fault conditions. As a further backup, an instantaneous voltage relay (#27) set to operate at about 70 percent volt-

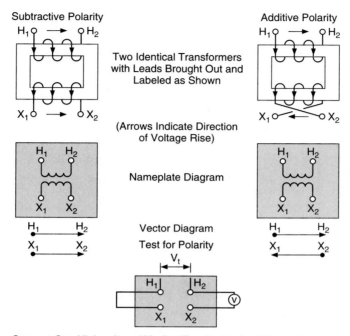

Subtractive Polarity

H_1 → H_2

X_1 → X_2

Additive Polarity

H_1 → H_2

X_1 ← X_2

Two Identical Transformers with Leads Brought Out and Labeled as Shown

(Arrows Indicate Direction of Voltage Rise)

Nameplate Diagram

Vector Diagram
Test for Polarity

V_t

Connect One High-voltage Winding Terminal to the Adjacent Low-voltage Terminal. Apply Test Voltage to High-voltage Winding. If High and Low Voltages Subtract, Transformer is Subtractive Polarity. If High and Low Voltages Add, Transformer is Additive Polarity.

FIGURE 8.15 Single-phase transformer polarity.

age with its contacts bypassing the series contacts of #32 and #81 provides further security for this sensitive relay combination.

Surge protection, consisting of lightning arresters and surge capacitors, is recommended for all medium-voltage motors. Such protective devices should be located at the motor terminals, with no more than a few feet (less than 10) of lead length. More than this distance from the motor terminals will negate the effect of the surge capacitors. Sometimes lightning arresters are mounted at the motor distribution bus for a bus serving several smaller motors to save the expense of many lightning arresters and relieve the congestion in a small motor terminal box. Surge capacitors *must* be mounted at the motor terminals to be effective.

Never ground motor neutrals. To do so seriously compromises the system ground overcurrent protection by providing a source of ground-fault current limited only by the subtransient reactance of the motor.

8.7.4 Generator Protection

Table 8.9 lists the usual protection for moderate-size industrial system generators up to about 50 MW. For smaller generators less than about 5 MW, some of this protection, such as the field ground detection relay, is omitted. For some generators, particularly

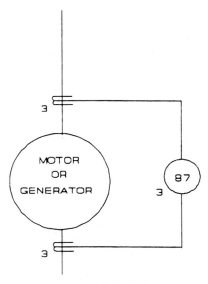

FIGURE 8.16 Six-CT rotating machine phase differential protection (relay and CT connections). 87 = high-speed differential relay.

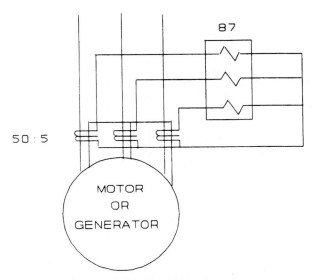

FIGURE 8.17 Three-CT rotating machine phase differential protection (relay and CT connections). 87 = three-element instantaneous overcurrent relay set on 0.5A tap.

larger and unit-connected generators, additional system relays, per IEEE/ANSI Standard C37.102, may be required. Sensitive generator ground-fault differential relays, similar to those described for transformer ground-fault differential relaying, apply.

8.7.5 Tie-Line Protection

Medium-voltage tie lines require instantaneous primary protection not provided by time-delay overcurrent relays set to be selective with downstream feeders and lacking instantaneous elements. Pilot wire differential relays meet this need. Figure 8.18 shows a simplified version of a pilot wire two-terminal protection arrangement. This protection is required to permit fast fault removal of tie-line faults, since backup overcurrent relaying may be set to operate at long times causing system instability. Pilot wire supervision relays monitor the condition of the pilot wire as well as providing transfer tripping functions to ensure that both ends of the line trip for all internal faults.

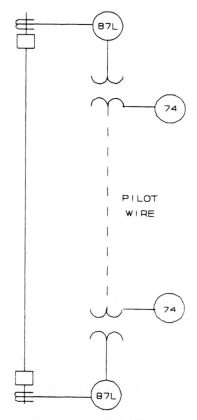

FIGURE 8.18 Two-terminal pilot wire line differential protection (simplified CT, pilot wire, and relay connections). 87L = pilot wire relay. 74 = line supervising relay.

8.7.6   Current and Voltage Transformer Accuracies

If feeders are a reasonable size compared to the available short circuit on any bus, saturation of CTs and PTs usually is not a problem. For feeders on a 13.8-kV bus with between 750 and 1000 MVA available short circuit the CT should be no smaller than 400 A to avoid saturation problems. Some higher-accuracy, specially constructed CTs may be available for unusual applications. Anticipate trouble with smaller CT ratios on high-fault-level buses. Good application practice is to make feeders no smaller than 10 percent of the incoming line rating and no larger than about 25 to 40 percent of the incoming line rating. While smaller CTs tend to saturate, larger feeders introduce overcurrent relay coordination problems. These problems usually require a compromise between coordination and protection. Check the burden on both CTs and PTs that look substantially loaded and compare it with the burden limits placed on the instrument transformers by the manufacturer. Procedures for making these examinations are given in the IEEE Red Book (IEEE Standard 141-1986).[6]

CT Accuracy Classifications.   CTs have relay classifications such as "C200." All classifications imply an error within 10 percent. C means calculated. 200 is the maximum voltage the CT can produce without saturation. T200 would mean the CT was tested and met this accuracy.

CTs have metering classifications such as "0.3B0.5." The 0.3 means an accuracy of 0.3 percent at a burden of 0.5 Ω.

Voltage Transformer Accuracies.   VTs have accuracies such as "0.3 thru Burden Z." Three accuracy classes apply to VTs. These classes specify the applicable *ratio correction factor* (RCF) and *transformer correction factor* (TCF) at certain burdens:

Burden W—12.5 VA at 120 V—0.10 burden power factor

Burden X—25.0 VA at 120 V—0.70 burden power factor

Burden Y—75.0 VA at 120 V—0.85 burden power factor

Burden Z—200.0 VA at 120 V—0.85 burden power factor

Three accuracy classes apply:

Accuracy class	RCF & TCF limit	Limit of PF (lagging load)
1.2	1.012—0.988	0.6—1.0
0.6	1.006—0.994	0.6—1.0
0.3	1.003—0.997	0.6—1.0

The designation of VT accuracies has the same meaning as for CT accuracies: The 0.3 is the percent accuracy at the designated burden with the limitations imposed by the standards.

8.8   LOW-VOLTAGE EQUIPMENT PROTECTION

For low-voltage systems, branch circuit or feeder overcurrent trip devices or fuses act as the primary protection for either the equipment or the circuit. Main fuses or main breaker overcurrent trip devices act as backup to downstream devices as well as primary protection for the immediate section of the circuit involved.

8.8.1 Low-Voltage Fuses

Low-voltage cartridge fuses are divided into classes by Underwriters Laboratory (UL). These classes specify dimensions, interrupting rating, and time-current characteristics:

UL Class H.   These fuses are non-current-limiting, with a 10,000-A ac interrupting rating, and include dual-element or time-delay fuses.

Classes G, J, L, CC.   These fuses are current-limiting, and do not have the same dimensions as Class H fuses, with various interrupting ratings up to 200,000 A ac.

Class K.   These fuses are current-limiting, with the same dimensions as Class H, having various characteristics and various interrupting ratings up to 100,000 A ac. There is no rejection feature when substituted for Class H fuses.

Class R.   These fuses are current-limiting, 200,000 A ac, with various characteristics. A rejection device prevents installation of any other fuse in equipments designed for Class R fuses but will fit in Class H or Class K fuse clips.

Class T.   These are dc fuses.

Limiters.   These function only for high values of fault current and will not operate at lower currents regardless of time.

Plug Fuses.   These fuses are limited to 125-V ac applications of 30 A or less. All new installations must use an S base, preventing installation of a larger fuse than initially installed.
   All fuses carry their rated current and operate at some value of overcurrent.

8.8.2 Low-Voltage Circuit Breakers

Low-voltage circuit breakers are divided into three categories: low-voltage power circuit breakers, insulated-case circuit breakers, and molded-case circuit breakers.
   Low-voltage power circuit breakers may be equipped with either dual-magnetic or solid-state trip devices. Most modern circuit breakers are equipped with rms sensing solid-state trip devices. These trip devices operate at their settings. Characteristics include long-time, short-time, instantaneous, and ground-fault trip elements which can be specified in various combinations and setting ranges. Current-limiting fuses can be specified to be incorporated with low-voltage power circuit breakers. Modern frame sizes range from 800 to 4000 A with interrupting ratings up to 200,000 A ac.
   Insulated-case circuit breakers may be equipped with either magnetic or solid-state trip devices, similar to low-voltage power-circuit-breaker trip devices. When specified without instantaneous tripping, a high set instantaneous is included to permit the circuit breaker to meet its specified rating. Frame sizes range from 800 to 4000 A ac with interrupting ratings up to 200,000 A at 240 V ac (150,000 A at 480 V ac).
   Molded-case circuit breakers may be equipped with thermal magnetic trip devices or, for some of the larger sizes, solid-state trip devices. All of the thermal magnetic trip devices are rated to carry the load and trip at some value of current higher than their rating. Some of the solid-state trip devices are fixed at 110 percent of their rating to permit the circuit breaker to carry the load. Others may be equipped with trip devices similar to low-voltage power circuit breakers arranged to trip at their settings. When specified as a molded-case switch, a high set instantaneous element is included.

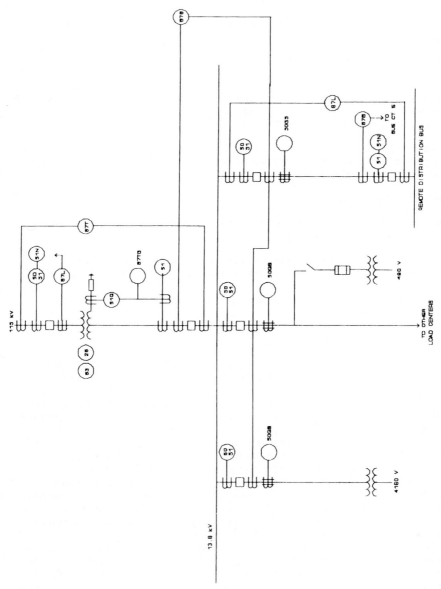

FIGURE 8.19 Radial system partial one-line diagram (major elements and protective functions).

8.30

8.9 SYSTEM CONFIGURATIONS

8.9.1 Radial System

A radial system is a single-source system supplying load through individual circuits. The one-line diagram in Chap. 7 (Fig. 7.7) depicts a radial system. Figure 8.19 shows a partial one-line diagram of a similar radial system showing the optimum phase and ground protection for each system component. Note that the differential zones overlap each other, providing primary short-circuit protection for each portion of the system except the feeders for which the instantaneous elements on the overcurrent relays provide this protection. Overcurrent relays and low-voltage trip devices provide backup protection throughout the system.

8.9.2 Primary Selective System

A primary selective system is one in which the primary source of electric energy may be one of two sources, usually feeders, each from a different distribution bus. Figure 8.20 shows only the major system elements, omitting the protective functions. Each system element requires the same type of protection as is required on a radial system. This type of system is often used for distribution of power in industrial plants not involving continuous processes.

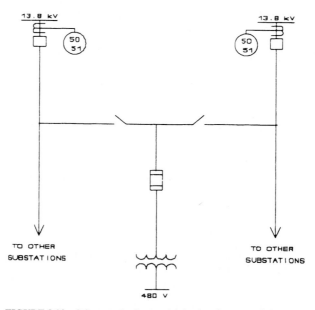

FIGURE 8.20 Primary selective system (major elements only).

8.9.3 Secondary Selective System

A secondary selective system contains at least two transformer secondary sources, each with a main circuit breaker, and at least one tie circuit breaker. Tie circuit breakers are normally open, with loads on each bus supplied by one transformer. Should one source fail, the tie circuit breaker may be closed, supplying the load from the other transformer. This type of system (main system elements are shown in Figure 8.21) applies to both medium- and low-voltage systems. With a two-transformer system, automatic throwover equipment, discussed in Sec. 8.96, may be applied. If the tie circuit breaker is closed, directional overcurrent relays (#67 and #67N) are required on each main-circuit-breaker protection package.

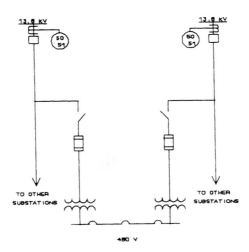

FIGURE 8.21   Secondary selective systems (major elements only).

8.9.4 Spot Network System

A spot network, when equipped with appropriate protective devices and properly operated, provides the ultimate in reliability; spot networks are commonly used by electric utilities. Industrial and utility spot networks operate differently, because different philosophies prevail.

Utility company philosophy—especially for underground systems—is based on the assumption that network faults, usually on underground cables, will burn themselves clear. For this reason, power is maintained to the network bus even under fault conditions. Industrial spot networks maintain power to the bus only under no-fault conditions. Phase and ground overcurrent relays (or trip devices) open the network transformer secondary protective device and directional relays provide selectivity.

Figure 8.22 shows a typical two-transformer utility spot network. The function of the network protector is to ensure that power is interrupted to the network bus only on the failure of the primary feeder. The network protector recloses automatically when the primary feeder is restored.

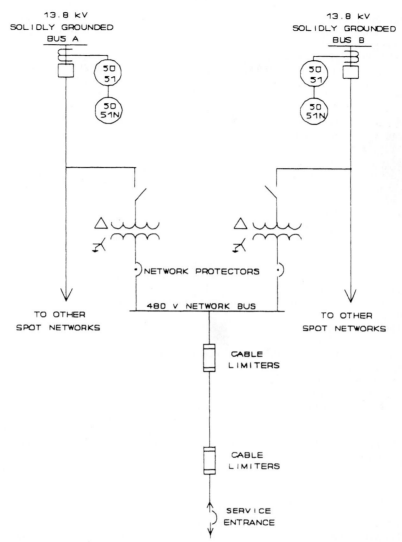

FIGURE 8.22 Utility two-transformer spot network one-line diagram. This diagram helps to understand the difference between utility and industrial practices. Network protectors operate only for medium voltage feeder faults and provide no overcurrent protection for the 480-V network bus. Although the system appears quite simple and uses fewer components than an industrial spot network, network protectors are actually complex and rather expensive.

The network protector is operated normally by master and phasing relays. The network master relay trips the network protector for power flow from the network into the transformer. This relay is sensitive to power flow as small as the transformer core losses. On three-phase and line-to-line medium-voltage feeder faults, the master relay operates. On line-to-ground faults, the medium-voltage feeder circuit breaker trips first; then the network master relay operates on the fault-current flow from the network or the transformer exciting current. The master relay also recloses the network protector if the source voltage is higher than the network bus voltage; the phasing relay interprets proper voltage relationships to prevent network protector pumping.

With this system configuration, network bus faults are presumed to be self-extinguishing; the network protector provides no overcurrent protection for the network bus. Network protector fuses are backup protection only, operating if the protector fails to open for a fault on its source side. Such fuses must be large enough to permit network cable limiters or network relays to operate before fuse operation. These fuses must also be sized to carry the maximum network load without blowing. Such fuses do not provide protection in accordance with the NEC[1] or any other accepted criterion.

Network protectors might not close into a dead network if there is no load connected to the system. For this reason, utility spot network systems sometimes include a 500-W line-to-ground resistance load on the phase containing the phasing relay to assure dead-bus network protector reclosing.

Figure 8.23 shows the industrial equivalent of the Fig. 8.22 utility spot network. The industrial version appears more complex because the NEC requires phase overcurrent and ground-fault protection for the network bus and because discrete elements are used for some of the functions incorporated in the utility network protector. Low-voltage power circuit breakers provide bus fault protection. Lockout devices prevent reclosing immediately following an overcurrent trip operation.

Because main secondary circuit breaker overcurrent trip devices for both transformers have identical settings, phase and ground directional overcurrent relays (#67 and #67N) must be provided for each incoming circuit to maintain selectivity for source side faults.

The relaying devices shown in Fig. 8.23 can sometimes be provided at less cost than the network protectors used by utilities, especially if standard devices intended for switchgear mounting are used. With the relaying shown, the industrial spot network (Fig. 8.23) can be reclosed on a dead network. The sensitive directional power relay (#32), when combined with the voltage relay (#62) used as a timer, accomplishes the reverse power function performed by a network master relay. The over/undervoltage relay (#27/59) provides the reclosing function upon restoration of the primary circuit. The voltage relay used as a timer supplies the additional function of tripping both transformer main breakers when a simultaneous undervoltage exists on both primary feeders. The synchronizing check relay (#25) prevents closing out-of-synchronism sources. The system shown can be easily modified to omit functions not required for some industrial systems.

If the 13.8-kV distribution system switchgear is reasonably close to the 480-V spot network, transfer tripping can be provided to trip all transformer main secondary breakers whenever the 13.8-kV circuit breaker is opened. If the distance from the spot network to the 13.8-kV circuit breaker is less than about 1500 ft, providing transfer tripping eliminates the need for the reverse power relay and timer.

The synchronizing check relay (#25) is not required if both sources are always synchronized. The control switch for the transformer main secondary breaker can be arranged to provide nonautomatic operation when in the normal-after-trip position, permitting secondary breaker automatic reclosing when in the normal-after-close position.

Figure 8.24 shows the closing and tripping circuits for the transformer main secondary circuit breakers of a two-transformer industrial spot network which includes the Fig. 8.23 relays.

8.9.5 Synchronizing Bus System

A synchronizing bus provides the means to connect several sources to a common bus through an impedance which limits the available short-circuit current to a value within

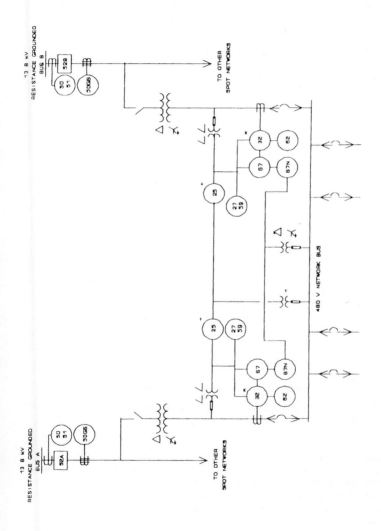

* Omit 32 if 52A or 52B trips respective secondary main breakers
+ Omit 25 if sources A and B always in synchronism

FIGURE 8.23 Industrial two-transformer spot network one-line diagram. Industrial spot network uses more components than utility spot network. If conventional relays designed for switchgear mounting are used, overall cost can be less than the cost of utility network protectors. Some components shown perform some functions of utility network protectors; others perform *NEC* overcurrent protective requirements for the 480-V network bus.

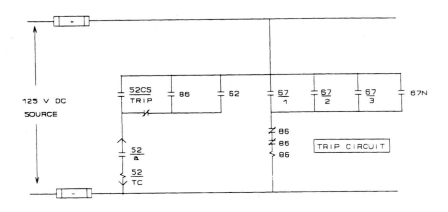

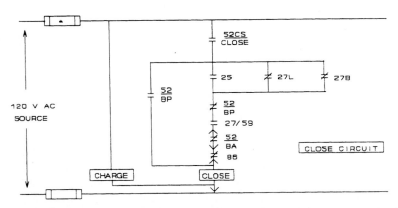

FIGURE 8.24 Control circuits for industrial spot network transformer main secondary circuit breakers. These diagrams show trip and close circuits. A reliable source of DC tripping power is required. Automatic reclosing on restoration of power is controlled by 27/59. Synchronism check relay 25 incorporates dead line 27L and dead bus 27B functions. Limit switches 52/BP permit test position operation.

the interrupting rating of the circuit breakers on any bus. Figure 8.25 shows such a system using reactors to limit the short-circuit current to each individual bus. Since the available short circuit on the synchronizing bus is high, no circuit breakers are directly connected to the bus. This type of bus is used in relatively large industrial installations with several sources of medium-voltage power and usually with in-plant generation.

High-speed differential relays protect all the system segments to help maintain system stability and isolate any faulted system component. Backup overcurrent relaying for the synchronizing bus may be difficult or impossible to set, even if directional overcurrent relaying is used; hence, duplicate differential relaying may be required for backup. Because of the voltage drops through reactors when carrying reactive power, synchronizing bus systems usually are operated in such a manner that mostly real power flows through the synchronizing bus.

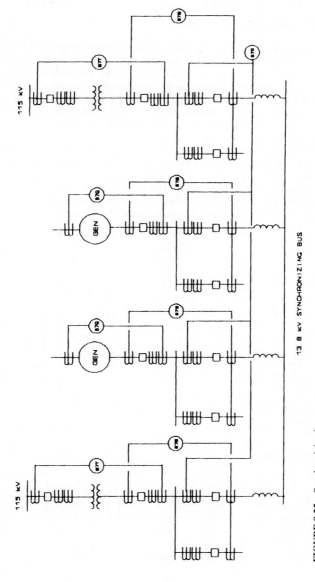

FIGURE 8.25 Synchronizing bus system.

8.37

High-voltage synchronizing bus applications may use transformers to limit the short-circuit current to the synchronizing bus. If the transformer reactance is high enough and circuit breakers are available with a sufficiently high short-circuit rating, circuit breakers can be connected to the synchronizing bus of such systems.

8.9.6 Other System Arrangements

Other system arrangements rarely used for industrial power systems and occasionally used for utility systems are the ring bus, breaker-and-a-half, and double bus systems.

Ring bus systems usually are limited to high-voltage applications because of the expense of primary protection. Occasionally, some medium-voltage applications lend themselves to this type of system. Figure 8.26 shows an arrangement of a ring bus. Relaying for this type of system consists of high-speed differential relays with backup duplicate relays since either directional or nondirectional overcurrent relays are difficult to set in a selective manner.

Breaker-and-a-half systems usually are used for large medium-voltage utility distribution substations. To save expense, only one or two extra circuit breakers are used for the middle breaker of this three-breaker, two-feeder system, as shown in Fig. 8.27.

Double bus systems were used for industrial systems in the early 20th century when circuit breakers were not as reliable as modern circuit breakers and required significant maintenance time. Figure 8.28 shows a typical double bus system.

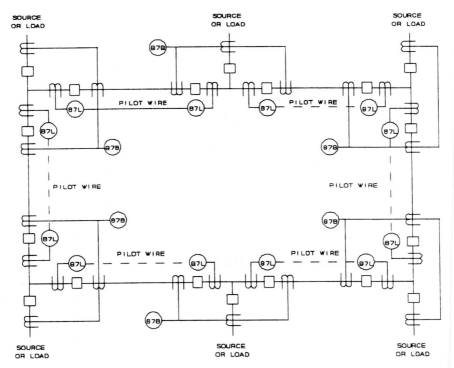

FIGURE 8.26 Ring bus system. 87B = high-speed bus differential relays. 87L = pilot wire differential relays.

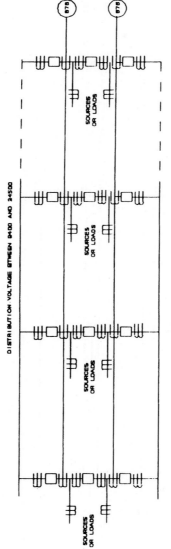

FIGURE 8.27 Breaker-and-a-half system. *Note*: Only primary bus differential protection shown. Load circuit and source protection varies with circuit and system configurations.

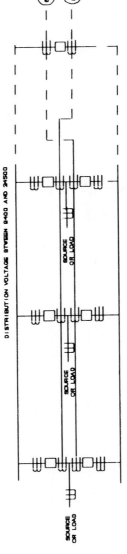

FIGURE 8.28 Double bus system. *Note:* Only primary bus differential protection shown. Load circuit and source protection varies with circuit and system configurations.

8.10 ADDITIONAL RELAY SYSTEMS

8.10.1 Breaker Failure Relaying

Some systems warrant breaker failure relaying. This type of relaying involves tripping backup circuit breakers if a breaker on a faulted circuit fails to open under fault conditions. A circuit-monitoring element transmits a signal to trip the backup circuit breaker when the breaker being monitored fails to operate correctly.

8.10.2 Automatic Throwover Systems

Throwover systems for medium- and low-voltage systems may be designed to automatically close a tie circuit breaker to an alternate source, which may either be another energized bus or an emergency generator. Transfer is usually delayed, but some boiler auxiliary buses employ a fast transfer arrangement if sufficient study shows that the motors on the bus will not be damaged by such a transfer.

The usual automatic throwover arrangement includes voltage detection, either single-phase (#27) or three-phase (#47 or #60). Loss of voltage from the usual power source is detected and throwover is prevented if the alternate source has no power. Overcurrent (#50/51 and #50/51N) lockout (#86) is included to prevent energizing a faulted bus from an alternate source. This feature can be performed by an overcurrent lockout device in low-voltage circuit breakers. Figure 8.29 depicts a typical medium-voltage automatic throwover system for a substation with large motors on each bus.

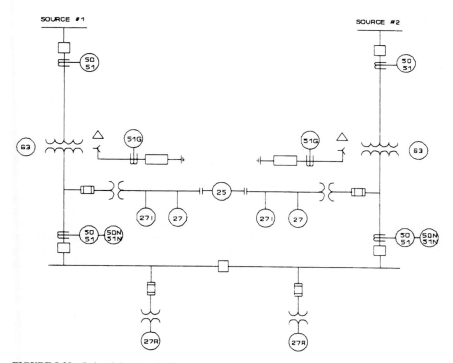

FIGURE 8.29   Industrial automatic throwover system.

Automatic throwback to the original source can be included. The installation requires a synchronizing check relay (#25) if the sources are momentarily paralleled during throwback.

Another function sometimes included is interlocking with an alternate transformer source primary breaker to further assure that the alternate source is energized. A load bus voltage relay (#27R) is sometimes included for motor buses to assure that the residual motor voltages have decayed sufficiently to allow safe reclosing. Position (or cell) switches on drawout circuit breakers prevent a throwover if the alternate source circuit breaker is not connected to the circuit. Automatic emergency generator starting may be included.

8.10.3 Automatic Synchronizing Systems

Most generators, either small or large, include automatic synchronizing relay systems, including speed-matching relays (#15) as well as synchronizing relays (#25). Manual override and dead-bus closing are usually included with such systems.

8.10.4 Automatic Reclosing Relays

Industrial systems, including cable distribution systems and large motors, seldom employ automatic reclosing. Utility systems with overhead distribution systems subject to transient faults use this type of operation in distribution substations on circuits subject to such phenomena. Automatic reclosing relays (#79) can be either single-shot or three-shot devices. The three-shot version can have various times, ranging from instantaneous to 30-s operation.

8.11 PHASE OVERCURRENT PROTECTION AND COORDINATION

The settings for protective devices in a radial circuit depend upon the characteristics of the equipment being protected allowing suitable margins of time between series protective devices to assure proper selectivity, as well as providing protection for the equipment concerned. An example of this procedure follows, using a feeder circuit from the one-line diagram in Fig. 7.7. Figure 8.30 shows the one-line diagram for the feeder considered, selecting the largest equipment connected to determine the relay and circuit-breaker settings as well as fuse sizes for the circuit.

Margins between relay times of operation at significant points should be about 0.3 s for electromagnetic relays. This margin may be decreased by another 0.1 s for static relays because of the rapid reset times of static relays. Since low-voltage circuit breakers have a characteristic curve which is a band between minimum and maximum tripping time, as long as these bands do not overlap, selectivity is assured. Much the same philosophy applies to fuses, although good practice suggests about a 0.2-s margin between fuses and relays of circuit breakers.

The procedure for a study of the circuit shown in Fig. 8.30 is as follows.

1. Construct the through-fault withstand curve for the largest transformer on a feeder. Determine currents and times for the maximum current for a 480-V fault and for the current at the transition point from the mechanical to the thermal curve. Table 8.10 shows the calculations involved following the procedures outlined in Table 8.3.

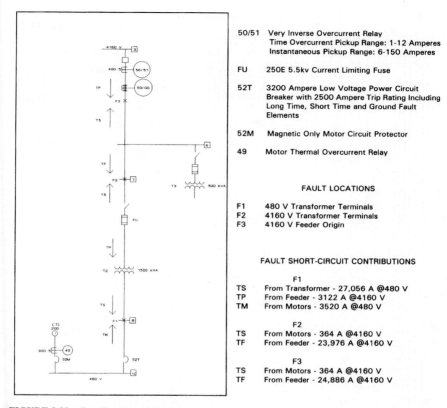

50/51   Very Inverse Overcurrent Relay
        Time Overcurrent Pickup Range: 1-12 Amperes
        Instantaneous Pickup Range: 6-150 Amperes

FU   250E 5.5kv Current Limiting Fuse

52T   3200 Ampere Low Voltage Power Circuit
        Breaker with 2500 Ampere Trip Rating Including
        Long Time, Short Time and Ground Fault
        Elements

52M   Magnetic Only Motor Circuit Protector

49   Motor Thermal Overcurrent Relay

                FAULT LOCATIONS

F1   480 V Transformer Terminals
F2   4160 V Transformer Terminals
F3   4160 V Feeder Origin

        FAULT SHORT-CIRCUIT CONTRIBUTIONS

                    F1
TS   From Transformer - 27,056 A @480 V
TP   From Feeder - 3122 A @4160 V
TM   From Motors - 3520 A @480 V

                    F2
TS   From Motors - 364 A @4160 V
TF   From Feeder - 23,976 A @4160 V

                    F3
TS   From Motors - 364 A @4160 V
TF   From Feeder - 24,886 A @4160 V

FIGURE 8.30   One-line diagram for protective device setting determination.

TABLE 8.10   Current and Time Calculations for Fig. 8.30 1500-kVA Transformer Through-Fault Withstand Curve

	Current	Time
Maximum current	$\dfrac{0.58 \times 208\text{A}}{0.0575} = 2100 \text{ A}$	2 s
Transition (mechanical)	$0.7 \times 2100 \text{ A} = 1470 \text{ A}$	$\dfrac{(2100 \text{ A})^2 \times 2''}{(1470 \text{ A})^2} = 4.08''$
I_0 Calculation	$I_o = \dfrac{0.58 \times 208 \text{ A}}{0.04} = 3016 \text{ A}$	
Transition (thermal)	1470 A	$\dfrac{3016^2 \times 2''}{1470^2} = 8.41''$

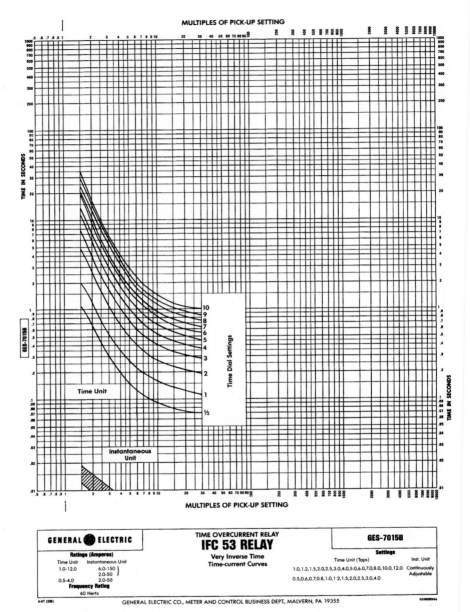

FIGURE 8.31 Time-current characteristic curves for very inverse electromagnetic overcurrent relays.

TABLE 8.11 Total Clearing Times and Margins at 1500-kVA Transformer Withstand Curve Transition Currents

250E EJO-1 Type 9F62 Fuse		
Current (A)	Total clearing time	Margin
2100	0.35 s	1.65 s
1470	1.5 s	2.58 s

TABLE 8.12 Relay-Fuse Coordination Data for Feeder Phase Overcurrent Relays and 1500-kVA Transformer Primary Fuse

Relay—fuse coordination		
	Time	
Current (A)	Relay	Fuse
2400	3.0 s*	0.13 s
6240	0.85 s*	0.03 s

*Add five cycles (0.083 s) for circuit-breaker operating time.

2. Check that the selected fuse for the transformer operates at least faster at the maximum and transition currents than the times indicated on the withstand curve for these currents. (The time-current characteristic for the maximum total clearing time is shown in Fig. 8.3.)

3. Determine the 4160-V circuit current for a fault (F_1) on the 480-V secondary bus. Set the instantaneous element of the overcurrent relay (50) at twice this value to detect faults in the primary winding of the transformer only (Set at 2×3122 A$\times5/400 = 78$ A).

4. Set the pickup of the time-overcurrent element (51) of the feeder relays at about twice the ampacity of the cable (425 A). ($425 \times 5/400 = 5.3$ A, set at 10-A tap, 800 primary amperes; the National Electrical Code,[1] Article 240-100, permits a setting of six times the cable ampacity.)

5. Set the time dial of the time-overcurrent element (51) to operate at least 0.3 s longer than the fuse maximum clearing time at three times relay pickup and at the instantaneous setting of the relay ($3 \times 800 = 2400$ A, $78 \times 400/5 = 6240$ A; set time dial #5). The time-current characteristic curve of very inverse relays selected appears in Fig. 8.31; the relay fuse coordination data is shown in Table 8.11. The fuse curves are Figs. 8.2 and 8.3.

6. Check to be sure that the overcurrent relay operates in less time than the cable short-circuit heating limit for faults at the source to the feeder at three times relay pickup, the instantaneous setting and the maximum short-circuit current. The cable short-circuit heating limits are shown in Fig. 8.6; Table 8.12 shows the cable short-cicuit heating-limit data.

7. Compare the time-current characteristic of the largest feeder in the load center substation with the primary fuse characteristic to assure that a reasonable setting can be made for the main secondary circuit breaker. The largest feeder in the example is the motor controller for the 200-hp motor. The controller manufacturer should provide the installation with a controller which is capable of being set to allow the motor to start and also meet the requirements of the NEC.[1]

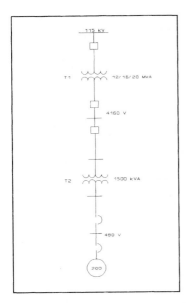

MOTOR STARTING FOR 200 HP MOTOR		
CALCULATION	Z_{pu} 10 MVA BASE	VOLTAGE DROP %
(UTILITY) $\dfrac{10}{5000}$	0.002	0.02%
(T1) $0.08 \times \dfrac{10}{12}$	0.067	0.81%
(T2) $0.0575 \times \dfrac{10}{1.5}$	0.383	4.63%
(200 HP MOTOR) $\dfrac{227}{1450} \times \dfrac{10}{0.2}$	7.828	94.54%
TOTAL	8.280	100%

FIGURE 8.32   Approximate motor-starting calculation for example.

In the example, the motor being controlled is a high-efficiency motor with a full-load ampere rating of 227 A and a locked rotor current of 1450 A. The locked rotor current in this case is about 6.4 times full-load current (1450/227 ≅ 6.4). The instantaneous on the magnetic-only circuit breaker should be set above the maximum peak current the motor will draw when initially started. Recent data has shown that, for high-efficiency motors, this value is between 2.5 and 2.83 times the locked rotor current. Setting the instantaneous trip below this value risks nuisance tripping when attempting a starting operation or a failure to be able to start at all. In the case of the 200-hp motor, a conservative value for the motor used in the example would be about three times locked rotor current or about 19.2 times full-load current (see Fig. 8.32).

The 1993 National Electrical Code[1] specifies the maximum setting for a motor controller magnetic-only circuit breaker as 13 times full-load current [Article 430-52(a) exception #2(d)]. Such a setting would risk the chance of requiring several attempts to start the motor. NEMA Standard MG1-1987[7] limits motors of this size to "two starts in succession (coasting to rest between starts) with the motor at the ambient temperature or one start with the motor initially at a temperature not exceeding its load operating temperature" [MG-1 12.50.1 (3.)]. The NEC provides an opportunity for the user to take exception to the NEC requirements by obtaining permission from the "authority having Jurisdiction" for enforcing NEC requirements (Article 90-4 Enforcement). It appears that the user's choice is either the obtaining of an exception or the risk of damaging a motor by unauthorized repeated starting attempts.

The manufacturer of the motor controller has the responsibility of supplying a controller which can be set to meet the NEC requirements and will permit starting the motor on the first attempt. The "standard" circuit breaker for a 200-hp motor is the 400-A circuit breaker (see Fig. 8.33 for time-current characteristic). The adjustable instantaneous

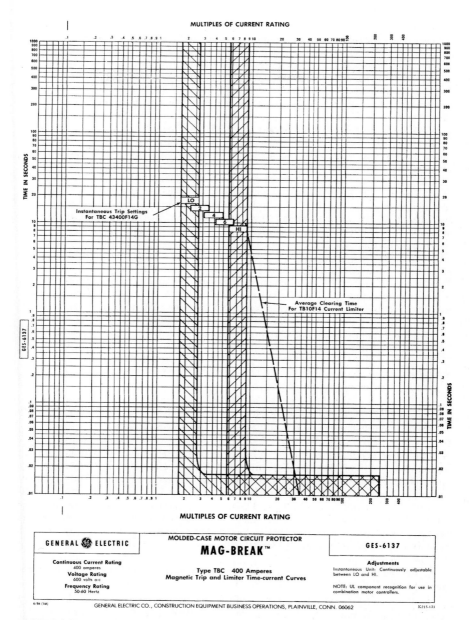

FIGURE 8.33 Time-current characteristic curve for 400-A molded-case motor circuit protector.

settings on this circuit breaker range from 1000 to 3300 A (4.4 to 14.5 times full-load current). The next largest size circuit breaker is the 600-A circuit breaker (see Fig. 8.34 for time-current characteristic). The adjustable instantaneous settings on this circuit breaker range from 3000 to 6000 A (13.2 to 26.4 times full-load current). The 3000-A setting almost meets the NEC requirement, and a higher setting to permit the motor to start can be obtained. The 600-A circuit breaker is somewhat larger than the 400-A circuit breaker and may not fit in the same standard enclosure. The example shown uses the larger circuit breaker with an instantaneous setting of three times locked rotor current ($3 \times 1450\,A = 4350\,A \cong 19$ times full-load amperes).

The controller manufacturer usually selects the thermal overload heater for the controller. In this case, the usually recommended heater for motors with full-load amperes between 216 and 231 is a heater with a rating of 269 A. This rating meets the NEC requirements for motors with a service factor of not less than 1.15 or a temperature rise of not over 40°C [NEC article 430-32(1) and 430-34]. An examination of the time-current characteristic in Fig. 8.35 reveals that the maximum persistent current this heater can withstand is 10 times its rating ($10 \times 269 = 2690$ A). This heater is not suitable for the application, indicating that a different type of thermal overload applies.

With 300:5 CTs and the relay shown in Fig. 8.36, the CT secondary current for the motor full-load of 227 A is $227 \times 5/300 = 3.7833$ A. The recommended heater for motors with secondary full-load currents between 3.75 and 4.12 A is rated at 4.68 A. This heater meets the NEC requirements mentioned previously. The CT secondary current for the instantaneous trip of 4350 A is $4350 \times 5/300 = 72.5$ A. From the characteristic curve in Fig. 8.36, the relay is capable of handling currents up to about 25 times its rating. The maximum current at the instantaneous setting of the circuit breaker is $72.5/4.68 = 15.49$ times the heater rating. An approximate motor-starting calculation in Fig. 8.32 shows this motor may take up to 10.25 s to start. Since locked rotor current (1450 primary amperes $\times$ 5/300 $\times$ 1/4.68 $\cong$ 5 times heater rating) may persist for almost this long, the slow time form of this relay has been selected.

The selected ratings for the motor controller are:

600-A motor circuit protector with adjustable instantaneous range from 3000 to 6000 A

Thermal overload relay with heater rated 4.68 A, slow time form

Examining the characteristic curve for the 600-A instantaneous-only motor protector in Fig. 8.34 reveals that on the 4 setting (4300 A), the circuit breaker may trip at any current between 5 and 8 times rating (3000 to 4800 A). On the 6 setting (5100 A), it may trip at any current between 6 to 8.5 times rating (3600 to 5100 A). The 6 setting should be sufficient to avoid false tripping at 4350 A (a conservative estimate). The operating time of the motor controller motor thermal relay at the maximum current for the circuit-breaker instantaneous trip compared to the minimum melting time of the transformer primary fuse is shown in Table 8.13.

8. Set the transformer main secondary circuit breaker in accordance with the criteria in Table 8.14.

9. Check to be sure that the main secondary circuit breaker trips before the primary fuse minimum melting time at the maximum long time delay for the short time delay pickup and at the maximum through-fault for the short time delay. (See Table 8.15.)

From Table 8.16, it can be seen that for certain high-magnitude secondary faults not detected by feeder breakers (bus faults), both the main breaker and the primary fuse will operate. This compromise is one that is usually accepted for this type of substation.

10. The settings determined by the preceding procedure may be verified by drawing time-current curves either by superimposing the individual device time-current curves

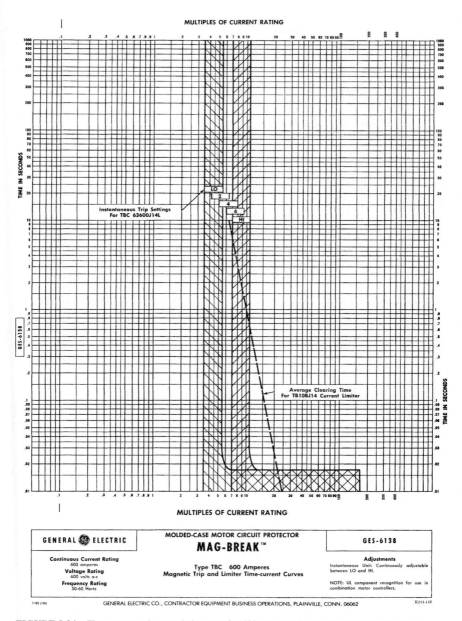

FIGURE 8.34   Time-current characteristic curve for 600-A molded-case motor circuit protector.

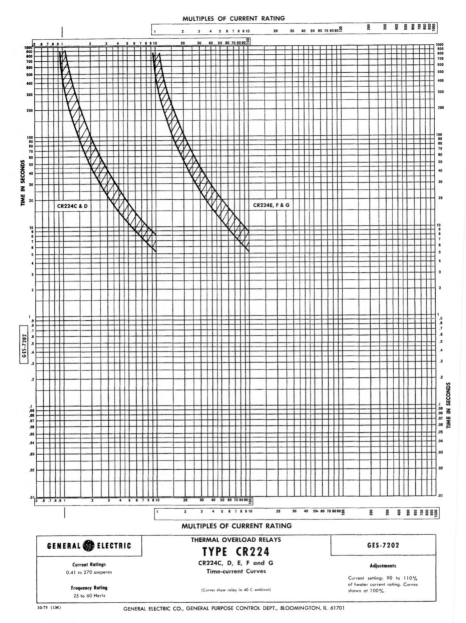

FIGURE 8.35 Time-current characteristic curve for thermal overload relays with maximum current rating of 10 times nominal rating.

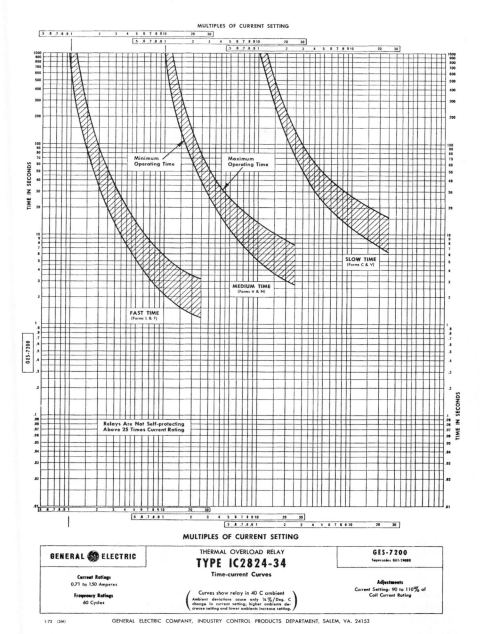

FIGURE 8.36 Time-current characteristic curve for thermal overload relays with maximum current rating of 25 times nominal rating.

TABLE 8.13   500-MCM Cable Short-Circuit
Heating Limit Protection Data

500-MCM cable protection		
	Time	
Current (A)	Cable	Relay
2400	>10 s	2.0 s
6240	>10 s	0.6 s*
24886	2.5 s	0.1 s†

*Add five cycles (0.083 s) for circuit-breaker oper-
ating time.

†Relay + circuit-breaker operating time = 1 + 5
cycles = 0.1 s.

TABLE 8.14   Motor Protection and Transformer
Primary Fuse Coordination Data for Example

Motor circuit breaker (See Fig. 8.34)	
Instantaneous setting	Maximum current
6	5100 A

Motor thermal relay (See Fig. 8.36)	
Calculation	Maximum time
$5100 \times 5/300 = 85$ A	20 s
$85/4.68 = 18$ times rating	

Fuse minimum melt (See Fig. 8.2)	
Calculation	Minimum time
$5100 \times 0.48/4.16 \times 0.58 = 341$ A	>1000 s

on a single curve with a selected current base or using a computer program to accomplish this function. Computer programs are available from many of the vendors of short-circuit programs. Figure 8.37 shows a set of curves for the sample system described.

Care must be taken to use this verification only for radial systems with a single source, because multisource systems are not easily represented on time-current curves. An example of multisource system coordination is given in the ground-fault example in Sec. 8.12.

NOTE: The classic 16 percent current curve separation, usually difficult to obtain for low-voltage devices, is sometimes ignored by protection engineers when employing electronic trip devices. As long as the curves do not overlap, selectivity exists.

TABLE 8.15 1500-kVA Transformer Main Secondary Circuit-Breaker Setting Criteria

Main secondary circuit breaker, 3200-A frame, 2400-A rating plug; long-time, short-time, ground-fault trip elements (see Fig. 8.4 for characteristic)

Criterion	Calculation	Setting
	Long-time pickup setting	
About 1.5 times maximum transformer current	1.5×1804 A = 2706 A	1.1X = C (2640 A)
	Time band	
Longer than maximum thermal relay time at maximum instantaneous operation of motor protector	5100/2640 = 1.93 20 s	1 (23 s)
	Short-time pickup and time-band settings	
1.1 times maximum trip instantaneous operating current of motor protector	1.1×5100 A = 5610A 5610/2640 = 2.34C	2.5C (6600 A) intermediate band

TABLE 8.16 Main Secondary Circuit-Breaker Setting and Primary Fuse Coordination Data

Current	Time
Maximum long-time-delay pickup at short-time-delay pickup (see Fig. 8.4)	
$1.1 \times 2.5 \times 2640$ A = 7260	20 s
Fuse minimum melt (see Fig. 8.2)	
$7260 \times 0.48/4.16 = 838$ A	30 s
Maximum short-time delay at maximum short-circuit current (see Fig. 8.3)	
27,056 A	0.32 s
Fuse minimum melt (see Fig. 8.2)	
$27,056$ A $\times 0.48/4.16 = 3118$ A	0.06 s

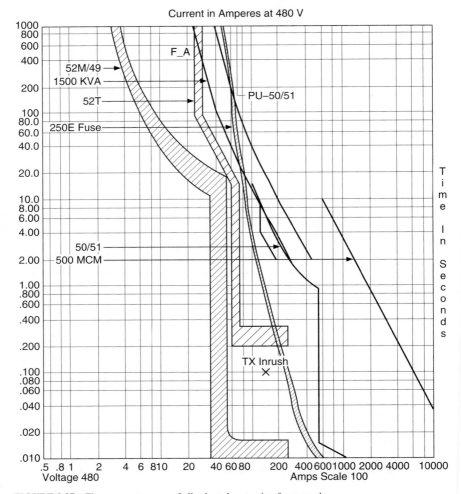

FIGURE 8.37 Time-current curves of all selected protection for example.

8.12 MULTISOURCE GROUND OVERCURRENT COORDINATION

Figure 8.38 shows the ground-fault current flow in different circuits for the fault location shown with three generators in parallel with a utility. Inverse time relays (see Fig. 8.39 for characteristic curves) are used, because the magnitude of ground currents in the different circuits varies with the location of the fault. The sequence of tripping selected should allow about 0.3 s between relay operation for any fault location.

Table 8.17 shows good coordination for faults on the utility transformer secondary system, but faults on the furthest feeder bus with the same settings are shown in Table 8.18.

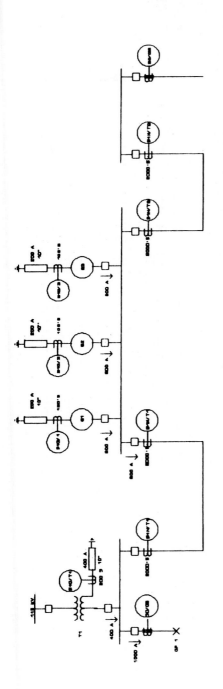

51G/T1   Inverse Time Overcurrent Relay, 1-12 A Taps

51N/T1,T2   Inverse Time Overcurrent Relay, 0.5-4 A Taps

51G/G1,G2,G3   Inverse Time Overcurrent Relay, 1-12 A Taps.

50/GS   Instantaneous Ground Sensor Overcurrent Relay, 0.5-4 A Taps

FIGURE 8.38 Multiground source system one–line diagram showing ground-fault current flow for fault GF_1.

8.55

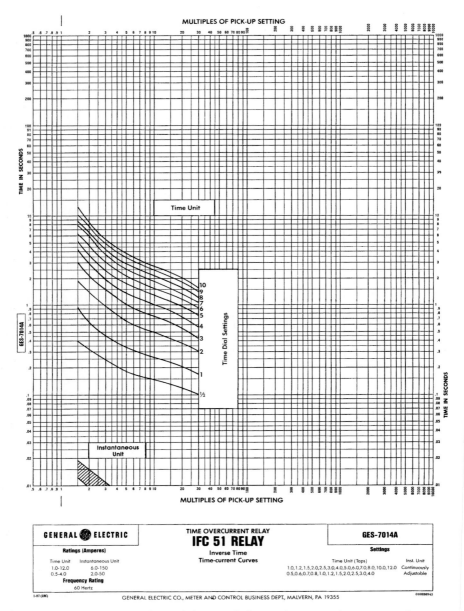

FIGURE 8.39 Time-current characteristic curves for inverse electromagnetic overcurrent relays.

TABLE 8.17 Initially Selected Ground-Fault Relay Settings

Ground-fault relay coordination for transformer bus feeder fault GF $_1$

Circuit	CT ratio	Relay pickup (A)		Circuit current (A)	Times pickup	Time dial	Seconds
		Secondary	Primary				
Feeder	50:5	0.5	15	1000			Instantaneous
Tie 1	2000:5	0.5	200	600	3	1	0.5
Transf.1	200:5	5	200	400	2	1	0.8
Gen.1	100:5	5	100	200	2	1.5	1.1
Gen.2	100:5	5	100	200	2	1.5	1.1
Gen.3	100:5	5	100	200	2	1.5	1.1

TABLE 8.18 Relay Operating Times for GF $_2$ Fault Using Initially Selected Relay Settings

Relay operating times for furthest bus feeder fault (settings from Table 8.12)

Circuit	CT ratio	Relay pickup (A)		Circuit current (A)	Times pickup	Time dial	Seconds
		Secondary	Primary				
Feeder	50:5	0.5	15	1000			Instantaneous
Tie 2	2000:5	0.5	200	1000	5	1	0.35
Tie 1	2000:5	0.5	200	400	2	1	0.7
Transf.1	200:5	5	200	400	2	1	0.8
Gen.1	100:5	5	100	200	2	1.5	1.1
Gen.2	100:5	5	100	200	2	1.5	1.1
Gen.3	100:5	5	100	200	2	1.5	1.1

The sequence of operation with the settings initially selected is not satisfactory because, for a GF2 fault, the transformer main secondary ground relay operates too soon. The fault GF-2 location is shown in Fig. 8.40.

Order	Circuit	Ground fault 1	Ground fault 2
1	Feeder	Instantaneous	Instantaneous
2	Tie 2	—	0.35 s
3	Tie 1	0.5 s	0.7 s
4	Transformer	0.8 s	0.8 s
5	Generators	1.1 s	1.1 s

The relay operating times given in Table 8.13 do not provide selectivity in accordance with the usual criteria (a 0.3-s margin at maximum fault current). Increasing the time dial setting of the transformer ground relay will result in settings which meet the usual criteria as shown in the tabulation given in Table 8.14 and the tabulation following:

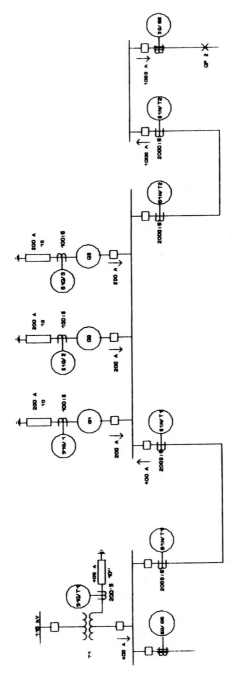

FIGURE 8.40 Multiground source system one-line diagram showing ground-fault current flow for fault GF_2.

Order	Circuit	Ground fault 1	Ground fault 2
1	Feeder	Instantaneous	Instantaneous
2	Tie 2	—	0.35 s
3	Tie 1	0.5 s	0.7 s
4	Transformer	1.1 s	1.1 s
5	Generators	1.1 s	1.1 s

The final selected relay settings are shown in Table 8.19.

TABLE 8.19 Relay Operating Times for GF$_2$ Fault Using Modified Time Dial Setting for Transformer Main Secondary Ground Relay

Relay operating times for furthest bus feeder fault (modified settings)

Circuit	CT ratio	Relay pickup (A) Secondary	Primary	Circuit current (A)	Times pickup	Time dial	Seconds
Feeder	50:5	0.5	15	1000			Instantaneous
Tie 2	2000:5	0.5	200	1000	5	1	0.35
Tie 1	2000:5	0.5	200	400	2	1	0.7
Transf.1	200:5	5	200	400	2	1.5	1.1
Gen.1	100:5	5	100	200	2	1.5	1.1
Gen.2	100:5	5	100	200	2	1.5	1.1
Gen.3	100:5	5	100	200	2	1.5	1.1

8.13 REFERENCES

1. ANSI/NFPA Standard 70 National Electrical Code (NEC), National Fire Protection Association, 1996.

2. ANSI Standard C2 National Electrical Safety Code (NECS), Institute of Electrical and Electronics Engineers, 1993

3. IEEE Standard C37.41-1988, *IEEE Standard Design Tests for High-Voltage Fuses, Distribution Enclosed Single-Pole Air Switches, Fuse Disconnecting Switches, and Accessories,* Institute of Electrical and Electronics Engineers.

4. ANSI/IEEE Standard C57.109-1985, *IEEE Guide for Liquid-Immersed Transformer Through-Fault-Current Duration,* Institute of Electrical and Electronics Engineers.

5. IEEE Standard C37.2-1991, *IEEE Standard Electrical Power System Device Function Numbers,* Institute of Electrical and Electronics Engineers.

6. IEEE Standard 141-1986, *IEEE Recommended Practice for Electric Power Distribution for Industrial Plants* (IEEE Red Book), Institute of Electrical and Electronics Engineers.

7. NEMA Standard MG1-1987, *Motors and Generators,* National Electrical Manufacturers Association.

CHAPTER 9
SYSTEM AND EQUIPMENT GROUNDING

9.0 INTRODUCTION

Grounding is one of the most important design aspects of an electric power system. Proper grounding contributes to the safety of equipment and personnel as well as to proper operation of control and other sensitive circuits. The primary reasons for grounding equipment and neutral points of a power system are:

- To limit fundamental-frequency electric potential between all uninsulated conducting objects in a local area
- To limit touch and step potentials
- To limit overvoltages on equipment and circuits for various operating conditions
- To limit system unbalances
- To provide for relaying and subsequent isolation of faulted equipment and circuits when phase-to-ground faults occur
- To provide four-wire, three-phase power supply (low-voltage)

Any potential difference, if sufficiently high, between uninsulated equipment or equipment and ground may damage equipment. Similarly, such potential difference may set up a shock hazard to personnel. Next to safety, isolating the faulted equipment or circuit as soon as possible is important so that any damage to the faulted and other nearby equipment is avoided or minimized. The grounding of neutral points limits the magnitude of an overvoltage as well as its propagation into other phases and parts of the equipment by holding the neutral potential close to ground potential.

This chapter covers the grounding of medium- and low-voltage systems that are found in industrial and commercial facilities. The grounding practices may be conceptually divided into three different classifications: system, equipment, and static and lightning grounding. System grounding deals with grounding of the neutral of the power supply equipment. For all practical and safety purposes, the neutral is considered as a live conductor and has insulation strength requirements. Equipment grounding deals with grounding of the uninsulated metallic or nonconducting parts of the equipment. Static grounding deals with safely discharging the built-up charge on the equipment or some other object touching the equipment. A person touching the keyboard of a computer and resetting the computer is a familiar example of the effect of static discharge. The lightning grounding serves a similar purpose, except the impulse is caused by lightning and the discharge current and energy involved are much larger.

Another important conceptual distinction should be understood before dealing with the details of grounding. It is not uncommon among power engineers to interchangeably use the terms earth, ground, and neutral. Using these terms interchangeably may not create a serious problem when the actual scenario being discussed is well understood by all the parties involved. However, the three terms are distinct and describe different aspects of grounding. *Earth* is the natural earth and it has a finite resistance expressed as ohm-meters. Depending upon the type, composition, and moisture content of the earth, the resistance may vary from a few tenths to thousands of ohm-meters. A *ground* is basically a zero potential electrical network or connection. By definition, the voltage difference between different points of a ground is within safe limits. A ground is usually connected to earth and/or neutral wire at one or more points. Whenever large currents (such as line-to-ground fault currents or lightning surges) pass through grounding points to earth, a high voltage (due to finite resistance and inductance of grounding and earth) may exist until the current ceases to flow. This is called the *ground potential rise* (GPR). If the GPR is high, this may cause other problems. Proper design of grounding should limit this GPR. The concept of *neutral* grounding was introduced earlier in this section.

A detailed discussion on grounding practices is available in several references, such as 6, cited at the end of this chapter. Further, the National Electrical Code (NEC) and other regulations spell out certain minimum grounding requirements applicable to industrial and commercial installations.

9.1 SYSTEM GROUNDING

In industrial and commercial power systems, the system should operate grounded. There are, however, other methods of operation that have been used historically and have operated satisfactorily for several years but do not have all of the advantages of the operation of a grounded system. Systems can be defined as:

- *Ungrounded* (although these are grounded through the distributed capacitance of the system)
- *Solidly grounded* (needed in four-wire systems with line-to-neutral loads)
- *Resistance-grounded* (either low- or high-resistance)
- *Reactance-grounded* (a special case that is not found in industrial and commercial systems)

There are advantages and disadvantages with each of these methods of grounding (Table 9.1), but for industrial and commercial power systems, the resistance-grounded system has the most advantages.

9.1.1 Ungrounded Systems

Historically, ungrounded systems were used initially because the first line-to-ground fault would not cause interruption of power. It did change the voltage to ground of the other two phases to full line-to-line voltage to ground. Figure 9.1(*a*) shows the phasor relationships to ground on an ungrounded system. When a second line-to-ground fault occurs on another feeder, it causes a phase-to-phase fault that interrupts two circuits instead of one.

It was discovered that there were multiple motor failures on this type of system. When the system was analyzed, it was determined that the distributed capacitance in parallel with a fault in an inductive device or an arcing fault could be a resonant circuit

TABLE 9.1 Summary of Advantages and Disadvantages of Grounded and Ungrounded Systems

	Grounded system	Ungrounded system
Advantages	• Fault detected and cleared as soon as possible • Safety • Balanced voltage • Ground current provides an easy means for fault detection • Resistance-grounding limits fault currents	• Single line-to-ground fault does not cause interruption of supply • Lower short-circuit current contribution to ground faults
Disadvantages	• L-G fault causes interruption of load • Local faults may trip the main supply breaker • Higher L-G fault-current levels if not resistance-grounded • May have marginally higher cost	• Equipment phases may be subjected to line-to-line outages • Phase-to-phase fault current may not be high enough for overcurrent relay tripping • Susceptible to arcing fault burn-downs and remote fault initiation due to overvoltages • Susceptible to more multiple faults • Locating faults difficult • Not as safe due to line-to-line voltage being impressed on phases to ground • Regular and fast response maintenance needed • Ground-fault levels high and may cause damage

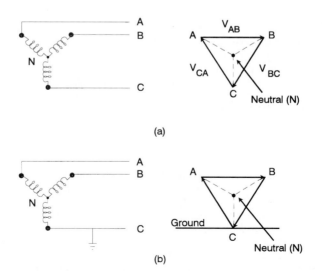

FIGURE 9.1 Voltages to ground under steady state conditions.

9.3

where the voltage could be several times normal (five or more times rated system voltage). These transient overvoltages occurred on low-voltage systems and caused multiple motor failures. To eliminate this problem, grounding systems were developed and promoted.

Many ungrounded systems were the result of delta-connected low-voltage windings. The transient overvoltage problem was eliminated by grounding one phase of the delta winding. This eliminated the high transient overvoltages, but still impressed full line-to-line voltage on the other two phases.

A simple ground-fault detection system for ungrounded systems consists of three lamps connected phase-to-ground. When all three lamps have the same brightness, there are no grounds (faults) present on the system. When one lamp goes out and the other two are bright, that phase which has the lamp out has a ground fault. These lamps are a resistance that is in parallel with the distributed capacitance so it tends to damp some of the transient overvoltages. The lamps indicate which phase is grounded but do not indicate where on the system the ground has occurred.

The operation of ungrounded systems requires that, at the indication of the first ground (fault), maintenance should locate the fault and correct it. This is one of the major disadvantages to ungrounded systems: finding the fault.

9.1.2   Solidly Grounded Systems

Solidly grounded systems are grounded by connecting the neutral of the system to ground or by connecting one phase to ground. With solid grounding, overvoltages on the system are eliminated. One phase connected to ground does impress full line-to-line voltage on the other two phases, but it eliminates the problem of transient overvoltage. Four-wire systems with line-to-neutral loads must be solidly grounded. Solidly grounded systems do not limit ground-fault current. If a ground fault occurs in a motor or generator, only the fault impedance limits the fault current. Burning of the laminations in the machine core can result in major damage that would require restacking the core of the machine. This would increase the cost of repair. For this reason, medium-voltage systems are resistance-grounded to limit the damage to machines connected to it.

It has been the practice to solidly ground low-voltage systems. Until ground sensor devices were available to detect low ground-fault currents, it was necessary to solidly ground low-voltage systems to have enough ground-fault current to trip circuit breakers or operate fuses in motor controllers. With the availability of ground-fault sensors, it is now possible to limit ground faults through the use of high-resistance grounding and still sense and operate a circuit breaker or an alarm to indicate the faulted circuit. These low-current faults will not blow fuses. See Figure 9.2 and Table 9.2.

9.1.3   Resistance-grounded Systems

When the source of the power in a system is a wye-connected transformer or generator, the neutral connection of the source is grounded through a resistor. Resistance-grounding is classified as low-resistance or high-resistance.

If a system neutral is not available, then a grounding transformer can be used to derive the system neutral. This transformer can be either a zigzag connection, or a three-phase distribution-type transformer (Fig. 9.3) that has its primary winding connected in wye and the secondary connected in delta. The normal impedance to positive sequence current is high, but the impedance to zero sequence current is low so that any ground-fault current will flow through the transformer.

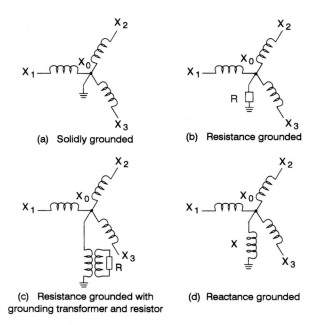

(a) Solidly grounded

(b) Resistance grounded

(c) Resistance grounded with grounding transformer and resistor

(d) Reactance grounded

FIGURE 9.2 Different grounding types.

TABLE 9.2 Characteristics of Different System Grounding Methods

	Solid grounding	Resistance-grounding		Reactance-grounding
		Low	High	
Phase-to-ground fault current in percentage of three-phase fault current	Varies, may be 100% or greater	5–20%	<1%	25%
Overvoltages	Not excessive	Not excessive	Not excessive	May be high
Automatic segregation of faulted circuit	Yes Trip	Yes Trip	No Alarm	Yes Trip

Resistance-grounded systems have all of the advantages and none of the disadvantages of other methods of grounding, with the exception of allowing the system to operate with one phase grounded. This, however, can be done with a high-resistance system. Medium-voltage systems use low-resistance grounding. Low-voltage systems use high-resistance grounding.

A low-resistance-grounded system is one that limits the fault current of a line-to-ground fault to a value in hundreds of amperes. Because sensitive fault detection is available, the lower ground fault minimizes the damage at the fault. Ground sensor relaying (Fig. 9.4) is sensitive and can detect a ground fault of 10 to 15 A and instantaneously trip the circuit breaker. If all circuits on the medium-voltage system have ground sensor relaying, the ground resistor can be rated as low as 200 A. If there are

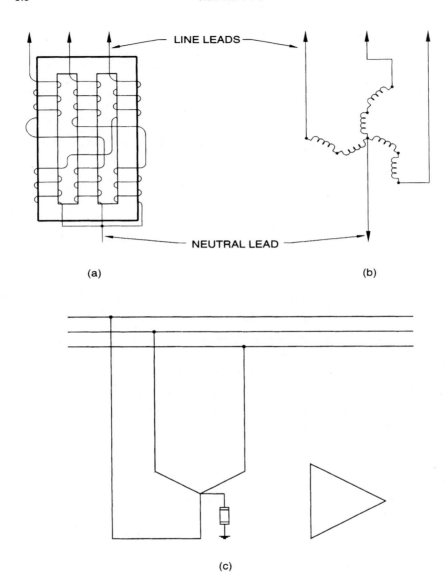

FIGURE 9.3 Grounding transformer connections.

A B C N

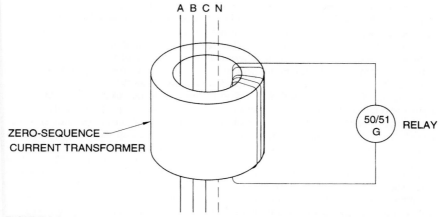

ZERO-SEQUENCE
CURRENT TRANSFORMER

50/51
G   RELAY

FIGURE 9.4 Ground sensor relay system showing a zero sequence (doughnut) current transformer and overcurrent relay.

multiple sources, i.e., more than one transformer and/or generator, the resistors will be in parallel so a larger ground-fault current will result.

If the feeders are using less sensitive residually connected overcurrent relays (Fig. 9.5), the grounding resistor needs to be higher current or 400 A. If the feeder current transformers are 400/5 A, the residual connected relay has a 0.5-A tap; the lowest current that can be detected is 40 A. With the impedance of the fault, the current could be limited to this value, even though the grounding resistor is rated at 400 A. The 400 A is the initial current flow and it decreases with time. Residually connected relays must have time delay, because there may be a false indication when the circuit is energized due to the unbalanced magnetization in the three current transformers.

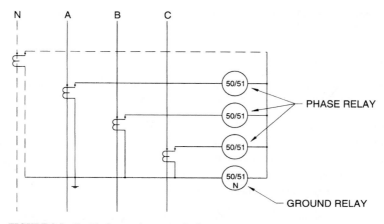

FIGURE 9.5 Residual ground connected relay.

High-resistance grounding of low-voltage systems enables operation with one phase grounded, which will allow a ground fault to flow of the same order of magnitude as the current that would flow through the distributed capacitance of the system. Thus, the grounding resistor must be: $R \leq X_{co}/3$. This resistor is in parallel with the distributed capacitance and, if its ohmic value is less than the reactance of the distributed capacitance, it becomes the dominant element in the zero sequence network. In low-voltage systems, the capacitive current in ungrounded systems is less than 2 A.

There are standard equipments that limit the ground current to 2 A. They come with a pulsing relay so that, if a ground fault occurs and an alarm is given, the pulsing relay can be activated and, with a clamp-on ammeter, the pulse of current (usually 6 A) can be detected on the faulted phase and traced down that phase to the fault. This allows location of the fault so that it can be removed and operation will not be interrupted.

It is possible to measure the capacitance charging current on a system. This is done by placing a variable resistor from one phase to ground and reducing this resistance to zero. The one phase is now grounded and the current to ground can be measured with an ammeter either in the circuit or with a clamp-on ammeter. If a section of a motor control circuit is available, one output phase from a motor controller can be used. The circuit breaker or fuse in the motor controller gives protection against short circuits. The measured current is usually 1 A or less.

The 2-A current is low enough so that the damage caused by leaving it on for a short time (hours) will not escalate to a larger fault, although if left on for a longer period, it could escalate to a phase-to-phase-to-ground fault depending upon the location of the fault. Good practice dictates that any fault be removed as soon as possible.

9.1.4 Reactance-Grounded Systems

A power system can also be grounded by placing a reactance in the neutral-to-ground circuit. As in the resistance-grounded system, the magnitude of the reactance determines the operating characteristics of the circuit. As indicated in Table 9.1, reactance-grounded systems are used in industrial power systems only under exceptional circumstances. Sometimes generator neutrals are reactance-grounded. In some European countries, transformer neutrals in transmission systems are reactance-grounded. In industrial and commercial power systems, resistance-grounded systems give the best service under both steady state and transient conditions. The resistance damps any transient that appears line to ground.

9.1.5 Other Considerations

In system grounding practices, the selection of grounding points is important. There are several factors to be considered in selecting the grounding points:

- Grounding at each voltage level
- Grounding at source, but not at load
- Grounding at each major source bus
- Number of sources
- Reference ground during isolated operation

At least one grounding point at each voltage level is necessary (Fig. 9.6). This is achieved because most of the transformers have delta (ungrounded on the high side)-wye (grounded on the low side) windings. Autotransformers have common neutrals. At

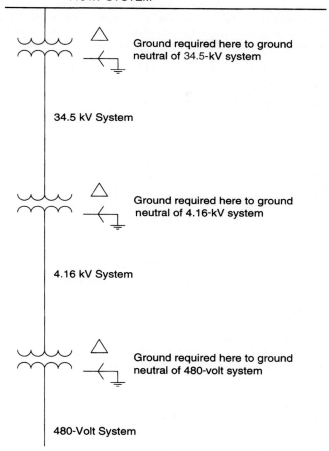

FIGURE 9.6 Common system grounding practice for industrial and commercial power systems.

every stepdown level, the transformer neutral is grounded on the low side. Where delta-delta windings are used, or when isolated operation creates a floating neutral condition, a grounding transformer is required.

Whenever impedance-grounding is used, the impedance in neutral should be applied properly. The five important factors are:

- Limiting of line-to-ground fault current
- Providing sufficient fault current for detection and relay operation
- Having proper thermal rating (I^2R)
- Having proper voltage rating
- Being within neutral BIL

EXAMPLE 9.1 *Consider a 50-MVA, 13.8-kV, wye-connected (with grounded neutral) generator within a cogeneration plant which has $x_d'' = 0.15$ and $x_d' = 0.23$ p.u. reactances. For the purpose of this calculation, assume that positive, negative, and zero sequence impedances are all equal. This generator is connected to the utility transmission system through a delta-wye transformer with wye being on the high-voltage side.*

1. Compute fault current for a line-to-ground fault at the generator terminal.
2. Calculate the grounding resistance needed to limit this fault current to 5 A.
3. If the resistance is connected to the secondary of a grounding transformer with a ratio of 12 kV to 240 V, compute the value of the resistance to limit the fault current to 5 A on the generator.

Solution:

1. Select the machine rating of 50 MVA and 13.8 kV (line-to-line voltage) as base quantities.

$$I_{base} = \frac{50 \times 10^6}{\sqrt{3} \times 13.8 \times 10^3} = 2091.8 \text{ A}$$

$$Z_{base} = \frac{13.8 \times 10^3}{\sqrt{3} \times 2091.8} = 3.81 \ \Omega$$

Note that (kV^2/MVA base) will also give the base ohms directly.
Line-to-ground fault current

$$I_f = \frac{3V}{Z_1 + Z_2 + Z_0}$$

$$= \frac{3}{0.1 + .1 + .15} = 6.7 \text{ p.u.}$$

$$= 13{,}945.7 \text{ A}$$

2. Total impedance needed to limit the fault current to 5 A or 2.39×10^{-3} p.u. is given by

$$Z = \frac{3}{2.39 \times 10^{-3}} = 1255.1 \text{ p.u.}$$

Thus, the grounding resistance (R_g) needed is given by

$$Z_1 + Z_2 + Z_0 + 3 R_g = 1255.1 \text{ p.u.}$$

Note that the generator sequence impedances Z_1, Z_2, and Z_0 are reactances and the grounding impedance is resistance. For all practical purposes, the total of these three impedances (0.84 p.u.) is very small as compared to the grounding resistance R_g, especially when expressed in the form of $3R_g + j(X_1 + X_2 + X_0)$. Hence, let us ignore the generator impedances in this calculation.

$$3 R_g = 1255.1 \text{ p.u. approximately}$$

$$R_g = 418.4 \text{ p.u.}$$

$$= 1593.5 \ \Omega$$

3. A grounding transformer with a ratio of 12 kV to 240 V may be used, so that a lower voltage rating and a smaller value of grounding resistance is possible.

The resistance value on the secondary side may be obtained directly by using the formula

$$R_{g,\text{low}} = R_{g,\text{high}} * \left(\frac{V_{\text{low}}}{V_{\text{high}}}\right)^2$$

$$R_{g,\text{low}} = 1593.5 * \left(\frac{240}{12 \times 10^3}\right)^2$$

$$= 0.637 \ \Omega$$

The current rating of this grounding resistor on the secondary (low-voltage side) of the grounding transformer may be directly calculated by

$$I_{g,\text{low}} = I_{g,\text{high}} * \frac{V_{\text{high}}}{V_{\text{low}}}$$

$$= 5.01 * \frac{12 \times 10^3}{240}$$

$$= 250.5 \ \text{A}$$

Approximate (shortcut) calculation:

$$I_{\text{fault}} = \frac{I_{\text{full load}}}{Z \ \text{p.u.}} = \frac{2091.8}{0.15} = 13,945 \ \text{A}$$

$$R_g = \frac{E_t}{\sqrt{3} \times I_f} = \frac{13,800}{\sqrt{3} \times 5} = 1,593.5 \ \Omega$$

9.2 EQUIPMENT GROUNDING

Equipment grounding connects the noninsulated, non-current-carrying parts of all electrical equipment with the earth. Equipment is grounded to:

- Protect personnel from electric shocks.
- Prevent damage to equipment by providing a low-impedance path between a fault and the source of ground-fault current.
- Facilitate the operation of protective devices.
- Minimize the buildup of static charges.

The importance of eliminating shock hazard through proper equipment grounding and ground mat design cannot be emphasized enough. However, it should be noted that motor neutrals are never grounded.

A simplified illustration of an indoor electric shock hazard and its mitigation by using equipment grounding are shown in Fig. 9.7. Figure 9.7(a) represents a condition in which a solidly grounded ac power system is supplying a motor with an ungrounded

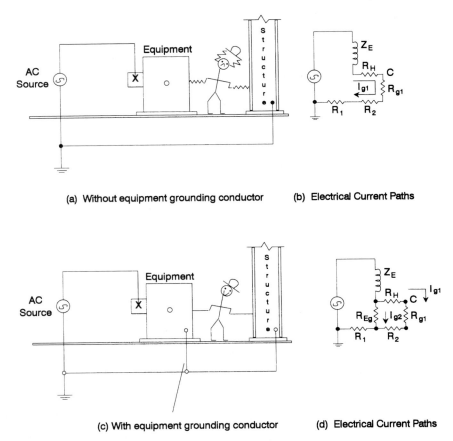

FIGURE 9.7   Principle of equipment grounding to mitigate electric shock hazard.

frame; a high-resistance ground fault occurs within the motor, and the fault current is not large enough to be detected by the motor overcurrent protection. If a person simultaneously touches the ungrounded motor enclosure and is in contact with the ground (such as touching the nearby grounded structural column), this person will become part of the electric circuit and receive an electric shock due to the potential difference between the motor enclosure and the steel column. The equivalent electric circuit is shown in Fig. 9.7(b). When the person touches either motor frame or column while holding the other, the touch potential is the line voltage. The current passing through the person is limited by the fault impedance (Z_E) because R_{g1} is very small and R_H is also small as compared to Z_E. In any case, the hazard of a high-voltage shock followed by a high current above the let-go current threshold is evident in this example.

In the equivalent circuit of Fig. 9.7(d), the equipment-grounding conductor resistance R_{Eg} is comparable to $R_{g1,}$ and resistance R_2 is very small in relation to R_H. Hence, the voltage differential is very small and almost all of the fault current flows through the equipment-grounding conductor (R_{Eg}).

All electric conductor housings, such as cable trays, junction boxes, conduits, equipment enclosures, and motor frames, should be connected to a properly grounded equipment-grounding system. The conductors should be properly sized so that they can carry phase-to-ground fault current. The grounding points should be such that step and touch potentials are within safe limits. The grounding conductor should be in the same raceway or conduit to minimize the reactance in the ground return path. The grounding conductor system requirements are given in ANSI/NFPA 70-1990 (NEC) and ANSI C2-1987 (NESC).

In substations, personnel can be subjected to shock hazards due to large potential gradients caused by ground-fault current returning to its source through the earth. Under very adverse conditions the earth potential gradient could be high enough to be hazardous for personnel walking in certain areas of the plant (step voltage hazard) and for personnel who come in contact with a grounded structure (touch voltage hazard) during a ground fault. IEEE Standard 80-1986 gives further information on substation grounding design requirements and detailed calculation procedures. ANSI/IEEE Standard 142-1991 provides details on preferred methods of grounding of different types of equipment. Further, all national and state electrical codes apply.

9.3 STATIC GROUNDING

The static grounding is used to discharge the static electricity to earth safely. The most common example of static electricity is the shock and spark experienced by people when they touch grounded and/or metallic objects in low humidity and heated buildings during the winter season. All objects above the earth have a capacitance with respect to ground. This is sometimes called *stray capacitance*. Hence, this capacitance can accumulate any charge that is generated by friction, dust, flow of liquids and gases, etc. The amount of charge and the resulting voltage is given by the basic equation

$$Q = CV \qquad (9.1)$$

where Q is the charge in coulombs
C is the capacitance in farads of the object with respect to ground
V is the voltage in volts across the capacitor

If the object is stationary, the voltage increases as the accumulated charge increases.

Static electricity may accumulate on belts, conveyors, drives, piping, paper machines, etc. If the voltage is high enough, arcing may result. This arcing to ground or other equipment in the presence of flammable or explosive material may cause fires and explosions. Motor-bearing failure due to static-induced current flow is well known. Thus, control of static charge is very important. The objectives of static-charge grounding are

- Protection of personnel
- Loss of production due to equipment outage
- Deterioration in product quality
- Loss of facility due to explosion of dust or liquid fumes

Usually the capacitance of objects is small . Thus, even for small values of static charge, very high voltage ($V = Q/C$) is possible. However, when the object is grounded through a low-value resistance, the charge leaks through the resistance and the voltage decays to zero. If the capacitance gets charged and discharged continuously (as in an ac circuit), then a leakage current flows through the resistance continuously.

If an object is placed in an electrostatic field (object 2 with capacitance C_{22} in Fig. 9.8), then, due to coupling capacitance (C_{12}), this second object acquires a voltage unless it is grounded. The electrostatically induced voltage (for the ungrounded condition) is given by:

$$V_2 = \left(\frac{C_{12}}{C_{22} + C_{12}}\right)V_1 \qquad (9.2)$$

Grounding of this second object discharges the stray capacitance.

If the electronic equipment is floating, its self-capacitance to the ground plane and mutual capacitance to the human body will increase the potential difference between them. This is true especially for small equipment (Fig. 9.8).

The voltages on the two objects in Fig. 9.8 are:

$$V_1 = \frac{Q_1}{\{C_{11} + C_{12} * C_{22}/C_{12} + C_{22}\}}$$

$$V_2 = -V_1 * \left\{\frac{C_{12}}{(C_{12} + C_{22})}\right\} \qquad (9.3)$$

$$V_1 - V_2 = V_1 * \left\{\frac{C_{22}}{(C_{12} + C_{22})}\right\}$$

where V_1, V_2 = voltages of objects 1 and 2
$\quad Q_1, Q_2$ = charges on objects 1 and 2, in coulombs
$\quad\quad C_{12}$ = mutual capacitance between objects 1 and 2, in farads
C_{11}, C_{22} = self-capacitances of objects 1 and 2, in farads

The last equation shows that there could be a large voltage difference between person and equipment.

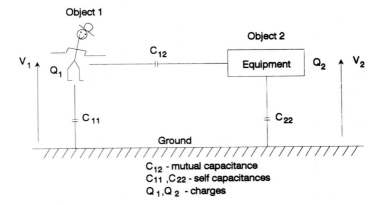

FIGURE 9.8   Mutual and self capacitances.

Bonding various parts of the equipment together and grounding the entire system is a common practice to reduce static charge buildup. Humidity control, electrostatic precipitators, and other neutralizers are other means to reduce static-charge buildup. Conductive floors, grounded mats, etc., are also used. Each static electricity generation case is a special situation. The source of static charge and the surrounding environment plays a big part in mitigation of this problem.

9.4 LIGHTNING PROTECTION GROUNDING

The purpose of lightning protection grounding is to provide a suitable path for lightning current to discharge to earth. The current magnitude of a stroke may range from 2 to 200 kA. The current buildup rate may be 10,000 A/μs; consequently, full current may build up within a 1- to 10-μs range. Because of fast rate of current rise, even with small inductance of 1 μH, high voltages such as 10 kV are possible.

The various considerations that are usually reviewed to determine the extent of lightning protection are:

- Personnel hazard
- Equipment damage
- Production loss
- Type of structure or equipment
- Isokeraunic level
- Number and severity of strokes per storm
- Cost of lightning protection
- Insurance premium

Equipment and structures for lightning protection purposes are divided into five classes:

First class: Needs very little or no additional protection; all grounded metal structures (except those with flammable material), water tanks, silos, and metal flagpoles.

Second class: Needs only addition of down conductor to connect metal roofs and metal clads to grounding electrodes.

Third class: Needs addition of conducting air terminals; includes metal-framed buildings with nonconducting facings.

Fourth class: Needs extensive lightning protection; includes nonmetallic buildings, high stacks, chimneys.

Fifth class: Needs full lightning protection; includes special and unique buildings, tanks and tank farms, power plants, substations, transmission lines.

Three different types of lightning protection are used in practice. They are

- Air terminal method
- Grounded masts
- Lightning (surge) arresters

The first two types of lightning protection avert direct strokes and the third type provides a conductive path to ground for the lightning impulse. The air terminal method uses a network of bare grounded conductors on the roof with the vertical rods sticking out and connected to the roof conductors. The grounded masts method uses either tall ground masts near the structure or ground wire suspended over it. This method is used in substations and on overhead transmission lines. Installing lightning arresters protects expensive and exposed equipment, such as transformers, generators, underground cables, etc., from the incoming impulse.

A lightning protection system must be connected to an adequate grounding system in order to be effective. The grounding system used for lightning protection of a structure should always be connected to other plant grounding subsystems (equipment, surge protection, and so on). This is to ensure that all grounded parts will remain at approximately the same potential during a lightning stroke.

The grounding conductor used to connect a surge arrester to the ground electrode should be as straight and short as practicable to reduce its surge impedance, thus minimizing the voltage drop in the conductor when the arrester operates. A local grounding electrode should be provided for grounding surge arresters, and the local electrode should be connected to the station grounding system. The following characteristics should be considered in selecting material for grounding:

- Electrical conductivity
- Fusing temperature
- Mechanical strength
- Corrosion resistance

Copper is the most commonly used metal for grounding in the United States, because of its high electrical conductivity, high fusing temperature, and high corrosion resistance in many soils. However, sufficient attention should be given to the corrosion effect of dissimilar metals. For example, when buried copper and steel are connected, steel will be consumed. Careful evaluation of all the factors involved should be made before aluminum is used for grounding. Use of proper sizes of grounding conductors and connections is also important for proper functioning of lightning protection grounds.

9.5 CONNECTION TO EARTH

Low-impedance (resistance) connection to remote earth is necessary for effective grounding of equipment, circuits, and facilities. A grounding resistance less than 5 Ω is considered a requirement for smaller substations and industrial plants. For larger plants, a resistance of 1 Ω or less is desired. Commonly used grounding electrodes include ground rods, groundmats, water pipes, and concrete-encased electrodes. The grounding resistance of an electrode depends upon

- Resistance of the metal electrode
- Contact resistance between soil and electrode
- Resistance of soil

The first two items for a properly sized electrode constitute a small fraction of an ohm. Hence, soil resistivity plays a big part in selecting the number of electrodes, rod or wire size, and type of design. All the grounding requirements for safety apply.

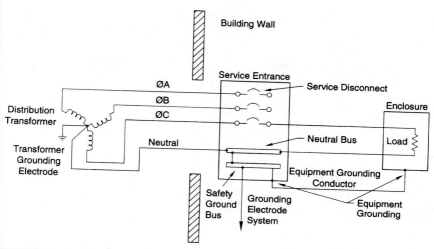

FIGURE 9.9   Grounding system with one equipment ground.

Figure 9.9 shows how each equipment ground should be connected to a system ground at the service entrance. The total facility ground scheme should include one system of ground conductors bonded together, which is then earthed at multiple points (Fig. 9.7). See also Fig. 9.10.

9.6 GROUNDING RESISTANCE MEASUREMENT

The grounding resistance is that resistance between the grounding electrode and remote earth. The remote earth is defined as a point beyond which the total ground resistance of the electrode under consideration does not change. Even though it is possible to compute the total electrode ground resistance, there are many assumed conditions, especially with respect to soil surrounding the electrode. Thus, measurement is usually a requirement not only after construction but also during regular intervals after commissioning. Measurements need not be made to a very high degree of accuracy (± 10 percent is acceptable). But proper and consistent methods should be used.

The following equipment is used to make these measurements:

- Instrument (commercially available)
- Two test electrodes, in addition to the electrode being tested
- Flexible single conductor #14 or larger
- Alligator clips
- Notebook
- Lineman's gloves (optional)

There are three commonly used methods for measuring this resistance:

OUTSIDE

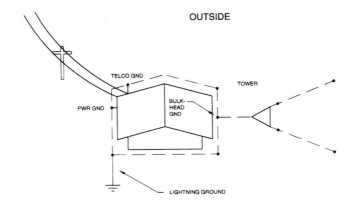

INSIDE

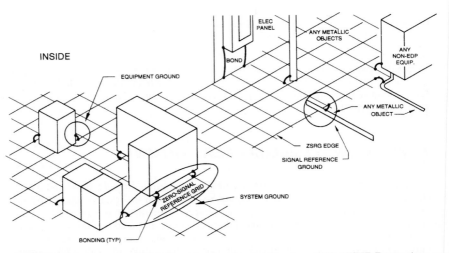

FIGURE 9.10 Types of grounds. (*Reproduced with permission from IEEE, "IEEE Transactions on Industry Applications", vol. IA-21, no. 6, Nov. 1985. © 1985 by IEEE.*)

1. Fall of potential method
2. Two-point method
3. Three-point method

A comparison of these three methods is shown in Table 9.3. More detailed discussion may be found in several references, including Ref. 7.

When inspecting the grounding and bonding system of an installation, check the following points:

1. Ground connections should remain intact at outdoor antennas and wires.

2. The dc resistance of the bond between the bulkhead panel at the building entrance and the ground electrode should be less than 0.01 Ω. This step will help to detect problems with corrosion, oxidation, and loose connections.

3. Bends in grounding and bonding conductors should have a radius exceeding 8 in.

4. No soldered connections should be used in the ground system.

5. The ground resistance to earth should be less than 5 Ω.

6. Surge protectors should be bonded to the nearest ground conductor.

7. Equipment ground conductors should be checked for continuity and arcing.

8. Raceway and cable tray joints should be checked for corrosion, loose connections, and broken conductors or straps.

9. Ground straps should not be exposed to the elements.

10. Bolts should be tight.

TABLE 9.3 Comparison of Ground Electrode Resistance Measurement Methods

	Fall of potential	Two-point method	Three-point method
Number of auxiliary electrodes			
Potential	1 variable		1 fixed
Current	1 distant	1 (water pipe)	1 distant
Measurement output	Plot of voltage vs. distance	Resistance value	3 resistances
Determination criteria	Leveling of potential	Actual value	Computed value
Used for	Large ground mat Small ground mat Single rod	Single rod near residence	Not suitable for large ground mats

11. Copper conductors should be inspected for discoloration due to heating.

9.7 REFERENCES

1. ANSI/IEEE Standard 142-1991 *IEEE Recommended Practice for Grounding of Industrial and Commercial Power Systems,* Institute of Electrical and Electronics Engineers.

2. ANSI/IEEE Standard 602-1986, *IEEE Recommended Practice for Electric Systems in Health Care Facilities,* Institute of Electrical and Electronics Engineers.

3. ANSI/IEEE Standard 1100-1992, *IEEE Recommended Practice for Powering and Grounding Sensitive Electronic Equipment,* Institute of Electrical and Electronics Engineers.

4. National Electrical Safety Code, ANSI/IEEE Standard, Institute of Electrical and Electronics Engineers, 1993.

5. National Electrical Code, National Fire Protection Association, Battery March Park, Quincy, Mass., 1987.

6. ANSI/IEEE Standard 80-1986, *IEEE Guide for Safety in AC Substation Grounding,* Institute of Electrical and Electronics Engineers, 1991.

CHAPTER 10
OVERVOLTAGES AND SURGE VOLTAGE PROTECTION

10.0 INTRODUCTION

Sudden overvoltages may be classified into different categories: surges, impulses, and fundamental-frequency overvoltages. Surges, an increase in voltage for a few milliseconds, are the result of switching circuit elements. Surges may be due to switching capacitive circuits (transmission lines, cables, capacitor banks) or the result of chopping current in an inductive circuit (blowing fuses and opening circuit breakers or contactors). Circuit breakers and contactors normally interrupt current at a current zero, but at times with low values of current they can force the current to zero at a time other than a normal current zero. This phenomenon is called *current chopping*. Impulses are rapid increases in voltage (in microseconds) on a power system due to lightning striking overhead conductors. The high current flowing down a conductor can induce large currents in adjacent conductors. These are not related to system frequency.

Other types of overvoltages have time frames in terms of several cycles of 60 Hz, i.e., seconds to minutes. These are usually called temporary (short-time) overvoltages. Overvoltages beyond minutes are usually steady state in nature.

The term *power quality,* which concerns the quality of voltage at a user's site, has come into use in the last few years. This subject is addressed in Chap. 11. When electric equipment is subjected to abnormal voltages, whether transient, temporary, or sustained, deterioration of electrical insulation (dielectric material) occurs. The factors governing this deterioration mechanism include the type of dielectric material and its physical properties, overvoltage exposure (magnitude, waveshape, duration, repetition rate, cumulative duration), ambient condition (temperature, humidity, moisture, dust, ions), and operating conditions. The physics of this deterioration and failure is complex. The cumulative effect of the deterioration is insulation failure and consequent short circuit. Once this failure process starts, it feeds on itself until either the protection system disconnects the equipment or the equipment fails catastrophically. Thus, the insulation failure is not only a function of the magnitude of overvoltage but also depends upon the cumulative effect of different exposure durations. The relationship is itself nonlinear and can only be empirically estimated.

The solution to the overvoltage-related failure problem is to limit the magnitude and duration of overvoltage exposure of equipment. This is achieved by the application of surge arresters (most of the time), capacitors, gaps, etc. Different parts of the system help to provide protection as well as dissipation of energy from overvoltage surges. Successful application of surge protection requires a certain basic understanding of:

- Surge characteristics
- Insulation withstand levels
- System and equipment behavior and response
- Mitigation (protective) device characteristics
- Individual equipment protection
- Coordinated application of the protective devices in the system

These topics are covered in the rest of this chapter.

In view of the many different sources of overvoltages and complicated insulation behaviors for these overvoltages, appropriate application of surge protective devices is necessary to obtain a cost-effective and reliable operating power system. Application of the surge protective devices has to be coordinated on a systemwide basis. This overall coordination is called *insulation coordination.* One of the important concepts in the selection of insulation and protection levels is that total or absolute protection from failures is neither possible nor cost-effective. Hence, a probabilistic approach consistent with cost and effect of failure is used. An overview of various steps in overvoltage protection and their interrelationship may be depicted as shown in Fig. 10.1. The entire process may take many iterations until acceptable protection at affordable cost is achieved. In industrial and commercial power systems, the approach is straightforward, based on standards[1,2] and experience.

Three important factors for surge protection emerge from this introductory discussion:

- Overvoltage magnitude and exposure time
- Cumulative exposure
- Probability of failure

Thus, the objective of surge protection may be said to be to minimize the probability of failure at the lowest possible cost for anticipated overvoltage exposure of the equipment and system over the life cycle.

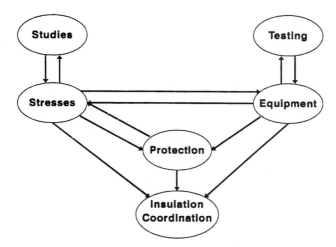

FIGURE 10.1   Relationship of various steps in surge or overvoltage voltage protection.

10.1   OVERVOLTAGES

The first step in surge voltage protection is identification of likely sources—the types, magnitudes, durations, and repetitive rates of overvoltage within a system. A working knowledge of the propagation of these overvoltages in different equipment and circuits is necessary to properly use this information. Different types of overvoltages include

- Fundamental frequency
- Temporary overvoltages
- Switching surges
- Lightning impulses

   The source, the type, the magnitude, and the duration of these overvoltages are different. Consequently, the remedy and protection requirements for these overvoltages are also different. Other than lightning, the other types of overvoltages are caused by the response of the resistance, inductance, and capacitance of the equipment and/or system to some sudden changes within the system. These sudden changes usually are the opening or closing of switches or circuit breakers, fuse blowing, faults, harmonic resonances, etc.

10.1.1   Switching of R, L, C Circuits

The magnitude of overvoltage for a given switching operation depends on the point on the voltage or current wave where switching occurs. Voltage and current waveforms in circuits with resistive, inductive, capacitive, and combinations of these elements are shown in Fig. 3.25(*a*)–(*c*) in Sec. 3.10.
   These waveforms for current and voltage were obtained through EMTP simulations (see Sec.3.10 for a description of EMTP). The equations corresponding to these simulations are given in Table 10.1.
   In Fig. 3.25(*c*), the effect of different types of loads on overvoltage are shown. In all these cases, the switch is being closed at time $t = 0$ or when the voltage is at its peak value and the current is zero. Current cannot be changed instantaneously in an inductance. This is simulated by the (di/dt) term in the equations in Table 10.1. Because current in an inductor lags the voltage across it by 90°, the initial current is zero. There are no step changes in current. The current in a resistor is always in phase with applied voltage. When the load is purely resistive, there is no overvoltage. The final steady state current magnitude in the load is dependent upon the applied voltage, source reactance X_s, and the load R.
   When a capacitor is switched [third item in Fig. 3.25(*c*)] at voltage peak, a high overvoltage results. The actual overvoltage magnitude is dependent upon the relative values of source reactance X_s and the capacitor. The capacitor is assumed to be completely discharged. Because there is no resistance in this circuit, the voltage oscillation continues. In practice, the circuit resistances damp this oscillation. If the capacitor is left with a charge and if the applied voltage is of the opposite polarity at the instant of closing the switch, an even higher transient overvoltage may result.
   In the next item in Fig. 3.25(*c*), a load consisting of an inductor and capacitor is being switched. The overvoltage is shown to be less than in the previous case of capacitor switching. However, if there is a resonance, an even higher overvoltage may result.
   In the last item in Fig. 3.25(*c*), the resistance in the load reduces the overvoltage as well as damps the oscillations to reach a steady state.

TABLE 10.1 Equations Corresponding to Five Elementary Circuits in Figure 3.25(a)–(c) (appropriate initial or preswitching conditions need to be assumed)

Type of load	Equation(s)
Resistor	$V_s(t) = L_s \cdot \dfrac{di}{dt} + R \cdot i(t)$
Inductor	$V_s(t) = L_s \cdot \dfrac{di}{dt} + L \cdot \dfrac{di}{dt}$
Capacitor	$V_s(t) = L_s \cdot \dfrac{di}{dt} + \dfrac{1}{C} \displaystyle\int i_c \cdot dt$
Inductor and capacitor in parallel	$V_s(t) = L_s \cdot \dfrac{di}{dt} + V(t)$ $V(t) = L \cdot \dfrac{di_L}{dt} = \dfrac{1}{C} \displaystyle\int i_c \cdot dt$ $i(t) = i_L(t) + i_c(t)$
Resistor, inductor, and capacitor in parallel	$V_s(t) = L_s \cdot \dfrac{di}{dt} + V(t)$ $V(t) = i_R \cdot R = L \dfrac{di_L}{dt} = \dfrac{1}{C} \displaystyle\int i_c \cdot dt$ $i(t) = i_R(t) + i_L(t) + i_c(t)$

In Fig. 3.25(a) and (b), the effect of two different instances of closing the switch for these five elementary circuits is shown. Refer to a basic network analysis textbook for more detailed technical analysis.

10.1.2 Fundamental-Frequency Overvoltage

As this name implies, the overvoltage is of fundamental frequency (150 or 60 Hz). The duration of these voltages is measured in seconds or minutes. There are two types of such overvoltages:

- Ferranti rise
- Ferroresonance

The Ferranti overvoltages occur when large leading current (such as current in capacitors, or charging current in long transmission lines or cables) flows through an inductive reactance circuit. For a given supply source voltage at one end, the voltage at the other end increases. The increase in voltage as shown in Fig. 10.2 is proportional to the circuit reactance and the capacitive leading current. This overvoltage will last until the net capacitive current is reduced. Usually 10 percent above the nominal volt-

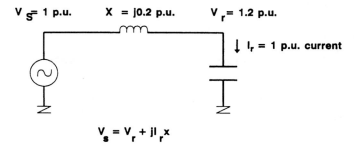

$$V_s = V_r + jI_r x$$

FIGURE 10.2   Ferranti voltage rise example.

age is acceptable for Ferranti-type overvoltages. These types of overvoltages typically occur during light (low) load operating conditions. The available remedy to reduce such overvoltages is to switch out capacitors and/or add shunt reactors to compensate the capacitive charging current. Transformer load tap changers (LTCs) in distribution and industrial systems reduce these overvoltages by decreasing the supply secondary bus voltage.

Ferroresonance overvoltages are due, as the name implies, to overexcitation of a transformer (due to switching, overvoltage, or unbalanced high voltage). The overexcitation saturates the transformer. This gives rise to harmonic currents which in turn may cause resonance between the series reactance in the system and the shunt capacitance of a line or cable. These are very subtle types of overvoltages and practicing engineers should be on the lookout for potential situations where such excessive overvoltages may develop. Single-phasing of a three-phase transformer that is protected with fuses is one situation in which ferroresonance can occur. See Fig. 10.3.

Transformers with the following types of windings may contribute to ferroresonance voltages, especially when they are terminated or switched with long cables, capacitors, etc.

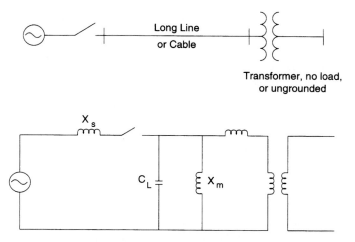

FIGURE 10.3   Typical elements in a ferroresonant circuit.

- Ungrounded wye-type
- Delta-winding
- Five-legged core-type

Mitigation for this problem includes avoiding this type of circuit configuration, grounding, using relays to alarm or trip under unbalanced voltage situations and using closure resistors in the switch.

10.1.3 Switching Surges

Switching surges are essentially overvoltages resulting from switching of equipment or facilities. As discussed earlier in Sec. 10.1.1, the instant of closing the switch, source characteristics, the resistances, inductances, and capacitances in the circuits determine the magnitude and duration of these overvoltages.

Circuit breakers, contactors, switches, and other types of interrupters are used to connect and disconnect equipment and circuits to the power system. Closing or opening these devices can generate overvoltages. On alternating-current circuits, these devices interrupt current at a current zero. That is, when the contacts part, an arc will be established to carry the current until the current would normally reach zero at the moment that it goes from positive to negative or vice versa. At the time that the current reaches zero, if there is enough dielectric strength (transient recovery voltage greater than the restrike voltage) built up between the contacts, the current will not be able to continue but will be interrupted. If there is not enough dielectric strength between the contacts to stop the conduction, the current will continue to flow until the next current zero or until enough dielectric strength is established. The reestablishment of current is called *reignition* or *restrike*. This current at the time of restrike is limited only by the impedance of the source so that a higher rate of change of current through the inductance of the circuit will cause an overvoltage.

Overvoltages occur on energizing capacitive circuits. Some examples are energizing long transmission lines, energizing lines terminated with transformers with little or no load, and energizing a capacitor bank with trapped charge. The initial capacitive current is higher than normal and when it flows through the inductance in the circuit, it causes overvoltage. The inductance in a transmission line not only includes the source inductance, but the line inductance also. A capacitor bank with an opposite, fully trapped charge will have a current that is twice the normal and, hence, the voltage generated can be twice the rated voltage if energized at the peak of the voltage wave.

Because a current cannot be established or reduced to zero instantaneously in an inductive circuit, any interruption of an inductive circuit can generate overvoltages if the current is forced to zero before a natural current zero. The level of overvoltages depends upon the point of switching on the voltage wave and any current chopping. The interrupting media determine the characteristic of the interrupter and include: air (disconnect switches, air magnetic circuit breakers, and air blast breakers), oil, SF_6, and vacuum. *Plasma* is a term that is used to denote the ionized interrupting media that may include material from the device's contacts. For example, in vacuum-based circuit switchers or circuit breakers, low levels of current are chopped (interrupted before a natural current zero). The results are a high rate of change of current ($L \cdot di/dt$, the circuit inductance times the rate of change of current), which results in an overvoltage. Interruption of current with an air disconnect switch takes longer than a circuit breaker or circuit switcher; thus, there is a possibility of interrupting the current at a current zero before enough dielectric strength has built up between the contacts so that there is a reignition or restrike. This restrike can result in a high $L \cdot di/dt$ with the resulting high voltage.

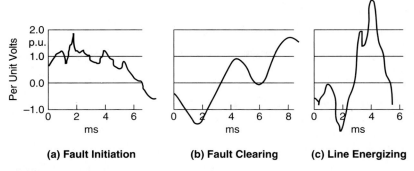

(a) Fault Initiation　　**(b) Fault Clearing**　　**(c) Line Energizing**

FIGURE 10.4　Examples of switching surge waveshapes.

Switching surges have a great variety of shapes, magnitudes, and duration, corresponding to the great variety of initiating events. For a particular event, the overvoltage magnitude and duration are determined by both the system and the characteristics of the switching device. Some examples of switching surge waveshapes are shown in Fig. 10.4. The shape can be unipolar, oscillatory, or quite irregular, and this shape may be superposed on power frequency voltage or other temporary overvoltages. Conditions producing the highest peaks are relatively rare. Voltage magnitudes are represented by frequency distribution curves.

Some of the highest switching surges are recorded when interrupting inductive currents before their natural zero. This *chopping* is most likely to occur for small currents, such as transformer magnetizing currents.[5] Applications involving vacuum interrupters are known to cause *current chopping*. The energy involved is small enough to be readily absorbed by surge arresters normally connected near the terminals of important transformers or in some cases incorporated in the vacuum circuit breaker. The use of preinsertion resistors (i.e., the switch or circuit breaker is opened with a resistance in parallel and this resistance is shorted out after several cycles) also assists in rendering these overvoltages harmless.

The interruption of capacitive currents, such as load currents of capacitor banks or charging currents of long unloaded transmission lines, can lead to a high voltage. The extent of switching overvoltage problems dictating the design of a distribution system can be determined only through studies. Determination of magnitudes of switching surge overvoltages is done by existing computer programs such as the electromagnetic transients program (EMTP) or by using an analog-type simulator called the transient network analyzer (TNA). By a systematic variation of relevant parameters, a complete statistical distribution, including the maximum values, can be obtained by using these simulation tools.

The purposes of switching transient investigations, in general, are:

1. To identify the type of transient overvoltages which may occur due to switching operation
2. To determine the magnitude, duration, and frequency of overvoltages
3. To determine any abnormal transient overvoltages by the inception and clearing of faults
4. To select the appropriate corrective measure to mitigate excessive transients

5. To determine alternative operating procedures, if necessary, to minimize or avoid unacceptable transient duties

The options available to reduce overvoltages include:

1. Temporary insertion of resistance between circuit elements (preinsertion resistors in circuit breakers)
2. Use of a zero voltage control switch
3. Use of inrush control reactors or tuning reactors
4. Use of surge capacitors or filter banks
5. Use of damping resistors
6. Use of surge arresters
7. Use of proper switching sequences and minimum and less frequent switchings

In industrial and commercial power systems, the experience and standards set the application of surge protective equipment.

10.1.4 Lightning-Induced Overvoltages

Overvoltage protection in power systems at voltages of 230 kV and below is based more on lightning requirements than switching surges. Some terms used in describing lightning effects in the power system arena are:

Flash Encompasses the entire electrical discharge from cloud to stricken object

Stroke The high-current components in a flash

Flashover An electrical discharge from an energized conductor to a grounded support

Tripout A flashover of an equipment circuit that does not clear itself

Several strokes may be present in a single flash. A lightning flashover may clear by itself as compared to a tripout which means the equipment or facility was disconnected from the power system.

The lightning and the resulting current and voltage surges can cause considerable damage to equipment and facilities if not properly protected. The three aspects of lightning that are applicable to equipment protection are:

- Keraunic level
- Stroke magnitude
- Rise time

Each location or area is said to have a certain keraunic level, or isokeraunic level (as it is usually called). The level represents the average number of thunder-days per year, i.e., the average number of days per year on which thunder will be heard. The keraunic level may be determined by referring to the isokeraunic maps (Figs. 10.5 and 10.6) in which lines of constant keraunic level are plotted. These lines of constant keraunic level are known as *isokeraunic lines.*

While the isokeraunic map provides a broad picture of storm frequency in a given area, it does not provide an accurate estimate of the number of lightning strokes to ground in a given locality—an important quantity in lightning protection design.

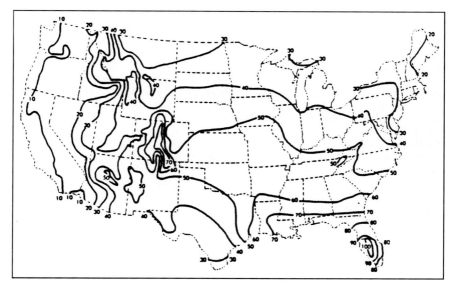

FIGURE 10.5   Isokeraunic map showing mean annual days of thunderstorm activity within the continental United States.

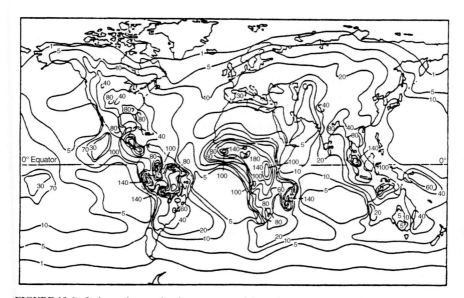

FIGURE 10.6   Isokeraunic map showing mean annual days of thunderstorm activity around the globe.

Effective lightning location systems have been developed in recent years. These systems can accurately locate and plot ground stroke data for a given area, providing better data for the engineer to work with. Depending upon the additional cost involved in obtaining this particular data, the engineer has to decide to use either interpolated data from an isokeraunic map or particular data collected for the locality.

The electrical characteristics (polarity, crest magnitude, and waveshape) of stroke currents vary over wide ranges, which is typical of natural events. The return stroke current magnitude ranges from 2 to 200 kA, with some extremely intense strokes exceeding 300 kA. The current magnitude increases rapidly to its peak value in 0.5 to 10 μs. The waveshape is in the form of an impulse. This is a greatly simplified explanation of the lightning mechanics.

In summary, any practical lightning protection system must deal with a large amount of energy in an extremely short period of time. Also, because of the fast rate of change of current, high voltages can develop across equipment. Thus, proper selection and application of surge arresters, shielding, grounding, and other lightning protection means are very important.

10.2 INSULATION COORDINATION

Insulation coordination is an overall process of protecting the power system and the equipment within the system from overvoltage that would cause an insulation failure.

The job of practicing engineer is to check whether or not an adequate margin of protection (often referred to as the *protective margin*) exists between the insulation withstand characteristic of the equipment and the protective characteristic of the surge arrester protecting that particular equipment. Protective levels and margins of different equipment, electrically connected, should be properly coordinated. The protective margin, in percent, is defined as:

$$\left\{ \frac{\text{Equipment withstand level}}{\text{surge arrester protection level}} - 1 \right\} \times 100 \qquad (10.1)$$

The three insulation withstand levels defined by IEEE/ANSI standards for equipment and the three corresponding protective levels for surge arresters are shown in Table 10.2.

TABLE 10.2 Minimum Protection Margin

Equipment withstand	Surge arrester protective level	Minimum protective margin*
Chopped-wave withstand (CWW)	Front-of-wave sparkover (FOW)	20%
Basic impulse insulation level (BIL)	Lightning protection level (LPL)	20%
Switching surge level (SSL)	Switching surge protective level (SPL)	15%

*Calculated by using Eq. (10.1). Minimum here means greater than or equal to. These margins were based on silicon carbide arresters. With zinc oxide arresters a higher margin is usually provided.

The following definitions may be found in references 1 and 3.

Chopped-wave withstand (CWW)   A voltage impulse that is terminated intentionally by spark over a gap. Impulse voltage above CWW of the equipment may damage insulation.

Basic impulse insulation level (BIL)   A reference impulse insulation strength expressed in terms of the crest value of withstand voltage of a standard full-impulse voltage wave.

Switching surge level (SSL)   Switching surge is a transient wave of overvoltage caused by a switching operation. SSL is a peak switching surge voltage above which the equipment insulation may be damaged.

Front-of-wave sparkover   The impulse spark over voltage with a wavefront that rises at the uniform rate and causes sparkover on the wavefront.

Lightning protection level   The maximum lightning impulse voltage expected at the terminals of a surge protective device under specified conditions of operation.

Based on the aforementioned three withstand and protective values, the three types of protective margins are shown in Fig. 10.7.

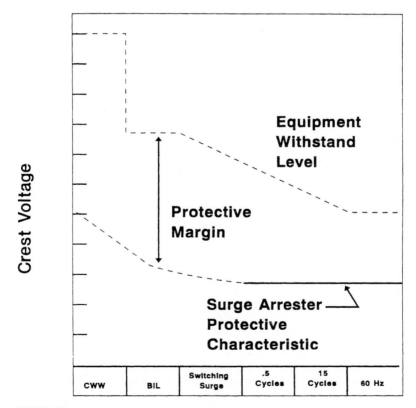

FIGURE 10.7   Typical insulation coordination curve.

Switching surge overvoltage severity in distribution and industrial systems is usually not important, because insulation designed to withstand lightning overvoltages will usually withstand switching surge overvoltages. Hence, only the first two types of protective margins are used in distribution and industrial insulation coordinations. The equipment withstand levels and surge arrester protective levels are obtained from manufacturers' catalogs and other literature.

10.3   SURGE VOLTAGE PROTECTION

As discussed earlier, surge arresters are used to protect equipment insulation from being subjected to overvoltages.[4] The protection level is such that the surge arresters keep these overvoltages at least 15 to 20 percent below the withstand voltage of insulation. About 40 percent margin is commonly used for applying zinc oxide arresters.

In the past, valve-type arresters were most widely used. These arresters consisted of a gap in series with nonlinear silicon carbide resistance disks. For certain applications in industrial and commercial-type systems, this type of arrester was not suitable. The gap sparkover characteristic greatly exceeded the arrester discharge voltage characteristic. But the silicon carbide did not possess a sufficient nonlinear property to be useful without the gap.

Recently developed metal oxide arresters, particularly zinc oxide (ZnO) arresters, have more ideal surge protective characteristics. For all new power systems and replacements, metal oxide arresters should be used. Only applications of metal oxide arresters will be presented here.

10.3.1   Surge Arrester Selection

The surge arrester selection is based on three primary considerations:

- Arrester rating
- Arrester class
- Arrester location

The three key variables defining the metal oxide arrester rating are maximum continuous operating voltage (MCOV), protective level, and energy absorption capability. MCOV is the continuous 60-Hz voltage which the arrester can withstand without overheating. The protective level is the maximum equivalent front-of-wave voltage rating based on so-called duty cycle tests defined in the standards. The third variable is the amount of energy that the arrester can absorb without damage. For each arrester, in addition to the MCOV rating, a corresponding front-of-wave (impulse) and switching surge voltage ratings are given in the manufacturer's catalog. Before selecting an arrester, its rating (from the manufacturer's catalog) should be confirmed to be adequate for the application under consideration. There should not be any encroachment of protective margins for different types of overvoltage; minimum margin levels should always be maintained. The lower the arrester rating, the better the protection of insulation (higher margin). However, the MCOV, and perhaps the energy absorption capabilities required by the particular application, may require a higher-rating surge arrester.

The voltage on unfaulted phases for a line-to-ground fault (temporary overvoltage) and duration of the fault is one of the important requirements for selecting an MCOV rating requirement. The line-to-ground voltage on the unfaulted phases is:

- Line-to-line voltage on ungrounded or resistance-grounded systems
- Line-to-ground voltage (or very close to it) for solidly grounded systems (usually a multiplication factor between 1.0 to 1.4 is used)

The use of other types of grounding lie in between these two extreme situations. In order to facilitate proper application, a *coefficient of grounding* is used as a measure of system grounding effectiveness. If the arrester rating satisfies the ungrounded requirement, it automatically satisfies the grounded case. For metal oxide arresters, as this sustained voltage is reduced, the time capability increases. The selection of a surge arrester of suitable (or acceptable) rating is influenced by:

- Effectiveness of grounding
- Underground equipment
- Maximum line-to-ground voltage and its duration
- Reflections (separation, lead length, etc.)
- Contamination
- Emergency operating modes
- Protective margins
- Energy discharge capability
- Other factors such as generator overspeeds, load rejection, and resonances

There are three classes of arresters and, typically, they are used as follows:

Station class	Equipment rating of 7.5 MVA or higher
Intermediate class	Used in substations of 1–20 MVA
Distribution class	Used in distribution-class equipment, rotating machine, and dry-type transformers
Riser class	For underground distribution

In actual practice, considerable overlap of applying these three classes exists. For example, station class arresters are used for nonshielded installation where intermediate class may suffice. The use of one higher-class arrester gives additional capability to cover uncertainties entailed in nonshielded installations. Similarly, intermediate class is used where distribution class arresters may be sufficient. The difference in cost between the different types of arresters is small when compared to their effectiveness and the number used in industrial systems. **Station-type arresters should be used to protect motors, generators, and transformers in industrial systems.**

The third important consideration in surge arrester application is its location with respect to the equipment. The ideal location is at the terminals of the equipment being protected. The arrester can be mounted on power transformers on a platform mounted on the case. Motors can have a large junction box that will hold lightning arresters and surge capacitors. However, the space, the need to protect more than one piece of equipment, and/or the cost require that the arrester be placed away from the equipment itself. As the arrester is moved away from the equipment, its effectiveness is greatly reduced as the cable length connecting the arrester to the protected equipment becomes a transmission line to the steep wavefront (high-frequency, megahertz). More specifics and recommended practices concerning arrester location are found in standards contained in reference 1.

Technical Factors for Surge Arrester Specifications.   While each specification for any equipment is unique, some of the technical factors for surge arrester specifications are:

1. Voltage rating
   kV in rms (60 Hz)
   Switching surge overvoltage (often at 2 kA)
   Lightning discharge current (usually at 10 kA)
2. Type of equipment being protected
   Line
   Cable
   Transformer
   Generator
   Motor
   Switchgear
   Substations, etc.
   SF_6 substation
3. Type of arrester (mostly metal oxide type is preferred these days)
4. Location of surge arrester
5. Arrester class (station, intermediate, and distribution)
6. Other application factors
   Type of grounding
   Discharge current and energy
   Number of parallel stacks
   Shunt gaps
   Insulation coordination curves
7. Other environmental factors
   Contamination
   Ambient temperature (maximum, minimum)
   Altitude
   Mounting, etc.

Metal-oxide arrester ratings, as well as other tables, formulas, and values, are available in the relevant standards and manufacturers' data.

EXAMPLE 10.1   *Select a surge arrester for a 15-kV class transformer with*

   CWW = 110 kV
   BIL = 95 kV

1. Select initially a 9-kV (rms) metal oxide (8.62-kV MCOV) distribution arrester which has
   FOW = 27.6 kV
   LPL = 24.0 kV (at 10 kA)

2. Calculate margins

$$PR\,(1) = \left(\frac{CWW}{FOW} - 1 \right) \times 100 = 299\% \text{ for chopped wave}$$

$$PR\,(2) = \left(\frac{BIL}{LPL} - 1 \right) = 296\% \text{ for BIL}$$

These margins are more than 20 percent, as shown in Table 10.2.

3. Consider all the other factors discussed earlier in this section. If changes are needed, then select another surge arrester and repeat steps 1 through 3. For example, consider an ungrounded system. Then surge arrester MCOV = 15.24 kV (nearest line-to-line voltage) with rms rating = 18 kV is the initial selection. At this rating, the distribution class metal oxide arrester has:

FOW = 55.21 kV
LPL = 47.8 kV

This gives the following margins:
PR (1) = 100%
PR (2) = 99%

Even though these margins are acceptable, they are considerably lower than the first selection. Thus, it is easy to see that for proper application, all relevant factors should be considered. These calculations can be made through a graphical method, as shown in Fig. 10.8.

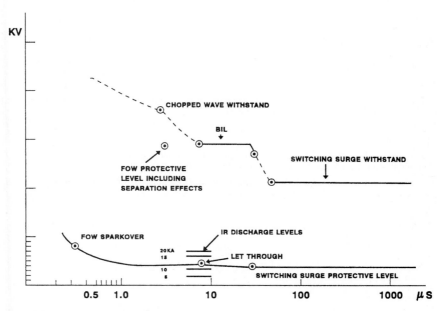

FIGURE 10.8 Insulation coordination through-curve.

On a resistance-grounded system a 15-kV arrester would give a little better protection and, under fault conditions, would exceed the MCOV for the short time the fault is on the line before it is interrupted.

A sample step-by-step approach is shown in reference 5.

10.3.2 Surge Capacitor Application

Surge capacitors are applied to alter the shape of a steep incoming wavefront, because the rate of rise of the surge voltage is limited by the charging rate of the capacitor. In practice, the surge capacitors are used with rotating machines, particularly motors:

Motors	Surge capacitors	Capacitance
Above 4000 V	Very high percentage	0.25 μF
4000 V	About 50%	0.5 μF
2300 V	Very low percentage	0.5 μF

A lightning arrester cuts off the top of an impulse but does nothing to reduce the steepness of the wavefront. Surge capacitors slope the front of the wave so that turn-to-turn voltages on machines are reduced. The surge capacitors should be connected as close to the equipment as possible. Connecting directly to the terminals is preferred. Each application should be analyzed carefully to assure that the equipment is protected properly. For example, it is a common practice to connect arresters on the line side of the feeder circuit breaker and capacitors at the motor terminals. In these circumstances, a careful analysis of the protective levels, the feeder lengths, capacitor waveshaping action, and the number of feeders during different operating modes should be performed. It is strongly recommended that the arrester and the surge capacitor be mounted at the terminals of the protected equipment.

10.4 REFERENCES

1. IEEE Standards Collection, *Surge Protection,* Institute of Electrical and Electronics Engineers, 1992.

2. IEEE Standard 399-1990, *IEEE Recommended Practice for Power System Analysis,* Chap. 11, Institute of Electrical and Electronics Engineers.

3. ANSI/IEEE Standards 100-1992, *Standard Dictionary of Electronics Terms,* Institute of Electrical and Electronics Engineers.

4. J. J. Burke, "Distribution Surge Arrester Application Guide," panel session, *1991 IEEE T&D Conference and Exposition,* Dallas, TX, Sept. 23–27, 1991.

5. ANSI/IEEE Standards 62.2-1987 (reaffirmed 1994), *IEEE Guide for the Application of Gapped Silicon-Carbide Surge Arrestors for Alternating-Current Systems,* Institute of Electrical and Electronics Engineers.

CHAPTER 11

POWER QUALITY—
VOLTAGE FLUCTUATIONS
AND HARMONICS

11.0 INTRODUCTION

The electric power generation, transmission, and distribution systems serving our modern civilization are based on an alternating current and voltage which are sinusoidal in nature. All the physical phenomena on which the energy system is based depend on an almost perfect sine wave of voltage and current. When loads that consist of inductances, capacitances, and resistances in any combination are connected to this system, the sine wave is preserved and the system components are said to be linear. Nonlinear loads that distort the current cause a voltage drop in the system that is not sinusoidal and, thus, also distort the voltage. The invention of semiconductor devices increased the use of nonlinear loads. Many of these loads in the first and second generation of their designs were also sensitive to variations in the voltage that were impressed upon them. This gave rise to the consideration of *power quality*.

Power quality is a term that is used to describe the quality of voltage and current that a facility has. The *voltage* is the responsibility of the power producer and is dependent on the generation and transmission system. The *current* is dependent on the requirement of the load; that is, if the load is linear, it will take current that is sinusoidal. If the load is nonlinear, it will take current that is not sinusoidal.

Most of the electric energy that is consumed is generated by synchronous generators. These machines are designed to produce a sine wave of voltage, which then carries the energy throughout the electric system, the transmission system, the distribution system, and down to the load.

The load determines the characteristic of the current that the voltage will carry through the system. If the load is linear, such as constant speed induction or synchronous motor, resistance heating, or incandescent lighting, then the current will also be sinusoidal. If, however, the load is nonlinear, such as fluorescent lighting, adjustable-speed motors (induction, synchronous, or dc), induction heating, resistance welding, electric arc furnaces, or static power converters, then the current will be nonsinusoidal. This nonsinusoidal current flowing through the impedance of the electric system will produce a nonsinusoidal voltage drop that will distort the sinusoidal voltage that the power producer is generating.

There are phenomena that accrue on the electric power system that vary the magnitude of the sinusoidal voltage that is produced by the generators. These phenomena include outages, sags, swells, surges, and impulses that vary the quantity or magnitude of the voltage that is delivered to the user. These phenomena are, in most cases, beyond the control of the power producer; however, the system can be designed to minimize the effect of these phenomena. Likewise, the user can design an electric power system to minimize the effect of nonlinear loads on the producer's system that affects other users.

11.1 VOLTAGE FLUCTUATIONS

Voltage fluctuations can be minimized by good engineering design of generation, transmission, and distribution systems. Since most of these phenomena are a result of natural occurrences, some of them will affect user loads. Proper use of lightning arresters and overhead ground wires on transmission and distribution systems will protect them. Generation stations with backup equipment can become very reliable.

11.1.1 Outages

An *outage* is a complete interruption of voltage for times from 30 cycles up to minutes or hours. It is caused by faults on lines that cause circuit breakers, reclosures, or fuses to interrupt the service. In some cases, this can be for days, as in the case of major system problems or natural disasters. The result of an outage is the shutdown of motors, static power converters, mercury vapor lighting, and processes that have undervoltage protection.

In most cases, motors should drop off-line to prevent reclosing of the voltage when it is out of phase with the internal voltage of the motor. Small induction motors lose their excitation within two or three cycles. Larger motors have a larger magnetic structure where the magnetic fields decay over a longer period. Figures 7B.3 through 7B.7 in Chap. 7 show the decay rate of synchronous and induction machines of different ratings. Large medium-voltage motors that are controlled by circuit breakers have undervoltage relays that will take them off-line if there is an outage, to keep from having excessive torques if the voltage were to come on that is out of phase with the motor voltage.

Adjustable-speed drives that have static power supplies controlling them will drop off line unless there is a concerted effort to keep them on-line and sense the voltage of the motor so that the inverter output voltage is synchronized with the residual motor voltage. If there is a complete outage, it is preferable to have the motor stop and be started manually or in sequence if it is associated with a process where it must coordinate with other drives.

11.1.2 Undervoltage

Undervoltage is a drop in voltage below the normal 5 percent that is allowed for normal operation of equipment. Most equipment will operate in the range of ± 10 percent for short periods of time without damage. If operation is for extended periods (hours) at 10 percent low voltage, there will be loss of efficiency when the equipment is not operating at its rating. Chapter 5 discusses the effect of low voltage in the operation of various equipment. Starting large motors may cause the voltage to drop close to 10 percent. Good system design would not allow the voltage to drop below 5 percent for motor starting.

11.1.3 Voltage Sags

A *sag* is defined as a drop in voltage for a few cycles (1 to 30). This is caused by a fault on the power system in a remote location. Figure 11.1 shows a drop in the bus voltage

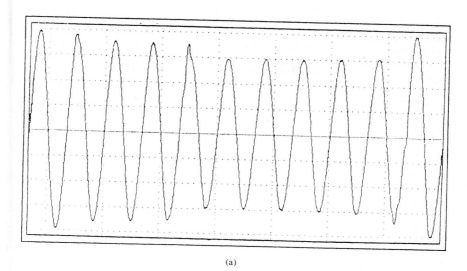

(a)

(b)

FIGURE 11.1   Voltage sag caused by (*a*) fault on a subtransmission line and (*b*) transformer fault on adjacent feeder.

for a remote fault on a subtransmission system. It can also be the result of a lightning strike that operates a circuit breaker or a reclosure which recloses instantaneously (3 to 10 cycles). Motors—both constant-speed and adjustable-speed ac motors—may drop out. Motor contactors drop out with approximately 65 percent voltage. Adjustable-speed ac drives have some power supplies that will drop out with very short drops in voltage. This type of voltage drop may cause electronic equipment to burp, which is indicated by a sharp dip in lighting output. Many office machines, computers, commercial cash registers, gasoline pump systems, etc., are affected by sags. To eliminate the results of a sag, uninterruptable power supplies (UPS) can be applied to isolate the sensitive equipment from the sag. The UPS is an inexpensive device that keeps the electronic equipment on-line and eliminates the loss of data.

It is not economical to provide UPSs for motor drives, either constant-speed or adjustable-speed, unless a particular drive is essential to operation and cannot stand a momentary outage. Many times, it is the control of a drive or process that is most sensitive to sags, and a UPS for that part of the system would be economical. The same is true on many control systems that regulate and govern large processes. These control systems should be furnished with UPSs to ensure that proper safety for personnel and equipment are observed for shutdown due to a sag or outage.

Some lighting will be extinguished if the voltage drops below a certain value. High bay mercury lamp ballast takes 5 to 10 minutes to come up to full illumination after an outage.

11.1.4 Overvoltage

An *overvoltage* is an increase in voltage above the standard voltage tolerance of ±5 percent which will last for seconds to minutes. Overvoltages can be a result of erroneous system operation, line-to-ground faults, or switching of major system components. Overvoltages should persist only for short periods until voltage-regulating equipment can correct the problem.

11.1.5 Voltage Swells

A *swell* (Fig. 11.2) is an increase of voltage for a short period of time (1 to 30 cycles). The cause can be changes in system loading (a large load is dropped), switching on of a capacitor bank, a fault on the system that depresses one phase and increases the voltage on the other two phases, etc. The lights blinking is an indication of this.

11.1.6 Voltage Surges

A *voltage surge* (Fig. 11.3) is an increase in voltage of a duration of milliseconds, less than a cycle. It is caused by switching in the circuit. It can be either a capacitive or an inductive circuit that has a forced current zero. Most circuit interrupters will interrupt

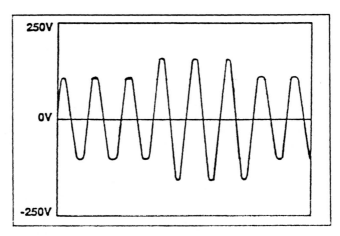

FIGURE 11.2   A voltage swell.

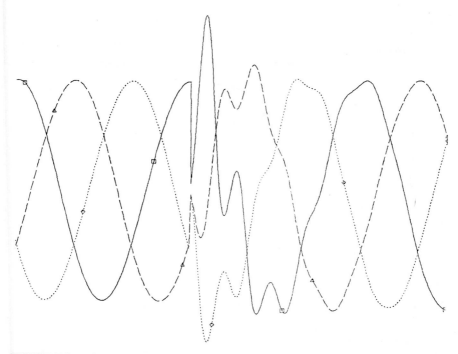

FIGURE 11.3   Voltage surge caused by capacitor switching.

the current at a natural current zero. However, there are times when interrupting particularly small currents such as transformer magnetizing currents, that the current will be
forced to zero and the rate of change of the current in the inductance of the circuit will
cause a *switching surge.*

$$V = L \frac{di}{dt} \tag{11.1}$$

Switching surges are limited to values of approximately four times or less of normal
circuit voltage. Switching devices are tested and rated so that switching surges are less
than the four-times value.

Sensitive equipment that cannot withstand this surge voltage, needs to be protected
by metal oxide varistors (MOVs). The MOV needs to be rated to absorb the energy in
any surge that might be on the circuit. The cost of MOVs is low enough so that furnishing a large energy-absorbing device is cheap insurance for the protection of the equipment.

11.1.7   Voltage Impulses

Voltage impulses (Fig. 11.4) are high voltages that have a duration on the order of
microseconds. They are caused by lightning and some switching. Protection from them
is accomplished with lightning arresters and proper equipment design. The basic

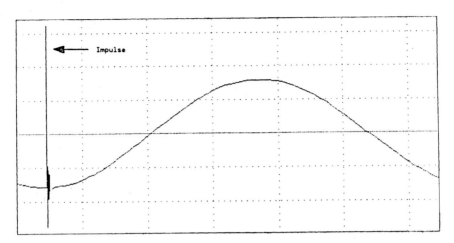

FIGURE 11.4   Voltage impulse.

impulse level (BIL) rating of the equipment is the rating that protects against impulses. Proper design of transmission and distribution systems will protect the user from most of these impulses.

There is another class of impulses that is very hard to detect, because its duration is in the nanosecond time frame. Low currents that are forced to zero by interrupters can produce very high values of voltage that last only for nanoseconds. An example of this type of impulse is an oil interrupter operating on a small (1- to 2-A) current. Measurements have been made on 600-V circuits that had oil interrupters feeding a transformer that interrupted the magnetizing current and resulted in voltages 10 to 12 times normal. Duplicating this can be done with the proper instrumentation and equipment. These fast impulses will not travel far, because they are attenuated by the circuit impedance. They have low energy, but they will stress insulation.

11.1.8   Voltage Flicker

Voltage flicker is the term used to define the fluctuation of voltage caused by loads that vary in frequency up to 8 Hz. It is based on the human eye's response to variation in lighting. Figure 5.12 shows the human-eye response to the lumen output of an incandescent lamp, where the lumen output varies as the voltage impressed upon the filament. Where the other phenomena are the results of their action on the utility system and are not associated with what occurs within the user's facility, voltage flicker is the result of a user's load's action on the utility system. The loads that are most associated with voltage flicker include electric melting arc furnaces, resistance welding, and hot rolling mills. There are other loads that cause a voltage drop on the system, but they are not repetitive to the extent that they become irritating.

There are techniques that may be used to minimize the effect of these load changes. The electric melting arc furnace is one of the largest loads, and the voltage drop is mainly the result of the reactive power changes. A static VAR compensator will operate fast enough to counteract the voltage drop fluctuation. Figure 11.5 shows the repetitive current operation of an arc furnace. This current set is repetitive at a frequency of approximately 5 Hz.

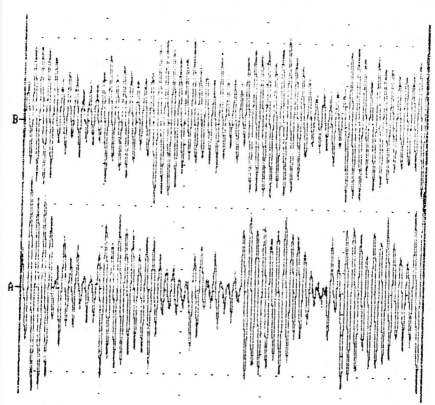

FIGURE 11.5 Repetitive nature of a large arc furnace current.

11.1.9 Summary of Voltage Fluctuations

Although at this time there are not formal definitions of the magnitude and time duration of the aforementioned terms, Table 11.1 lists the terms and approximate time durations and magnitudes.

TABLE 11.1 Summary of Terms and Definitions for Voltage Fluctuations

Phenomenon	Duration	Voltage (per unit)
Outage	>0.5 s	0.0
Undervoltage (rms)	>0.5 s	<0.86
Voltage sags (rms)	0.02 s–0.5 s	<0.86
Overvoltage (rms)	>0.5 s	>1.055
Voltage swells (rms)	0.02 s–0.5 s	>1.055
Surges (peak)	<0.01 s	>2.0
Impulses (peak)	1 μs–5.0 μs	>2.0
Flicker	continuous	±0.008

11.2   EQUIPMENT DESIGN TO WITHSTAND VOLTAGE FLUCTUATIONS

The importance of power quality has been emphasized in the last two decades of the 20th century because of the use of much more sensitive equipment. Electronic devices use low signal levels to operate and, thus, are more sensitive to power quality. The Computer Business Equipment Manufacturer's Association (CBEMA) developed a curve that has been used as a standard to the environment in which equipment must operate. In general, it can provide a good indication of what the environment is without the need for extensive study. Figure 11.6 shows this curve. A later study,[1] completed over a period of four years, shows the events (see Fig. 11.7) that lie outside of the

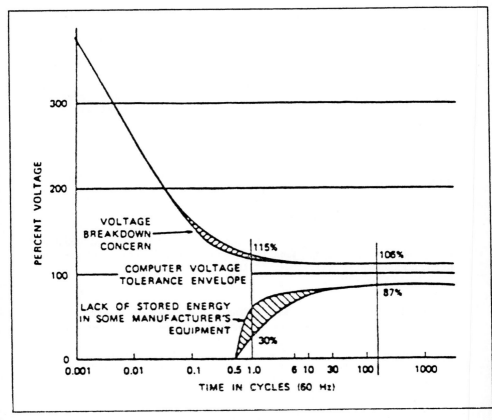

FIGURE 11.6 Computer Business Equipment Manufacturer's Association (CBEMA) Curve of Equipment Susceptibility.

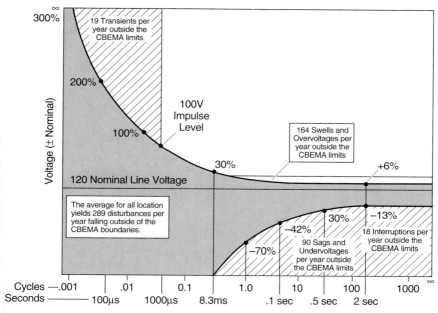

FIGURE 11.7   Events outside the CBEMA curve.

CBEMA curve. There are committees within the IEEE that are working on defining the environment that equipment must withstand as a result of the phenomena that this equipment sees in actual practice.

There are several auxiliary equipments that can be used to mitigate most of the phenomena. These are furnished by several different manufacturers and should be considered whenever sensitive electronic equipment, control, or data handling, is essential to the operation of a process or facility.

Except for large installations, the equipment itself should exceed the CBEMA curve. The use of the *switched-mode power supplies* on most electronic equipment can be designed to ride through most of the disturbances seen. Figure 11.8 shows a typical switched-mode power supply. The size of the dc link capacitor determines whether or not the equipment will ride through short time sags and low voltage. For large computers that are essential to the facility operation, a UPS should be considered. Figure 11.9 shows such a system, using a battery as the stored energy source. A motor-generator with a flywheel can electrically isolate the sensitive loads from the main power supply. An M-G set has enough inertia to ride through short outages and low voltage.

Most of the phenomena previously listed are the results of natural causes. Several studies have been done to determine the cause of the sustained outages. Adverse weather caused 75 percent of outages.[2]

Users can protect themselves by ensuring that the specifications are written so that the majority of disturbances do not affect the equipment being purchased. Working with the vendor can help in providing the correct power conditioning equipment.

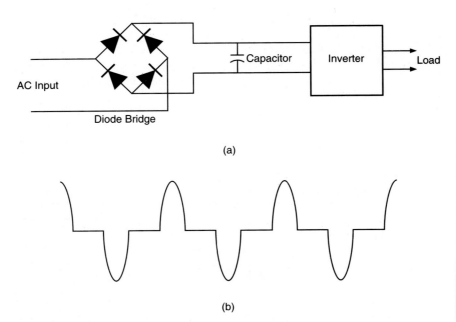

(a)

(b)

FIGURE 11.8   (*a*) Typical switched-mode power supply (SMPS). (*b*) Current wave of SMPS.

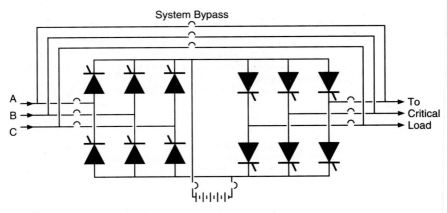

FIGURE 11.9   Typical UPS system with battery energy storage.

11.3 HARMONICS

When nonlinear devices are connected to the system, the fundamental sinusoidal character of the current flowing through the system is changed. This is the problem that must be faced with nonlinear static power converters (diode rectifiers and thyristor power supplies), loads that have discontinuous currents, and loads where the impedance in the positive direction of current flow is different from the impedance in the negative direction of current flow. Besides the static power converters, nonlinear devices include electric metal-melting arc furnaces where the carbon electrodes in contact with steel have dissimilar impedances between the positive and negative flows of current, resistance welding where the copper electrodes and the steel being welded have dissimilar impedances between the positive and negative flows of current, current that magnetizes iron cores, and current that is modulated by the action of thyristor switches. These types of loads are nonlinear. Because static power converters are used more extensively for controlling loads, and they produce nonlinear currents, they will be discussed in detail.

11.3.1 Static Power Converters

Static power converters come in many forms for many applications. Rectifiers, inverters, cycloconverters, single-phase, three-phase, single-way, double-way, bridge, star connection, and six-pulse are all terms that describe different circuit arrangements of static power converters. They all have one thing in common: They are nonlinear. They require current from the power system that is nonsinusoidal. However, they are governed by the same basic laws that allow an analysis of the effects of the nonlinearity.

Consider a basic three-phase bridge rectifier, shown in Fig. 11.10(a), that is furnishing a constant dc current to an inductive (motor) load. This constant current is switched sequentially among the three phases of the ac system. The resulting theoretical square current wave, shown in Fig. 11.10(b), is 120 electrical degrees of positive amplitude, repeated 60 degrees later with 120 degrees of negative amplitude. This configuration can be simulated by a series of sine waves which include all of the odd harmonics, except the triplens (3, 9, 15, etc.).

Thus, through Fourier analysis, the nonlinear currents can be described by sine waves of different frequencies, and the characteristics of the power system can be used to predict the phenomena to be expected with these nonlinear currents. The terms in the Fourier series in Fig. 11.10(b) and (c) are the same, except the 5th, 7th, 17th, 19th, etc., terms have opposite signs in the two circuits. By combining the two circuits with equal loads, it becomes a 12-pulse circuit, where the lowest theoretical harmonic current is the 11th.

The theory to which most static power converters conform states that the harmonic current will be of the order of:

$$h = kq \pm 1 \tag{11.2}$$

where h = harmonic order
  k = any integer 1, 2, 3,...
  q = the pulse number of the circuit

and the magnitude of the harmonic will be:

$$I_h = \frac{I_1}{h} \tag{11.3}$$

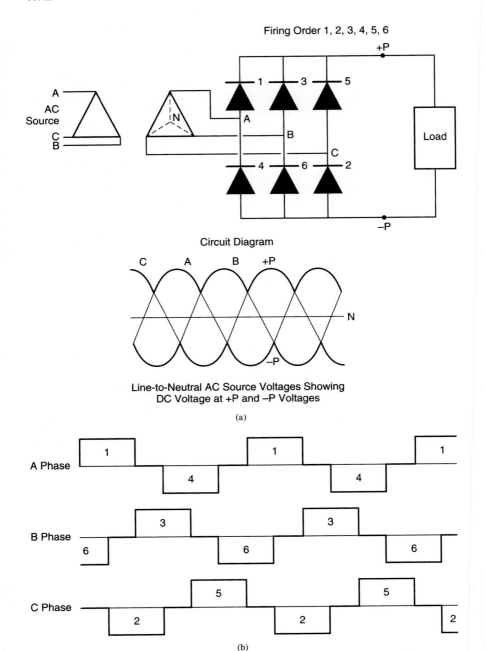

FIGURE 11.10 (a) ANSI circuit 25, delta-delta, three-phase bridge, six-pulse circuit, voltage with respect to neutral; (b) theoretical current reflected to the ac line using ANSI circuit 25, numbers represent element carrying current.

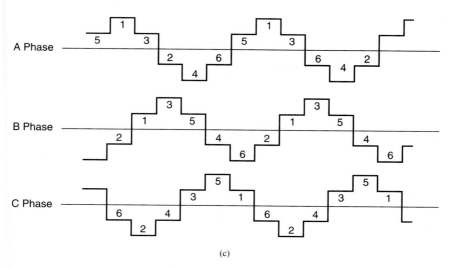

(c)

FIGURE 11.10 (Continued) (c) Theoretical current reflected to the ac line using ANSI circuit 24, delta-wye, numbers represent converter element carrying current.

where I_h = harmonic current magnitude
  I_1 = fundamental current magnitude
  h = harmonic order

Thus, the common three-phase bridge circuit, Fig. 11.10, will have the theoretical values of harmonic currents shown in Table 11.2.

  These are theoretical values for a square wave. Real-life circuits have inductance to retard commutation of current between the phases, so that edges are rounded and the values of the harmonic currents are reduced. Practical values for medium-voltage systems are more like those shown in Table 11.3.

TABLE 11.2 Theoretical Values of Harmonic Currents in a Six-Pulse Bridge Circuit

Harmonic	5	7	11	13	17	19	23	25
Current (per unit)	0.200	0.143	0.091	0.077	0.059	0.053	0.043	0.04

TABLE 11.3 Values of Harmonic Currents in a Six-Pulse Circuit with $X_c = 0.12$ and an Alpha = 30°

Harmonic	5	7	11	13	17	19	23	25
Current (per unit)	0.192	0.132	0.073	0.057	0.035	0.027	0.020	0.01

11.3.2 Resonant Circuits

Power systems are made up of inductances, capacitances, and resistance. Inductance and capacitance in either series or parallel combinations set up resonant circuits at some frequency. If we had only the fundamental frequency to consider (50 or 60 Hz), we could design our power systems to always avoid resonance at these frequencies. However, the nonlinearity of static power converters and other nonlinear loads produces currents of many frequencies. One of these may be near the resonant frequency of circuit components.

The characteristic of a parallel resonant circuit, shown in Fig. 11.12(*a*), is high impedance to the flow of current at the frequency of resonance. A series-resonant circuit, shown in Fig. 11.12(*b*), is low impedance to the flow of current at the frequency of resonance. A harmonic filter is designed as a series-resonant circuit. It is also called a *trap* because it tends to trap or control the flow of harmonic currents.

Again, looking at fundamentals, we know that:

$$V = IZ \tag{11.4}$$

where V = voltage
$\quad I$ = current
$\quad Z$ = impedance

For any given I, V is proportional to Z, which is important in parallel resonance where Z is very high, then V is high. That characteristic of a parallel-resonant circuit (also called a *tank circuit* in radio terminology) gives rise to an oscillating current of the frequency of parallel resonance if an outside current excites the circuit. It takes only a small current to excite a large current that will oscillate between the energy storage of the capacitor and the energy storage of the inductance in the power system. The magnitude of this oscillating current is limited only by the resistance in the circuit. Therefore, a circuit with a high X/R ratio will have a relatively high oscillating current. A circuit with a low X/R ratio will have a smaller oscillating current. Figure 11.11 shows the effect of the circuit resistance or load in limiting the amplification of the parallel-resonant circuit.

Likewise,

$$I = \frac{V}{Z} \tag{11.5}$$

For any given voltage, I is inversely proportional to Z, which is important in series resonance where Z is very low and I is very high. For normal static converter operation, the circuit parameters of inductance, capacitance, and resistance do not combine to be resonant at any characteristic harmonic. The capacitance of transformers, cables, and distribution lines is small and with normal transformer, cable, or distribution line inductance, the resonant (natural) frequencies are high. The natural frequencies of power circuits are in the kilohertz range. However, when power capacitors are added to the system for power factor improvement or voltage control, the resonant circuits can fall into the range of frequencies normally encountered with static power converters. The frequency of resonance is:

$$f_r = \frac{1}{2\pi} \sqrt{\frac{1}{LC}} = f_1 \sqrt{\frac{X_c}{X_L}} = f_1 \sqrt{\frac{\text{MVA}_{SC}}{\text{MVAR}_C}} \tag{11.6}$$

where f_r = resonant frequency
$\quad L$ = circuit inductance, henries

C = circuit capacitance, farads
f_1 = fundamental frequency, Hz
X_C = capacitive reactance, ohms
X_L = inductive reactance, ohms
mVA_{sc} = short-circuit capacity of the system in mVA
mVAR_c = capacitor value in mVAR

The equations are the same for either parallel or series resonance, depending on the point from which the circuit is viewed.

Figure 11.12(a) and (b) shows the equivalent parallel or series circuit, respectively, when viewed from the converter as a source of harmonic currents. An example of a parallel circuit is the system inductance in parallel with power factor improvement capacitors. An example of a series circuit is the inductance of a stepdown transformer in series with capacitors connected and switched with motors connected to the secondary of the transformer.

Figure 11.13 illustrates the relationship between system short-circuit capacity and power factor improvement capacitors with respect to harmonic resonant frequencies. There are some quick rules of thumb that can be used to determine if there might be a problem. The first is the resonant value of the combination of system impedance and the capacitor bank size:

$$f_p = \sqrt{\frac{\text{MVA}_{sc}}{\text{MVAR}_c}} = \sqrt{\frac{X_c}{X_{sys}}} \tag{11.7}$$

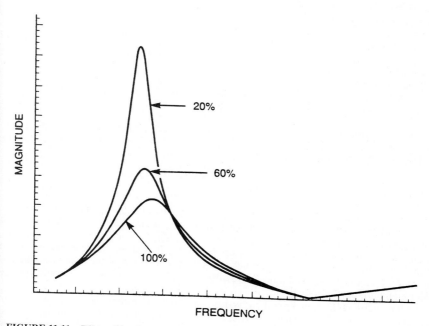

FIGURE 11.11   Effect of loading on the magnitude of parallel-resonant current.

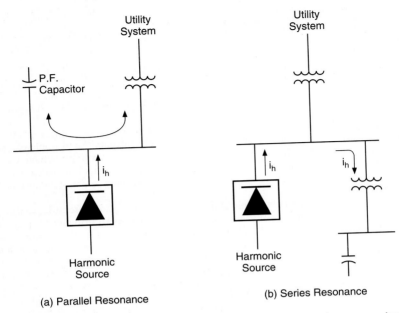

(a) Parallel Resonance

(b) Series Resonance

FIGURE 11.12 (*a*) Parallel and (*b*) series circuits of inductance and capacitance as seen by a static power converter.

where f_p = per-unit parallel resonant frequency
X_c = capacitor bank reactance (per-unit or ohms)
X_{sys} = system reactance (per-unit or ohms)

A second rule of thumb involves the size of the static power converter with respect to the size of the electrical system feeding the converter. The term *short-circuit ratio* (SCR) has been used to describe this and is defined as:

$$\text{SCR} = \frac{\text{MVA}_{\text{short circuit}}}{\text{MW}_{\text{converter}}} \qquad (11.8)$$

If the converter is small compared to the system capacity, the per-unit harmonic currents will be small and the system impedance will be low, so any harmonic voltage will be insignificant.

If the SCR is above 20 and the f_p is above 8.5, the probability of problems is low. If the SCR is below 20 and if the parallel resonance is near one of the converter characteristic harmonics, there is a high probability of producing excessive harmonic voltage and high harmonic currents.

An example will show how the physics of the power systems leads to establishing the parallel resonance at or near the first two converter characteristic harmonics. A system normally operates at a 0.80 power factor, and it requires power factor improvement capacitors to change to a 0.95 power factor. If we were to assume a 36-mW load on a

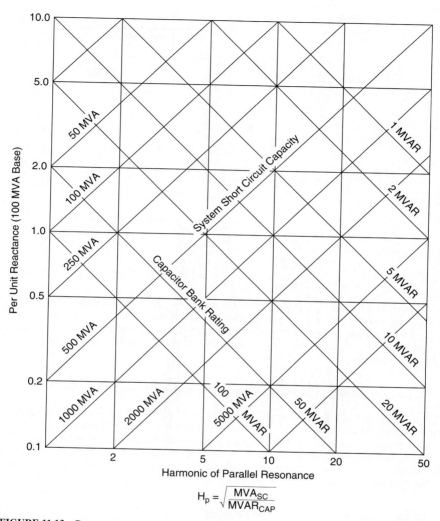

FIGURE 11.13 Reactance versus frequency plot of system inductance and power factor improvement capacitors.

500-mVA short-circuit duty system, we can calculate the amount of capacitors needed to raise the 0.80 power factor to a 0.95 power factor.

$$kVAR = kW[(\tan \cos^{-1} \phi_1) - (\tan \cos^{-1} \phi_2)] \qquad (11.9)$$

$$MVAR = 36 \text{ MW} (\tan 36.7° - \tan 18.19°) \qquad (11.10)$$

$$MVAR = 36(0.75 - 0.329) = 15.2 \text{ mVAR} \qquad (11.11)$$

TABLE 11.4 System Resonant Characteristics

System	Capacitor	First parallel resonance	
MVA_{sc}	$MVAR_{cap}$	f_p (Hz)	Per-unit frequency
1000	20.4	420	7
500	10.2	420	7
250	5.1	420	7
125	2.5	420	7
1000	30	346	5.77
500	15	346	5.77
250	7.5	346	5.77
125	3.75	346	5.77
1000	40	300	5
500	20	300	5
250	10	300	5
125	5	300	5

So the parallel resonance is:

$$f_p = \sqrt{\frac{MVA_{sc}}{MVAR_c}} = \sqrt{\frac{500}{15.2}} = 5.7 \text{ per-unit frequency} \qquad (11.12)$$

Table 11.4 lists several system short-circuit values and capacitor values that give resonances near the first two characteristic harmonics of six-pulse static power converters.

Table 11.4 shows that most power factor improvement capacitors will place the natural frequency of a system close to the characteristic frequencies of six-pulse static power converters.

11.3.3 Controlling the Flow of Harmonic Currents

The harmonic currents from any load can be controlled by filters tuned to the characteristic harmonic currents of the load. A filter or trap is a series-resonant circuit that consists of reactors in series with capacitors. The capacitors can be, and normally are, the power factor improvement capacitors in the circuit. They need to be sized and have a voltage rating that will let them carry, not only the normal fundamental current, but also the harmonic current that they will trap.

11.3.4 Harmonic Filter Design Process

On the one-line diagram on Fig. 11.14, using a 10 MVA base, the calculations of the harmonic filter attached to the 13.8-kV bus are as follows:

$$X_s = \frac{MVA_{base}}{MVA_{sc}} = \frac{10}{1500} = 0.0067 \text{ p.u. } \Omega \qquad (11.13)$$

$$X_{T1} = Z \frac{MVA_{base}}{MVA_{trans}} = 0.08 \frac{10}{15} = 0.0533 \text{ p.u. } \Omega \qquad (11.14)$$

The 13.8-kV bus short circuit is then:

$$MVA_{sc} = \frac{MVA_{base}}{Z} = \frac{10}{0.0067 + .0533} = 166.7 \text{ MVA}_{sc} \qquad (11.15)$$

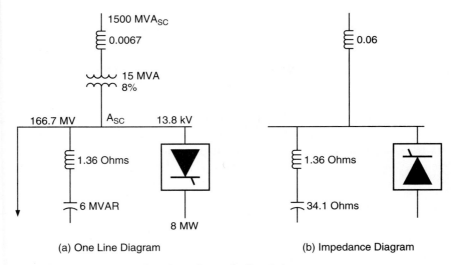

(a) One Line Diagram   (b) Impedance Diagram

FIGURE 11.14   One-line and impedance diagram for filter design.

The 12-mW load operates at a 0.78 power factor.

$$\text{MVA}_{\text{load}} = \frac{\text{MW}}{\text{PF}} = \frac{12}{0.78} = 15.385 \text{ MVA} \tag{11.16}$$

$$\text{MVAR} = \text{MW} (\tan \cos^{-1} \phi) = 12 \, (0.802) = 9.627 \text{ MVAR} \tag{11.17}$$

The power triangle is, then, as shown in Fig. 11.15.

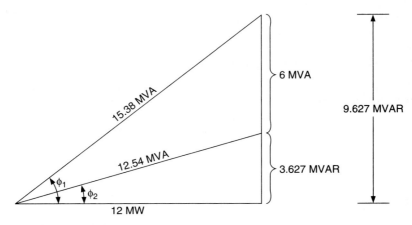

FIGURE 11.15   Power triangle of filter design.

To correct to 0.95 PF:

$$\text{MVAR} = \text{MW tan cos}^{-1}\phi = 12\ (0.3287) = 3.944\ \text{MVAR} \tag{11.18}$$

$$\text{MVAR}_{\text{capacitor}} = 9.627 - 3.944 = 5.683\ \text{MVAR} \tag{11.19}$$

Use a 6.0-MVAR capacitor, which gives a power factor of 0.9572. For 6.0 MVAR, use 10 to 200 kVAR, 7960 volt units per phase.

5th-Harmonic Filter Design. For a 6-MVAR capacitor:

$$X_C = \frac{kV^2}{\text{MVAR}} = \frac{13.8^2}{6} = 31.74\ \Omega \tag{11.20}$$

$$C = \frac{1}{2\pi f X_C} = 83.57\ \mu\text{fd} \tag{11.21}$$

$$X_L = \frac{X_C}{h^2} = \frac{31.74}{5^2} = 1.27\ \Omega \tag{11.22}$$

$$L = \frac{X_L}{2\pi f} = \frac{1.27}{2\pi 60} = 3.37\ \text{mH} \tag{11.23}$$

Capacitor current (filter current) is then:

$$I_C = \frac{V}{X_C - X_L} = \frac{13,800}{(31.75 - 1.27)\ \sqrt{3}} = 261\ \text{A} \tag{11.24}$$

Fundamental voltage across the capacitor is then:

$$V_{C1} = I_{C1}X_{C1} = 261 \times 31.74 = 8284\ \text{V} \tag{11.25}$$

$$\frac{8240}{7960} = 1.04\ \text{p.u. V} \tag{11.26}$$

This voltage is too high for the application. The capacitor must also account for the harmonic voltages that add to the fundamental voltage.

Filter Capacitor Voltage. Assume that the total static power converter load is 8 MW. Also assume that 1 mW is equal to 1 MVA.

$$V_C = I_{C1}X_{C1} + I_{C5}X_{C5} \tag{11.27}$$

$$I_1 = \frac{kW}{\sqrt{3}(kV)} = \frac{8000}{\sqrt{3}(13.8)} = 335\ \text{A} \tag{11.28}$$

I_5 (the 5th-harmonic current) is approximately 0.2 of I_1.

$$I_5 = 335 \times 0.2 = 67\ \text{A} \tag{11.29}$$

Then:

$$V_C = (261 \times 31.74) + 67\left(\frac{31.74}{5}\right) = 8709 \text{ V} \tag{11.30}$$

On the 7960-V capacitor unit:

$$V_C = \frac{8709}{7960} = 1.094 \text{ V p.u.} \tag{11.31}$$

This is an excessive voltage! At this point, a higher-voltage capacitor must be used. The standard unit voltage of 8660 V will be used. To obtain the desired mVAR, 11 units per phase will be used. Then:

$$X_C = \frac{kV^2}{MVA} = \frac{(8660 \times \sqrt{3})^2}{6.6} = 34.09 \ \Omega \tag{11.32}$$

$$X_L = \frac{X_C}{h^2} = \frac{34.09}{5^2} = 1.36 \ \Omega \tag{11.33}$$

$$I_1 = \frac{V_1}{X_{C1} - X_{L1}} = \frac{7960}{34.09 - 1.36} = 243 \text{ A} \tag{11.34}$$

The voltage in this case is the circuit voltage that is impressed upon the filter.

$$V_C = I_1 X_{C1} + I_5 X_{C5} = (243 \times 34.09) + \frac{67 \times 34.09}{5} = 8741 \text{ V} \tag{11.35}$$

The per-unit voltage on the capacitor is then:

$$V_C = \frac{8741}{8660} = 1.009 \text{ p.u.} \tag{11.36}$$

This is well within the capacitor's rating. The arithmetic sum of the fundamental and the harmonic voltage to which the filter is tuned is conservative. Although the filter will also absorb other harmonic currents which will have corresponding harmonic voltages, that may or may not add arithmetically to the total voltage. These currents do add to the heating effect on the inductor and must be taken into account.

Filter Inductor Current. The filter inductor must withstand the heating effect of the total rms current, fundamental plus all the harmonic current. The rms current of the fundamental and the tuned harmonic is:

$$I_{rms} = \sqrt{I_1^2 + I_5^2} = \sqrt{243^2 + 67^2} = 252 \text{ A} \tag{11.37}$$

Other harmonic currents will flow into the filter depending upon the impedance ratios between the system impedance and the filter impedance. A good factor to use to increase the reactor current rating to account for these harmonic currents is 1.2. Therefore:

$$I_{rms} = 1.2 \, I_{rms} = 1.2 \times 252 = 302 \text{ A} \tag{11.38}$$

Use a 300-A-rated inductor. As long as the filter size is large enough to improve the power factor similar to the preceding example, the tuning point will be broad enough that taps on the reactor should not be needed. The example in Fig. 11.16 shows the broad tuning point of a large filter with its trapping a large portion of the 11th-harmonic current.

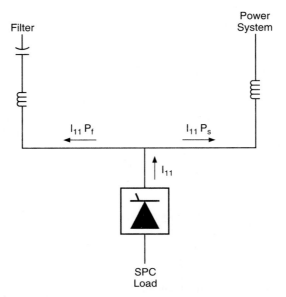

FIGURE 11.16   Division of harmonic current flow.

The 11th-harmonic current division between the filter and the system can be calculated by using the impedance ratios. The impedances must be changed to reflect the
higher frequency of the 11th harmonic: 660 Hz currents. In this case, the actual ohms
will be used to calculate the per-unit current flowing into the filter or the system.
Because the X/R ratio is over 5 for the elements being considered, using only the reactances values makes the calculations easier and does not degrade the results. The system X/R ratio with medium voltage has a minimum of 8. The reactor X/R ratio varies
between 30 and 60 and the capacitor X/R ratios are 500 to 5000.
    Calculating the system ohms from the per-unit value:

$$X_s (\Omega) = \frac{X_{s\,(p.u.)}kV^2}{MVA_{base}} = \frac{0.06 \times 13.8^2}{10} = 1.1426 \tag{11.39}$$

At 660 Hz, the value is 11 times as much, or:

$$X_s = 1.1426 \times 11 = 12.569\ \Omega$$

$$X_C = \frac{34.1}{11} = 3.10\ \Omega$$

$$X_L = 1.36 \times 11 = 14.96\ \Omega \ @ \ 11\text{th harmonic}$$

$$I_s = \frac{X_C + X_L}{X_s + (X_C + X_L)} \tag{11.40}$$

$$I_s = \frac{-3. + 14.96}{12.569 + (-3.1 + 14.96)} = 0.4855 \text{ p.u.} \tag{11.41}$$

$$I_f = \frac{X_s}{X_s + (X_C + X_L)} = \frac{12.569}{24.429} = 0.5145 \text{ p.u.} \tag{11.42}$$

Notice that X_L is the opposite sign from X_C.

Thus, the filter will trap 51.5 percent of any 11th-harmonic current in addition to the 5th-harmonic current.

Assume 30 A of 11th-harmonic current. Then:

$$I_s = 30 \times 0.4855 = 14.565 \text{ A} \tag{11.43}$$

$$I_f = 30 \times 0.5145 = 15.435 \text{ A} \tag{11.44}$$

Even though the filter was tuned to the 5th harmonic, 51.45 percent of 11th-harmonic current is trapped by the filter. The filter is large and so has a relatively wide tuning point. The size was picked to improve the power factor; thus, it has the ability to absorb a larger part of the higher harmonics.

An example of how effective filtering can be is shown in the one-line diagram of Fig. 11.17. The system consists of a large plant with a 1000-mVA short circuit on its 13.8-kV bus. The static power converter load is 30 mVA of six-pulse converters. It is proposed to place 30 mVAR of capacitors on the system to improve the power factor and filter the harmonic currents from the static power converter. Five cases will be investigated to see how effective the filtering is.

Case 1 is with the capacitors on the bus, with no tuning reactors in series.

Case 2 is with 30 mVAR of the capacitors tuned to the 5th harmonic.

Case 3 is with half of the capacitors tuned to the 5th and half to the 7th harmonics.

Case 4 is with one-third of the capacitors tuned to each of the 5th, 7th, and 11th harmonics.

Case 5 is with one-fourth of the capacitors tuned to each of the 5th, 7th, 11th, and 13th harmonics.

Since the harmonic currents will divide in accordance with the impedance ratios between the filters and the system inductance, there are two terms that will be used to determine the split in the currents. The term ρ_f is the per-unit current that will flow into the filter and ρs is the per-unit current that will flow into the system. The equations for these two terms are:

$$\rho_f = \frac{Z_s}{(Z_s + Z_f)} \tag{11.45}$$

$$\rho_s = \frac{Z_f}{(Z_s + Z_f)} \tag{11.46}$$

By using the values in Table 11.5, we can determine the characteristics of the different filter arrangements and see how effective each is in filtering the harmonic currents.

Case 1 concerns the 30-mVAR capacitor bank without any tuning filter. The first parallel-resonant frequency is at 346 Hz. It shows that 19.6 p.u. of any 346-Hz current present will oscillate between the energy storage of the capacitor, which has an angle of 92°,

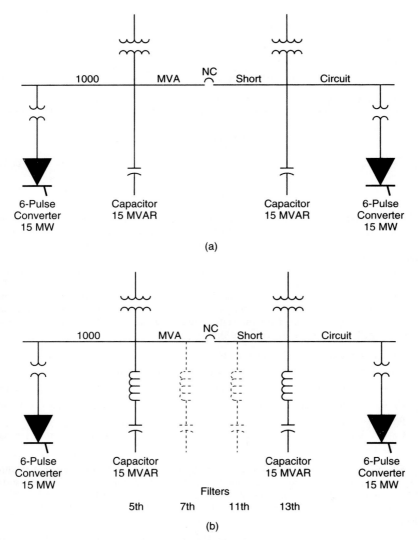

FIGURE 11.17   One-line diagram of system used in filter effectiveness example. (*a*) No filters; (*b*) with filters.

and the energy storage of the inductance of the power system, which has an angle of −85°. There is 177° difference in the angles between these currents that signifies that it is an oscillating current. Since there is normally no 346-Hz current present, this is no problem. However, there is 5th-harmonic (300-Hz) current present and there is 3 p.u. 5th-harmonic current oscillating between the two energy storage elements. Again, the angle between the currents is about 177°. One p.u. of the 300-Hz current is flowing back into the system. There is 2.1 p.u. 7th-harmonic current also oscillating. Only 0.4 p.u. of the 11th harmonic and 0.2 p.u. of the 13th harmonic is oscillating.

TABLE 11.5 Effectiveness of Filters on Power Systems with Static Power Converters

Case no.	Parallel Resonant Frequency (Hz)	Per-unit current of characteristic harmonic										Harmonic filter
		60-Hz		300-Hz		420-Hz		660-Hz		780-Hz		
		ρ_f	ρ_s	ρ_f	ρ_s	ρ_f	ρ_s	ρ_f	ρ_s	ρ_f	ρ_s	
1	346	19.6/92	19.7/-85	3.0/170	4.0/-6.9	3.1/5.5	2.1/-172	1.4/1.3	0.4/-175	1.2/1.0	0.2/-175	None
2	227	13/86.2	13/-89	1.0/2.4	0.04/-89	0.6/1.0	0.4/-1.2	0.5/0.3	0.5/-0.3	0.5/0.2	0.5/-0.2	5
3	238	13/92.3	13/-83	1.0/4.8	0.08/-79	1.0/5	0.1/-79	0.6/0.6	0.4/-1	0.6/0.4	0.4/-0.6	5 & 7
4	248	13/90.7	13/-85	1.0/2.2	0.04/-88	1.0/1	0.1/-107	1.0/0.5	0.0/-107	0.8/0.2	0.2/-0.9	5,7,&11
5	254	13/94	13/-82	1.0/9.5	0.2/-88	1.0/6	0.1/-88	1.0/1.4	0.1/-94	1.0/1	0.1/-86	5,7,11,13

Case 2 concerns the capacitor bank tuned to the 5th harmonic. The first parallel-resonant point is now at 227 Hz. Again, the angle between the two elements is 175°, close to 180°. Almost all of the 5th-harmonic current is flowing into the filter. There is also 60 percent of the 7th, 50 percent of the 11th, and almost 50 percent of the 13th. So with just one fairly large 5th-harmonic filter, most of the characteristic harmonic currents are seeing the filter as a lower impedance than the system and so these harmonic currents are being controlled in their flow into the system.

Case 3 has the capacitor divided equally in two filters. Again the 5th- and 7th-harmonic current is almost all going into the filters. Sixty percent 11th and almost 60 percent 13th is going into the filters. The angle difference is no longer near 180°, so the currents are no longer oscillating. Notice that the first parallel resonance has increased to 238 Hz. As the 5th-harmonic filter gets smaller, the first parallel resonance moves closer to the 5th harmonic.

Cases 4 and 5 continue the effectiveness of the filtering. However, an economic evaluation needs to be done to see just how much money needs to be spent on the filtering system when one large filter does such a good job.

The total value of the capacitors is determined by the amount of VARs needed to improve the power factor to an optimum level. How those capacitors are used in the filters is dependent upon the effectiveness that needs to be accomplished. The preceding example shows how effective it is to use the total value of capacitance in one filter. In any actual system, the size of the filters should not be equal, but should be sized according to the amount of harmonic current that must be filtered. The ratio among the filters should be approximately as the ratios of their harmonic order squared. That is, the square of the 5th is 25, and the square of the 7th is 49. The 7th-harmonic filter should be half the size of the 5th. The 11th filter should be about half the size of the 7th, etc.

11.3.5 Reducing Harmonic Currents through Multipulsing

In the 1930s, the main users of static power converters were the electrometallurgical and the electrochemical industries. These industries were using mercury arc rectifiers to produce dc current for their processes. The loads were relatively large compared to others during that period. The harmonics currents that were produced caused problems in the communication circuits (telephones) at that time. A concerted effort was undertaken by the Edison Electric Institute (EEI) representing the electric utility industry; Bell Laboratories, representing the telephone industry; and the major electrical manufacturers. From this work, the technique of increasing the pulses of a rectifier installation was developed. The theory follows the characteristic of rectifier circuits:

$$h = kq \pm 1 \qquad (11.47)$$

Likewise, the theoretical magnitude of the harmonic current based on a 120° square wave is:

$$I_h = \frac{1}{h} \qquad (11.48)$$

where h = harmonic order
k = any integer, 1, 2, 3,...
q = pulse number of the rectifier circuit
I_h = magnitude of the harmonic current

By this multipulsing technique, the lower harmonics and most of the others were eliminated. Table 11.6 shows the results of multipulsing.

TABLE 11.6 Theoretical Harmonic Currents Present with Multipulsed Static Power Converters

| Static power converter pulse number | | | | | | | | Frequency 60-Hz base | Theoretical current magnitude |
6	12	18	24	30	36	42	48		
5								300	0.200
7								420	0.143
11	11							660	0.091
13	13							720	0.077
17		17						1020	0.059
19		19						1140	0.053
23	23		23					1380	0.043
25	25		25					1500	0.040
29				29				1740	0.034
31				31				1860	0.032
35	35	35			35			2100	0.032
37	37	37			37			2220	0.029
41						41		2460	0.027
43						43		2580	0.024
47	47		47				47	2820	0.023
49	49		49				49	2940	0.021

Although a 12-pulse converter shows no 5th- and 7th-harmonic currents, because of unbalances in the circuit, there will be 10 to 15 percent of what would be expected; e.g., instead of 20 percent 5th-harmonic current, there may be 2 percent.

Figure 11.18 shows the transformer winding configuration for a 24-pulse arrangement. With the 7.5° phase shift connected, only two transformer designs are needed. Note also that each transformer has a delta winding to furnish a low-impedance path for the 3d-harmonic magnetizing currents. Figures 11.19 through 11.23 show the current waveform of individual converters of a 24-pulse installation and the total current going to the converters. These converters had a 15° phase shift between the individual converter transformers. The balance among the four units is good, as well as the ratio of the transformer turns.

When more than one six-pulse bridge circuit is needed to furnish the total power to a load, it is easy to change the transformer connections to give the installation a multipulse arrangement. The amount of phase shift between bridges is determined by:

$$P_{shift} = \frac{360°}{6k} \qquad (11.49)$$

where P_{shift} = degrees of phase shift between six-pulse units
k = number of six-pulse units

Table 11.7 lists the phase shift between transformers for various numbers of six-pulse units. The cancellation of the lower harmonic currents is based on the units being

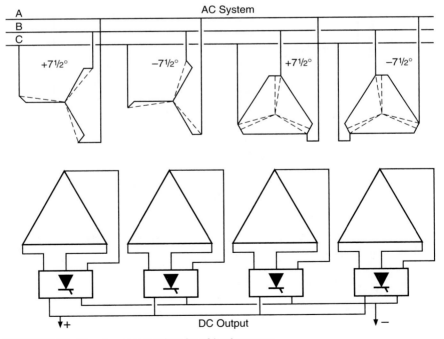

FIGURE 11.18   Transformer connections for a 24-pulse system.

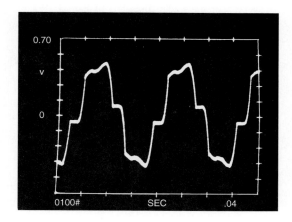

5th = 18.45   11th = 3.48

7th =   8.14   13th = 2.19

FIGURE 11.19   Converter 1 current from a hexagonal-delta transformer. Harmonic current percentage is: 5th = 18.45, 7th = 8.14, 11th = 3.48, 13th = 2.33, 17th = 1.33, 19th = 0.97, 23d = 0.91, 25th = 0.64.

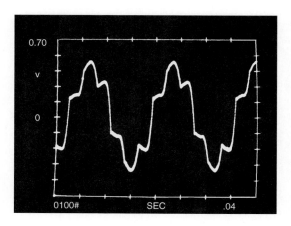

5th = 19.90   11th = 4.33

7th =   8.69   13th = 2.33

FIGURE 11.20   Converter 2 Current from a zigzag-delta transformer. Harmonic current percentage is: 5th = 19.90, 7th = 8.69, 11th = 4.33, 13th = 2.33, 17th = 1.59, 19th = 1.13, 23d = 1.08, 25th = 0.74.

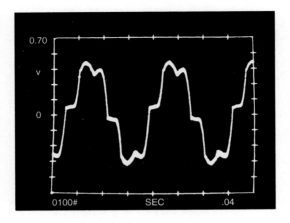

$$5\text{th} = 19.28 \qquad 11\text{th} = 3.19$$
$$7\text{th} = 7.34 \qquad 13\text{th} = 2.24$$

FIGURE 11.21   Converter 3 current from a hexagonal-delta transformer. Harmonic current percentage is: 5th = 19.28, 7th = 7.34, 11th = 3.19, 13th = 2.24, 17th = 1.42, 19th = 1.13, 23d = 0.78, 25th = 0.43.

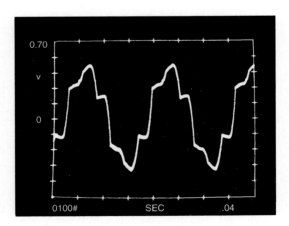

$$5\text{th} = 18.04 \qquad 11\text{th} = 3.49$$
$$7\text{th} = 9.23 \qquad 13\text{th} = 2.42$$

FIGURE 11.22   Converter 4 current with a zigzag-delta transformer. Harmonic current percentage is: 5th = 18.04, 7th = 9.23, 11th = 3.49, 13th = 2.42, 17th = 1.19, 19th = 1.15, 23d = 095, 25th = 0.73.

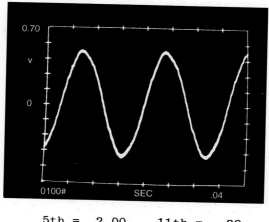

5th = 2.00   11th = .29
7th = .60   13th = .23

FIGURE 11.23 Total current from converters 1–4, each phase-shifted 15° from the others through transformers with ±7.5° for a balanced 24-pulse system. Harmonic current percentage is: 5th = 2.00, 7th = 0.60, 11th = 0.29, 13th = 0.23, 17th = 0.34, 19th = 0.23, 23d = 0.66, 25th = 0.52.

TABLE 11.7 Transformer Phase Shift for Various Multipulse Systems

Pulse number	6	12	18	24	30	36	42	48
No. of units	1	2	3	4	5	6	7	8
Phase shift (in degrees)	0	30	20	15	12	10	8.57	7.5

equally loaded, e.g., a large load being fed by several converter units all equally loaded. The impedance differences between transformers will prevent perfect cancellation.

Some installations have many drives, but they are not equally loaded and do not necessarily run at the same time. An example is a steel mill where there are several auxiliary drives that are not equal in rating and do not necessarily operate at the same time. In this case, if half of the horsepower of the drives had a delta-delta transformer and half a delta-wye transformer, there would be some cancellation of the 5th-, 7th-, etc., harmonic currents, even though they would not be completely canceled. A rule of thumb would be that 25 percent of the harmonic currents would be present instead of being canceled. Although this is not perfect, it is better than if all of the drives had the same transformer connection.

11.4   EFFECTS OF HARMONIC CURRENTS AND VOLTAGES

The effect that harmonics have on various power system components, loads, and adjacent facilities can be broken down into these categories:

- Rotating machines
- Power system components
- Loads
- Power factor
- Communications

11.4.1   Rotating Machines

Balanced harmonic currents in a three-phase system can be given sequence designations. Table 11.8 lists the sequences of harmonics.

When viewed from the rotor of a machine, the negative and positive sequence harmonics appear as a zero sequence on the rotor. The 5th, negative sequence, is rotating in the opposite direction from the rotor and that additional rotation makes it appear as a 6th harmonic on the rotor. Likewise, the 7th harmonic is rotating faster than the rotor and appears as a 6th harmonic on the rotor. Thus, both the 5th- and 7th-harmonic currents on the stator appear on the rotor across the air gap as a 6th-harmonic current induced onto the rotor. These harmonic currents then reduce the negative sequence current capability of the machine. Basically, they are adding to the heating of the rotor. As long as the voltage distortion is within the limits of the standard (5 percent voltage distortion), the additional heating is within the tolerance of the machine.

The second effect of the harmonics on the machine is how they affect the torque of a motor. The negative sequence harmonics subtract from the torque and the positive sequence harmonics add to the torque. These harmonics result in pulsating torques. Again, as long as the distortion is within limits, the torque effect is insignificant.

The current that a machine draws from the system is nearly sinusoidal. The harmonic currents from other loads flowing through the impedance of the system cause a voltage drop that distorts the voltage waveform. It is this distorted voltage that affects motors. Generators are affected by the distorted current they must produce for the load.

TABLE 11.8   Sequences of Harmonics

Harmonic	Sequence	Harmonic	Sequence	Harmonic	Sequence
1	Pos	10	Pos	19	Pos
2	Neg	11	Neg	20	Neg
3	Zero	12	Zero	21	Zero
4	Pos	13	Pos	22	Pos
5	Neg	14	Neg	23	Neg
6	Zero	15	Zero	24	Zero
7	Pos	16	Pos	25	Pos
8	Neg	17	Neg	26	Neg
9	Zero	18	Zero	27	Zero

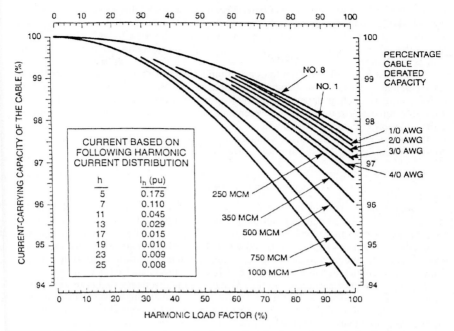

FIGURE 11.24 Cable derating vs. harmonics from a six-pulse harmonic current distribution.

11.4.2 Power System Components

When conductors have to carry harmonic currents or fundamental currents that are distorted by harmonic currents, the "skin effect" of the harmonic currents increases the effective resistance of the conductor. The larger the cable, the more effect the frequency has on the internal eddy current (skin effect) losses. The actual calculation of the effective resistance is complicated by the proximity effect of other conductors, conductor size, permeability of the conductor, etc. Smaller cables have less of a derating factor than larger ones. Several texts give the calculating methods for determining the effective resistance for various frequencies. The result of harmonics currents in conductors is additional losses.

Figure 11.24 indicates the derating factor for cables when there is a six-pulse static power converter on the system. If the static power converter is 50 percent of the load, then the cable needs to be derated 1 percent if it is a 250-MCM cable.

Transformers are complex devices that have losses that are divided into load losses and no-load losses. The load I^2R losses will constitute 75 to 85 percent of the total, and about 75 percent of these are not frequency dependent. The remainder varies with the square of the frequency. The no-load losses (core loss) constitute between 15 and 25 percent of the total loss and, depending upon the flux density, these losses vary with frequency $f^{3/2}$ to f^3. From this, the reactance increases directly with the frequency (inductance constant). It can be seen that the harmonic X/R ratios will be less than the fundamental-frequency X/R ratio. The additional resistance at the higher frequencies damps oscillations and reduces the amplification factors in parallel resonances.

Reactors can be air core or iron core. The air-core reactor losses increase with frequency similar to the losses in conductors. Some manufacturers use small conductors in parallel to reduce this effect. Iron-core reactors have losses similar to transformers, where the X/R ratios are reduced at the higher frequencies. Again, the higher resistance damps oscillations.

Transmission and distribution lines are complex circuits that include all three impedance elements: resistance, inductance, and capacitance. The lines represent resonant circuits at some frequency, depending on the length and construction of the line. Using the long-line equations to represent these elements in studies will bring out the resonant circuits and their effect on harmonics.

Capacitor reactance decreases with frequency. As a result, the capacitor acts as a sink to harmonic currents. These harmonic currents will increase the voltage

$$X_c = \frac{1}{2\pi f C} \tag{11.50}$$

across the capacitor tending to overvoltage the capacitor. The capacitor is rated for 110 percent voltage, or 120 percent including harmonic voltage. It is the peak voltage and not the rms voltage that tends to break down the capacitor. The 135 percent capability of the capacitor in its kVAR rating allows it to absorb the harmonic current without stress, except for the increased voltage associated with the additional harmonic currents. There is also an increase in the reactive kVAR that is produced by the capacitor.

The main effect of harmonics on the capacitor is the resonance that is set up with the system inductance. This resonance can result in a relatively high oscillating current if the resonance is at a frequency of a harmonic current that is present on the circuit. This high current can cause the capacitor fuses to fail, thus removing part of the capacitor bank from service. It can unbalance the three-phase currents.

Switchgear will have additional losses due to the harmonic currents, but otherwise will not be affected.

Some relay characteristics will be changed with distorted currents. Tripping devices on low-voltage circuit breakers reacting to rms currents will take into account the effect of these harmonics. Some devices look at peak currents. These devices will give erroneous trip signals if the circuit breaker should react to the rms current. Many newer circuit protective devices take into account the distorted currents that loads have.

11.4.3 Loads

The distorted voltage caused by distorted currents flowing through the impedance of the system force harmonic currents to flow in loads. If the loads are for heat, this distorted current will add to the heating. If heating is harmful, then the effect is detrimental. If the voltage distortion is great enough to change the zero crossing of the voltage wave, then those devices that depend upon the zero crossing of the voltage will be affected.

Electronic loads that use a switched-mode power supply will be isolated from the distortion by the dc link in this power supply. If the dc link capacitor is large, it will swamp out the effect of the distortion on the input. There are devices that depend upon zero crossings that can be affected by voltage distortion. Carrier signals that control hot water heaters, clocks, and similar loads can be affected by the harmonic currents and voltages.

11.4.4 Power Factor

In Chap. 6, power factor and VAR control were discussed. What was not discussed was the effect of harmonics on power factor. Power factor can be divided into two parts: displacement power factor and distortion power factor. *Displacement power factor* is what is normally referred to as power factor. It is the relationship between the relative displacement of the current with respect to the voltage. It is defined as:

$$PF_{displacement} = \frac{kW}{kVA} = \frac{V_1 I_1 \cos \phi}{V_1 I_1} = \cos \phi \tag{11.51}$$

This does not take into account the rms value of current and voltage that increases the kVA in the circuit. The *total power factor* is defined as:

$$PF_{total} = \frac{V_1 I_1 \cos \phi}{V_{rms} I_{rms}} = \frac{kW}{kVA} \tag{11.52}$$

Then, if we assume that the system is stiff—that is, the power system has a low impedance and the amount of voltage distortion is low such that V_1 is equal to V_{rms}—then the *distortion power factor* can be:

$$PF_{distortion} = \frac{V_1 I_1}{V_{rms} I_{rms}} = \frac{I_1}{I_{rms}} \tag{11.53}$$

$$V_1 = V_{rms}$$

Thus, the distortion power factor is the ratio of the fundamental current to the rms current.

The quality of the voltage and current has been defined as a *distortion factor* (DF) or *total harmonic distortion* (THD). The relationship between THD and the distortion power factor can be derived. The THD, as it pertains to current and the rms current, is:

$$THD = \frac{\sum_2^H \sqrt{I_h^2}}{I_1} = \frac{\sqrt{I_2^2 + I_3^2 + I_4^2 + \dots + I_H^2}}{I_1} \tag{11.54}$$

$$I_{rms} = \sqrt{I_1^2 + I_2^2 + I_3^2 + I_4^2 + \dots + I_H^2} \tag{11.55}$$

Substituting in the expression for distortion power factor:

$$PF_{distortion} = \frac{I_1}{I_{rms}} = \frac{I_1}{\sqrt{I_1^2 + I_2^2 + I_3^2 + I_4^2 + \dots + I_H^2}} \tag{11.56}$$

By dividing each term by I_1, we can reduce the equation to:

$$PF_{distortion} = \frac{1}{\sqrt{1 + THD^2}} \tag{11.57}$$

Table 11.9 can be constructed to show this relationship. It can be seen that the distortion power factor decreases rapidly as the THD increases. The phasor diagram of the total power factor is then as shown in Fig. 11.25.

The displacement power factor can be improved by providing the capacitor to furnish the VAR needed by the load. The distortion power factor can be improved only by providing the harmonic current needed by the load through passive or active filters. By reducing the load current distortion, it becomes closer to the fundamental waveshape; thus, the harmonics do not flow into the power system.

TABLE 11.9 THD vs. Distortion Power Factor

THD	PF	THD	PF
0.10	0.995	0.70	0.819
0.20	0.981	0.80	0.781
0.30	0.978	0.90	0.743
0.40	0.928	1.00	0.707
0.50	0.894	1.10	0.673
0.60	0.867	1.20	0.640

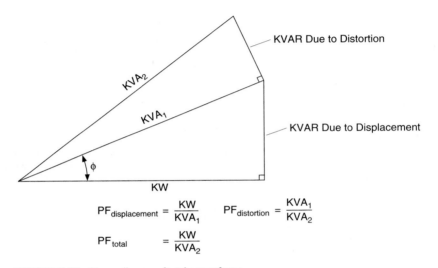

$$PF_{displacement} = \frac{KW}{KVA_1} \qquad PF_{distortion} = \frac{KVA_1}{KVA_2}$$

$$PF_{total} = \frac{KW}{KVA_2}$$

FIGURE 11.25 Phasor diagram of total power factor.

11.4.5 Communications

Noise induced into communications circuits was one of the first major problems that arose from harmonics currents in power systems. Where telephone circuits parallel power lines, distorted currents in the power system can induce unequal voltages in the two conductors of the telephone circuit. This noise can disrupt communications. In the 1920s and 1930s, when many of the telephone circuits were open wire, this was a major problem. Work was done to investigate this problem and standards were set up to control the inductive coupling between the power systems and the telephone system. The telephone industry developed multiconductor twisted-pair cable that eliminated a large part of the problem. The use of fiber optic cables on long distance service eliminates this inductive interference. There are still cases where the telephone noise problem is evident.

An analytical approach to handling the problem has been developed. The term *telephone influence factor* (TIF) is used to define the response of the human ear and the telephone equipment to different frequencies. Figure 11.26 is the 1960 TIF weighting curve.

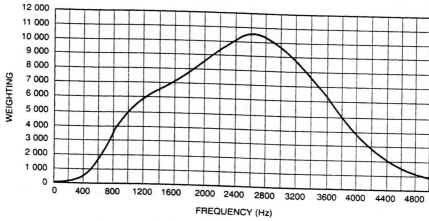

FIGURE 11.26   1960 TIF weighting values.

Earlier curves varied, depending on the evolution of the telephone equipment. It can be seen that the frequencies between 1600 and 3500 Hz are the most sensitive. This is the 25th to 60th harmonic on 60 Hz or 30th to 70th harmonic on 50 Hz.

Table 11.10 lists the single-frequency weighting values based on the curve of Fig. 11.24.

The term *I*T product* is used to determine the quality of the interference. The *I*T* product is calculated by:

$$I*T \text{ product} = \sqrt{\sum_{H=1}^{H=50} [(I_h)(\text{TIF})_h]^2} \qquad (11.58)$$

The current of each of the harmonics is obtained from measurements. The telephone company makes these measurements. From the measurements, the *I*T* product can be calculated. If there needs to be a reduction in the *I*T* product, filters can be designed to reduce the harmonic currents that are causing the most problems. *I*T* products less than

TABLE 11.10   Single-Frequency Weighting Values

Freq	TIF	Freq	TIF	Freq	TIF	Freq	TIF
60	0.5	1,020	5,100	1,860	7,820	3,000	9,670
180	30	1,080	5,400	1,980	8,330	3,180	8,740
300	225	1,140	5,630	2,100	8,830	3,300	8,090
360	400	1,260	6,050	2,160	9,080	3,540	6,730
420	650	1,380	6,370	2,220	9,330	3,660	6,130
540	1,320	1,440	6,650	2,340	9,840	3,900	4,400
660	2,260	1,500	6,680	2,460	10,340	4,020	3,700
720	2,760	1,620	6,970	2,580	10,600	4,260	2,750
780	3,360	1,740	7,320	2,820	10,210	4,380	2,190
900	4,350	1,800	7,570	2,940	9,820	5,000	840
1,000	5,000						

10,000 will probably not cause problems unless there is extreme exposure between the power system and communication circuits. An extreme exposure example occurs in rural areas where the power or telephone circuits use a ground return path for one conductor. In these cases, $I*T$ products as low as 1500 may be required.

There is another curve that is called the *C-message weighting curve*. Instruments have been developed that measure the noise based upon this curve. It is similar to the curve in Fig. 11.26, except it is plotted on log-log paper. Table 11.10 lists the single-frequency weighting values.

In some cases, the commutation notch from static power converters can cause noise if the natural frequency of the power system is excited by the notch.

11.5   HARMONIC STANDARDS

The first harmonic standards were established by the British Electricity Board in Britain as G5/1. Since it was established by the electrical producer, it was restrictive. When U.S. manufacturers had to meet what they thought was not a practical standard, work was started within IEEE-IAS to establish a standard that was based on the experience that the application engineers had in dealing with the problem. As a result, the first IEEE standard for harmonics was established as IEEE Standard 519-1981, *IEEE Guide for Harmonic Control and Reactive Compensation of Static Power Converters*. This standard established a limit of voltage distortion for power systems that would allow equipment connected to it to operate satisfactorily.

The need for changes in the document became apparent during the first meeting to review and revise the document. The revised document was cosponsored by both the Industry Application Society (IAS), representing the users, and the Power Engineering Society (PES), representing the electric utility industry. The original document was good because it brought to the attention of power engineers the problems associated with nonlinear harmonic-producing loads. However, it did not address the concern that many electric utilities had about one user using up all the capacity of their system to absorb harmonic currents. The voltage distortion criteria in the 1981 edition could not be used to distribute among users the ability of the utility system to absorb harmonic currents. This problem is addressed in the revised document.

There are two criteria that are used to evaluate harmonic distortion. The first is a limitation in the harmonic current that a user can transmit into the utility system. The second criteria is the quality of the voltage that the utility must furnish the user. The interrelationship of these criteria shows that the harmonic problem is a system problem and not tied just to the individual load that requires the harmonic current.

Table 11.11 lists the harmonic current limits based on the size of the user with respect to the size of the power system to which the user is connected. The ratio of I_{sc}/I_L is the short-circuit current available at the point of common coupling (PCC) to the load current associated with the demand of the facility. Thus, as the size of the user load decreases with respect to the size of the system, the percentage of harmonic current the user is allowed to inject into the utility system becomes larger. This protects other users on the same feeder as well as the utility that is required to furnish a certain quality of power to its customers.

The second limitation specifies the quality of voltage that the utility must furnish the user. Table 11.12 lists the amount of voltage distortion that is acceptable from the utility to the user. To meet the power quality values listed in Table 11.12, cooperation among all users and the utility is needed to ensure that no single user deteriorates the power quality beyond the values in Table 11.12.

TABLE 11.11 IEEE-519 Standard Current Distortion Limits for General Distribution Systems, (120–69,000 V)

I_{sc}/I_L	Maximum harmonic current distortions in % of I_1					
	Individual harmonic order (odd harmonics)					
	<11	$11 \leq h < 17$	$17 \leq h < 23$	$23 \leq h < 35$	$35 \leq h$	TDD
<20	4.0	2.0	1.5	0.6	0.3	5.0
20<50	7.0	3.5	2.5	1.0	0.5	8.0
50<100	10.0	4.5	4.0	1.5	0.7	12.0
100<1000	12.0	5.5	5.0	2.0	1.0	15.0
>1000	15.0	7.0	6.0	2.5	1.4	20.0

Even harmonics are limited to 25 percent of the odd harmonic limits above.

Current distortion that results in a direct current offset (e.g., half-wave converters) are not allowed.

where I_{sc} is the short-circuit current at the PCC

I_L is the maximum demand load current (fundamental-frequency component at PCC).

All power generation equipment is limited to these values of current distortion regardless of the actual I_{sc}/I_L.

TABLE 11.12 Voltage Distortion Limits

Bus voltage at PCC	Individual voltage distortion (%)	Total voltage distortion THD (%)
69 kV and below	3.0	5.0
69.001–161 kV	1.5w	2.5
161.001 kV and above	1.0	1.5

NOTE: High-voltage systems can have up to 2.0 percent THD where the cause is an HVDC terminal that will attenuate by the time it is tapped for a user.

Two examples are given in the IEEE-519 standard to show how the limits are applied. One of them is shown in Fig. 11.27, which represents a utility distribution feeder to which four users are connected. Each user sees a different value of short circuit or system size. Table 11.13 lists the cases that were considered in meeting the IEEE-519 criteria on voltage and current distortion. Table 11.14 lists the voltage THD at the PCC for each user for each of the various cases.

Case A in Table 11.13 lists the voltage distortion if nothing is done. Case B shows the utility's effort to correct the problem without involving the users. In case C, the users who are causing the majority of the problem make a correction by furnishing harmonic filters on their own buses. Case D shows the utility's effort to correct the problem with a larger filter. Table 11.14 lists the voltage THD for each of these cases.

In case A of Table 11.14, the distortion for all users is above the 5 percent that is recommended by the IEEE-518 Standard. In fact, users 3 and 4 have distortions that could affect how their loads operate. This condition would not be allowed to persist. The utilities' efforts to correct the problem with case B did improve the voltage quality for users 1 and 2, but not for the other two.

Case C shows the biggest improvement for all users. The filters that users 3 and 4 installed corrected not only their problem, but also the overall problem. They had the largest nonlinear loads and, by placing the filters on their lower-voltage buses, the har-

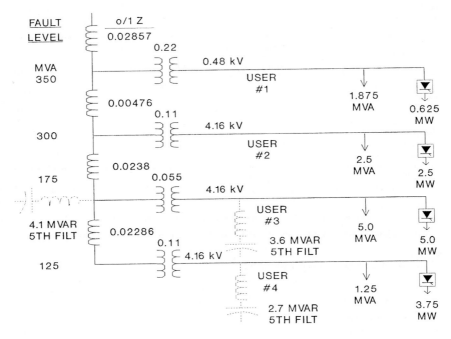

FIGURE 11.27   One-line diagram of distribution system with user's loads for example.

TABLE 11.13   Arrangement of Harmonic Filters for the Various Study Cases

Case number	Filter size (mVA)	Location of filter
A	None	—
B	4.1	At user 3, 13.8-kV bus
C	3.6	At user 3, 4.16-kV bus
	2.7	At user 4, 4.16-kV bus
D	5.8	At user 3, 13.8-kV bus

TABLE 11.14   Summary of Total Harmonic Voltage Distortion for Example

Case number	User 1	User 2	User 3	User 4
A	7.13	8.25	12.62	14.43
B	3.93	4.53	6.39	8.02
C	2.94	3.37	4.34	4.69
D	3.48	3.99	5.47	7.12

monic currents from their loads saw a much lower filter impedance for all of the generated harmonics than the impedance back into the utility system. The addition of the step-down transformer impedance to the total made the difference.

Case D was an attempt by the utility to increase the size of its filter (lower its impedance), to absorb more of the harmonic currents. It was not successful.

This example shows that the users who distort the current beyond that shown in Table 11.12 should make the correction by using harmonic filters at their own buses.

11.6 REFERENCES

1. Don O. Koval, *Frequency of Transmission Line Outages in Canada,* IEEE-IAS Conference Record, 94CH34520, Oct. 1994, pp. 2201–2208.

2. Douglas S. Dorr, *Point of Utilization Power Quality Study Report,* IEEE-IAS Conference Record, 94CH34520, Oct. 1994, pp. 2334–2344.

3. David E. Rice, *Adjustable Speed Drive and Power Rectifier Harmonics—Their Effect on Power Systems Components,* IEEE Paper No. PCIC-84-52, pp. 269–287.

4. P. G. Cummings, "Estimating Effect of System Harmonics on Losses and Temperature Rise of Squirrel-Cage Motors," *IEEE Transactions on Industry Applications,* Nov./Dec. 1986, pp. 1121–1126.

5. W. C. Ball and C. K. Poarch, "Telephone Influence Factor and Its Measurement," *AIEE Trans. on Communications and Electronics,* vol. 9, Jan. 1961, pp. 659–664.

6. E. W. Kimbark, *Direct Current Transmission, vol. 1,* Wiley Interscience, New York, 1971.

7. B. R. Pelly, *Thyristor Phase-Controlled Converters and Cyclo-converters,* John Wiley, New York, 1971.

8. AIEE Committee Report, "Inductive Coordination Aspects of Rectifier Installations," *AIEE Transactions,* vol. 65, 1946, pp. 417–436.

9. R. P. Stratford, "Rectifier Harmonics in Power Systems," *IEEE Transactions on Industrial Applications,* vol. IA-16, no. 2, Mar./Apr. 1980, pp. 271–276.

10. R. P. Stratford, "Harmonic Pollution on Power Systems—A Change in Philosophy," *IEEE Transactions on Industrial Applications,* vol. IA-16, no. 5, Sept./Oct. 1980, pp. 617–623.

11. R. P. Stratford, "Analysis and Control of Harmonic Current in Systems with Static Power Converters," *IEEE Transactions on Industry Applications,* vol. IA-17, no. 1, Jan./Feb. 1981, pp. 71–81.

12. IEEE Standard 100-1992, *IEEE Recommended Practices and Requirements for Harmonic Control in Electric Power Systems,* Institute of Electrical and Electronics Engineers.

CHAPTER 12
ELECTRIC LIGHTING

12.0 INTRODUCTION

Good lighting improves work quality, production efficiency, personnel safety, and installation security. Such improvements are achieved by careful consideration of six major subjects:

1. Illuminance—light on the task
2. Light source—lamp, ballast, fixture (luminaire)
3. Light quality—color, glare, stroboscopic effects
4. Physical environment—temperature, reflectance, zonal cavity
5. Electrical environment—voltage sags and surges, harmonics
6. Energy codes—state and federal

12.1 GLOSSARY

Ballast A device used with an electric-discharge lamp to obtain the necessary circuit conditions (voltage, current, and wave form) for starting and operating.

Candela The SI unit of luminous intensity. One candela is one lumen per steradian (a solid angle subtending an area on the surface of a sphere equal to the square of the sphere radius).

Coefficient of utilization (CU) The ratio of the luminous flux (lumens) from a luminaire calculated as received on the work plane to the luminous flux emitted by the luminaire's lamps alone.

Color The characteristics of light by which a human observer may distinguish between two structure-free patches of light of the same size and shape.

Correlated color temperature (CCT) The absolute temperature in kelvins (K) of a blackbody whose chromaticity most nearly resembles that of the light source. A blackbody is an ideal body of surface that completely absorbs all radiant energy falling on it with no reflection. Higher CCT indicates cooler (bluer) light source.

Color rendering index (CRI) A measurement of the color shift an object undergoes when illuminated by a light source, as compared to a reference source at the same color temperature measured on a scale from 0 to 100. Objects and people under natural daylight (CRI = 100) generally appear more true to life.

Efficacy (luminous)   The luminous efficacy of a light source is the quotient of the total luminous flux emitted by the total lamp power input expressed in lumens per watt.

Fixture   (See **luminaire**)

Fluorescence   The emission of light (luminescence) as the result of, and only during, the absorption of radiation of other (mostly shorter) wavelengths. A fluorescent lamp is a low-pressure mercury electric-discharge lamp in which a fluorescing coating (phosphor) transforms some ultraviolet energy generated by the discharge into light.

Footcandle (fc)   The illuminance on a surface one square foot in area on which there is a uniformly distributed flux of one lumen.

Glare   The sensation produced by luminance within the visual field that is sufficiently greater than the luminescence to which the eyes are adapted to cause annoyance, discomfort, or loss in visual performance or visibility.

Halogen   See definition included under **Metal halide lamp.**

High-pressure sodium lamp (HPS)   A high-intensity discharge lamp in which light is produced by radiation from sodium vapor operating at a partial pressure of about 1.33×10^4 Pa (100 torr). Includes clear and diffuse-coated lamps.

Illuminance, $d\rho/dA$ The density of luminous flux incident on a surface; it is the quotient of the luminous flux by the area of the surface when the latter is uniformly illuminated. See Table 12.1 for the units of measurement used. Lux is the SI unit of measurement for illuminance.

Incandescence   The self-emission of radiant energy in the visible spectrum due to the thermal excitation of atoms or molecules. An incandescent filament lamp is one in which light is produced by a filament heated to incandescence by an electric current.

Instant-start fluorescent lamp   A fluorescent lamp designed for starting by a high voltage without preheating of the electrodes.

Lamp   A generic term for a man-made light source. By extension, sources that radiate in regions of the spectrum adjacent to the visible.

Light   Radiant energy capable of exciting the retina producing a visual sensation.

Low-pressure sodium lamp (LPS)   A discharge lamp in which light is produced by radiation from sodium vapor operating at a partial pressure of 0.13 to 1.3 Pa (10^{-3} to 10^{-2} torr).

Lumen (lm)   SI unit of luminous flux. Radiometrically, it is determined from the radiant power. Photometrically, it is the luminous flux emitted within a unit solid angle (one steradian) by a point source having a uniform luminous intensity of one candela.

TABLE 12.1   Illuminance Conversion Factors

Units	Footcandles	Lux	Phot	Milliphot
Footcandles	1	0.0929	929	0.929
Lux (lx)	10.76	1	10,000	10
Phot (ph)	0.00108	0.0001	1	0.001
Milliphot	1.076	0.1	1,000	1

1 lm = 1/683 light-watt
1 fc = 1 lm/ft^2
1 Ws = 10^7 ergs
1 ph = 1 lm/cm^2
1 lx = 1 lm/m^2 = 1 metercandle

Lumen maintenance Lumen output (usually expressed in percent of rated lumen output) at a specific percent of rated lamp life.

Luminaire A complete lighting unit consisting of a lamp or lamps together with the parts designed to distribute the light, to position and protect the lamps, and to connect the lamps to the power supply.

Mercury lamp A high-intensity discharge (HID) lamp in which the major portion of light is produced by radiation from mercury operating at a partial pressure in excess of 1.013×10^5 Pa (1 atmosphere). Includes clear, phosphor-coated (mercury-fluorescent), and self-ballasted lamps.

Metal halide lamp A high-intensity discharge (HID) lamp in which the major portion of light is produced by radiation of metal halides and their products of disassociation possibly in combination with metallic vapors such as mercury. Includes both clear and phosphor-coated lamps. [Halide is a binary compound of a halogen (fluorine, bromine, iodine, or astatine forming part of group VIIA of the periodic table and existing in the free state normally as diatomic molecules) with a more electropositive element or radical.]

PAR lamp Parabolic aluminized reflector lamp.

Rapid-start fluorescent lamp A fluorescent lamp for operation with a ballast that provides a low-voltage winding for preheating the electrodes and initiating the arc without a starting switch or the application of high voltage.

Reflectance of a surface or a medium, $\rho = \Phi_r/\Phi_i$ The ratio of reflected flux to incident flux.

12.2 ILLUMINANCE

The units of measurement for illuminance (light on the task) are lux (lm/m^2) or foot-candles (lm/ft^2). 1 lx (SI unit) = 10.76 fc. See *IESNA Lighting Handbook*,[1] Fig. 11-1, for recommended illuminance for specific visual tasks in any of the nine illuminance categories listed therein. See Tables 12.2 through 12.5 of this handbook for an abbreviated summary of the values listed in Ref. 1.

Recommended illuminance is only one of the factors considered in a lighting installation. Those unfamiliar with lighting could tend to select the often inappropriate highest values in the ranges. Advice of an expert in lighting should be consulted before making any design decisions.

12.3 LIGHT SOURCES

The selection of a light source involves evaluating the lamp, ballast, and fixture types best suited to meet the objectives of illuminance, first cost and operating cost, maintainability, esthetics, and physical space constraints. Three major lamp types available are (1) incandescent (Fig. 12.1), (2) fluorescent (Fig. 12.2), and (3) high-intensity discharge (HID) (Fig. 12.3). Fluorescent and HID lamps require a ballast to control the voltage and current applied to the lamp. Fluorescent lamp ballasts consist of a transformer core and coil plus capacitor and electronics where required. HID lamp ballasts may be reactor, autotransformer, constant wattage (regulator or lead circuit autotransformer type). Section 12.4 discusses ballasts, and Sec. 12.5 discusses luminaires.

TABLE 12.2 Illuminance for Industrials—Indoors

Area/activity	lx	fc
Garages		
Repair	500–1000	50–100
Active traffic areas	100–200	10–20
Loading platforms	200	20
Machine shops and assembly areas		
Rough bench/machine work, simple assembly	200–500	20–50
Medium bench/machine work, moderately difficult assembly	500–1000	50–100
Difficult machine work, assembly	2000–5000	200–500
Fine bench/machine work, assembly	5000–10000	500–1000
Receiving and shipping	200–500	20–50
Warehouses, storage rooms		
Active	100–200	10–20
Inactive	50–100	5–10

TABLE 12.3 Illuminance for Industrials—Outdoors

Area/activity	lx	fc
Storage yards		
Active/inactive	200/100	20/10
Open construction plants		
Chemical or power plants, refineries	200/100	20/10
Parking areas		
Open—high activity/medium activity	20/10	2/1
Covered—parking, pedestrian areas	50	5
Entrances—day/night	500/50	50/5

12.3.1 Incandescent Lamps

Incandescent lamps operate by heating a filament in a translucent envelope producing both light and heat. This type of lighting usually involves the least initial investment, lowest efficacy (energy efficiency), shortest life, highest maintenance, and most heat generation of the lamp choices. Incandescent lamps provide some of the best color distribution in the higher wavelengths (orange and red). Table 12.6 compares the approximate characteristics of incandescent lamps to other types of lamps.

Incandescent lamps usually employ a tungsten filament and are either vacuum type (below 40 W for general service) or gas filled using a combination of argon and nitrogen or nitrogen alone. Figure 12.4 shows the effect of voltage and current variations on incandescent filament lamps.

Three categories of incandescent lamps are available: (1) standard, (2) energy efficient, and (3) halogen. Energy-efficient lamps are lower-wattage lamps providing about the same lumen output as a standard lamp of a high watt rating. Changing either the filament or the reflector (in PAR lamps) accomplishes this energy saving. Halogen lamps use a halogen gas to allow the filament to burn more intensely without sacrificing life. The halogen vapor combines with particles of tungsten that have evaporated from the filament and redeposits them on the filament. Halogen-IR lamps redirect infrared heat

TABLE 12.4 Illuminance for Offices

Activity	Examples	lx	fc
Visual tasks only occasionally performed	Lobbies, reception areas, corridors, stairs, washrooms, circulation areas	100–200	10–20
Visual tasks of high contrast or large size	Reading newsprint, types originals, 8–10-point print, impact printing (good ribbon), ballpoint pen, felt-tip pen. Conference rooms, library areas, general filing.	500–1000	50–100
Visual tasks of medium contrast or small size	Mail sorting, reading thermal printing, xerography, 6-point print, drafting with high-contrast media, photographic work (moderate detail), writing (#3 pencil and softer).	500–1000	50–100
Visual tasks of low contrast or very small size	Drafting (low-contrast media), charting, graphing, reading poor thermal paper quality, writing (#4 pencil).	1000–2000	100–200

TABLE 12.5 Illuminance for Stores

Category	Activity	Merchandise		Displays	
		lx	fc	lx	fc
Discount/mass merchandising	*High* Readily recognizable merchandise usage. Rapid evaluation and viewing time. Merchandise attracts and stimulates impulse buying.	1000	100	5000	500
Family stores	*Medium* Merchandise is familiar type or usage.	700	70	3000	300
Upscale stores	*Low* Infrequent buying. Assistance and time required to buy.	300	30	1500	150

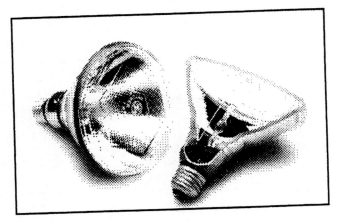

FIGURE 12.1 Incandescent lamps.

FIGURE 12.2 Compact fluorescent lamps.

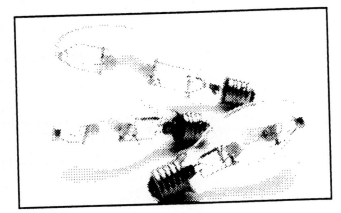

FIGURE 12.3 High-intensity discharge lamps.

TABLE 12.6 Approximate Range of Lamp Characteristics

Lamp type	W	Avg. lm/W Initial	Avg. lm/W Mean*	% lm maintenance	Rated avg. life† (h)	Warm-up/ restrike max/min (min)
LP sodium‡	10–180	100–180			10000– 18000	7–15/ immediate
HP sodium	35–1000	64–140	58–126	90	24000+	3–4/1–2
Metal halide	175–1000	80–115	57–92	71–83	10000– 20000	2–4/10–15
Energy-saving metal halide	32–175	78–95	63–76	77–80	10000– 18000	2–4/10–15
Deluxe HP sodium	70–400	51–93	41–86	80–92	10000	3–4/1–2
Fluorescent	4–215	34–90	27–82	66–92	10000– 20000+	Immediate
Mercury vapor	50–1000	32–63	25–43	57–84	16000– 24000+	5–7/3–6
Incandescent	100–1500	17–24	15–23	90–95	750–2000	Immediate

*Mean lm = initial lm × % lm maintenance at 40% (metal halide or fluorescent) or 50% (mercury) rated life. Efficacy = "efficiency" = lm/W (LPW). Highest mean LPW = lowest cost of light.

†Average rated life = number of hours a group of lamps operates before 50% have failed. Lower mean LPW plus longer life means higher cost of light.

‡Monochromatic, hence seldom used in U.S. except in warehouses and for parking lots.

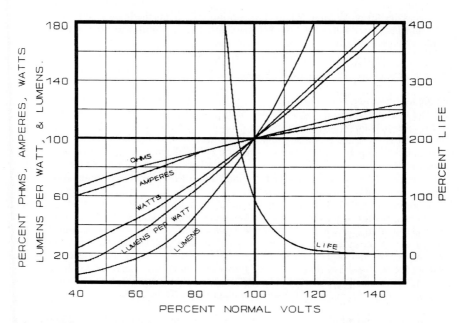

FIGURE 12.4 Effect of voltage and current variation on the operating characteristics of incandescent lamps in general lighting circuits.

with a special coating to heat the filament, producing more visible light and saving energy, as well as reducing air-conditioning load. Less power input is necessary to keep the filament hot. A comparison of the three categories of lamp for a specific PAR38 lamp appears below:

Lamp type	W	lm	LPW	Life
Standard	75	765	10.2	2000
Energy efficient	65	675	10.4	2000
Halogen	60	1150	19.2	3000

lm = initial rated lumens.
LPW = lumens per watt.
Life = rated average life in hours.

12.3.2   Fluorescent Lamps

Fluorescent lamps operate with a low-pressure gas discharge source. Light is produced when a coating of fluorescent powders on the glass envelope is activated by ultraviolet energy generated by a low-pressure mercury arc. Fluorescent lamps must operate with a series current-limiting device (ballast). Three types of starting are used for fluorescent lamps: preheat, rapid, and instant start. Preheat starting uses an automatic switch starter to initiate the arc by boiling electrons off the lamp electrodes and is still used for smaller lamps similar to desk and bathroom-fixture type lamps. Preheating accelerates loss of the emissive coating, resulting in shorter lamp life than that of the same lamp operated on rapid-start ballasts. Instant start jolts the lamp electrodes with a high voltage to avoid the necessity for preheating. Rapid start continuously heats the lamp electrodes. Tables 12.7 and 12.8 give some of the typical characteristics for rapid- and instant-start lamps. Table 12.9 gives this data for energy-saving lamps.

12.3.3   Compact Fluorescent Lamps

Compact fluorescent lamp installation configurations are three: dedicated, self-ballasted, and modular. Dedicated systems are similar to full-size fluorescent lighting systems with separate ballasts hard-wired within a luminaire. This section concerns only screwbase self-ballasted and modular compact fluorescent lamps. A self-ballasted lamp is one in which the lamp and ballast form an integral unit and cannot be separated. A modular compact fluorescent product consists of a screwbase ballast with a replaceable lamp. Since ballast life is approximately five times lamp life, modular systems allow lamp replacement without ballast replacement.

High lamp system efficacies (lm/W) and long lamp life make compact fluorescent lamps an attractive replacement for incandescent lamps. Lamp system efficacies range from about 40 to 60 LPW for 13- to 28-W lamp systems. These efficacies, four times or more than for incandescent lamps, approach those available using standard fluorescent systems 32 W and higher. Lamp life is about 10 times that of incandescent lamps. Ballast life is about five times compact fluorescent lamp life. Correlated color temperature (CCT) and color rendering index (CRI) are respectively acceptable for most applications. With electronic ballasts instead of magnetic ballasts, compact fluorescent lamps provide even higher efficacies.

The drawbacks of compact fluorescent lamp systems are minor for most legitimate applications: low power factor (0.50 to 0.6) for most systems; high total harmonic distortion with electronic ballasts (as high as 100 to 140 for some systems); difficulty fit-

TABLE 12.7 Typical Characteristics for Rapid-Start Hot-Cathode Fluorescent Lamps

Designation	Length mm	Length in	Base	Approx. lamp W	Rated life (h)
Lightly loaded (430-ma) lamps					
T-12	900	36		32.4	18,000
	1,200	48		41	20,000
T-8	900	36	Medium bipin	25	20,000
	1,200	48		32	20,000
	1,500	60		40	20,000
Medium loaded (800-ma) lamps					
T-12	600	24		37	9,000
	900	36		50	
	1,200	48		63	
	1,500	60	Recessed double contact	75.5	12,000
	1,800	72		87	
	2,100	84		100	
	2,400	96		113	
Highly loaded (1500-ma) lamps					
T-10	1,200	48		105	
	1,800	72		150	
	2,400	96		205	
T-12	1,200	48		116	
	1,800	72	Recessed double contact	168	9,000
	2,400	96		215	
PG-17	1,200	48		116	
	1,800	72		168	
	2,400	96		215	

Designations: T means tubular, PG means power grooved ("Power-Groove" is a GE trademark). The number following the letter designation gives the number of ⅛ inches in the tube diameter. T-12 indicates a tubular lamp 1½-in in diameter.

TABLE 12.8 Typical Characteristics for Instant-Start Hot-Cathode Fluorescent Lamps

Designation	Length mm	Length in	Base	Approx. lamp W	Rated life (h)
T-6	1,050	42		25.5	
	1,600	64		38.5	
T-8	1,800	72		38	7,500
	2,400	96		51	
T-12	600	24		21.5	
	900	36		30.	7,500–9,000
	1,050	42	Single pin	34.5	
	1,200	48		39	
	1,500	60		48	
	1,600	64		50.5	7,500–12,000
	1,800	72		57	
	2,100	84		66.5	
	2,400	96		75	12,000

Designations: T means tubular, PG means power grooved ("Power-Groove" is a GE trademark). The number following the letter designation gives the number of ⅛ inches in the tube diameter. T-12 indicates a tubular lamp 1½-in in diameter.

TABLE 12.9　Typical Characteristics for Energy-Efficient Fluorescent Lamps

	Length			Approx. lamp	Rated life
				Rapid start	
Designation	mm	in	Base	W	(h)
T-12 (38 mm)	900	36	Medium bipin	25	18,000
	1,200	48		34–35	20,000
	2,400	96		95	12,000
PG-17 (54 mm)	2,400	96	Recessed double contact	185	12,000
	1,200	48		95	12,000
			Preheat start		
T-12 (38 mm)	1,200	48	Medium bipin	34	15,000
T-17 (54 mm)	1,500	60	Mogul bipin	82–84	9,000
			Instant-start		
T-12 (38 mm)	1,200	48		30–32	9,000
T-8 (25 mm)	2,400	96	Single pin	60	12,000
				40–41	7500

Designations: T means tubular, PG means power grooved ("Power-Groove" is a GE trademark). The number following the letter designation gives the number of $\frac{1}{8}$ inches in the tube diameter. T-12 indicates a tubular lamp 1½-in in diameter.

ting unusually shaped lamps in existing luminaires; and electromagnetic interference with electronic ballasts (may affect hearing aids, communications systems, VCRs, etc.). Special high-power-factor systems are available at extra cost. Harmonic distortion can be tolerated on systems where the lighting load is not the major load. Location may solve the electromagnetic interference problem. Considering all these factors, the energy savings make compact fluorescent systems attractive as a replacement for many incandescent systems, even though the initial cost is higher than for incandescent lamps.

12.3.4　High-Intensity Discharge (HID) Lamps

Three types of HID lamps are mercury, metal halide, and high-pressure sodium. Because of its applications, the low-pressure sodium lamp is often included in this group, even though it is a low-pressure gaseous discharge lamp. LP sodium lamps find limited applications in the United States because they are monochromatic.

　　All HID lamps operate using a light-producing element consisting of a wall-stabilized arc discharge contained within a refractory envelope with wall loading in excess of 3 W/cm². Table 12.6 shows some of the ranges of characteristics for the different types of HID lamps in common use. The following sections discuss construction and arc production.

HID Mercury Lamps.　Mercury lamps operate by preheating a small amount of easily vaporized argon gas, which produces the original arc when subjected to the circuit voltage. This arc vaporizes the mercury, which produces the light. The lamp has a double envelope. The outer lamp envelope contains an inert gas and acts as a filter for ultraviolet light wavelengths as well as shielding the inner envelope from wide temperature variations. If the outer envelope is broken, the lamp should not be operated because of the danger of ultraviolet radiation as well as the fire hazard. Long life is a beneficial characteristic of mercury lamps. Time to reignite, high lamp lumen depreciation, and lack of red color rendition are detrimental characteristics. Use of phosphor on the outside lamp envelope improves color rendition of mercury vapor lamps.

HID Metal Halide Lamps.   Metal halide lamps are similar in construction and application to mercury lamps, except that, in addition to mercury and argon, the arc tube contains various metal halides which produce the light. The main advantage of these lamps is better color rendition and higher efficacy. This advantage is obtained at the expense of reduced lamp life and higher lamp cost, although the total cost of light is usually lower with metal halide lamps due to their higher lm/W output.

HID High-Pressure Sodium (HPS) Lamps.   HPS lamps are similar in construction to mercury lamps, except the inner envelope is made of a high-temperature ceramic to contain the very corrosive sodium instead of the quartz material used in the inner envelope of mercury and metal halide lamps. The main advantages of these lamps are shorter reignition time, higher efficacy (lm/W), and better lumen maintenance. The main disadvantage is color-rendering ability. HPS lamps will usually provide the lowest cost of light due to their high efficacy, long lamp life, and high lumen maintenance.

Low-Pressure Sodium (LPS) Lamps.   LPS lamps produce a spectrum consisting primarily of yellow. Because of the limited application of this type of lamp and the special disposal requirements due to the large amount of sodium used, these lamps are not used much in the United States and are not manufactured by many producers.

12.4   BALLASTS

Every type of arc discharge lamp requires a current-limiting device. Since the arc produced has a negative resistance characteristic, the lamp would destroy itself if current were not limited during discharge. Other tasks performed by ballasts include providing starting voltage, maintaining operating voltage, and providing power factor correction. Ballasts must be matched to the lamp selected. Incandescent lamps do not require ballasts.

12.4.1   Fluorescent Lamp Ballasts

Three available types of fluorescent lamp ballasts are low-power-factor reactor, instant-start, and rapid-start. Early fluorescent lamps all used simple, low-power-factor, reactor-type ballasts. These ballasts are available for use with a manual switch which, when closed, energizes the lamp electrodes in series for preheating. Opening this switch after the preheating time permits the lamp to operate. A starter is usually employed to automatically open the switch at the proper time.

   Instant-start lamps have high-power-factor circuits which use an autotransformer and capacitor combined to provide the functions necessary for starting and operation. An instant-start ballast imposes a high voltage on the lamp electrodes, preheating them fast enough to start the lamp within a very short time.

   Rapid-start ballasts utilize either special windings on the ballast transformer or a separate transformer to preheat the lamp electrodes within a short time using lower voltages yet allowing the lamps to operate within about one second. Rapid start is the predominant method of present-day fluorescent lamp operation. Figure 12.5 shows some of the typical fluorescent lamp and ballast circuits.

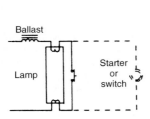

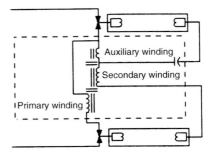

(a) Preheat, low power factor circuit

(b) Instant start, high power factor lamp circuit for use with slim-line indoor units

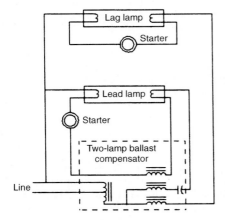

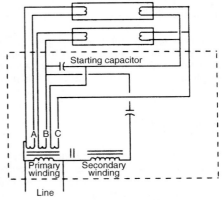

(c) Preheat, lead-lag, high power factor lamp circuit

(d) Rapid start, high power factor for high output, indoor and low temperature applications

FIGURE 12.5   Typical fluorescent lamp and ballast circuits.

12.4.2 HID Mercury Lamp Ballasts

At least nine different circuits are available to operate mercury lamps, providing the proper ballast characteristics. These are:

   Normal-power-factor reactor
   High-power-factor reactor
   Normal-power-factor high-reactance autotransformer
   High-power-factor high-reactance autotransformer
   Constant-voltage autotransformer
   Two-lamp lead-lag circuit

Two-lamp series (isolated) constant wattage

Constant current series regulator

Two-level mercury ballasts

Figure 12.6 shows some of the ballast circuits listed here. In addition, some mercury lamps are self-ballasted with a current-limiting tungsten filament and do not require an external ballast. These self-ballasted lamps are installed as incandescent replacements and are not as efficient as the lamp with a separate ballast.

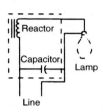

(a) High power factor, reactor mercury lamp ballast

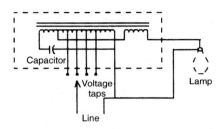

(b) High power factor, autotransformer mercury lamp ballast

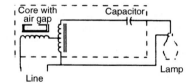

(c) Constant wattage autotransformer ballast for mercury lamps or peak-lead ballast for metal halide lamps

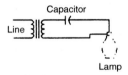

(d) Constant wattage (isolated circuit) mercury lamp ballast

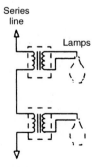

(e) Constant current series regulator ballast for mercury lamps

FIGURE 12.6   Typical circuits for HID lamps.

12.4.3   HID Metal Halide Lamp Ballasts

Metal halide lamps normally use a lead-peaked autotransformer ballast circuit specifically designed for this type of lamp. Certain metal halide lamps can be operated on mercury lamp ballasts with the manufacturer's concurrence.

12.4.4   HID High-Pressure Sodium Lamp Ballasts

High-pressure sodium lamps require a high-voltage pulse to break down the amalgam of sodium and mercury in a xenon gas atmosphere into charged particles and start the lamp. A special electronic pulse generator produces up to a 4000-V peak for 1000-W lamps and about a 2500-V peak for other types of lamps. Four different types of ballast regulate the current flow for voltages within the lamp operating range to maintain specified lamp wattages for the line voltage. These four ballast types are lag (or reactor), magnetic regulator (or constant-wattage), lead-circuit, and electronically regulated reactor.

Lag (or reactor) ballasts are similar to the simple mercury lamp ballasts shown in Fig. 12.6 and provide good wattage regulation for up to ±5 percent line voltage variation. Regulation is poor but losses are low, as is the price.

A *magnetic regulator (or constant-wattage) ballast* consists of a voltage-regulating circuit feeding a current-limiting reactor. This type of ballast has good wattage regulation for a voltage range of ±10 percent, but both losses and price are high.

A *lead-circuit ballast* is similar to a constant-wattage autotransformer mercury ballast. It maintains good wattage regulation for voltages of ±10 percent and is intermediate in both losses and price.

An *electronically regulated reactor ballast* adds a circuit to sense and regulate lamp wattage. It provides almost perfect wattage regulation for both line and lamp voltage changes up to ±10 percent line voltage variation. It is the most costly ballast and has the lowest losses.

12.4.5   Electronic Ballasts

Electronic ballasts convert a line frequency source into a high-frequency power supply for fluorescent lamps. Operating lamps at a high frequency (20 to 60 kHz) improves the lamp efficacy to 114 percent of rated LPW and higher. Figure 12.7 shows an electronic ballast block diagram for a circuit design which accomplishes this improvement.

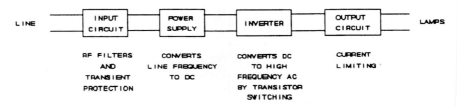

FIGURE 12.7   Block diagram for electronic ballast.

12.4.6 Ballast Losses and Lighting Circuit Power Requirements

High magnetic ballast losses compared to the total circuit watts consumed by a fluorescent lamp circuit, ranging from as much as 25 percent for small preheat-start lamps to between 10 and 20 percent for most other lamps, led the industry to develop better ballasts now on the market. The standard ballast of the past is the comparison ballast to which other ballasts are contrasted. Standard magnetic ballasts (not now available) have been replaced by high-efficiency magnetic ballasts and electronic ballasts for most fluorescent applications.

Table 12.10 tabulates the standard magnetic ballast losses and total circuit watts consumed. Higher losses are associated with smaller preheat lamps and instant-start lamps. Lower losses are associated with highly loaded rapid-start lamps.

High-efficiency magnetic ballasts used with standard-efficiency rapid-start lamps save between 7 and 9 percent of the lighting circuit power requirements. The higher values go with the higher-rated lamps. When used with high-efficiency rapid-start lamps, the power savings can be greater than 20 percent when compared to standard-efficiency lamps with standard-efficiency ballasts.

Electronic ballasts with high power factors (>0.9) and low total harmonic distortion (less than 10 to 15 percent) provide power savings of 20 to 24 percent when used with standard-efficiency rapid-start lamps and greater than 35 percent when used with high-efficiency lamps. Electronic ballasts that meet the criteria set forth tend to be more expensive than those which do not, but the power savings more than make up for this in the life of an installation.

Table 12.11 compares the lamp circuit power requirements for several types of electronic ballasts with the standard magnetic ballast for both standard lamps and high-efficiency lamps. Data for Tables 12.10 and 12.11 were adapted from Refs. 1 and 2.

Table 12.12 (adapted from Ref. 2) gives the approximate ballast losses and circuit requirements for magnetic ballasts that Fig. 12.6 illustrates. Losses represent a significant portion of the lighting circuit power: about 10 percent for higher lamp ratings and between 15 and 25 percent for lower ratings. Electronic ballasts for this type of lamp have been slow to develop and are rare in today's market.

12.4.7 Estimating Plant Lighting Load Power Requirements

Reference 3 assumes 3.1 W/ft^2 for the lighting load in the manufacturing area of a typical industrial plant. With the improvements in lighting circuit efficiencies made since this 1951 estimate, more practical values are 1 to 3 W/ft^2 for manufacturing or offices. Lower values apply to general lighting circuits using newer high-efficiency lighting. Higher values apply to special areas, particularly where incandescent lighting is required.

12.5 LIGHT QUALITY—COLOR, GLARE, STROBOSCOPIC EFFECT

Color and glare are important factors in almost any lighting installation. Stroboscopic effects are important only in installations where rotating machines are installed and rotating parts are visible.

TABLE 12.10 Approximate Ballast Losses and Fluorescent Lamp Circuit Power Requirements

Lamp type	Operating current (ma)	Each lamp	Single-lamp circuit		Two-lamp circuit	
			Ballast	Total	Ballast	Total
Standard magnetic ballasts						
Watts consumed — Preheat start						
14T12	380	14	5	19	5	33
15T12	325	14.5	4.5	19	9	38
20T12	380	20.5	5	25.5	10	51
Rapid-start (lightly loaded)						
40T12	430	41	13	54	13	95
Rapid-start (medium-loaded)						
48T12	800	63	22	85	30	146
72T12		87	19	106	31	200
96T12		113	27	140	21	252
Rapid-start (highly loaded)						
48T12	1500	110	20	140	30	250
72T12		160	20	180	30	360
96T12		215	30	245	30	460
Instant-start						
48T12	425	39	12	59	32	110
72T12		55	27	82	34	144
96T12		75	27	102	34	184
High-efficiency magnetic ballasts (standard lamps)						
Rapid-start						
40T12	430	34	7	41	18	86
Rapid-start (slimline)						
96T12	425	75			10	160
Rapid-start (medium-loaded)						
96T12	800	113			11	237
High-efficiency magnetic ballasts (energy-saving lamps)						
Rapid-start						
40T12	460	34	7	41	18	86
Rapid-start (slimline)						
96T12	440	60			8	128
Rapid-start (medium-loaded)						
96T12	840	95			17	207

TABLE 12.11 Comparison of Electronic and Reactive Ballast Fluorescent Lamp Circuit Power Requirements

	Two-lamp 40T12 rapid-start		
Ballast	Lamp	Efficacy LPW	Total circuit watts consumed
Standard magnetic	Standard cool white	60	96
Energy-efficient magnetic	Standard cool white	65	88
Electronic	Standard cool white	78	73
Electronic	Energy-saving cool white		60
Electronic (dimming)	Standard cool white		77
Electronic (dimming)	Energy-saving cool white		64

TABLE 12.12 Approximate Magnetic Ballast Losses and HID Lamp Circuit Power Requirements

Lamp type	Lamp watts	Ballast watts	Total watts
Mercury	100	20	120
	175	30	205
	250	40	290
	400	50	450
	1000	75	1075
Metal halide	175	35	210
	250	50	300
	400	60	460
	1000	100	1100
	1500	125	1625
High-pressure sodium	100	35	135
	150	50	200
	250	60	310
	400	75	475
	1000	100	1100

12.5.1 Color

Three basic concepts characterize the nature of light sources:

- *Correlated color temperature (CCT).* The absolute temperature in kelvins (K) of a blackbody whose chromaticity most nearly resembles that of the light source. Figure 12.8 shows the CCT of various light sources.
- *Color rendering index (CRI).* The measurement of the color shift an object undergoes when illuminated by a light source, compared to a reference source at the same color temperature. CRI is measured on a scale from 0 to 100 established by CEI (Commission Internationale de l'Eclairage, International Commission on Illumination. Objects and people generally appear more true to life under natural daylight, which has a CRI of 100. Table 12.13 gives some selected values of some light sources.

- *Color preference index (CPI).* This is sometimes called the flattery index. While this index is often mentioned, there seems to be no definite measurement system extant. Preference is subjective and relates to color combinations and contrast as well as illuminance and reflected color. Lightness contrast between object and background colors, some variety, and personal taste all affect color preference.

Color is thoroughly discussed in section 5 of the *IESNA Lighting Handbook.*[1] A general observation reveals that incandescent light sources produce a predominance of reds with very few blues. LPS lamps produce mainly yellows. HID and fluorescent lamps produce fewer reds than sunlight; hence, some installations may mix incandescent and

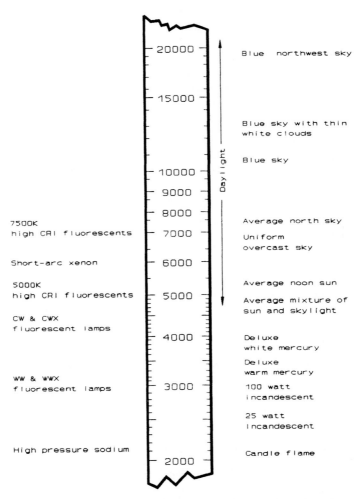

FIGURE 12.8 Correlated color temperature in kelvins of several light sources including daylight.

TABLE 12.13 Selected Color Rendering Index (CRI) Values of Several Light Sources

Correlated color temperature (CCT) of each source also shown

Lamp designation	CCT (K)	CRI
Fluorescent		
Warm white	3020	52
Warm white deluxe	2940	73
White	3450	57
Cool white	4250	62
Cool white deluxe	4050	89
Daylight	6250	74
Mercury		
Clear	5710	15
Improved color	4430	32
Metal halide, clear	3720	60
High-pressure sodium	2100	21
Low-pressure sodium	1740	−44
DXW tungsten halogen	3190	100

discharge-type lamps to provide a color rendition closer to that from natural light sources. Metal halide lamps produce better color rendition than any other type of discharge lamp but still lack as much red production. Color is an important factor in any installation, and a study of the *IESNA Lighting Handbook* as well as consultation with a lighting expert are recommended.

12.5.2 Glare

Luminaires too bright for the physical environment produce glare: either *discomfort* glare or *disability* glare, or both. Glare can be either *direct* or *reflected*. Reducing glare involves positioning the luminaires, as well as selecting the correct luminance for the installation.

IESNA Lighting Handbook, Chapter 3, recommends the following steps to reduce direct glare:

1. Reduce the luminance of light sources or lighting equipment.
2. Reduce the area of high luminance causing the glare situation.
3. Increase the angle between the glare source and the line of vision.
4. Increase the luminance of the area surrounding the glare source and against which it is seen.

12.5.3 Stroboscopic Effects

Both fluorescent and high-intensity discharge lamps emit light, depending on the peak power input. When operating on a 60-Hz system, 120-cycles-per-second flicker results. This flicker (not usually noticeable to the human observer) can produce ghost images of rapidly moving objects, particularly machines with rapidly moving, visible rotors.

Stroboscopic effects are particularly noticeable in metal halide lamps and not so noticeable in fluorescent lamps. Metal halide lamps lack a phosphor coating present on the inside of the glass envelope of fluorescent lamps. This coating creates a time delay in the flicker, which somewhat modifies its effects. To somewhat mitigate the stroboscopic effect, metal halide lamp installations operating on three-phase systems usually wire adjacent fixtures to different phases and sometimes use lead-lag ballasts.

12.6 PHYSICAL ENVIRONMENT— TEMPERATURE, REFLECTANCE, ZONAL CAVITY

The physical environment affects the lamp operation plus the lighting component selection and installation layout. Some subjects discussed in Sections 12.3, 12.4, and 12.5 contain information used to determine factors for lighting calculations in Sec. 12.10, performed either manually or with computer programs.

12.6.1 Ambient Temperature

For incandescent and high-intensity discharge lamps, very little light output loss accompanies a reasonable range of temperatures (-20 to $80°C$). Ballasts must be operated within the temperature ranges of the applied components. Special ballasts and lamps are available for many special environments. Extremes (the top of a boiler, for example) may require special or remote mounted ballasts. Fluorescent lamps lose most of their ability to produce light at temperatures below freezing.

Most light sources produce heat. In designing air-conditioning systems, this heat must be considered. Refer to the *ASHRAE Handbook*[4] (American Society of Heating, Refrigerating and Air-conditioning Engineers) for more information on this subject.

12.6.2 Reflectance

Walls, floors, and ceilings all reflect light. Each of these room components has a reflectance. Photometric data from a luminaire manufacturer usually contains a table relating reflectance values to the coefficient of utilization (CU). Lighting design calculations can be performed either manually or with a computer program (available from a lamp manufacturer). The CU quantity is shown in Secs. 12.6.3 and 12.10.2 of this chapter. Reflectances are either estimated or measured. If unknown, the usual assumed values are 20 percent for the floor reflectances and 30 percent for the ceiling and wall reflectances. Table 12.14 gives approximate reflectance values for office spaces.

TABLE 12.14 Reflectance Values for Office Areas

Surface	Reflectance
Ceiling	80–90% (less for acoustic materials)
Walls	40–60%
Furniture, office machines, & equipment	25–45%
Floors	20–40%

12.6.3 Zonal Cavities

For indoor rooms, divide the volume of the room into three zones and define each zone as follows:

Ceiling cavity	The space between the ceiling and the luminaire plane (bottom of the luminaires)
Room cavity	The space between the luminaire and the work plane
Floor cavity	The space between the work plane and the floor plane

Each of these cavities has a special ratio calculated from its zone dimensions. These cavity ratios ultimately determine the *coefficient of utilization* (CU). The cavity ratio formula is:

$$\text{Cavity ratio} = 5h \times \frac{(\text{room length} + \text{room width})}{(\text{room length} \times \text{room width})}$$

In this formula, h is the height of the particular cavity under consideration. If the luminaires are recessed in a dropped ceiling, the ceiling cavity height is zero and so is the ceiling cavity ratio (CCR). If the floor is the work plane, the floor cavity height is zero and so is the floor cavity ratio (FCR). In such a case, the room cavity ratio (RCR) determines the CU. Equations for each cavity ratio are:

$$\text{Room cavity ratio (RCR)} = 5h_{rc} \times \frac{(\text{room length} + \text{room width})}{(\text{room length} \times \text{room width})}$$

$$\text{Ceiling cavity ratio (CCR)} = 5h_{cc} \times \frac{(\text{room length} + \text{room width})}{(\text{room length} \times \text{room width})}$$

$$\text{Floor cavity ratio (FCR)} = 5h_{fc} \times \frac{(\text{room length} + \text{room width})}{(\text{room length} \times \text{room width})}$$

Luminaire manufacturers provide photometric data for each luminaire design available. Using the ratios calculated in the preceding equations, charts provided in the photometric data determine the CU. Photometric data also provides candela distribution charts and tables from which to obtain additional data required for lighting calculations.

12.7 ELECTRICAL ENVIRONMENT—VOLTAGE SAGS, SURGES, HARMONICS

Every electrical power system is subject to voltage sags and surges, as well as interruptions and outages. Lighting systems operated outside the normal voltage ranges specified in ANSI Standard 84.1 for extended times invite poor performance. Every power system which serves nonlinear loads is subject to the harmonics produced by such loads. Examples of nonlinear loads are electronic ballasts, computer power supplies, heat pumps, adjustable-speed drives, and switched power supplies.

12.7.1 Voltage Variations

Incandescent lamps produce reduced lumen output at low voltages. At high voltages, lamp life decreases appreciably. Fluorescent and high-intensity discharge lamps will extinguish at voltages somewhat less than 90 percent of rated voltage. Most ballasts are constructed to ride through dips of 10 percent for 4 seconds. If the voltage dip is more severe than 10 percent, the lamp usually extinguishes. The approximate restarting times for different types of lamps are given in Table 12.15.

TABLE 12.15 Approximate Lamp Restarting Times for Fluorescent and HID Lamps

Lamp	Restarting time
Fluorescent (rapid-start)	1 s
Fluorescent (instant-start)	≤1 s
Fluorescent (preheat-start)	A few s
Mercury	3–7 min
Metal halide	Up to 15 min
Sodium	3–4 min

12.7.2 Harmonics

Third harmonics produced by fluorescent lamps have long plagued power systems. Balancing loads among phases does nothing to alleviate this problem since triplen harmonics from each phase add in the neutral conductor and, in some installations, total more than the rated line current in the phase conductor. Three of the solutions proposed have been:

1. Increase the ampacity of the neutral conductor.
2. Divert the lighting load to other distribution panels which supply three-phase power to nonlighting loads, sizing the neutral for the lighting load.
3. Supply the lighting load from two wye-connected transformers with reverse phase sequences (A-B-C and A-C-B) and a common neutral, providing a six-phase lighting supply. These solutions may not appeal to many office building designers but may be less expensive than 3d-harmonic filters or special ballasts.

Switched power supplies and other nonlinear loads produce many harmonics. These harmonics can be alleviated by judicial use of either passive or active filters. These filters may be incorporated in the offending equipment or located on the distribution bus.

12.8 ENERGY MANAGEMENT AND LIGHTING CONTROL

Efficient management of electric loads contributes not only to the profitability of any enterprise but also to the optimum planning of supply systems providing reliable electric power to any consumer. Recently enacted codes and standards help the consumer not only to better control the electric load but also to design better installations providing better-lighted and more comfortable work and living spaces. In response to this situation, manufacturers have made a range of advances, products, and systems to make more efficient operation a reality.

12.8.1 Codes and Standards

State and federal codes, those already existing and those to come, establish minimum lumens per watt, work area illuminance limitations, building lighting load limits, lamp labeling, plus other lighting system energy-saving requirements. The Federal Energy Policy Act of 1992 sets labeling requirements for most lamps to be published by April 1994, with efficiency standards to be in effect by October 1995. Revised standards are to be published by April 1997.

California's Title 24 Lighting Code mandates a minimum level of illumination and a maximum allowed watts per square foot, and gives incentives which allow engineers to offer creative solutions to the problems encountered. Allowed lighting power density is calculated by one of three methods: (1) complete building, (2) area category, and (3) tailored. The following lists are a few of the allowed power densities applicable to industrial and commercial installations in W/ft^2:

Type of use	Allowed lighting power
Complete building method	
General work buildings	1.2
Storage buildings	0.8
Office buildings	1.5
Area category method	
Storage	0.6
Office	1.6
Precision work	2.0

Increased lighting power density is permitted if lighting controls are employed and for special occupancies.

The Title 24 Code applies to all new structures as well as retrofits and additions. The code concerns lighting controls that must be installed, lighting circuitry, and lighting control devices. It gives prescriptive requirements and three different methods of calculating the allowed power density. Some public utilities require customer compliance with a utility-generated energy code. Several utilities have demand-side management programs which attempt to influence utility customers to achieve the most cost-effective overall supply of electricity.

Lights produce heat which affects both the air-conditioning and heating loads at any installation. A combined effort by the Illuminating Engineering Society of North America (IESNA) and the American Society of Heating, Refrigeration and Air Conditioning Engineers (ASHRAE) has produced ASHRAE/IES Standard 90.1 which sets minimum requirements for energy-efficient design of new buildings, criteria for determining compliance, and design guidance.

12.8.2 Lighting Control Strategies

Lighting control involves ensuring that the required light is available for the time it is required. At least four strategies can be used to help control electric demand:

- Scheduling and use of daylight
- Lumen maintenance
- Load shedding and peak shaving
- Esthetic controls

Scheduling predicted times involves ensuring that the lights are turned on when required and turned off when not required. This can be done with timers arranged to do the job. Large banks of lamps provide the most effective means of reducing load at scheduled times. Unscheduled events, such as periodic out-of-town absences, can be accommodated by the use of motion detectors to control lighting. When sufficient daylight is available during certain hours of the day, photosensor controls can dim the lights at the appropriate time.

Lumen maintenance is the control of the amount of light required, using dimming controls and photosensors. An installation is designed for a certain minimum illuminance and photosensors are used to dim the light to that value. As the lamps age and lumens decrease, the photosensors detect and control the power required to maintain the required illuminance.

Load shedding involves shutting off nonessential loads during a peak electric demand period. This is accomplished with load controllers or, in case of emergency, by protective relays installed in the main power distribution switchgear.

Esthetic control changes the light to the required amount for a specific application which requires different amounts of light for a number of different time periods. This may be a repetitive process or a nonrepetitive process.

12.8.3 Lighting Control Techniques

Three major lighting control techniques are:

1. Switching or dimming control
2. Local or central control
3. Degree of automation

Switching or Dimming Control. Switching can be done manually with wall switches or with relays and a control system. Two-level switching in private offices gives the occupants the freedom to change the lighting, considering both available daylight and task requirements. For multiballasted systems, split-wiring multilamp luminaires can provide up to four different levels of illumination using arrangements discussed in Chap. 31 of Ref. 2. Dimming control provides continuous variability and is particularly suited for applications where daylight is available.

Local or Central Control and Automation. Lighting can be controlled locally, centrally, or by a combination of both. A local control system divides the lighting system into small, individually controlled zones. A central control system is usually accomplished with a building control system, which controls all energy-consuming system elements including air conditioning and heating. Some central control systems include complete electrical control and protection accomplished with a central computer. Such systems often supervise the entire power system protection components automatically. Completely automated systems require careful and judicious monitoring to prevent system catastrophes.

12.8.4 Lighting Control Equipment

Manual Control. Each area should have separate control with similar work areas grouped together on one circuit. High-level areas and luminaires along window walls should be on separate circuits. Single or two-lamp luminaires should be placed on alternate circuits to provide half or full light. Three- and four-lamp luminaires should be on a separate circuit from outside lamps and wired to provide multilevel lighting.

Timing and Sensing Devices. Simple timers, mechanical timers, and electronic astronomical clock timers provide a wide choice of control devices. For electronic timers, a battery backup should be provided to allow for power outages.

Photosensors. Careful placement of photosensors is a definite requirement for optimum operation. Outdoor photosensors usually are oriented toward the north in the northern hemisphere and toward the south in the southern hemisphere. Calibration of these sensors is critical to their correct application.

Motion Sensors. Motion detectors sense occupancy by audio, ultrasonic, passive infrared, or optical means. They switch on the lights as an occupant enters and switch them off at a predetermined time after the occupant leaves. These sensors often do not detect passive activities such as word processing, reading, or telephoning. This may cause nuisance light switching.

Central Processors. Central processors may be incorporated in building management systems and central computers. Local and distributed processors can be installed if a central processor does not seem appropriate for any building or plant area.

12.9 FIXTURES (LUMINAIRES)

Selection of luminaires requires consideration of many items:

- Luminaire characteristics
- Environmental conditions
- Electrical considerations
- Mechanical considerations
- Codes and standards

12.9.1 Luminaire Characteristics

Figure 12.9 shows three different types of reflector designs for individual lamps. *Parabolic reflectors* are used when a well-defined beam is required. Typical applications include automobile headlights, product display tables, and any area in which a focused light source is required. *Elliptical reflectors* are used where the light source is required to be unobtrusive, such as in a cocktail lounge. *Hyperbolic reflectors* are used where light is to be spread over a wide area, such as in high bay industrial lighting.

Fluorescent lamps sometimes use twin parabolic reflectors to concentrate the light on a specific area. For office spaces and similar occupancies with low ceilings, fluorescent lamps usually are installed in trougher type enclosures, using louvers or prismatic lens–type diffusers.

Replacing and cleaning lamps and luminaires must be considered in luminaire selection. Other factors include lamp position (some lamps can be mounted in only one position, either base up or base down). Ventilation and air circulation bear consideration, as well as factors discussed in other sections of this chapter.

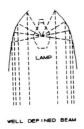
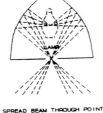

FIGURE 12.9 Luminaire reflector designs.

12.9.2 Environmental Considerations

Ambient temperature, as well as wet, hazardous, or corrosive locations, either require special luminaires or must have other provisions to mitigate the environment to allow the installation of standard components.

12.9.3 Electrical Considerations

The quality of the lighting circuits (harmonics, voltage) must be considered. Circuits with adjustable-speed drives, welders, and similar qualities that affect loads should be isolated from lighting circuits.

12.9.4 Mechanical Considerations

Coordination with other installed services, such as power cable, steam, and air lines, is required to ensure that the luminaires as well as the illuminance quality are not compromised in the installation. Proper mounting and feeder wire supports should be installed.

12.9.5 Codes and Standards

National Electrical Code[5] Article 410 applies to all industrial locations in the United States. Local codes may contain additional provisions required. The Canadian Electrical Code, usually enforced by the serving utility, is patterned after the NEC but contains some differences. IEEE, IES, NEMA, and ANSI Standards contain provisions which affect any lighting installation. The references at the end of this chapter list a few of these standards.

12.10 DESIGNING AND INSTALLATION

Two widely used methods of designing an indoor installation are the zonal cavity method and the point method.

The *zonal cavity method* is an offspring of the lumen method (rarely used today). This method considers overall illumination of office and manufacturing or process areas. The *point method* considers the requirements of a specific area, taking into consideration all sources of light, as well as the specific installation under consideration. This section does not contain a detailed description of the calculations involved. Complete discussions with sample calculations are given in Refs. 6, 7, and 8. These references also contain other procedures for outdoor lighting, roadway lighting, sports arena lighting, and special applications.

The five steps listed as follows abbreviate the zonal cavity design procedure. Most of the data given in this chapter is condensed from Refs. 1 and 6. The procedure outlined in this section is meant only to give an idea of the procedures involved in installation design. Refer to the references cited for an exact procedure.

The five basic steps in a lighting installation design are:

- Select a lamp, ballast, and luminaire.
- Determine the number of luminaires required.
- Determine the spacing of the luminaires.
- Check the level of illumination.
- Compare an alternate system to that originally considered.

12.10.1 Lamp, Ballast, and Luminaire Selection

Consider the information given in the appropriate six sections of this chapter and consult with both a lamp and a luminaire manufacturer.

12.10.2 Number of Luminaires

The formula for determining the number of luminaires is:

$$N = \frac{E \times \text{area}}{\text{LL} \times \text{CU} \times \text{LLD} \times \text{LDD} \times \text{BF}}$$

Quantity	Definition	Source
N	Number of luminaires	
E	Maintained level of illumination in lx (or fc)	Refs. 1, 2, 3 or Tables 13.1–13.4
Area	Area of the space lighted in m² (or ft²)	
LL	Initial rated lumens	Lamp manufacturer or Table 13.5
CU	Coefficient of utilization	Photometric data or luminaire manufacturer or Refs. 1–3
LLD	Lamp lumen depreciation factor (%lumen maintenance ÷ 100)	Lamp manufacturer or Table 13.5
LDD	Luminaire dirt depreciation factor	Luminaire manufacturer Typical: 0.94–0.97 (light) 0.74–0.88 (heavy)
LLF	Light-loss factor	LLF = LLD × LDD
BF	Ballast factor	Usually 1.0 for operation under rated conditions

12.10.3 Luminaire Spacing

Square spacing is preferred but not essential. Luminaires near a wall are placed at a distance equal to half the spacing in that direction.

$$\text{Spacing} = \sqrt{\frac{\text{area}}{N}}$$

12.10.4 Illumination Levels

Illumination levels are checked by a series of equations and methods detailed in Refs. 1 and 2. These methods are beyond the scope of this handbook.

12.10.5 System Comparisons

Lamp and luminaire manufacturers provide computer programs for use by customers to perform many of the calculations required for a successful lighting installation. A knowledge of the terms used in this chapter is necessary to use these programs. The references given at the end of this chapter provide help to perform lighting calculations and give details concerning all the subjects briefly described in this chapter.

12.11 LIGHTING IN THE FUTURE

Light transmission and distribution by fiber optic cable appears to be the future for lighting. One of the first applications installed using such a system is the relighting of four clock faces on the 24th floor of the Manhattan Consolidated Edison Company headquarters building. This system consists of 24 tiny metal halide lamps connected to optical fiber cable replacing 860 60-W incandescent lamps and using 83 percent less energy. Other new fiber optic applications include automobile headlights, decorative lighting, underwater lighting, neon replacement, and sign lighting.

Fiber optic construction can be *end dedicated,* where most of the light leaves the delivery end of the fiber, or side *dedicated,* with light coming out of the entire fiber length. The flexibility of fiber optic lighting promises revolutionary ways of lighting not only industrial and commercial plants and offices but also homes and other personal spaces.

12.12 REFERENCES

Many of the data included in this chapter were condensed from Refs. 1 and 6, as well as some of the papers listed in "Further Reading." References 1 and 6 are particularly valuable to understanding the application and design of lighting systems.

1. *IESNA Lighting Handbook,* 8th ed., Mark S. Rea, ed., Illuminating Society of North America, New York, 1993.
2. Joseph Murdock, *Illumination Engineering From Edison's Lamp to the Laser,* Macmillan, New York, 1985.
3. D. L. Beeman, ed., *Industrial Power Systems Data Book,* Section .7103, p. 2, General Electric Company, 1951.

4. *ASHRAE Handbook,* American Society of Heating, Refrigerating and Air-conditioning Engineers, Atlanta, GA, 1994.

5. NFPA 70, *National Electrical Code 1993,* National Fire Protection Association, Quincy, MA, 1992.

6. John P. Frier and Mary E. Gazley Frier, *Industrial Lighting Systems,* McGraw-Hill, New York, 1980.

7. R. B. Yarbrough, *Electrical Engineering Reference Manual,* 5th ed., Professional Publications, Inc., Belmont, CA, 1990.

8. A. R. Borden IV, "Lamp Specification: A Key to Good Lighting," *Consulting-Specifying Engineer,* March 1994, pp. 50–56.

FURTHER READING

Books

Mark W. Earley, editor-in-chief, *National Electrical Code Handbook,* National Fire Protection Association, 1992.

Consensus Standards

IEEE 1100-1992, *IEEE Recommended Practice for Powering and Grounding Sensitive Electronic Loads,* Institute of Electrical and Electronics Engineers.

IEEE 1250-1995, *IEEE Guide on Service to Equipment Sensitive to Momentary Voltage Disturbances,* Institute of Electrical and Electronics Engineers.

ANSI C84.1-1990, *Voltage Ratings for Electric Power Systems and Equipment (60 Hz),* American National Standards Institute.

Conference Papers

Samir Datta, "Power Pollution Caused by Lighting Control Systems," *IEEE 0-7803-0453-5/91,* Institute of Electrical and Electronics Engineers, Piscataway, NJ, 1991.

A. C. Liew, "Excessive Neutral Currents in Three-phase Fluorescent Lighting Circuits," *IEEE CH2499-2/87/000,* pp. 1364–1371, Institute of Electrical and Electronics Engineers, Piscataway, NJ, 1987.

R. R. Verderber, "Harmonics from Compact Fluorescent Lamps," *IEEE 0-7803-0453-5/91,* Institute of Electrical and Electronics Engineers, Piscataway, NJ, 1991.

CHAPTER 13
TESTING SYSTEM COMPONENTS

13.0 INTRODUCTION

Continuous service from an electrical power system depends upon adequate equipment with protective devices and relays properly set and in good working condition. Good maintenance programs, including recommended testing and calibration, assures that such is the case.

This chapter deals with field testing of power system components. Requirements for testing components are given in the manufacturers' instruction books, industry standards, and several textbooks. Two industry standards covering maintenance and testing exclusively are NFPA Standard 70B,[1] *Recommended Practice for Electrical Equipment Maintenance* and IEEE Standard 902, *Guide for Maintenance, Operation and Safety of Industrial and Commercial Power Systems*[2] (to be issued shortly). Other IEEE standards in the C37 and C57 series for individual devices and equipment include maintenance and field testing procedures therein. Every standard includes explicit safety procedures for conducting tests. In addition to the procedures outlined in Chap. 14, the actions indicated in these references require strict compliance in order to safely accomplish the tasks undertaken.

This chapter discusses test procedures for some of the more important electric power system components and includes sample test report forms which can be modified as needed for any specific power system. The test reports are not necessarily complete for every circumstance encountered and should be supplemented by reference to the manufacturer's instruction book for the specific equipment involved. Low-frequency dielectric tests are discussed separately and are not shown on any of the sample test reports.

Specific test procedures apply to each system element, however, all test procedures have the following steps in common:

1. Isolate the element under test, *observing all safety rules particularly lockout and tagout rules.*

2. Conduct a visual inspection of the element under test, noting nameplate information, physical condition, ambient conditions, instrument or meter readings, and protective device settings.

3. Clean equipment under test using a vacuum cleaner and lint-free cloth. *Do not use compressed air.*

4. Perform mechanical adjustments necessary and replace worn or broken parts, putting equipment as close to as-delivered condition as possible.

5. Perform electrical tests in the order required by the applicable standards, making sure that if a hipot test is to be conducted that it is the last test performed.

Special troubleshooting procedures, such as infrared inspection, wire checking, x-ray inspection, etc., are not discussed in this chapter.

13.1 TRANSFORMERS

13.1.1 Liquid-Filled Transformers

ANSI/IEEE Standard C57.12.90[3] is the test code for liquid-immersed distribution, power, and regulating transformers which specifies factory tests to be performed by the manufacturer as well as field tests. NFPA 70B, *Recommended Practice for Electrical Equipment Maintenance,*[1] discusses maintenance procedures but does not discuss tests in the detail included in ANSI/IEEE Standard C57.12.90.

   Following the testing procedure outlined in this chapter introduction, the field-testing sequence would be:

1. Consult the manufacturer instruction book and other maintenance recommendations available from the test equipment manufacturer or appropriate industry standards.
2. Isolate the transformer, observing lockout and tagout rules, as outlined in the NESC[4] and OSHA[5] rules.
3. Make a visual inspection, noting any irregularities (correct if possible), and clean the transformer.
4. Record gage readings.
5. Sample liquid per ASTM D923[6] and conduct appropriate tests (see sample test report form in Sec. 13.10).
6. Perform insulation power factor test.
7. Perform other recommended tests per manufacturer instruction book (refer to sample test report in Sec. 13.10 for tests and limits).
8. Perform optional tests as desired.
9. Hipot tests are *not* recommended but if they are performed, adhere strictly to the procedures and limits discussed in ANSI/IEEE Standard C57.12.90. Perform this test after all other tests have been performed.

13.1.2 Dry-Type Transformers

ANSI/IEEE Standard C57.12.91[7] is the test code for dry-type distribution and power transformers which specifies factory tests to be performed by the manufacturer, as well as field tests. NFPA 70B, *Recommended Practice for Electrical Equipment Maintenance,*[1] discusses maintenance procedures but does not discuss tests in the detail included in ANSI/IEEE Standard C57.12.91.

   Following the testing procedure outlined in this chapter introduction, the field testing sequence would be:

1. Consult manufacturer instruction book and other maintenance recommendations available from the test equipment manufacturer or appropriate industry standards.

2. Isolate the transformer, observing lockout and tagout rules as outlined in the NESC[4] and OSHA[5] rules.

3. Make a visual inspection, noting any irregularities.

4. Record gage readings.

5. Clean the transformer using a vacuum cleaner. *Do not use compressed air.*

6. Perform insulation resistance tests with a megohmeter.

7. Perform other recommended tests per manufacturer instruction book (refer to sample test report in Sec. 13.10 for tests and limits).

8. Perform optional tests as desired.

9. Hipot tests are *not* recommended, but if they are performed, adhere strictly to the procedures and limits discussed in ANSI Standard C57.12.91.[7] Perform this test after all other tests have been performed.

13.2 SWITCHGEAR

13.2.1 Low-Voltage Switchgear

Observe all safety precautions including lockout and tagout rules.

ANSI Standard C37.51,[8] ANSI/IEEE Standard C37.20.1,[9] NEMA Standard SG 5,[10] and UL 1558[11] cover low-voltage switchgear testing. In addition to the circuit-breaker tests covered in Sec. 13.3.2, these standards contain warnings not to subject the switchgear or circuit breakers to greater than 75 percent of the dielectric (hipot) factory test nor to periodically perform hipot tests without first restoring the switchgear and circuit breakers to the same condition they were in when first received from the factory.

In addition to circuit-breaker testing, inspect the general condition of the switchgear, vacuum cleaning the bus, looking for evidence of tracking or other possible insulation failures, inspecting for loose connections, and subjecting the main bus to resistance measurements with a megohmeter. Check the integrity of all interlocks (both mechanical and electrical). Check the drawout mechanism for all circuit breakers.

Test and calibrate all relays and instruments. Check wiring to assure conformance of control circuits to the latest revision of equipment wiring diagrams.

13.2.2 Medium-Voltage Switchgear

Observe all safety precautions including lockout and tagout rules.

ANSI/IEEE Standard C37.20.2[12] and UL Standard 1670[13] cover medium-voltage metal-clad switchgear testing. In addition to the circuit-breaker tests covered in Secs. 13.3.3 and 13.3.4, these standards contain warnings not to subject the switchgear or circuit breakers to greater than 75 percent of the dielectric (hipot) factory test nor to periodically perform hipot tests without first restoring the switchgear and circuit breakers to the same condition they were in when first received from the factory.

In addition to circuit-breaker testing, inspect the general condition of the switchgear, vacuum cleaning the bus, looking for evidence of tracking or other possible insulation failures, inspecting for loose connections, and subjecting the main bus to resistance measurements with a megohmeter. Check the integrity of all interlocks (both mechanical and electrical). Check the drawout mechanism for all circuit breakers.

Test and calibrate all relays and instruments. Check the wiring to assure conformance of control circuits to the latest revision of equipment wiring diagrams.

13.2.3 High-Voltage Switchgear

Observe all safety precautions including lockout and tagout rules.

High-voltage switchgear and circuit breakers require special procedures outlined in the manufacturers' instruction books. The procedures generally follow those outlined in the previous sections with additional precautions appropriate for the equipment.

13.3 CIRCUIT-INTERRUPTING ISOLATING AND FAULT-DETECTION EQUIPMENT

13.3.1 Molded- and Insulated-Case Circuit Breakers

NEMA Standards AB 1,[14] AB 3,[15] AB 4,[16] and UL Standard 489[17] cover molded-case circuit breakers. Section 13.10 contains a sample molded-case circuit-breaker maintenance and test record. After the visual and mechanical inspections have been completed and the circuit breaker has been cleaned, the trip elements should be tested. These tests are usually accomplished using a high-current test set. Test sets come in a variety of ratings ranging from 750 A output for testing smaller circuit breakers (up to 125 A) to outputs up to 60,000 A (for testing the complete range of circuit-breaker ratings). All of these test sets come with ammeters and timers. Complete instructions on the use of the test set are supplied with each test set.

Some molded-case circuit breakers have solid-state trip devices. Special test sets for these units are available from the circuit-breaker manufacturer and other vendors. Usually, testing trip units requires only disconnection of the trip device from the current sensor. Be sure to reconnect the trip device to the current sensors before reenergizing the circuit to prevent damaging overvoltages which may result from open-circuited current sensors in a loaded primary circuit.

13.3.2 Low-Voltage Power Circuit Breakers

ANSI Standards C37.16[18] and C37.17,[19] NEMA Standard SG 3[20] and UL Standard 1066[21] cover low-voltage power circuit breakers. Section 13.9 contains a sample low-voltage power-circuit-breaker maintenance and test record. After the visual and mechanical inspections have been completed and the circuit breaker has been cleaned, test the trip elements. These tests are usually accomplished using a high-current test set. Test sets come in a variety of ratings ranging from 750 A output for testing smaller circuit breakers (up to 125 A) to outputs up to 60,000 A (for testing the complete range of circuit-breaker ratings). All of these test sets come with ammeters and timers. Complete instructions for using the test set are supplied with each test set.

Most low-voltage power circuit breakers have solid-state trip devices. Special test sets for these units are available from the circuit-breaker manufacturer and other vendors. Usually, testing trip units requires only disconnection of the trip device from the current sensor. Be sure to reconnect the trip device to the current sensors before reenergizing the circuit to prevent damaging overvoltages which may result from open-circuited current sensors in a loaded primary circuit.

13.3.3 Medium-Voltage Air Magnetic Circuit Breakers

ANSI Standard C37.06,[22] ANSI/IEEE Standard C37.09,[23] and NEMA Standard SG 4[24] cover medium-voltage power circuit breakers. Section 13.10 contains a sample medium-voltage air magnetic power-circuit-breaker maintenance and test record. After the visual and mechanical inspections have been completed and the circuit breaker has been cleaned, complete the verification tests, low-resistance contact test, and the control power voltage verification. Check and calibrate, as necessary, all meter and instrument tests and calibrations, and test the relays to obtain results for the test report, as described in Sec. 13.10.

13.3.4 Medium-Voltage Vacuum Circuit Breakers

ANSI Standard C37.06,[22] ANSI/IEEE Standard C37.09,[23] and NEMA Standard SG 4[24] cover medium-voltage power circuit breakers. Section 13.9 contains a sample medium-voltage vacuum power-circuit-breaker maintenance and test record. After the visual and mechanical inspections have been completed and the circuit breaker has been cleaned, complete the verification tests, low-resistance contact test, and the control power voltage verification. Check and calibrate, as necessary, all meter and instrument tests and calibrations, and test the relays.

Test the integrity of the vacuum interrupters with a hipot set in accordance with the manufacturer's instructions.

13.3.5 Medium-Voltage Fused Motor Starters

NEMA Standard ICS 1.3,[25] UL Standard 347,[26] and UL Standard 508[27] cover industrial control equipment. No sample test form is included in Sec. 13.10. The tests performed on this type of equipment generally follow the same guidelines as for medium-voltage switchgear equipment.

13.3.6 Medium-Voltage Interrupter Switches

ANSI Standard 37.20.3[28] and ANSI/IEEE Standard 37.20.4[29] cover interrupter switches for use in metal-enclosed switchgear. Section 13.10 contains a sample maintenance and test record. After completing the visual and mechanical items and cleaning the equipment, perform the insulation resistance tests and the low-resistance contact test.

13.3.7 Protective Relays

ANSI/IEEE Standard C37.90[30] covers the general requirements for protective relays and relay systems associated with electric power systems. This standard includes no testing instructions. NFPA Standard 70B, Sec. 18-10.3,[1] contains some testing instructions for electromagnetic relays. IEEE Standard 242 (Buff Book), Sec. 15.5,[31] contains detailed test instructions for induction disk overcurrent relays. UL recognizes several protective relays in a "UL Recognized Component Directory" under UL category codes NKCR2, NRGU2, and XCFR2.

Protective relays fall under at least two distinct categories: electromagnetic and solid-state (or static). Most solid-state relays include self-testing diagnostics and do not require periodic testing by the same procedures that electromagnetic relays require. In either case, the manufacturer's instruction book informs the user of any test procedure required.

Section 13.10 contains a sample test record for electromagnetic, induction disk overcurrent relays. Two of the important warnings associated with relay testing are:

1. Never open circuit current transformers on an energized circuit because damaging high voltages in the current transformer secondary circuit may result from such action.

2. Never hipot (or resistance test with a megohmeter) a solid-state relay or a static relay which contains solid-state components because such components will be damaged by this action.

These two warnings apply, in addition to the usual safety precautions which component testing demands.

Manufacturers' instruction books usually contain testing instructions applying to the specific relay under consideration. In addition, test equipment manufacturers offer test sets with various degrees of flexibility. These test sets include:

Equipment	*Application*
General-purpose test table	Testing Wh meters, demand meters, indicating graphic instruments, transducers, small transformers, ammeters, voltmeters, etc.
Universal protective relay test set	Overcurrent, current-phase balance, directional overcurrent, voltage-restraint overcurrent, under- or overvoltage, thermal, percentage differential, and various ac and dc auxiliary relays.
Phase shifter	Used in conjunction with aforementioned test set for testing relays requiring phase angle control including reverse power, loss of excitation, impedance, mho, ohm, and various other distance relays.
Frequency relay test set	Under- and overfrequency or other frequency-responsive devices: transducers, synchronizers, meters, etc.
Protective relay test table	Large test console designed for laboratory use, containing facilities to test virtually all types of solid-state and electromechanical protective relays.

This list does not cover all of the test equipment available but gives only an indication of the range of available products.

Some relay systems include testing facilities in their basic installation, including the necessary special instruments. Check the instruction books of the installed equipment for complete testing instructions.

13.4 BATTERY SYSTEMS

ANSI/IEEE Standard 450[32] covers testing of stationary control batteries in substations. "Maintaining Stationary Batteries"[33] contains several tests to assure that stationary batteries can meet the application requirements of an installation. Additional information on battery testing appears in two papers: "Battery Diagnostic Testing for Improved Reliability"[34] and "Selection, Use and Care of Station Batteries for Paper Mill Service."[35] The battery system maintenance and test record included in Sec. 13.10 is the minimum that should be done for a battery system.

UL Standard 1236[36] covers battery chargers. The battery charger should be maintained and tested in accordance with the manufacturer's instructions. Periodic visits to the battery installation should be made for visual inspections. A failure of any station battery means a loss of the complete protection system for an entire plant and cannot be tolerated for any length of time.

13.5 CABLES

ANSI/IEEE Standards 48[37] and 400[38] and IEEE Standard 62[39] cover test procedures and field testing of high-voltage cable terminations and conductors. Hipot testing of cables requires strict safety precautions. Such procedures are contained in the instructions for the test sets used. The criteria for testing are contained in the referenced standards. Consultation with the cable manufacturer is advised.

13.6 MOTORS

NEMA Standard MG 1[40] covers motors and generators. For field tests, this standard refers to IEEE Standards 112,[41] 113,[42] 115,[43] and others. These tests are not usually applied on a routine basis and require the supervision of a qualified motor tester. Consultation with the motor manufacturer is advised.

13.7 ADJUSTABLE-SPEED DRIVES

Most adjustable speed (variable-frequency) drives using solid-state components include self-testing diagnostic programs. Field tests on this equipment are not required. Consult the manufacturer for advice on field testing.

13.8 CAPACITORS

NEMA Standard CP 1[44] and NFPA 70B[1] cover capacitors. No electrical tests are recommended for shunt capacitors, but extreme care should be used to ensure that the capacitors are completely discharged and grounded before attempting any cleaning or close inspection.

13.9 LOW-FREQUENCY DIELECTRIC TESTING

Dielectric testing involves both megohmeter testing and hipot testing. Megohmeters give a rough idea of the condition of a circuit as far as short circuits to ground or another phase. These instruments impose a dc voltage on the insulation under consideration. Five-hundred-volt megohmeters apply to low-voltage control circuits. Primary circuits require 1000- or 2500-V megohmeters. The dc voltage imposed on tested circuits by megohmeters seldom causes insulation degradation.

Transformers, switchgear, motors, and other electrical equipment, except cable, employ multiple-material insulation systems, in many cases using air gaps of varying dimensions as part of the insulation system. Factor dielectric tests (hipot tests) employing ac test voltages assure that such insulation systems are adequate for the service indicated.

For almost all types of equipment, the recommended value of an initial field hipot acceptance test per the relevant IEEE standard is an ac test at 75 percent of the factory hipot test value. Any subsequent ac hipot testing or periodic ac hipot testing is recommended at lower values (60 to 65 percent of the factory test voltages). Dc hipot testing is not recommended since it may produce questionable data unless conducted under precisely the same conditions each time. Repeated hipot tests may seriously degrade equipment insulation. Periodic hipot testing is not recommended.

Cable testing is an entirely different story. The significant value of capacitance to ground in most cable systems precludes the use of ac hipot testing because the testing equipment size required would be quite large. The solid nature of cable systems lends itself to dc hipot testing.

Two common test procedures which overcome some of the disadvantages of dc insulation testing are the *dielectric absorption ratio method* and the *step voltage method*. The dielectric absorption ratio method uses the ratio of two time insulation resistance readings. The ratio of a 10-min resistance measurement to a 1-min resistance measurement is called the *polarization index*. The magnitude of this ratio depends on the test circuit time constant, the presence of trapped charges on the insulation, the condition of the insulation, and the dc test voltage level. (The insulation circuit to be measured should be grounded for 30 min or longer prior to the actual test.) Values of these ratios are significant in a relative sense and must be related to prior ratios taken under similar conditions.

The *step voltage method* may be a nondestructive way to detect impending insulation failures. This test increases the dc test voltage in uniform steps and current measurements are taken after each specific time period. The resistance of certain "weak" insulation areas decreases rapidly as the electrical stress is increased beyond a critical limit. The point is detected by a noticeable increase in the current with an incremental voltage change, or by a negative slope in a plot of resistance values versus dc voltage. This test is potentially more dangerous than the dielectric absorption ratio test, because higher voltages are used for longer periods of time.

When using a dc test for cable circuits, all other electrical equipment should be disconnected, including transformers, lightning arresters, surge capacitors, etc. Some switchgear manufacturers sanction the use of dc tests for cable circuits. The test equipment is connected to the cable circuit with a ground and test device. This connection increases the cable circuit by conductor length from the cable side of the circuit breaker to the test set. Make sure all potential transformers, surge suppressors (or arresters), etc., are isolated from the test circuit.

13.10   SAMPLE TEST RECORD INDEX

The sample test records in this section show the type of information usually appearing in such records. They do not contain any entries for hipot tests, as indicated in the introduction to this chapter. If such tests are conducted, include an item to document the voltage and duration of these tests.

LIQUID FILLED TRANSFORMER MAINTENANCE AND TEST RECORD
(Refer to manufacturer instruction book)

Location _____   Inspection Date _____
Equipment#_____   Last Inspection Date _____
Installation Date _____

NAMEPLATE DATA

Manufacturer _____   Taps: ☐ Primary   ☐ Secondary
Class _____   Tap Range: ☐ 2 - ±2½%   ☐ Other   ☐ None
Serial # _____   Winding B.I.L.: HV _____   LV _____
kVA _____ Temp. Rise _____   Surge Arresters: ☐ Primary   ☐ Secondary
Voltages: Arrester Rating: _____   _____
Primary _____Connection_____   Bushing Current Transformers:
Secondary_____Connection_____   ☐ Primary ☐ Secondary ☐ Neutral
Tertiary _____Connection_____   Number: _____ _____ _____
Phases _____ Frequency _____   Ratio: _____ _____ _____
Insulating Liquid: ☐ 10C Oil ☐ Silicone   Instruction Book:_____
             ☐ Other _____   Specials: _____

SETTINGS

Voltage Taps: Bushing Current Transformers
HV   ☐ 1   ☐ 2   ☐ 3   ☐ 4   ☐ 5   HV   _____
LV   ☐ 1   ☐ 2   ☐ 3   ☐ 4   ☐ 5   LV   _____
                                                 Neutral _____

READINGS

Ambient Temperature ☐ ___°C ☐ ___°F   Pressure/vacuum ____ psi (+ or -)
Liquid Temperature ☐ ___°C ☐ Max ___°C   Liquid level _____
Winding Temperature ☐ ___°C ☐ Max ___°C   Surge arrester operations _____

INSPECTION AND MAINTENANCE CHECK LIST

☐ Paint condition   ☐ Primary and secondary connections clean
☐ Visual leaks   and tight
☐ Relief diaphragm set   ☐ Excessive or irregular vibration or noise
☐ Gasket and seal condition   ☐ No evidence of overheating (discoloration)
☐ Bushing condition (cracks, chips, leaks)   ☐ No moisture in junction & control boxes
☐ No loose hardware   ☐ Auxiliary contacts properly adjusted
☐ Check protective devices, alarms and   ☐ Fan operation, direction, balance, lube
   indicators for circuit integrity and operation ☐ Ground pad connection to ground

TESTS

☐ Insulating Liquid (Sampling per ASTM-D923)
   Dielectric   ASTM D877   ASTM D1816   Acidity ASTM D974 and D1534
   Limit   Unprocessed oil   Used oil   Limit - < 0.15 mg of KOH to neutralize
   ≥ 28 kV _____ _____   1 gram of oil _____
   Water Content ASTM D1533   Visual Appearance ASTM D1524
   Limit   < 30 ppm _____   Unusual changes indicate trouble
   Interfacial Tension ASTM D971   Color ASTM D1500
   Limit   ≥ 25 dynes/cm _____   Limit - ≤ 1 to 2 depending on transformer age

☐ Insulating Liquid (Optional tests)

	Limits
Gas Content ASTM D1827, D2945, D3612	Check manufacturer _____
Specific Gravity ASTM 1298	0.91 _____
Power Factor (of liquid) ASTM D924	1.0 _____
PCB in Oil by Gas Chromatography	Check Federal Regulation 40 CFR, Part 761
Pour Point ASTM D97	-40°C _____
Flash Point ASTM D92	140°C _____
Viscosity ASTM D445, D2161	<12 cSt at 40°C (mm²/s) _____
Oxidation Inhibitor Content ASTM D1473, D2668	0.3% by wgt _____
Oxidation Stability ASTM D2112, D2440	0.25% sludge _____
Foaming Characteristics of (Lubricating) Oil ASTM D892	_____

☐ Insulation Resistance

Equipment Temperature ____°C ____°F
H with X & Y grounded _____ meg Ω
X with Y & H grounded _____ meg Ω
Y with H & X grounded _____ meg Ω
(Y = tertiary)
Approximate minimums

Winding kV Class	Resistance Meg Ω
1.2	600
2.5	1000
5.0	1500
8.7	2000
15.0	3000

☐ Core Ground
(Lift core ground and megohm core to ground)
This test depends on transformer
construction and may require lowering
liquid level.

Equipment Temperature ____°C ____°F
_____ meg Ω

☐ Insulation Power Factor Test
☐ Polarity and Phase Relation Test*
☐ No Load Losses and Excitation Current*

* Optional tests
Procedures for these optional tests are described
in ANSI/IEEE Std C57.12.90 *IEEE Standard Test
Code for Liquid-Immersed Distribution, Power and
Regulating Transformers.* Refer to manufacturer's
instruction book.

☐ Ratio Test†

HV Winding H1-H2   _____ kV

Tap	LV Nominal	LV Actual
1	_____ kV	_____ kV
2	_____ kV	_____ kV
3	_____ kV	_____ kV
4	_____ kV	_____ kV
5	_____ kV	_____ kV

HV Winding H2-H3   _____ kV

Tap	LV Nominal	LV Actual
1	_____ kV	_____ kV
2	_____ kV	_____ kV
3	_____ kV	_____ kV
4	_____ kV	_____ kV
5	_____ kV	_____ kV

HV Winding H3-H1   _____ kV

Tap	LV Nominal	LV Actual
1	_____ kV	_____ kV
2	_____ kV	_____ kV
3	_____ kV	_____ kV
4	_____ kV	_____ kV
5	_____ kV	_____ kV

Limit 0.5 % of nominal ratio

† For LTC transformers, check all taps

Signed _____   Date _____

DRY TYPE TRANSFORMER MAINTENANCE AND TEST RECORD
(Refer to manufacturer instruction book)

Location _____   Inspection Date _____

Equipment# _____   Last Inspection Date _____

Installation Date _____

NAMEPLATE DATA

Manufacturer _____

Class _____   Taps: ☐ Primary   ☐ Secondary

Serial # _____   Tap Range: ☐ 2 - ±2½%   ☐ Other   ☐ None

kVA _____ Temp. Rise _____   Winding B.I.L.: HV _____   LV _____

Voltages: Surge Arresters: ☐ Primary   ☐ Secondary

Primary _____ Connection _____   Arrester Rating: _____   _____

Secondary _____ Connection _____

Phases _____ Frequency _____

Transformer Type: ☐ Open Dry ☐ Cast Coil

SETTINGS

Voltage Taps: ### READINGS

HV   ☐ 1   ☐ 2   ☐ 3   ☐ 4   ☐ 5   Ambient Temperature ☐ ____°C   ☐ ____°F

LV   ☐ 1   ☐ 2   ☐ 3   ☐ 4   ☐ 5   Winding Temperature ☐ ____°C ☐ Max ____°C

INSPECTION AND MAINTENANCE CHECK LIST

☐ Paint condition   ☐ Check that frame connected to ground
☐ No loose hardware   conductors
☐ Check protective devices, alarms and   ☐ Check that all accessible electrical
   indicators for circuit integrity and operation   connections are tight
☐ Remove all dust and dirt from exposed bus and
   from open windings (open dry transformers)
   with vacuum cleaner

TESTS

☐ Insulation Resistance   ☐ Insulation Power Factor Test*
   Equipment Temperature ____°C ____°F
   H with X grounded _____ meg Ω   ☐ Polarity and Phase Relation Test*
   X with H grounded _____ meg Ω
                                               ☐ No Load Losses and Excitation Current*

Winding kV Class	Resistance Meg Ω
1.2	600
2.5	1000
5.0	1500
8.7	2000
15.0	3000

Approximate minimums (heading above table)

* Optional tests

Procedures for these optional tests are described in ANSI/IEEE Std C57.12.91 *IEEE Standard Test Code for Dry-Type Distribution and Power Transformers.* Refer to manufacturer's instruction book.

☐ Ratio Test

HV Winding H1-H2   _____ kV
Tap   LV Nominal   LV Actual
1   _____ kV   _____ kV
2   _____ kV   _____ kV
3   _____ kV   _____ kV
4   _____ kV   _____ kV
5   _____ kV   _____ kV

HV Winding H3-H1   _____ kV
Tap   LV Nominal   LV Actual
1   _____ kV   _____ kV
2   _____ kV   _____ kV
3   _____ kV   _____ kV
4   _____ kV   _____ kV
5   _____ kV   _____ kV

Limit 0.5 % of nominal ratio

HV Winding H2-H3   _____ kV
Tap   LV Nominal   LV Actual
1   _____ kV   _____ kV
2   _____ kV   _____ kV
3   _____ kV   _____ kV
4   _____ kV   _____ kV
5   _____ kV   _____ kV

Signed _____   Date _____

MOLDED CASE CIRCUIT BREAKER MAINTENANCE AND TEST RECORD

This form is suggested record keeping of maintenance procedures and tests. For information on how to perform tests, consult circuit breaker manufacturer instructions.

Circuit Breaker Location: _____ Date installed: _____

Manufacturer & Breaker Catalogue No. _____

Continuous Current Rating: _____ Interrupting rating: _____

INSPECTION AND MAINTENANCE CHECK LIST

Comments

☐ **Overheating** (Check for hot surfaces, discoloration)
☐ **Verify application & rating**
☐ **Case Intact** (Check for breaker case cracks)
☐ **Exercise Mechanism** (Operate ON & OFF several times)
☐ **Push-to-trip** (Trip breaker, reset and repeat)
☐ **Clean Breaker** (Use lint-free dry cloth or vacuum cleaner)

VERIFICATION TESTS

Insulation resistance - if less than one megohm, investigate cause.

Phase to ground, breaker closed

Phase Resistance

Phase to phase, breaker closed

A	B	C

Between line & load terminals, breaker open

A-B	B-C	C-A
A	B	C

Watts Loss (Breaker Resistance)

	Millivolt Drop	Service Current		Measured Watts Loss	Typical Maximum Watts Loss Value
Left Pole	_____	x _____	=	_____	_____
Center Pole	_____	x _____	=	_____	_____
Right Pole	_____	x _____	=	_____	_____

Overload Tripping At 300% rated current

	Tripping Time Measured Time	Acceptable Time		Tripping Time Measured Time	Acceptable Time
Left Pole	_____	_____	Right Pole	_____	_____
Center Pole	_____	_____			

Magnetic Tripping   ☐ Pulse Method   ☐ Run-up Method

Tripping Current   Left Pole _____   Center Pole _____   Right Pole _____

Rated Current Hold-in Test   ☐ Passed   ☐ Failed

Date Tested _____   **Signed** _____

LOW VOLTAGE POWER CIRCUIT BREAKER MAINTENANCE AND TEST RECORD

This form is suggested record keeping of maintenance procedures and tests. For information on how to perform tests, consult circuit breaker manufacturer instructions.

Circuit Breaker Location: _____ Date installed: _____
Manufacturer & Breaker Catalogue No. _____
Continuous Current Rating: _____ Interrupting rating:_____
Trip Device Type and Setting: _____

INSPECTION AND MAINTENANCE CHECK LIST

Comments

☐ **Overheating, tracking, connections** (Check for hot surfaces, discoloration)
☐ **Verify application & rating**
☐ **Appearance** (Clean, rust free, exterior paint, damage)
☐ **Exercise Mechanism** (Operate ON & OFF several times, check mechanism charging, tripping and trip free, latching, excessive wear)
☐ **Push-to-trip** (Trip breaker, reset and repeat)
☐ **Clean Breaker** (Use lint-free dry cloth or vacuum cleaner)
☐ **Clean and Check Arc Chutes**
☐ **Examine Primary Contacts** (check wipe and general condition)
☐ **Alignment, Clearances, Lubrication**
☐ **Check Racking Mechanism**
☐ **Check Auxiliaries** (shunt trip, undervoltage device, overcurrent lockout device, etc.)

VERIFICATION TESTS
Insulation resistance - if less than one megohm, investigate cause.

Phase Resistance

Phase to ground, breaker closed	A	B	C
Phase to phase, breaker closed	A-B	B-C	C-A
Between line & load terminals, breaker open	A	B	C

Watts Loss (Breaker Resistance)

	Millivolt Drop	Service Current	Measured Watts Loss	Typical Maximum Watts Loss Value
Left Pole	_____ x	_____ =	_____	_____
Center Pole	_____ x	_____ =	_____	_____
Right Pole	_____ x	_____ =	_____	_____

Trip Devices (See separate sheet)

Date _____   **Signed** _____

LOW VOLTAGE POWER CIRCUIT BREAKER TRIP DEVICE TEST RECORD
(For static trip devices - may be adapted for magnetic trip devices)
Refer to Manufacturer's Time-Current Curves for Test Limits

Breaker I.D. ______________________ Breaker Frame ______________________
Mfg. Type ______________________ Trip Device Type ______________________
S/N ______________________ Sensor Rating ______________________

SETTINGS									
Condition			Long Time		Short Time		Inst.	Ground Fault	
	Sensor Rating (Amps)	Rating Plug (Amps)	Pickup Setting (Amps)	Band I^2t In or Out	Pickup (Amps)	Band I^2t In or Out	(Amps)	Pickup (Amps)	Band I^2t In or Out
	S		X						
As Found									
As Left									

LONG TIME PICKUP (PU) TEST

Mode	PU	Test Current	Phase A		Phase B		Phase C		Test Limits in Seconds		
			Secs	Band	Secs	Band	Secs	Band	Min	Int	Max
No Trip		0.88xPU									
Trip		1.18xPU									

LONG TIME DELAY

PU	Test Current	Phase A		Phase B		Phase C		Test Limits in Seconds	
		Secs	Band	Secs	Band	Secs	Band	Lower	Upper
	2.0xPU								
	3.0xPU								

SHORT TIME PICKUP (PU) TEST

Mode	PU (Amperes)	Test Current (Amperes)	Trip Time (Seconds)		
			Phase A	Phase B	Phase C
No Trip < 1 Second		0.873 x PU			
Trip < 1 Second		1.133 x PU			

SHORT TIME DELAY

PU	Test Current	Phase A		Phase B		Phase C		Test Limits in Seconds		
Amperes	> 2 x PU	Secs	Band	Secs	Band	Secs	Band	Band	Lower	Upper
								Min		
								Int		
								Max		

INSTANTANEOUS PICKUP (PU) TEST

Mode	PU (Amperes)	Test Current (Amperes)	Phase A	Phase B	Phase C
No Trip		0.873 x PU			
Trip		1.133 x PU			

GROUND FAULT PICKUP (PU) TEST

Mode	PU (Amperes)	Test Current (Amperes)	Trips
No Trip		0.873 x PU	
Trip		1.133 x PU	

GROUND FAULT DELAY TEST

PU (Amperes)	Test Current (Amperes)	Band	Time	Test Limits (Seconds)	
				Lower	Upper
	2.0 x PU	Min			
	2.0 x PU	Int			
	2.0 x PU	Max			

Trip Annunciators OK: Overload _____ Short-Circuit _____ Ground _____
Phase Sensor Check OK: Phase A_____ Phase B _____ Phase C _____ Breaker Trip Test OK _____

Signed _____ Date: _____

MEDIUM VOLTAGE VACUUM POWER CIRCUIT BREAKER MAINTENANCE AND TEST RECORD

This form is suggested record keeping of maintenance procedures and tests. For information on how to perform tests, consult circuit breaker manufacturer instructions.

Circuit Breaker Location: _____ Date installed: _____

AMBIENT   Temperature ____ °C _____ °F Environment Conditions _____

NAMEPLATE INFORMATION
Manufacturer _____   Feeder _____
Model Number _____   Serial Number _____
Maximum Voltage _____   Continuous Current _____
Operating Mechanism Type _____   Interrupting Time _____
Closing Control Voltage _____   Tripping Control Voltage _____

INSPECTION AND MAINTENANCE CHECK LIST
Comments

☐ Check for loose or damaged parts and vacuum clean breaker   _____
☐ Check mechanical and electrical connections tight   _____
☐ Check for overheating, corona, insulation breakdown   _____
☐ Check ratchet mechanism by manually charging closing springs   _____
☐ Perform slow closing operation   _____
☐ Check close coil linkage and coil shaft for free movement   _____
☐ Check trip coil linkage and coil shaft for free movement   _____
☐ Record primary contact wipe   Phase A _____ Phase B _____ Phase C _____
☐ Record primary contact gap   Phase A _____ Phase B _____ Phase C _____
☐ Record contact erosion   Phase A _____ Phase B _____ Phase C _____
☐ Record operation counter reading _____

VERIFICATION TESTS
Insulation resistance - if less than one megohm, investigate cause.

Phase Resistance

Phase to ground, breaker closed

A	B	C

Phase to phase, breaker closed

A-B	B-C	C-A

Between line & load terminals,
breaker open

A	B	C

Secondary control circuit @500 V _____

VACUUM INTERRUPTER INTEGRITY TEST
Test Set Manufacturer _____
Model _____
Breaker open @ _____ kV _____ seconds

LOW RESISTANCE CONTACT TEST (Micro-ohms)
Phase A _____ Phase B _____ Phase C _____

CONTROL POWER
Rated _____
Voltage at fuse terminals _____

TIMING (Optional)
Requires travel recorder and oscillograph.
Check manufacturer for timing limits, operating speeds and other test details.

Signed _____   Date _____

MEDIUM VOLTAGE AIR MAGNETIC POWER CIRCUIT BREAKER MAINTENANCE AND TEST RECORD

This form is suggested record keeping of maintenance procedures and tests. For information on how to perform tests, consult circuit breaker manufacturer instructions.

Circuit Breaker Location: _____ Date installed: _____

AMBIENT   Temperature _____°C _____°F Environment Conditions _____

NAMEPLATE INFORMATION

Manufacturer _____   Feeder _____

Model Number _____   Serial Number _____

Maximum Voltage _____   Continuous Current _____

Operating Mechanism Type _____   Interrupting Time _____

Closing Control Voltage _____   Tripping Control Voltage _____

INSPECTION AND MAINTENANCE CHECK LIST   Comments

☐ Check for loose or damaged parts and vacuum clean breaker   _____

☐ Check mechanical and electrical connections tight   _____

☐ Check for overheating, corona, insulation breakdown   _____

☐ Check ratchet mechanism by manually charging closing springs   _____

☐ Perform slow closing operation   _____

☐ Check close coil linkage and coil shaft for free movement   _____

☐ Check trip coil linkage and coil shaft for free movement   _____

☐ Check arc chutes and arcing assembly

☐ Record primary contact wipe   Phase A _____   Phase B _____   Phase C _____

☐ Record primary contact gap   Phase A _____   Phase B _____   Phase C _____

☐ Record arcing contact wipe   Phase A _____   Phase B _____   Phase C _____

☐ Record operation counter reading _____

VERIFICATION TESTS

Insulation resistance - if less than one megohm, investigate cause.

                                     Phase Resistance

Phase to ground, breaker closed   _____   _____   _____

                                 A   B   C

Phase to phase, breaker closed   _____   _____   _____

                                 A-B   B-C   C-A

Between line & load terminals,   _____   _____   _____

breaker open   A   B   C

Secondary control circuit @500 V   _____

LOW RESISTANCE CONTACT TEST (Micro-ohms)

Phase A _____   Phase B _____   Phase C _____

CONTROL POWER

Rated _____

Voltage at fuse terminals _____

TIMING (Optional)

Requires travel recorder and oscillograph. Check manufacturer for timing limits, operating speeds and other test details.

           Signed _____   Date _____

PROTECTIVE RELAY TEST RECORD

This form is suggested record keeping of periodic tests. For information on how to perform tests, consult relay manufacturer instructions.

Relay Location: _____ Date installed: _____

AMBIENT CONDITIONS
Temperature ____°C ____°F Environment Conditions _____

NAMEPLATE INFORMATION

PHASE RELAY		GROUND/NEUTRAL RELAY	
Mfg:		Mfg:	
Type/Model:		Type/Model:	
Range:		Range:	
Characteristic:		Characteristic:	
CT Ratio:	PT Ratio:	CT Ratio:	PT Ratio:

RELAY SETTINGS

	Instantaneous				Target				Tap Setting				Time Dial			
	A	B	C	G	A	B	C	G	A	B	C	G	A	B	C	G
Specified																
As Found																
As Left																

AS FOUND TEST OPERATIONS - CURRENT IN AMPERES

INSTANTANEOUS (pickup and dropout)								TARGET (pickup and dropout)							
A		B		C		G		A		B		C		G	
pu	do	pu	do	pu	do	pu	do	pu	do	pu	do	pu	do	pu	do

AS LEFT TEST OPERATIONS - CURRENT IN AMPERES

INSTANTANEOUS (pickup and dropout)								TARGET (pickup and dropout)							
A		B		C		G		A		B		C		G	
pu	do	pu	do	pu	do	pu	do	pu	do	pu	do	pu	do	pu	do

AS FOUND TEST OPERATIONS - CURRENT IN AMPERES, TIME IN SECONDS

TIME ELEMENT							
A		B		C		G	
x pu	time	x pu	time	x pu	time	x pu	time

AS LEFT TEST OPERATIONS - CURRENT IN AMPERES, TIME IN SECONDS

TIME ELEMENT							
A		B		C		G	
x pu	time	x pu	time	x pu	time	x pu	time

Signed _____   Date _____

INTERRUPTER SWITCH MAINTENANCE AND TEST RECORD

This form is suggested record keeping of maintenance procedures and tests. For information on how to perform tests, consult interrupter switch manufacturer instructions.

Switch Breaker Location: _____ Date installed: _____

AMBIENT   Temperature ____°C _____°F Environment Conditions _____

NAMEPLATE INFORMATION

Manufacturer _____   Feeder _____
Model Number _____   Serial Number _____
Maximum Voltage _____   Continuous Current _____
Operating Mechanism Type _____

INSPECTION AND MAINTENANCE CHECK LIST

Comments

☐ Check for loose or damaged parts and vacuum clean switch   _____
☐ Check mechanical and electrical connections tight   _____
☐ Check for overheating, corona, insulation breakdown   _____
☐ Check operating mechanism, arc chute alignment, contact pressure _____
☐ Record primary contact wipe   Phase A _____ Phase B _____   Phase C _____
☐ Record primary contact gap   Phase A _____ Phase B _____   Phase C _____

VERIFICATION TESTS

Insulation resistance - if less than one megohm, investigate cause.

Phase Resistance

Phase to ground, switch closed

A	B	C

Phase to phase, switch closed

A-B	B-C	C-A

Between line & load terminals,
switch open

A	B	C

LOW RESISTANCE CONTACT TEST (Micro-ohms)

Phase A _____ Phase B _____ Phase C _____

Signed _____ Date _____

BATTERY SYSTEM MAINTENANCE AND TEST RECORD

This form is suggested record keeping of maintenance procedures and tests. For information on how to perform tests, consult battery manufacturer instructions.

Battery Location: _____ Date installed: _____

AMBIENT Temperature ____°C ____°F Environment Conditions _____

NAMEPLATE INFORMATION
Manufacturer _____ Model No. _____
Model Number _____ Volts _____ Amperes _____

VISUAL INSPECTION   Comments
☐ Check for visual damage, corrosion, bus link
  integrity   _____
☐ Check charging system per manufacturer's
  recommendations   _____

RECORD INDIVIDUAL CELL CONDITIONS
(Continue Table for All Cells)

Cell No.	Voltage	Electrolyte Specific Gravity	Level

13.11 REFERENCES

1. NFPA Standard 70B, *Recommended Practice for Electrical Equipment Maintenance,* National Fire Protection Association, 1 Batterymarch Park, P.O. Box 9101, Quincy, MA 02269-9101.

2. IEEE Standard 902, *Guide for Maintenance, Operation and Safety of Industrial and Commercial Power Systems* (Yellow Book - work in progress), Institute of Electrical and Electronics Engineers, P.O. Box 1331, Piscataway, NJ 08855-1331.

3. ANSI/IEEE Standard C57.12.90, *Standard Test Code for Liquid-Immersed Distribution, Power and Regulating Transformers and Guide for Short-Circuit Testing of Distribution and Power Transformers,* Institute of Electrical and Electronics Engineers, P.O. Box 1331, Piscataway, NJ 08855-1331.

4. ANSI Standard C2, *National Electrical Safety Code,* Institute of Electrical and Electronics Engineers, P.O. Box 1331, Piscataway, NJ 08855-1331.

5. *Federal Register* (53 FR 1546), U.S. Government Printing Office, Washington, DC 20402 (Telephone: 202-783-3238).

6. ASTM D923, *Standard Test Method for Sampling Electrical Insulating Liquids,* American Society for Testing and Materials, 1916 Race Street, Philadelphia, PA 19103.

7. ANSI/IEEE Standard C57.12.91, *Test Code for Dry-Type Distribution and Power Transformers,* Institute of Electrical and Electronics Engineers, P.O. Box 1331, Piscataway, NJ 08855-1331.

8. ANSI Standard C37.51, *Metal-Enclosed Low-Voltage AC Power-Circuit-Breaker Switchgear Assemblies—Conformance Test Procedures,* American National Standards Institute, 11 W. 42nd St., New York, NY 10036.

9. ANSI/IEEE Standard C37.20.1, *Standard for Metal-Enclosed Low-Voltage Power Circuit Breaker Switchgear,* Institute of Electrical and Electronics Engineers, P.O. Box 1331, Piscataway, NJ 08855-1331.

10. NEMA Standard SG 5, *Power Switchgear Assemblies (1000 Volts or Less Including Network Protectors as Well as DC Low Voltage Power Circuit Breakers up to 3200 Volts),* NEMA Publication Distribution Center, P.O. Box 338, Annapolis Junction, MD 20701-0338.

11. UL Standard 1558, *Metal Enclosed Low Voltage Power Circuit Breaker Switchgear,* Underwriters Laboratory, P.O. Box 75330, Chicago, IL 60675-5330.

12. ANSI/IEEE Standard C37.20.2, *Standard for Metal-Clad and Station-Type Cubicle Switchgear,* Institute of Electrical and Electronics Engineers, P.O. Box 1331, Piscataway, NJ 08855-1331.

13. UL Standard 1670, *Medium Voltage Switchgear over 1000 Volts* (tentative title pending publication), Underwriters Laboratory, P.O. Box 75330, Chicago, IL 60675-5330.

14. NEMA Standard AB 1, *Molded Case Circuit Breakers,* NEMA Publication Distribution Center, P.O. Box 338, Annapolis Junction, MD 20701-0338.

15. NEMA Standard AB 3, *Molded Case Circuit Breakers and Their Application,* NEMA Publication Distribution Center, P.O. Box 338, Annapolis Junction, MD 20701-0338.

16. NEMA Standard AB 4, *Guidelines for Inspection and Preventive Maintenance of Molded Case Circuit Breakers Used in Commercial and Industrial Applications,* NEMA Publication Distribution Center, P.O. Box 338, Annapolis Junction, MD 20701-0338.

17. UL Standard 489, *Molded Case Circuit Breakers and Circuit Breaker Enclosures,* Underwriters Laboratory, P.O. Box 75330, Chicago, IL 60675-5330.

18. ANSI Standard 37.16, *Standard for Low-Voltage Power Circuit Breakers and AC Power Circuit Protectors—Preferred Ratings, Related Requirements and Application Recommendations,* American National Standards Institute, 11 W. 42nd St., New York, NY 10036.

19. ANSI Standard 37.17, *Standard for Trip Devices for AC and General Purpose DC Low-Voltage Power Circuit Breakers,* American National Standards Institute, 11 W. 42nd St., New York, NY 10036.

20. NEMA Standard SG 3, *Low Voltage Power Circuit Breakers,* NEMA Publication Distribution Center, P.O. Box 338, Annapolis Junction, MD 20701-0338.

21. UL Standard 1066, *Low Voltage AC and DC Power Circuit Breakers Used in Enclosures,* Underwriters Laboratory, P.O. Box 75330, Chicago, IL 60675-5330.

22. ANSI Standard C37.06, *Standard for AC High-Voltage Circuit Breakers Rated on a Symmetrical Current Basis—Preferred Ratings and Related Required Capabilities,* American National Standards Institute, 11 W. 42nd St., New York, NY 10036.

23. ANSI/IEEE Standard C37.09, *Standard Test Procedure for AC High-Voltage Circuit Breakers Rated on a Symmetrical Current Basis,* Institute of Electrical and Electronics Engineers, P.O. Box 1331, Piscataway, NJ 08855-1331.

24. NEMA Standard SG 4, *Alternating-Current High Voltage Circuit Breaker,* NEMA Publication Distribution Center, P.O. Box 338, Annapolis Junction, MD 20701-0338.

25. NEMA Standard ICS 1.3, *Preventive Maintenance of Industrial Control and Systems Equipment,* NEMA Publication Distribution Center, P.O. Box 338, Annapolis Junction, MD 20701-0338.

26. UL Standard 347, *High Voltage Industrial Control Equipment,* Underwriters Laboratory, P.O. Box 75330, Chicago, IL 60675-5330.

27. UL Standard 508, *Industrial Control Equipment,* Underwriters Laboratory, P.O. Box 75330, Chicago, IL 60675-5330.

28. ANSI/IEEE Standard C37.20.3, *Metal Enclosed Interrupter Switchgear,* Institute of Electrical and Electronics Engineers, P.O. Box 1331, Piscataway, NJ 08855-1331.

29. ANSI Standard 37.20.4, *Standard for Indoor AC Medium Voltage Switches in Metal-Enclosed Switchgear,* American National Standards Institute, 11 W. 42nd St., New York, NY 10036.

30. ANSI/IEEE C37.90, *IEEE Standard for Relays and Relay Systems Associated with Electric Power Apparatus,* Institute of Electrical and Electronics Engineers, P.O. Box 1331, Piscataway, NJ 08855-1331.

31. ANSI/IEEE Standard 242, *IEEE Recommended Practice for Protection and Coordination of Industrial and Commercial Power Systems,* Institute of Electrical and Electronics Engineers, P.O. Box 1331, Piscataway, NJ 08855-1331.

32. ANSI/IEEE Standard 450, *IEEE Recommended Practice for Maintenance, Testing and Replacement of Large Lead Storage Batteries for Generating Stations and Substations,* Institute of Electrical and Electronics Engineers, P.O. Box 1331, Piscataway, NJ 08855-1331.

33. M. W. Migliaro, "Maintaining Stationary Batteries," *IEEE Transactions on Industry Applications,* vol. IA-23, no. 4, July/August 1987, pp. 765–772.

34. A. L. Lamb and J. H. Bellack, "Battery Diagnostic Testing for Improved Reliability," *IEEE Industry Application Society Annual Meeting Conference Record,* CH2272-3, 1986, pp. 1019–1023.

35. E. C. Korbeck and J. W. Blankley, "Selection, Use and Care of Stationary Batteries for Paper Mill Service," *IEEE Transactions on Industry and General Applications,* vol. IGA-7, no. 6, November/December 1971, pp. 742–749.

36. UL Standard 1236, *Battery Chargers,* Underwriters Laboratory, P.O. Box 75330, Chicago, IL 60675-5330.

37. ANSI/IEEE Standard 48, *IEEE Standard Test Procedures and Requirements for High-Voltage Alternating-Current Cable Terminations,* Institute of Electrical and Electronics Engineers, P.O. Box 1331, Piscataway, NJ 08855-1331.

38. ANSI/IEEE Standard 400, *IEEE Guide for Making High-Direct-Voltage Tests on Power Cable Systems in the Field,* Institute of Electrical and Electronics Engineers, P.O. Box 1331, Piscataway, NJ 08855-1331.

39. IEEE Standard 62, *IEEE Guide for Field Testing Power Apparatus Insulation,* Institute of Electrical and Electronics Engineers, P.O. Box 1331, Piscataway, NJ 08855-1331.

40. NEMA Standard MG 1, *Motors and Generators,* NEMA Publication Distribution Center, P.O. Box 338, Annapolis Junction, MD 20701-0338.

41. IEEE Standard 112, *Standard Test Procedure for Polyphase Induction Motors and Generators,* P.O. Box 1331, Piscataway, NJ 08855-1331.

42. IEEE Standard 113, *IEEE Guide on Test Procedures for DC Machines,* P.O. Box 1331, Piscataway, NJ 08855-1331.

43. IEEE Standard 115, *IEEE Test Procedures for Synchronous Machines,* P.O. Box 1331, Piscataway, NJ 08855-1331.

44. NEMA Standard CP 1, *Capacitors,* NEMA Publication Distribution Center, P.O. Box 338, Annapolis Junction, MD 20701-03382.

CHAPTER 14
SAFETY

14.0 INTRODUCTION

Safety concerns everyone: labor (both employees and contractors), management (both on-site and off-site), and visitors (both scheduled and casual). Safety considerations include protection of life and plant property, and uninterrupted productive output. Protection of life is the paramount safety consideration and deserves the highest priority over all other considerations, including economy. (The cost of one preventable accident can far exceed the price of compliance with safe practices.) A lost life cannot be recovered. No compensation can adequately alleviate the effects of individual pain and suffering.

Major factors influencing safety include site conditions, adequate equipment, and work rules. Safe behavior or design implies the exercise of common sense in the design, operation, maintenance, and testing of electrical systems. The application of common sense, however, presumes a complete knowledge of all aspects of safe practices as well as an intimate knowledge of the electric system and its protecting and protected equipments. Every involved person in plant activities cannot be expected to possess this knowledge or to exercise safe practices all the time without the help of strictly enforced work rules.

14.1 PROTECTION OF LIFE

14.1.1 Site Conditions

Housekeeping, lighting, access, and working space all affect safety. Codes such as the *National Electrical Code* (NEC),[1] *National Electrical Safety Code* (NESC),[2] and the *California Electrical Code*[3] require compliance in some of these areas.

Good housekeeping—maintaining clear access aisles with no storage of any kind blocking the front or rear of metal-enclosed switching equipments—seems like a common-sense requirement. Good housekeeping rules are not always voluntarily observed, which makes it necessary to include them in many codes such as the NEC.

Any kind of work requires adequate lighting. This is as true for switching operations performed in an enclosed room as it is for maintenance and testing operations performed on-site or in a testing laboratory. Some codes (the *California Electrical Code,* for example) specify not only minimum illumination for work areas, but also maximum consumed wattage to meet the required lighting levels. Chapter 12 discusses some of these requirements. Architects must allow adequate space for a safe electric system in any installation.

Major electrical switching and control centers should be accessible only to qualified personnel. Even with such limited access, fences around the operating faces of critical control boards sometimes are necessary to prevent accidental operation by plant visitors or unauthorized personnel.

The NEC requires warning signs on rooms housing electric equipment, as well as requiring guards to prevent accidental contact with live parts. Both the NEC and the NESC specify minimum live conductor clearances, as well as front and rear electric equipment access aisles.

14.1.2   Adequate Equipment, Tools, and Clothing

Protecting life means preventing hazardous events such as fires and explosions resulting from overloads and short circuits. This requires installation of equipment capable of interrupting overloads and short circuits in the minimum time, as required by code. The purpose of the NEC as stated in Article 90.1 and repeated in the NESC, is "the practical safeguarding of both persons and property from the hazards arising from the use of electricity." Chapter 7 discusses short circuits, and Chap. 8 discusses protective device settings.

For areas used by the public and unqualified employees, greater safeguards than those required by code may be indicated. Special occupancies, such as hazardous areas, paint booths, computer rooms, health care facilities, and similar areas, require special design and safety considerations, some of which are covered by the NEC and other applicable codes.

Operating personnel safeguards include adequate enclosures, safety key or electrical interlocking, high-speed fault detection, clear operating instructions for unusual system occurrences, adequate training and intimate knowledge of the electric system, well-maintained electric equipment, effective equipment and system grounding (see Chap. 9).

Maintenance, testing, and troubleshooting personnel require the proper tools to perform these functions. Proper tools vary with the sophistication of the installed electric equipment. A study of manufacturers' instruction books may reveal the need for tools not normally thought to be required. Metal ladders or step stools should never be used in an electrical area.

Proper dress for persons working around electric equipment includes plastic or fiber safety helmets and safety glasses. Do not allow loose key chains, tool pouches, or pieces of wire to hang from the body. Remove rings and metal watchbands. Do not wear loose clothing that might catch in rotating machines or on control-switch handles.

14.1.3   Work Rules

The NESC devotes all of Part 4 to work rules for utility workers. The Occupational Safety and Health Administration (OSHA) of the U.S. Department of Labor periodically issues work rules that apply to all industrial installations. Among these work rules is required compliance with the NEC and lockout/tagout rules. The "Rule on Lockout/Tagout" is published in the *Federal Register* (53 FR 1546), January 2, 1990.[4] A few rules other than those published in formal work rules are:

1. Always have a definite written plan for work to be accomplished.

2. Always follow prescribed tagout/lockout rules.

3. Assume every conductor is energized unless you have personal knowledge that it is not energized and is grounded.

4. Never bypass mechanical or electrical safety interlocks.

5. Do not reclose an automatically tripped circuit breaker or replace a blown fuse until after the reason for the operation has been determined. *Never* assume false relay operation. *Always* record time and identification of relay or circuit-breaker trip device targets after any automatic operation.

6. Avoid instantaneous automatic reclosing on any circuit or service supplying motors of about 100 hp or larger.

7. *Never* touch the outer insulation surface of an energized, unshielded insulated conductor. Examples of this type of conductor are metal-clad switchgear or metal-enclosed equipment, insulated buses, and transformer primary switch throat connections. (The air surrounding such conductors is part of the conductor insulation system.)

8. Never leave a CT secondary open-circuited on an energized circuit.

9. Use extreme caution and care with electric equipment located in confined or hazardous and damp or humid areas. Water is the biggest enemy of electric systems.

These are examples of the many rules that may not be documented, but should be observed when working with electric equipment.

14.2 PROTECTION OF PLANT PROPERTY

All of the safety aspects mentioned in Sec. 14.1 apply to this section, with the addition of three management responsibilities concerning operating instructions, adequate system protection, and maintenance and testing.

14.2.1 Operating Instructions

Skillful operators—persons authorized to perform switching operations—enhance the chances of minimizing plant property damage when emergency situations such as faults occur. Such operators should be well trained, have an intimate knowledge of the plant electric system, and know the limitations of the system-protective devices. Operators should be instructed about what to do in the case of an outage in a certain part of the plant and how to restore power in as short a time as possible.

14.2.2 Adequate System Protection

Adequate system protection not only means the capability to interrupt overloads and short circuits; it also means appropriate speed of interruption and automatic throwover equipment where necessary. An examination of each protective component should be made not only when the system is initially installed but also periodically (probably every five years or so). Such a power system study can assure that the proper equipment settings or ratings are applied, indicate a need to correct system protection to account for power system changes, and discover new opportunities to apply recently developed technologies. Chapter 3 discusses power system studies.

14.2.3   Planned Maintenance and Testing

A good power system with appropriate protective functions installed may fail to perform as expected if it is not periodically maintained and tested. Manufacturers' instruction books supplemented by industry recommendations such as IEEE maintenance guides and other handbooks offer recommendations. Chapter 15 discusses maintenance, and Chap. 13 discusses testing.

The best maintenance programs are well-planned, periodic ones that are not compromised in times of fiscal restraint.

14.3   PROTECTION AGAINST INTERRUPTION OF OUTPUT

While all of the ideas discussed in Secs. 14.1 and 14.2 tend to protect a system from unscheduled outages, interruption of output can best be avoided by frequent examinations of the power system to discover how it can be improved or updated. A few of the questions often asked in the examinations are:

1. Is a higher plant distribution voltage indicated?
2. Is high-speed differential protection required for any system element that does not have this protection installed?
3. Will automatic throwover for any particular bus increase plant reliability?
4. Are the economies of cogeneration applicable to this plant?
5. Is another utility source indicated?
6. Is the present system being operated in the most efficient manner?
7. Should the system be changed to a more complex system, which may be more reliable theoretically but more difficult to operate?

14.4   SUMMARY

High safety standards require continuous monitoring. Such standards include at least the following items:

- High-quality electric system components
- Proper system design and rating selection
- Adequate short-circuit and overload interrupting capabilities
- Metal-clad or metal-enclosed construction with proper safety interlocks
- Simple, easy-to-operate system design
- Adequate, effective relaying protection
- Adequate grounding, both equipment and system
- Appropriate, scheduled maintenance and testing
- Safe operating practice code (work rules)
- Adequate record logs

14.5 REFERENCES

1. ANSI/NFPA 70, *1993 National Electrical Code,* National Fire Protection Association, Batterymarch Park, Quincy, MA 02269.

2. ANSI/IEEE C2-1993, *1993 National Electrical Safety Code,* Institute of Electrical and Electronics Engineers, 445 Hoes Lane, P.O. Box 1331, Piscataway, NJ 08855-1331.

3. *California Administrative Code, Title 24, Part 3—California Electrical Code,* California Building Standards Commission, 428 Jay Street, Suite 450, Sacramento, CA 95814.

4. *Federal Register* (53 FR 1546), U.S. Government Printing Office, Washington, DC 20402 (Telephone: 202-783-3238).

CHAPTER 15
ELECTRICAL MAINTENANCE

15.0 INTRODUCTION

Once equipment and facilities are properly designed and installed, reliable and safe performance is ensured by proper operation and maintenance. The objectives of maintenance may include:

- Ensuring the safety of personnel and property
- Providing for operational readiness of emergency equipment
- Complying with codes, standards, and regulations
- Providing the highest quality product or service
- Maximizing availability
- Minimizing the number of outages
- Minimizing the duration of outages (downtime)
- Preserving and enhancing useful life
- Maximizing profit and minimizing the cost of service

Maintenance may be defined as actions taken to ensure that the equipment and facilities continue to be in good operating condition. Maintenance is commonly classified as *planned* (scheduled) and *emergency* (unscheduled). Planned maintenance is performed according to a predetermined plan. This plan would have been assembled to meet the general maintenance objectives just listed, taking into account the specific plant, equipment, and manpower constraints.

Planned maintenance may be divided into *preventive* and *corrective* maintenance. Preventive maintenance is regular maintenance performed before any failure or defect occurs. Corrective maintenance is undertaken to repair or correct a failure, but is not needed on an emergency basis. The process may not be halted immediately for repairs. Such outages are also sometimes called *postponable outages.* Emergency maintenance is self-explanatory.

Maintenance within a plant covers a large range of disciplines. This chapter is included in this book to introduce the subject of maintenance to the reader and provide a glimpse of the maintenance aspects of some major electric apparatus and equipment within a plant. Whether one is implementing a maintenance program or actually performing maintenance, other books, manufacturers' instructions, and standards and codes should be consulted. The need for proper and adequate training, including safety-related aspects for the people performing the maintenance work, cannot be emphasized enough. The next section briefly describes maintenance management. The remainder of this chapter acquaints the reader with the maintenance testing requirements and other related maintenance activities for certain major electrical equipment. Only electrical aspects of maintenance are discussed in this chapter. Maintenance disciplines such as mechanical aspects, process control, etc., are equally important. For example, lubrication is one of the most important preventive maintenance items.

15.1 MAINTENANCE MANAGEMENT

The scope of maintenance management encompasses engineering, technology, skilled labor, accounting, and administration. The actual method used for maintenance management depends upon the type of maintenance philosophy, the type of plant and/or service, the product, whether it involves an automated or labor-intensive process, the safety requirements, the laws and regulations, etc. We will briefly discuss in this section different types of maintenance approaches. References 1–12, listed at the end of this chapter, provide a simple and straightforward exposition on the topic of maintenance.

Historically, preventive maintenance has been a major maintenance approach. In this approach, outages are taken at regular intervals to test, inspect, and correct any problem discovered during this planned or scheduled outage. The extent of work performed at these scheduled outage periods varies. Longer and extensive scheduled outages are used to perform extensive maintenance. These are sometimes referred to as major overhauls. For example, a regular outage of one week every six months may be used to do routine maintenance work, and every two years a two-week outage may be taken to perform detailed inspection, testing, and repair. The preventive maintenance concept is easily understood, simple to implement, and used by a wide range of plants and facilities around the world.

The regularly scheduled maintenance assumes that failure characteristics are the same for a given category of equipment. The type of operation, the surrounding environment, the quality of equipment, and quality of maintenance are all factors which influence the failure rate. Because of the influence of these different variables, the equipment may fail before the next regularly scheduled maintenance period. As the common saying goes, failure may occur when the equipment is most needed. The emergency repair and/or lost production may be very expensive. Too-frequent preventive maintenance outages may cause failure because of mistakes made when putting the equipment back into service. Interchange of wiring, leaving a wrench or screwdriver inside, or improper setting are such examples. In order to overcome this drawback of preventive maintenance, several new maintenance concepts have been introduced, including reliability-centered maintenance, predictive maintenance, just-in-time maintenance, reliability and maintainability, and total productive maintenance.

In reliability engineering, the outage characteristics of equipment are represented generically by the so-called bathtub curve, as shown in Fig. 15.1. The life of any equipment could be divided into three major periods.

Period	Characteristics	Maintenance action
Immature period (shakedown)	High failure rate	Troubleshooting and repairs
Mature period	Low failure rate	Preventive or as-needed maintenance
Wearout period	High failure rate	Major repair or replacement

The failure during the mature period is a chance failure. The design and selection of equipment should be such that the failure rate is below the acceptable level during the

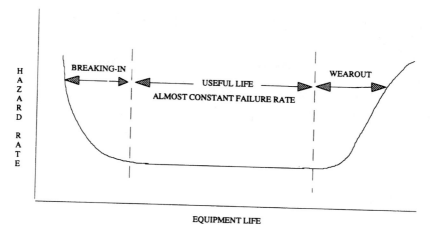

FIGURE 15.1 Bathtub curve for hazard rate.

mature operation period. This mature period is not constant for all equipment and operations. It is better to detect impending failure and repair the problem before the failure occurs. This is called *predictive maintenance.*

A predictive maintenance management philosophy uses data from the actual operating condition of the plant, systems, subsystems, and equipment to determine the need for maintenance. The maintenance can be scheduled for only the items needing repair and when actually needed, but before an outage occurs. This is considered to be more cost-effective. Data is collected through nondestructive testing techniques such as vibration monitoring, process parameter monitoring, and thermography. The analysis of this data can provide information on the performance of equipment. The monitoring, collection, and analysis of pertinent data are the main requirements of predictive maintenance. Software programs are available to set up a database for predicting when maintenance should be performed. Predictive maintenance should be considered to supplement the conventional preventive maintenance approach so that cost savings are realized, and, at the same time, improved reliability of plant and equipment is obtained.

The most recent maintenance concept is called *total production maintenance* (TPM) originated in Japan. The important features of TPM are:

- To maximize overall effectiveness of equipment and facility
- To establish a system of preventive maintenance for the entire life of the equipment, facility, or plant.
- To implement the plan in all disciplines and departments
- To involve all personnel, from top management all through to workers on the floor
- To use small groups for preventive maintenance

TPM may be described as maintenance performed on a companywide basis and organized into small group activities. As more automation is introduced into manufacturing, the equipment effectiveness (production and cost) increases in importance. TPM is described as being parallel to total quality control and fits with the zero-breakdown and zero-defects goal. Many articles, books, and seminars are available on this topic. Reference 1 is a concise and easy-to-read book introducing this topic.

The preceding discussion gives a general idea of maintenance management philosophies. Without getting involved in concepts and philosophies of maintenance management, we will itemize and describe the following six steps for organizing maintenance programs:

- Establishing responsibility
- Planning and scheduling work
- Preparing a work order
- Inspecting and testing
- Record keeping
- Evaluating effectiveness

15.1.1 Establishing Responsibility

The responsibility for maintenance within the company should be clearly defined and documented. The job descriptions for the maintenance manager (plant engineer) and the maintenance department staff need to be defined. The personnel should understand how they fit into the overall structure of the department and the company. Other departments within the company should be informed of the responsibilities assigned to the maintenance department so that maintenance personnel are properly utilized.

15.1.2 Planning and Scheduling

A successful maintenance department plans and schedules maintenance before the actual work needs to be performed, so that the production activity is affected as little as possible. Thus, a maintenance schedule(s) (which has been approved or accepted by other departments) should be prepared to meet the overall planned maintenance. Provision for unscheduled and emergency maintenance requirements must be made. The actual schedule varies with type of plant, process, equipment, age, etc. Thus, the frequency of maintenance will be constantly adjusted. The availability of skilled personnel within the plant and the cost of contracting outside help are factors which influence maintenance scheduling and costs. Preventive maintenance scheduling is defined as *what to maintain and how often.* It may include:

- Type of inspection (operational/shutdown/major overhaul)
- Type of tests
- Frequency of inspection
- Method
- Procedure
- References (drawings/codes/instructions)

15.1.3 Preparing Work Orders

Maintenance requests are processed by the maintenance department and work orders are issued. Typically work orders include information on when, where, and exactly what has to be done. A particular person with the proper skill should be assigned. Whether this is a routine, immediate, or emergency situation should be clearly communicated. Any safety-related work has to be expeditiously executed, whether a formal work order is prepared or not.

15.1.4 Inspection and Testing

Inspection and testing (see Chap. 13) is needed to check if the equipment has any defect and to identify the source of the problem, if any. Testing performed by qualified personnel using appropriate techniques at proper intervals or as needed (in a predictive maintenance approach) is important, so that repair or replacement of the defective equipment can be made before actual failure occurs.

15.1.5 Record Keeping

The record keeping, with the associated paper work, is usually disdained, especially by the skilled labor on the shop floor. The availability of microcomputers, electronic notepads, remote sensing, metering, and recording has made record keeping easy. It is important to select the appropriate software, database, and computer system for the type of maintenance used in the plant. The same computers can be used for planning and scheduling; to issue work order tickets; to monitor progress, stores, and parts inventory; and for information and document retrieval. It is cost-effective to have a computer system for an up-to-date record-keeping system. The initial effort to implement a computer-based maintenance management system may appear to be substantial, but in the long run it pays off.

Inspection and test data on all equipment is readily available for review, analysis, and evaluation. All data and reports are organized. Means for detecting problems and trends of performance of equipment, process, etc., can be available. If the inspection and maintenance data is not recorded properly, then one of the primary purposes of planned maintenance is lost.

15.1.6 Evaluating Effectiveness of Maintenance

Maintenance programs should be constantly evaluated for their effectiveness. The method of determining effectiveness depends upon the type of plant, facility, equipment, etc. The goals of maintenance stated at the beginning of Sec. 15.0 should be met at the lowest possible cost. Some of the items that can be readily identified for measuring effectiveness include:

- Comparing plant/facility maintenance needs with those of similar industries (benchmarking) or with the maintenance history of the facility
- Monitoring maintenance quality (frequency of recurring problems and corrective actions)
- Comparing maintenance costs to other plant costs or to last year's costs
- Reviewing operating costs and aiming for high availability, low downtime, and process efficiency
- Monitoring productivity
- Monitoring product quality
- Monitoring manpower turnover

- Providing alternatives for maintenance cost reduction
- Monitoring the safety record

The outcome of this evaluation should identify changes for recommended future action to improve effectiveness in areas where it is necessary.

15.2   TECHNICAL ACTIVITIES OF MAINTENANCE

Whereas maintenance management focuses on the cost, performance, and scheduling aspects, the technical activities of maintenance deal with the so-called nuts and bolts. The technical-related activities may be described in great detail. However, we can identify some important activities and briefly describe them here. These are:

- Plant equipment survey
- Critical equipment identification
- Materials and stores need
- Maintenance planning and scheduling
- Instructions and procedures documentation
- Education and training

15.2.1   Plant Equipment Survey

Accurate data about the electric power system that can be used as reference for maintenance and testing is essential. Typically, system diagrams are used to document equipment and circuit installation. Established IEEE or other standards for diagram symbols, device designations, and electrical symbols are used in these diagrams so that there is no confusion in the interpretation by maintenance personnel and others. The types of diagrams and drawings in common use are the following:

- System diagrams
  Control and monitoring system
  Lighting system
  Ventilation system
  Heating and air-conditioning system
  Emergency system
  Other systems
- Block diagram
- Process or flow diagram
- One-line (single-line) diagram
- Schematic (elementary) diagram
- Control sequence (truth-table) diagram
- Wiring (connection) diagram

System diagrams may interface with other diagrams, such as electrical, fire, and emergency power diagrams. It is important to know where these interfaces are, how they work, and how they can be coordinated in the maintenance program.

It may not be necessary to perform an equipment plant survey from scratch except for a new plant. It may only be necessary to reorganize and augment existing information. It may be cost-effective to start fresh rather than wasting time on figuring out what is available.

15.2.2   Critical Equipment Identification

Failure of electrical equipment and the electric power system may pose a serious threat to people and property. The type of critical equipment varies from plant to plant; therefore, a team consisting of personnel from different disciplines needs to identify and list the critical equipment (electric and other) vital to the operation of a facility. All the critical equipment and/or systems should be identified on the drawings. Maintenance personnel should know where the critical equipment is located and its function. Different priorities for the critical equipment may be developed.

15.2.3   Materials and Stores Need

This aspect of maintenance is primarily a management or organization control type of activity and deals with administration. However, the technical disciplines have an important part to play in this activity, including:

- Identifying spare parts and materials that need to be stored
- Identifying critical parts
- Defining turnaround times
- Identifying outside sources
- Performing quality inspection and approval of suppliers and parts
- Performing or witnessing acceptance tests
- Defining shelf life of parts and material
- Assisting store administration in optimizing costs
- Assisting in evaluation of reconditioning of equipment for restocking rather than purchasing new parts
- Substitution

This list illustrates that technical personnel have a major contribution to make in the area of materials and stores. The location of the store (central vs. decentralized), the type of control, the cost, procurement procedure, layout, etc., are more within the responsibility of administration.

15.2.4   Maintenance Planning and Scheduling

The purpose of maintenance planning is to:

- Set goals and priorities
- Schedule maintenance for the next period
- Determine material, labor, and other support requirements
- Minimize maintenance cost
- Coordinate with all departments/personnel concerned

- Develop a consistent work flow
- Provide for record keeping and evaluation
- Determine implementation (work orders, clearance procedures, etc.)

The maintenance goals and priorities should be set by the person in charge of maintenance. The maintenance schedule may be required for the day, week, or month. The daily schedule needs to be more detailed as compared to a semiannual schedule. The material, spare parts, and skilled and unskilled labor needs should be identified before commencing maintenance work. The need for outside assistance should be recognized as early as possible and arrangements for procuring such assistance should be made ahead of time. The source of assistance, on an emergency basis, should be identified. It is the maintenance planner's job to minimize the cost of maintenance. Coordination with all departments concerned will make the maintenance work go smoothly. Manufacturers' service manuals specify recommended frequency of maintenance and/or inspection. These time intervals are based upon standard operating conditions and environments. If these standard conditions change for the equipment, then the maintenance frequency should be modified accordingly.

Timely and high-quality preventive maintenance is very important for the reliability of equipment. A properly developed and implemented electrical preventive maintenance program minimizes equipment failure. Performing maintenance at too-frequent intervals is expensive, both in labor and material costs. A maintenance outage may introduce problems due to human error and this may cause a failure. There is an optimum interval between occurrences of scheduled preventive maintenance. Table 15.1 lists the data regarding percentage of failures since last maintenance. For example, 7.4 percent (under all electrical equipment classes combined) in Table 15.1 indicates that 74 pieces out of 1000 pieces of electrical equipment failed within 12 months of maintenance. As can be seen from this table, the following conclusions regarding maintenance frequency can be drawn.

- A one-year or less interval for scheduled maintenance of all electrical equipment combined is desirable.
- A one-year interval for circuit breakers is appropriate.
- A two-year interval for motors (dc motors may need more frequent maintenance compared to ac motors) should be sufficient (bearings may need more attention).
- A two-year interval for transformers is needed.

Of course, this interval needs to be adjusted for specific equipment, the type of duty, operating environment, and quality of maintenance. Use of predictive maintenance con-

TABLE 15.1 Percentage of Failure Caused Since Maintained

Failure (months since maintained)	All electrical equipment classes combined (%)	Circuit breakers (%)	Motors (%)	Open wire (%)	Transformers (%)
Less than 12 months ago	7.4	12.5*	8.8	0*	2.9
12–24 months ago	11.2	19.2	8.8	22.2*	2.6
More than 24 months ago	36.7	77.8	44.4	38.2	36.4
Total	16.4	20.8	15.8	30.6	11.1

*Small sample size; less than 7 failures caused by inadequate maintenance.

Source: Reprinted from ANSI/IEEE-Standard 493-1990, *Design of Reliable Industrial and Commercial Power Systems,* copyright © 1991, IEEE; with the permission of IEEE.

cepts in timing preventive maintenance was discussed earlier in this chapter. Quality of maintenance can be factored into the failure rate by using multipliers shown in Table 15.2. For example, consider a transformer maintained at less than a 12-months interval. The failure rate is 2.9 percent from Table 15.1. If the quality of maintenance was excellent, then the failure rate is 0.95 times 2.9 percent; i.e., there is a 5 percent reduction in the failure rate.

TABLE 15.2 Equipment Failure Rate Multipliers versus Maintenance Quality

Maintenance quality	Transformers	Circuit breakers	Motors
Excellent	0.95	0.91	0.89
Fair	1.05	1.06	1.07
Poor	1.51	1.28	1.97
All	1.0	1.0	1.0
Perfect maintenance	0.89	0.79	0.84

Source: Reprinted from ANSI/IEEE-Standard 493-1990, *Design of Reliable Industrial and Commercial Power Systems,* copyright © 1991, IEEE; with the permission of IEEE.

Whenever manpower constraints prevent the facilities manager from following the suggested maintenance frequency, procuring outside contractors is an option. If budgetary constraints make this an impossible task, then maintenance should be scheduled as close to the suggested interval as possible. Exceptions should not be made for maintaining equipment and facilities which serve critical loads and functions. The maintenance group or user should immediately report any defect beyond their repair capability to the proper authority. Records should be kept of all defects in the system and corrective actions taken to repair these defects.

15.2.5 Instructions and Procedures Development

Development of instructions, procedures, and methods is necessary to ensure that the maintenance performed is consistent with the safety and operating requirements of the equipment and the system. The maintenance department should have fully developed procedures and instructions for servicing all equipment and components within the plant. This includes shutdown procedures, safeguards, lack of equipment, interlocking of equipment, and alarms. Methods of recording data should be established. The role of microcomputers and databases for this purpose was discussed earlier.

15.2.6 Education and Training

The quality of maintenance is directly related to the capability of the maintenance staff, which in turn depends upon their education, training, and experience. With the increasing use of automation, microprocessor technology, and robots and with the stringent requirements of productivity and cost control, there is an ever-increasing need for education and training. There are strict environmental and public safety laws and regulations which need to be followed without exceptions. There is a high level of competency required from skilled labor through the ranks to the engineer level so that every person performs the maintenance tasks effectively. This is achieved through proper education and training.

Minimum-level-of-education requirements should be established for different categories of personnel. Education and training are achieved through many different approaches. These include:

- Apprenticeships
- On-the-job training
- Training specific to equipment, process, and operation
- Specialized indoctrination seminars
- Special training courses
- Continuing education courses
- Conferences and seminars
- Multidisciplinary training
- Computer-use courses
- Retraining programs
- Safety-related training
- Maintenance management systems

The actual type of training required is dependent on many factors. The training within the plant or from outside institutions specializing in a particular type of training can be cost-effective. Universities, manufacturers, and professional and trade associations provide specialized training. Steps needed in ensuring proper education and training include

- Determining the education and training needs for the plant and/or equipment
- Matching the education, training, and experience of personnel to the maintenance needs
- Identifying the additional training need
- Selecting a cost-effective method of providing needed training

The cost factor acts as an obstacle to education and training. However, the long-term cost of not providing the appropriate kind of training is usually greater than the short-term cost of training. The appropriate type of training for selective staff should be provided so that the money is not wasted. Training based on video instructions and courses is increasingly available on the market. These formats provide self-paced study without a long absence from the workplace, are easy to repeat, and no travel costs are involved. Some of these tutor the student and administer tests. Supervisors can monitor the progress. A training budget needs to be established and used.

15.3 CODES AND SPECIFICATIONS

Maintenance of electrical systems and equipment must adhere to the applicable codes and specifications. Manufacturers' maintenance instructions which accompany select electric components must be applied. Some of the applicable codes and specifications are:

- The National Electrical Code, ANSI/(NFPA 70). This is an American National Standard. This code is the most widely adopted set of electrical safeguarding practices. It defines approved types of conductors and equipment, acceptable wiring methods, mandatory and advisory rules, operating voltages, limitations on loading of conductors, required working spaces, methods of guarding energized parts, interrupting capacity requirements and splices, insulation resistance requirements, and grounding requirements.

- *Recommended Practice for Electrical Equipment Maintenance* (NFPA 780b).
- *IEEE Recommended Practice for Protection and Coordination of Industrial and Commercial Power Systems,* Chapter 15 (ANSI/IEEE Standard 242-1986).

15.4 MAINTENANCE OF MAJOR EQUIPMENT

Brief descriptions of maintenance of major equipment are presented in this section. This discussion is not intended to be inclusive of all electrical equipment and facilities, but only provides a cursory view of the maintenance items and procedures. For detailed procedures and practices, refer to other manuals and books, including Chap. 14, on safety, of this book.

15.4.1 Transformers

A transformer is a very simple, rugged device and is often ignored and forgotten until transformer failure occurs. Transformers are a vital link in the electric distribution system and should be given proper care and attention. Transformer maintenance schedules should be determined according to the critical or noncritical nature of the transformer and the load to which it is connected. Large power transformers are obviously more important than small lighting and distribution transformers; thus, they warrant more attention and care. Proper maintenance of transformers should involve routine inspection and testing, including transformer liquid testing, transformer winding insulation testing, and any other special maintenance that is recommended by the manufacturer of the transformer. A power transformer inspection and maintenance checklist includes the following:

General	Windings	Transformer oil
Alarms	Gas-analysis test	Color
Auxiliary equipment	Insulation resistance	Dielectric strength
Breather	Power factor	Neutralization number
External inspection		Power factor test
Grounding of neutral		
High and low voltage		
Lightning arresters		
Liquid level		
Load current		
Pressure-relief valve		
Relay and protection		
Tap changer		
Temperature		

The frequency of these checks varies from hourly (recorded values) to yearly.

15.4.2 Electric Motors

Ac motors, with reasonable care, give long continuous service. There is a tendency to neglect motor maintenance and, as a result, motor failures are frequent and repairs become a continuous and costly process. It is recommended that a preventive maintenance program be established to minimize failures. A squirrel-cage induction motor is the most widely used type of motor and it is the most rugged and least expensive of all types of induction motors. Periodic inspection and regularly scheduled preventive maintenance checks and services will enhance continuous operation without undue failures. The frequency of these inspections depends on the criticality of the service, the hours that equipment is in service, and the environment under which the equipment operates. The following safety precautions should be observed when working on electric motors:

1. The machine should be deenergized, tagged, and locked out before starting work.
2. Personal protective equipment such as goggles, gloves, aprons, and respirators should be worn if working with hazardous substances.
3. Great care should be provided in selecting solvents to be used for a particular task.
4. Adequate ventilation must be provided to avoid fire, explosion, and health hazards where cleaning solvents are used.
5. A metal nozzle used for spraying flammable solvents should be bonded to the supply drum and to the equipment being sprayed.
6. After tests have been made, discharge stored energy from windings by proper grounding before handling test leads.

Preventive maintenance will involve lubrication, cleaning, and checking for sparking brushes (dc or wound rotor machines), vibration, loose belts, high temperature, and unusual motor noises. Repair work on larger motors is normally limited to replacement or refinishing of bearings, commutators, collector rings, brushes, etc. Motor rewinding should not be attempted by the installation support groups since it is more economical to contract such work to commercial shops that specialize in motor rewinding. Some troubleshooting steps for ac induction motors are summarized in Tables 15.3 and 15.4.

15.4.3 Motor Controls

Control equipment should be inspected and serviced simultaneously with the motors. Overhaul procedures are less involved than those of motor overhauling. Most repairs can be made on-site. Motor starters represent one area in which simplicity of construction and wiring has been very much emphasized by the manufacturers. Improvements have resulted in starters that are simple to install, maintain, and operate. Typical motor control maintenance steps are given in Table 15.5. Some troubleshooting steps are summarized in Table 15.6.

TABLE 15.3 Ac Induction Motor Troubleshooting Summary

Cause	Action
Motor will not start	
Bearing stiff	Free bearings or replace.
Driven machine locked	Disconnect motor from load. If motor starts satisfactorily, check driven machine.
Failed starter capacitor	Isolate and discharge capacitor and check impedance. If opened or shorted, replace.
Faulty (open) fuses	Test fuses and circuit breakers.
Faulty control	Troubleshoot the control.
Grease too thick	Use special lubricant for special conditions.
Loose terminal-lead connection	Tighten connections.
Low voltage	Check motor-nameplate values with power supply. Also check voltage at motor terminals with motor under load to be sure wire size is adequate.
Open circuit in stator or rotor winding	Check for open circuits.
Overload	Reduce load
Overload control trip	Wait for motor to cool. Try starting again. If motor still does not start, check all the causes as outlined below.
Power not connected to motor	Connect power to control; check control sequence and power to motor. Check connections.
Short circuit in stator winding	Check for shorted coil.
Winding grounded	Test for grounded winding.
Wrong control connections	Check connections with control wiring diagram.
Motor noisy	
Electrical load unbalanced	Check current balance.
Motor running single-phase (3-phase type)	Stop motor, then try to start. It will not start on single-phase. Check for "open" in one of the lines or circuits.
Motor running at higher than normal temperature or smoking	
Belt too tight	Remove excessive pressure on bearings.
Electrical load unbalanced	Check for voltage unbalance or single-phasing.
Fuse blown, faulty control, etc.	Check for "open" in one of the lines or circuits.
Incorrect voltage and frequency	Check motor-nameplate values with power supply. Also check voltage at motor terminals with motor under full load.
Motor stalled	Remove power from motor. Check machine for cause of stalling.
Overload	Measure motor loading with watt-meter. Reduce load.
Rapid reversing service	Replace with motor designed for this service.
Restricted ventilation	Clean air passages and windings.
Rotor winding with loose connections	Tighten, if possible, or replace with another rotor.
Stator winding grounded/shorted	Use insulation testing procedures.

Note: These steps are not necessarily in any order of priority or sequence.

TABLE 15.4 Ac Induction Motor (Wound Rotor) Troubleshooting Summary

Cause	Action
Motor will not start	
Faulty connection	Inspect for open or poor connection.
Friction high	Make sure bearings are properly lubricated. Check bearing tightness, belt tension, load friction, alignment.
Load too great	Remove part of load.
Open circuit in one phase	Test, locate, and repair.
Short circuit in one phase	Open and repair
Voltage too low	Reduce the impedance of the external circuit.
Wrong direction of rotation	Reverse any two main leads of 3-phase motor. Single-phase, reverse-starting winding leads.
Motor will not come up to speed or fails to pull into step	
Excessive load	Reduce the load or may be a misapplication—consult manufacturer. Check operation of unloading device (if any) on driven machine.
Low voltage	Check the reason for low voltage and increase voltage.
Motor runs at low speed with external resistance cut out	
Brushes sparking	Check for looseness, overload, or dirt.
Brushes stuck in holders	Use right size brush; clean holders.
Control too far from motor	Bring control closer to motor.
Dirt between brush and ring	Clean rings and insulation assembly.
Eccentric rings	True up rings.
Incorrect brush tension	Clean brush tension and correct
Open circuit in rotor circuit (including cable to control)	Test to find open circuit and repair.
Rough collector rings	Sand and polish.
Wires to control too small	Use larger cable to control.
Motor pulls out of step or trips breaker	
Excessive torque peak	Check driven machine for bad adjustment, or consult motor manufacturer.
Line voltage too low	Increase if possible. Raise excitation.
Load fluctuates widely	See motor "hunts," below.
Power fails	Reestablish power circuit.
Motor "hunts"	
Fluctuating load	Correct excessive torque peak at driven machine or consult motor manufacturer. If driven machine is a compressor, check valve operations. Increase or decrease flywheel size.
Stator overheats in spots	
Open phase	Check connections and correct.
Rotor not centered	Realign and shim stator or bearings.
Unbalanced currents	Loose connections; improper internal connections.
All parts overheat	
Excessive room temperature	Supply cooler air.
Improper ventilation	Remove any obstruction and clear out dirt.
Improper voltage	See that nameplate voltage is applied.
Overload	See above

Note: These steps are not necessarily in any order of priority or sequence.

TABLE 15.5 Motor Control Preventative Maintenance Steps

Inspection items	Action
Exterior and surroundings	Check for dust, grease, oil; high temperature; rust and corrosion; mechanical damage; condition of gaskets.
Interior of enclosure, nuts, and bolts	Same as above. In addition, check for loosened nuts, bolts, or other mechanical connections.
Connections	Tighten main line and control conductor connections; look for discoloration of current-carrying parts.
Contactors, relays, solenoids	
Arc chutes	Check for breaks or burning.
Coils	Look for overheating, charred insulation, or mechanical injury.
Contact tips	Check for excessive pitting, roughness, copper oxide; do not file silver contacts.
Flexible leads	Look for frayed or broken strands; be sure lead is flexible, not brittle.
General	Check control circuit voltage; inspect for excess heating of parts evidenced by discoloration of metal, charred insulation, or odor; freedom of moving parts; dust, grease, and corrosion; loose connections.
Magnets	Clean faces; check shading coil; inspect for misalignment, bonding.
Springs	Check contact pressure.
Fuses and fuse clips	Check for proper rating, snug fit; if copper, polish ferrules; check fuse clip pressure.
Overload relays	Check for proper heater size; trip by hand; check heater coil and connection; inspect for dirt, corrosion.
Pushbutton station and pilot devices	Check contacts; inspect for grease and corrosion.
Dashpot-type timers and overload relays	Check for freedom of movement; check oil level.
Resistors	Check for signs of overheating; loose connections; tighten sliders.
Control operation	Check sequence of operation of control relays; check relay contacts for sparking on operation; check contacts for flash when closing; if so, adjust to eliminate contact bounce; check light switches, pressure switches, temperature switches, etc.

Note: These steps are not necessarily in any order of priority or sequence.

TABLE 15.6   Motor Control Troubleshooting Summary

Cause	Action
Contactor or relay does not close	
Coil open or shorted	Replace.
Interlock or relay contact not being made	Adjust or replace if badly worn.
Loose connection	Turn power off first, then check the circuit visually with a flashlight.
Low voltage	Check power supply. Wire may be too small.
Mechanical obstruction	With power off, check for free movement of contact and armature assembly.
No supply voltage	Check fuses and disconnect switches.
Overload relay contact open	Reset.
Pushbutton contacts not being made	Clean or replace if badly worn.
Wrong coil	Check coil number.
Contactor or relay does not open	
Gummy substance on pole faces	Clean with solvent.
Interlock or relay contact not opening circuit	Adjust contact travel.
Pushbutton not connected correctly	Check connections with wiring diagram.
Shim in magnetic circuit (dc only) worn, allowing residual magnetism to hold armature closed	Replace.
"Sneak" circuit	Check control wiring for insulation failure.
Worn or rusted parts causing burning	Replace parts.
Contacts weld shut or freeze	
Abnormal inrush of current	Use larger contactor or check for grounds, shorts, or excessive motor load current.
Foreign matter preventing contacts from closing	Clean contacts with approved solvent.
Insufficient contact spring pressure causing contacts to burn and draw arc on closing	Adjust, increasing pressure. Replace if necessary.
Low voltage preventing magnet from sealing	Correct voltage condition. Check momentary voltage dip during starting.
Rapid jogging	Install larger device rated for jogging service or caution operator.
Short circuit	Remove short-circuit fault and check to be sure fuse or breaker size is correct.
Very rough contact surface causing current to be carried by too small an area	Smooth surface or replace if badly worn.
Contact chatter	
Broken pole shader	Replace
Low voltage	Correct voltage condition. Check momentary voltage dip during starting.
Poor contact in control circuit	Improve contact or use holding circuit interlock (3-wire control).
Arc lingers across contacts	
Arc box might be left off or not in correct place	See that arc box is on contactor as it should be.
If blowout is series, it may be shorted	Check wiring diagram.
If blowout is shunt, it may be open circuited	Check wiring diagram.
If no blowout used, note travel of contacts	Increasing travel of contacts increases rupturing capacity.

TABLE 15.6 Motor Control Troubleshooting Summary (*Continued*)

Cause	Action
Excessive corrosion of contacts	
Chattering of contacts as a result of vibration outside the control cabinet	Check control spring pressure and replace spring if it does not give rated pressure. If this does not help, move control so vibrations are decreased.
High contact resistance because of insufficient contact spring pressure	Replace contact spring.
Abnormally short coil life	
Ambient temperature too high	Check rating of contact.
Dirt on contact surface	Clean contact surface.
Excessive jogging	Install larger device rated for jogging or caution operator.
Filing or dressing	Do not file silver-faced contacts. Rough spots or discoloration will not harm contacts.
Gap in magnetic circuit (ac only)	Check travel of armature. Adjust so magnetic circuit is completed.
High voltage	Check supply voltage and rating of controller.
Interrupting excessively high currents	Install larger device or check for grounds, shorts, or excessive motor currents. Use silver-faced contacts.
Loose connections	Clean and tighten.
Short circuits	Remove short-circuit fault and check for proper fuse or breaker size.
Sustained overload	Install larger device or check for excessive load current.
Weak contact pressure	Adjust or replace contact springs.
Panel and apparatus burned by heat from resistor	
Motor being started frequently	Use resister of higher rating.
Coil overheating	
Dirt or rust on pole faces increasing air gap	Clean pole faces.
Incorrect coil	Check rating and replace with proper coil if incorrect.
Overvoltage or high ambient temperature	Check application and circuit.
Shorted turns caused by mechanical damage or corrosion	Replace coil.
Undervoltage, failure of magnet to seal in	Correct pole faces.
Overload relays tripping	
Incorrect heater	Relay should be replaced with correct size heater unit.
Loose connection on load wires	Clean and tighten.
Sustained overload	Check for grounds, shorts, or excessive motor currents.
Overload relay fails to trip	
Mechanical binding, dirt, corrosion, etc.	Clean or replace.
Motor and relay in different temperatures	Adjust relay rating accordingly or make temperature the same for both.
Wrong heater or heaters omitted and jumper wires used	Check ratings. Apply proper heaters.

(*Continued*)

TABLE 15.6 Motor Control Troubleshooting Summary (*Continued*)

Cause	Action
Noisy magnet (humming)	
Broken shading coil	Replace shading coil.
Dirt or rust on magnet faces	Clean and realign.
Low voltage	Check system voltage and voltage dips during starting.
Magnet faces not mating	Replace magnet assembly or realign.

Note: These steps are not necessarily in any order of priority or sequence.

15.4.4 Switchgear

The switchgear items to be maintained include:

- Metal enclosures
- Busbar and terminal connections, including insulators
- Busways and ducts
- Circuit breakers, including insulation, main and arcing contacts, trip devices, auxiliary devices, and operating mechanisms
- Network protectors
- Auxiliary switchgear equipment, including fuses, capacitors, battery supply, instrument transformers, metering, alarms, indicators, and protective relaying

A periodic maintenance schedule must be established to obtain the best service from the switchgear. Annual checks should be made on all major switchgear devices after installation. After trends have been established regarding the equipment condition and reliability, the maintenance interval may be extended (18 to 36 months) in keeping with the operating conditions. Some troubleshooting procedures for switchgear equipment are given in Table 15.7. A permanent record of all maintenance work should be kept. The record should include a list of periodic checks and tests made (including date of test), condition of the equipment, repairs or adjustments performed, and test data that would facilitate performing a trend analysis. Maintenance personnel must follow all recognized safety practices and applicable nationally published standards. The following are some specific suggestions for dealing with switchgear maintenance:

1. Tools designed for slowly closing switchgear circuit breakers or other devices during maintenance are *not* suitable for use on an energized system. The speed necessary for device closing is as important as its speed in opening; therefore, a wrench or other manual tool is not fast enough.

2. Before working on a switchgear enclosure, verify that the enclosure is deenergized by checking for voltage using a voltage detector. The voltmeter with proper range can be used as a voltage detector. A Simpson multimeter or an amprobe (overhead extension type) are commonly used. Digital types of meters are easy to use. Make sure the instrument is rated for the voltage being tested.

3. Disconnect all draw-out or tilt-out devices, such as circuit breakers, instrumentation transformers, and control power transformers.

4. Do not set tools on the equipment while working. It is all too common to forget a wrench when closing up an enclosure. Don't take the chance.

5. *Never* rely upon the insulation surrounding an energized conductor to provide protection to personnel. Use suitable safety clothing and equipment.

6. Always use the correct maintenance forms and equipment when performing maintenance. The following should be available:

   a. Forms for recording the conditions "as found" and "work done."

   b. Control power connections, test couplers, and spare parts recommended by the manufacturer to facilitate repair and maintenance of each type of circuit breaker.

   c. Special tools, such as lifting mechanisms for removing and transporting power circuit breakers, relay test plugs for testing and calibrating protective relays, a low-resistance ohmmeter or bridge for measuring the resistance of contacts, ammeters, voltmeters, megohmmeters, low-voltage/high-current test sets for testing power circuit breakers, and other special test equipment.

   d. Manufacturers' instruction books regarding the maintenance of switchgear devices, such as circuit breakers, relays, bus bars, and meters.

TABLE 15.7   Switchgear Equipment Troubleshooting Summary

Cause	Action
Meters inaccurate	
C.T. circuit shorted or shorting strap left	Remove the short.
Dirt or dust may be impeding movement; particles may be adhering to the magnets	Clean, test, and calibrate meter.
Loose connections	Tighten, test, and calibrate meter.
Meter may be damaged—have a cracked jewel, rough bearing, bent disk or shaft, insufficient disk clearance, or damaged coils	Repair or replace damaged parts; test and calibrate meter.
Meters failing to register	
Blown potential transformer fuse	Ascertain reason and correct trouble. Replace blown fuses.
Broken wires or fault in connections	Repair break, correct fault.
C.T. circuit shorted or shorting strap left on	Remove the short.
Wedge or block accidentally left at time of test or inspection	Remove wedge or block, test and calibrate meter.
Damaged control, instrument transfer switch or test blocks	
Burned or pitted contacts from long use without attention or from unusual conditions	Dress or clean burned contacts, or replace with new contacts.

(Continued)

TABLE 15.7 Switchgear Equipment Troubleshooting Summary (*Continued*)

Cause	Action
Relays failing to trip breakers	
Contacts improperly adjusted	Readjust so that contacts close with proper amount of wipe.
Dirty, corroded, or tarnished contacts	Clean contacts with burnishing tool. Do not use emery or sandpaper.
Faulty or improperly adjusted timing devices	If timing device is of bellows or oil-film type, clean and adjust. If of induction-disk type, check for mechanical interference.
Improper application of target and holding coil	Target and holding coils should correspond with tripping duty of breaker to assure proper tripping.
Improper setting	Adjust setting to correspond with circuit conditions.
Open circuits or short circuits in relay connections	Check with instruments to ascertain that voltage is applied and that current is passing through relay.
Connections overheating	
Bolts and nuts in the connection joints not tight	Tighten all bolts and nuts. Too much pressure must be avoided.
Increase of current due to additional load that is beyond normal current rating of bars or cables	Increase the number or size of conductors. Remove excess current from circuit.
Failure in function of all instruments and devices having potential windings	
Blown fuse in potential transformer circuit	Replace fuses.
C.T. circuit shorted or shorting strap left	Remove the short.
Loose nuts, binding screws, or broken wire at terminals	Tighten all loose connections or repair broken wire circuits.
Open circuit in potential transformer primary or secondary	Repair open circuit and check entire circuit.
Breaker fails to trip	
Blown fuse in control circuit (where trip coils are potential type)	Replace blown fuse.
Damaged trip coil	Replace damaged coil.
Failure of latching device	Examine surface of latch. If worn or corroded, it should be replaced. Check latch wipe and adjust according to instruction book.
Faulty connections (loose or broken wire) in trip circuit	Repair faulty wiring. See that all binding screws are tight.
Mechanism binding or sticking due to lack of lubrication	Lubricate mechanism.
Mechanism out of adjustment	Adjust all mechanical devices, such as toggles, stops, buffers, opening springs, etc., according to instruction book.

Note: These steps are not necessarily in any order of priority or sequence.

15.4.5 Power Cables

Power cables are made up of three components: conductor, insulation, and protective covering. The single most important component of a cable is its insulation. The best way to assure continued reliability of a power cable is through visual inspection and electrical testing of its insulation. If touching, handling, or moving cables in manholes or at terminations is involved, then all circuits in the group to be inspected should be deenergized before the work is started.

Cables in manholes, ducts, or below-grade installations should be annually inspected for the following:

- Sharp bends in the cables
- Physical damage
- Excessive tension
- Cables lying under water
- Cable movement or dangling
- Insulation swelling
- Soft spots
- Cracked protective coverings
- Damaged fireproofing
- Poor ground connections or high impedance to ground
- Deterioration of metallic sheath bond
- Corrosion of cable supports or trays

Terminations and splices of nonlead cables should be squeezed (deenergized circuits) in search of soft spots and inspected for tracking or signs of corona. The ground braid should be inspected for corrosion and tight connections. Inspect the bottom surface of the cable for wear or scraping due to movement at the point of entrance into the manhole for spalling concrete or deterioration above ground. If the manhole is equipped with drains, these may require cleaning or, in some instances, it may be necessary to pump water from the manhole prior to entrance. Do not enter a manhole unless a test for dangerous gas has been made or adequate ventilation gives positive assurance that entry is safe. Potheads should be inspected for oil or compound leaks and cracked or chipped porcelains. The porcelain surfaces should be cleaned and, if the connections are exposed, their tightness should be checked. Since inspection intervals are normally one year or more, comprehensive records are an important part of the maintenance inspection. They should be arranged so as to facilitate comparison from one year to the next. Troubleshooting steps for overheated cables are listed in Table 15.8.

15.5 ELECTRICAL SAFETY

Electrical safety is an important part of maintenance. Chapter 14 is devoted to this topic. However, a brief discussion is presented in this section. The protection of human life is paramount. Electrical equipment can be replaced; lost production can be made up; but human life can never be recovered nor human suffering ever compensated. The principal personnel dangers from electricity are that of shock, electrocution, and/or severe burn from an electrical arc or its effects, which may be an explosion. The major con-

TABLE 15.8   Cable Overheating Troubleshooting Summary

Cause	Action
Cables in racks	
Cables closely spaced or in a location where heat is confined, such as near ceilings, etc	If constricted portion of cable run is short, fans can be set up to provide cooling.
Cables spaced horizontally affected by mutual heating	Increase space between cables. (For best cooling, minimum center-to-center distance between cables should be twice the cable diameter.)
External sources of heat	Reroute cable or remove heat source. Shield cables from heat or ventilate with fan.
Heat from lower cables in vertical racks rises and heats upper cables	Provide baffles to deflect rising warm air.
Cables in floor channels	
Mutual heating of cables that have been piled aimlessly in overcrowded floor channels	Rack cables systematically and maintain spacing necessary to minimize mutual heating.
Restriction of air circulation by solid covers on channels	Where practical, replace solid covers with perforated covers to increase air circulation.
Cables in tunnels	
External sources of heat	Force air circulation through tunnel. Insulate adequately from the external heat source.
Mutual heating of cables spaced too closely on rack	Space cables on racks to minimize mutual heating. Place cables near the floor.
Overloading	Reroute part of load from overloaded cables to cables carrying lighter loads.
Cables in underground ducts	
Overloading	Transfer load from overloaded cables to cables carrying lighter loads. Place power cables in outside ducts with most heavily loaded cables at the corners of bank. Install ventilating covers on congested manholes. (A fan to force air out through a ventilating cover may help.)
Cables buried in earth	
Overloading	Wetting dry soil improves its conductivity and may slightly improve cable capacity. Only real remedy is transferring portion of load to another circuit.
Aerial cables	
Cable in hot sun	If practical, shade from sun.
Overloads	Reduce load.
Cable risers	
Cable chosen for underground operation instead of in a conduit in air	Provide fans to cool risers during overload periods.
Exposure to sun	Shade risers, if possible.
Heating air rising and trapped at top of conduit	Provide ventilating bushing at top of conduit.

TABLE 15.8 Cable Overheating Troubleshooting Summary (*Continued*)

Cause	Action
All installations	
High current	Low-power-factor equipment, and low voltage at receiving end. Install capacitors to improve power factor.
Unbalanced currents	Unbalanced loading of phases and unbalanced arrangement of single-conductor cables in group. Raise voltage by means of taps on transformer or reduce voltage drop by moving single-conductor cables closer together. Move transformer closer to load. If load can be operated at two voltages, use higher value. Balance arrangement of single-phase loads to divide current equally among three conductors. With two or more single-conductor cables in parallel per phase, consideration must be given to phase arrangement of cables to prevent unbalanced currents.

Note: These steps are not necessarily in any order of priority or sequence.

tributors to work-related electrical accidents are unsafe conditions and unsafe practices. The most common unsafe conditions are damaged, defective, burned, or wet insulation or other parts; improperly guarded or shielded live parts; loose connections or loose strands; and equipment not grounded, or poor or inadequate grounding connections. Unsafe practices include failing to deenergize equipment, using tools or equipment too near bare energized parts, and misusing tools or equipment. In general, to improve safety to personnel and avoid accidents, special attention must be directed to the following:

1. *Be alert.*   Alertness is particularly essential on new assignments until safe habits are formed, but should never be relaxed because conditions often change.
2. *Be cautious.*   Caution should be exercised at all times.
3. *Develop safe habits.*   Safe habits result from repeated alertness and caution, and continuous adherence to the intent and the letter of the rules.
4. *Know your job.*   Have complete and thorough information before proceeding.
5. *Observe the rules.*   The rules and instructions applying to a variety of cases, both electrical and mechanical in nature, cover most of the common causes of accidents.

15.6 HOUSEKEEPING AND RECORD KEEPING

Housekeeping and record keeping are important activities of facilities management. The importance of maintaining manufacturers' data and instruction books is discussed in Chap. 18. Detailed data of equipment and their ratings should be recorded and maintained. A detailed maintenance record for major equipment should be maintained.

With the cost-effectiveness of personal computers, this record keeping can be easily implemented. If present record keeping is not satisfactory, a new program of record keeping could be implemented on microcomputers. A large and sudden expenditure is not needed. A gradual implementation of a well-developed program over a 3- to 5-year range may be suitable, depending upon the type and size of the facility.

Computer-based record keeping may be implemented by using spreadsheet software (Lotus, Symphony, Quattro, and other commercial software) and database software (e.g., dBase III or Paradox). Special-purpose maintenance and record-keeping software is also available. Trade magazines, such as *Plant Engineering,* publish lists of software and related products from time to time. Typical data items usually recorded are:

- General information
- Primary fuse
- Circuit breaker
- Relay (overcurrent, differential, voltage, frequency), type, and settings
- Transformer and its grounding
- Cable, lines, and feeders
- Equipment

On one-line diagrams, all equipment should be identified. Portions which do not apply are usually crossed out. For directional relays, tripping direction is marked. Typical data items recorded include the following:

- Customer
- Date
- Substation identification
- One-line drawing reference number
- Bus voltage
- Normal bus load (MW, MVAR)

The following four items are common to all data; hence, they will not be repeated for each item.

- Identification
- Manufacturers
- Model number
- Type

Other types of typical data recorded are shown in Tables 15.9 and 15.10.

The discussion in this section is mainly to introduce to the reader the importance of record keeping. The reader is encouraged to investigate the many options available before selecting any one system of record keeping, as well as deciding the extent and type of data recording.

TABLE 15.9 Example of Data Items Typically Recorded

Primary fuse data	Breaker data	Transformer data
Ampere rating	Interrupting time (number of cycles)	kVA rating
Volt rating	Interrupting rating (symmetrical/asymmetrical)	Type of cooling
Interrupting rating	Current and voltage rating	Primary & secondary voltage
	MOM rating (close & latch)	Impedance (%)
	CT & PT ratios	Type of connection (Y or Δ) and vector diagram
	BIL	Taps
		X/R (if available)
		Grounding type and its rating
		BIL

TABLE 15.10 Example of Data Items Typically Recorded

Cable/line/feeder data	Motors & rotating equipment	Relays
Number	Feeder connection I.D.	Protected equipment identification
Single-phase, 3-phase, or other configuration	Quantity	Relay type, model, characteristics
Length	Type (induction/synchronous/other)	Range of adjustments
Conductor size & type	hp/kW	PT/CT ratios
Insulation	Voltage rating	Range of adjustments
Conduit information	Reactance (locked rotor current)	Settings (current/voltage/time/time delay)
	X/R (if available)	Identification of tripped breaker
	Starting type	
	Normal starting time	
	Motor load, 50 hp or larger	
	Total motor load	

15.7 REFERENCES

1. S. Nakajima, *Introduction to Total Productive Maintenance,* Productivity Press, Cambridge, Mass., 1988.

2. R. K. Mobley, *An Introduction to Predictive Maintenance,* Van Nostrand Reinhold, New York, 1990.

3. A. S. Corder, *Maintenance Management Techniques,* McGraw-Hill Book Company (UK) Ltd., 1976.

4. A. S. Gill, *Electrical Equipment Testing and Maintenance,* Prentice Hall, Inc., Englewood Cliffs, N.J., 1982.

5. R. L. Dunn, "Maintenance of Continuous Processes," *Plant Engineering,* August 16, 1990, pp. 71–76.

6. W. M. Windle, "TPM: More Alphabet Soup or a Useful Plant Improvement Concept?" *Plant Engineering,* February 4, 1993, pp. 61–63.

7. R. L. Dunn, "How Much is Enough?" *Plant Engineering,* April 22, 1993, p. 67.

8. E. C. Wordehoff, "Using Maintenance Benchmarking: Can it Really Lead to Cost Reduction," *Plant Engineering,* October 8, 1992, pp. 114–115.

9. J. L. Foszcz, "Texas Instruments Goal is Maintaining Total," *Plant Engineering,* March 4, 1993, pp. 68–72.

10. M. H. Bos and M. V. Brown, "Get Both Operations and Maintenance Involved in Setting Workload Priorities," *Plant Engineering,* September 17, 1992, p. 142.

11. C. Poxton, "Infrared-Thermographic Inspection Improves Service Reliability," *Transmission & Distribution,* April 1992, pp. 32–35.

12. S. Shores and J. Hollis, "Reliability Based Maintenance," *Proceedings of Inter-RAMQ Conference for the Electric Power Industry,* August 25–28, 1992, pp. 280–285.

CHAPTER 16
ENGINEERING ECONOMICS

16.0 INTRODUCTION

Engineers are not only involved in the production of products, equipment, and services on a sound technical basis (including safety and reliability), but they also have to ensure that these have the lowest possible cost. Every engineer who is involved in a project, design, or manufacturing realizes that the final selection from competing alternatives is based on least cost. Note that, at this point, we have not defined what is the least cost or the method of comparing the alternatives. The cost may be the per-unit cost of a product or device, the capital cost, operating cost, etc. The least cost to whom is not necessarily self-evident unless it is specifically defined. The comparison of alternatives may be a simple evaluation of initial cost or a complex evaluation including probability of risk, social acceptance, etc. In any case, proving the cost-effectiveness is as important as the technical feasibility and reliability.

Before embarking on the task of economic evaluation, the overall nature of the problem needs to be reviewed and appropriate assumptions have to be made. Proper methods of economic evaluation should be selected. As compared to engineering systems or problems, economic systems or problems are difficult to define and solve. This is partly due to the human behavior and/or response factor which is inherent in the outcome of economics.

The economic problems associated with analyzing venture capital and repair/replace and marketing strategies are all somewhat different. The basic economic problem may be that of deciding how to optimize the allocation of severely restricted resources to promote the most rapid overall balanced economic growth. In an industrial firm, the economic problem of the moment may be how to best promote revenue growth while still maintaining an adequate profit margin. The analysis of larger economic problems of profitability of a business firm is much more difficult when compared to the day-to-day engineering problems. For engineers, one of the most frequently faced economic evaluation problems is that of selecting between alternate patterns of future capital investments. Engineers and managers are continually faced with choices between alternate designs, plants, and equipments being offered by various vendors. The capital investment decision problem is the single most important economic evaluation problem facing the engineering staff.

In this chapter, we will introduce the common terms used in engineering economics. Some of the methods used in the engineering economic analyses and comparison of alternatives will be presented. Some examples will be used for illustration of different methodologies.

This chapter deals only with engineering economics evaluation but not with tax- or accounting-related issues. When we use the term "engineering economics," there is an underlying assumption that, irrespective of the selected alternative (including doing nothing), the financial position of the business firm is not altered. This means that problems such as business diversification, major corporate strategic maneuvers, abandoning major parts of the business, or embarking on a single project which will increase total capitalization of the company are not part of the engineering economic evaluation.

16.1 TERMINOLOGY

Engineering economics, like any specialized field, has its own language. The commonly used terms will be described here. A broad cost structure can be represented as in Fig. 16.1. The percentages of each item may vary from project to project. Many cost items may have both fixed and variable cost components. It may not always be possible to clearly differentiate and assign these costs. Practicing engineers should bear in mind that the amount of effort to be expended in clearly identifying cost components should be commensurate with the importance of the particular cost item to the overall cost of the item as well as to the outcome of the economic analysis.

Amortization   The distribution of the initial cost by periodic charges to operational expenses. Home mortgage monthly payments to reduce debt and cover interest charges is a very good example of amortization.

Annuity   A series of equal (could be variable) payments occurring at equal intervals.

Book life   The expected (or average) service life of a plant or equipment.

Book value   The current value of a plant or equipment.

Breakeven analysis   Determining at what point the alternatives are economically equivalent as underlying parameters are varied.

Capital   The financial resources for establishing and operating a plant or equipment.

Capital recovery   Amount periodically charged or recovered from operations so that, ultimately, this total is equal to the amount of capital expended.

Direct cost   The direct costs of material and labor are those costs which are directly (and, of course, easily) identifiable and allocated to a specific product, operation, service, or project. For example, the steel used as raw material in an automobile is a

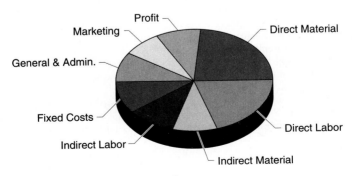

FIGURE 16.1   Components of operating costs.

direct material cost. The cost of labor involved in machining, painting, and packing a product or part is a direct labor cost.

Discount rate (present worth factor)   The interest rate used for evaluating the effect of the time value of money. This is also typically the cost of capital.

Escalation   Annual rate of increase of the price of a commodity, labor, or service.

Fixed cost   This type of cost does not vary with the quantity of output. Some examples of this type of fixed costs are mortgage payments, rent, insurance, real estate taxes, and administration. However, step changes are possible. For example, a company may close a plant, sell an extra building, or shut down heat and air-conditioning for unused parts of a building.

Installed cost   This is sometimes referred to as *capital cost.* The installed cost includes the purchase price, shipping, installation, construction, etc. In some instances, the cost of specialized training required to operate specific machinery may be included in the installed cost. All cost items which need to be capitalized may be included in the installed cost item. The income tax angle may also influence the number of items segregated and/or included in the installed cost. Installed cost is a one-time cost. It is also referred to as *first cost.*

Life cycle costs   As previously discussed, owning an equipment or facility involves capital cost (installed cost), operating and maintenance (O&M) cost, and disposal cost. The installed cost is a one-time cost, whereas the O&M cost is a recurring cost at the same or different levels and different frequencies. It is very common in practice to see a low-installed-cost alternative requiring a higher O&M cost and a higher-installed-cost alternative requiring a lower O&M cost. This is shown in Fig. 16.2. The life cycle cost is the total cost and has an optimum value. The purpose of the economic analysis is to find the alternative corresponding to this optimum or near this optimum, everything else being equal. The life cycle cost is a type of summation (discussed later) of capital and O&M cost, but is not a straight arithmetic sum.

Operating and maintenance costs   As evident from the name itself, these costs include material, labor, and overhead items. The direct material and labor costs may

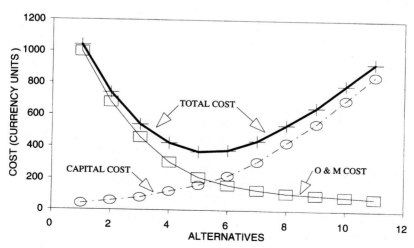

FIGURE 16.2   Life cycle cost.

be segregated separately from the indirect costs. Preventive maintenance and repair costs are part of O&M costs. O&M costs may include both fixed and variable costs. As can be seen from this description, the O&M cost could be a major component of the total cost of a product, equipment, or service.

Opportunity cost   Any project or product requires the investment of capital. The same capital may be invested in other products or other types of investments for a higher profit. This higher return represents opportunity cost. For example, an individual has $1000 in a simple passbook account at 5 percent interest. However, a certificate of deposit (say for one year) yields 6 percent. The additional 1 percent represents an opportunity cost with no additional risk because both accounts are insured. Of course, there is a penalty for early withdrawal for the certificate of deposit. Economic analysis has to deal with this possible penalty vs. the opportunity cost.

Overhead cost   This is also referred to as *overhead burden*. The costs included in this category may vary from company to company and include:

- indirect material cost
- indirect labor cost
- facilities cost
- insurance cost
- real estate and other business taxes
- income tax
- general and administrative costs

It is evident that the overhead cost may be used as a catchall category. The accounting department usually has a method in place to assign and categorize these costs. In the economic analysis, the engineers usually follow the accounting department's categories. Otherwise, it would be very difficult to properly assign various costs in a meaningful pattern.

The indirect material cost includes cost items such as paint used for a part, packaging material, etc. The labor used in a purchasing department, the loading dock, and delivery of parts are examples of *indirect labor*. The *facilities cost* would typically include the cost of owning or renting the space/building, water, electricity, etc. Some of these items may be included in the general and administrative overhead. In some companies, the labor overhead to account for Social Security taxes, medical insurance, and other employee benefits may be accounted as a separate category. The cost of marketing, if it is a major component, may be recognized as a separate item rather than being included as an overhead expense. The *cost of insurance* for the property, liability, etc., *taxes* of various kinds, and other *general and administrative costs* may be part of indirect costs.

Payback period   The time required (usually in months or years) to recover the initial capital, without regard to the time value of money.

Rate of return   The amount (expressed as a percent) earned (or saved) from an investment.

Salvage value   The net market value of a plant or equipment at the end of the estimated life. This can be viewed as the trade-in value of a piece of equipment. Sometimes there is a disposal cost associated with retiring or replacing a piece of equipment. Any disposal cost should be factored into the calculation of salvage value. The term "useful life" means the economic life. A piece of equipment may still be in working condition, but it may not be useful to the owner. It is possible to end up with a negative value for salvage value, implying a liability, such as environmental cleanup, decommissioning costs, or demolition costs, which is higher than the remaining value.

Service life   The number of years (or other suitable units of time) the plant or equipment is in service. Note the difference between book life and service life.

Sunk cost   The expenditures already incurred (past costs) that are unrecoverable. Usually, sunk costs are not considered to be important in making future decisions. However, there are exceptions. For example, sunk cost may be used as capital losses for income tax purposes and, thereby, may reduce the tax liability of the company.

Variable costs   Variable cost is directly proportional to output. For example, direct material and labor costs belong to this category. In an electric power company, the amount of fuel used is proportional to the amount of electric energy generated.

16.2   PRINCIPAL AND INTEREST

A given sum of money at the present time is called the *principal amount.* Let us say that you have $1000 to lend to your friend. At the end of one year, your friend will pay back $1050. The interest rate in this case is 5 percent per year. At the end of two years you will get $1100. The interest for the second year is also $50 on the original loan of $1000. This is called *simple interest.* The total amount is computed by

$$A = P\left(1 + \frac{ni}{100}\right) \tag{16.1}$$

where P = principal
  n = number of years
  i = simple interest rate in percent per year

In the preceding example, if the lender took $1050 from the first borrower and lent it again at the same 5 percent interest rate, he would have $1102.50 at the end of the second year. The additional $2.50 represents 5 percent interest on the $50. This type of interest computation is called *compound interest* and is computed by the formula

$$A = P\left(1 + \frac{i}{100}\right)^n \tag{16.2}$$

If the compounding period is less than one year, say m times a year, then the formula is,

$$A = P\left(1 + \frac{i}{m \times 100}\right)^{nm} \tag{16.3}$$

because interest rate i is expressed as percent per year.

One useful rule of thumb for compound interest calculations, an easy-to-remember formula, is called the Rule of 72. This states that the duration for an initial investment to double its value is obtained by

$$\text{Time} = \frac{72}{\text{interest rate}} \tag{16.4}$$

This obviously implies that if the time period is known, then the interest rate can be computed by

$$\text{Interest rate} = \frac{72}{\text{period}} \tag{16.5}$$

For example, at a 12 percent interest rate, the initial investment will double in six years. Given ten years, a 7.2 percent compounding rate is required for doubling the initial value. This rule applies to any compounding problem, such as load growth or sales growth.

EXAMPLE 16.1   *A person invests $1000 at a 10 percent interest rate for 10 years. Calculate the total amount at the end of 10 years (a) based on simple interest, (b) compounded annually, (c) compounded quarterly, and (d) compounded daily.*

(*a*) Based on simple interest, amount

$$A = 1000 \left(1 + \frac{10}{100} \times 10 \right)$$

$$= \$2000$$

(*b*) Based on annual compounding, amount

$$A = 1000 \left(1 + \frac{10}{100} \right)^{10}$$

$$= \$2593.74$$

(*c*) Based on quarterly compounding, amount

$$A = 1000 \left(1 + \frac{10}{4 \times 100} \right)^{40}$$

$$= \$2685.06$$

When there are multiple compounding periods in a year, an equivalent interest rate may be computed. This is called the *annual percentage rate* (APR) and is required to be posted by lending institutions (the Truth-in-Lending Law). The APR is calculated as follows:

$$\text{APR} = \left[\left(1 + \frac{i}{m \times 100} \right)^{m} - 1 \right] \times 100$$

Thus, the APR in this case is 10.38 percent, whereas the nominal interest rate is 10 percent.

(*d*) Based on daily compounding, amount

$$A = 1000 \left(1 + \frac{10}{365 \times 100} \right)^{3650}$$

$$= \$2717.91$$

$$\text{APR} = 10.52 \text{ percent}$$

16.3   TIME VALUE OF MONEY

The time value of money is the most fundamental concept in engineering economics. The basic concept is that a dollar has time-varying earning (buying) power, hence, value. Example 16.1 showed the impact of interest and compounding.

16.3.1   Present Worth

Consider the compound-interest-based formula given by Eq. (16.2). In this equation, the future amount A is based on the present principal P, interest rate i, and compounding periods of n. This equation may be rewritten to compute the present value (or present worth denoted by PW, or net present value NPV), as follows:

$$PW = \frac{A}{\left(1 + \dfrac{i}{100}\right)^n} \tag{16.6}$$

EXAMPLE 16.2   *A student needs $2000 by high school graduation. How much money should be deposited at the beginning of a four-year high school term in an account yielding 5 percent, compounded quarterly?*

$$PW = \frac{2000}{(1 + 5/(4 \times 100))^{16}} = \$1639.49$$

where $A = \$2000$
$\quad\ \ i = 5\%$
$\quad\ \ n = 4$
$\quad\ \ m = 4$

In this example, only one single sum of money was involved. However, in practice, the cash flow occurs at different instances and at different levels. The most general cash flow is shown in Fig. 16.3. The present worth for this cash flow is given by

$$PW = \frac{A_1}{(1 + i/100)} + \frac{A_2}{(1 + i/100)^2} + \cdots + \frac{A_n}{(1 + i/100)^n} \tag{16.7}$$

If it is necessary to compute the value of this cash flow at the end of n years,

$$A = A_1\left(1 + \frac{i}{100}\right)^{n-1} + A_2\left(1 + \frac{i}{100}\right)^{n-2} + \cdots + A_n \tag{16.8}$$

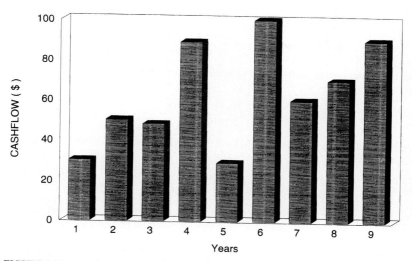

FIGURE 16.3   An example of cash flow.

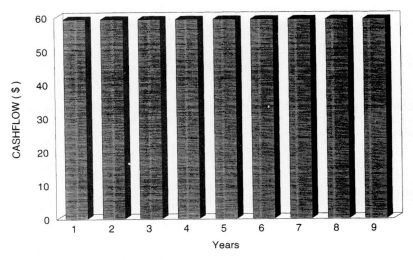

FIGURE 16.4   An example of uniform cash flow

In Fig. 16.3 and Eqs. (16.7) and (16.8), cash flow $A_1, A_2..., A_n$ were assumed to be at the end of the years (or start of the following year). This needs to be stated clearly. If all the cash flows occurred at the beginning of each year including the current year (year 0), the terms in Eqs. (16.7) and (16.8) will change by a factor of $(1 + i/100)$. Attention to this type of detail is important during the economic analysis.

The general cash flow diagram in Fig. 16.3 may be redrawn as shown in Fig. 16.4 for a uniform series of cash flow at regular intervals. These two cash flows are equivalent on a present worth basis.

Then Eq. (16.7) can be rewritten as

$$PW = A \left[\frac{1}{(1 + i/100)} + \frac{1}{(1 + i/100)^2} + \cdots \frac{1}{(1 + i/100)^n} \right] \qquad (16.9)$$

This equation is in the well-known geometric series form and can be rewritten as a sum of n terms:

$$PW = A \left[\frac{(1 + i/100)^n - 1}{(i/100)(1 + i/100)^n} \right] \qquad (16.10)$$

If the present worth value is known, then the regular periodic payment amount may be computed by

$$A = PW \left[\frac{(i/100)(1 + i/100)^n}{(1 + i/100)^n - 1} \right] \qquad (16.11)$$

The term within the parentheses in Eq. (16.11) is called the *capital recovery factor*. This fraction gives the equivalent uniform payment when multiplied by the principal.

EXAMPLE 16.3   *We will use the most familiar home mortgage monthly payment example to illustrate the aforementioned concept. Consider a homeowner who borrows*

$100,000 at a 10 percent fixed interest rate. The period for the loan is 30 years. (a) What is the monthly payment? (b) If the mortgage were for 15 years, what would be the monthly payment?

$$PW = 100,000$$

$$\text{Interest rate} = 10 \text{ percent per annum}$$

$$= \frac{10}{12} \text{ percent per period}$$

$$\text{Number of periods} = 360 \text{ months}$$

(a)

$$\text{Monthly payment}$$

$$= \frac{100,000}{[(1 + 10/(12 \times 100))^{360} - 1/(10/(12 \times 100))(1 + 10/(12 \times 100))^{360}]}$$

$$= \$877.57$$

(b)

$$\text{Monthly payment}$$

$$= \frac{100,000}{[1(+ 10/(12 \times 100))^{180} - 1/(10/(12 \times 100))(1 + 10/(12 \times 100))^{180}]}$$

$$= \$1074.61$$

Because of availability of spreadsheets on personal computers, these calculations are easy to make. The standard financial functions are available for calculations. A 15-year mortgage is shown in spreadsheet format in Table 16.1.

EXAMPLE 16.4 *Compute the present worth of the following series of cash flows. Assume each cash flow to be at the end of the year.*

Year	Amount
1	11,000
2	8,000
3	28,000
4	4,000
5	45,000
6	3,000
7	15,000
8	0
9	82,000
10	9,000

Use an interest rate of 10 percent per year. The solution is shown in spreadsheet Table 16.2.

It can be readily seen that the difference between the present values in the two columns is 10% which is the discount rate used in this example.

TABLE 16.1 Example of a 15-Year Mortgage*

Month	Principal	Interest	Balance	Month	Principal	Interest	Balance	Month	Principal	Interest	Balance
1	241.28	833.33	99,758.72	61	396.97	677.64	80,919.32	121	653.14	421.47	49,922.68
2	243.29	831.32	99,515.43	62	400.28	674.33	80,519.04	122	658.59	416.02	49,264.09
3	245.31	829.30	99,270.12	63	403.62	670.99	80,115.42	123	664.08	410.53	48,600.01
4	247.36	827.25	99,022.76	64	406.98	667.63	79,708.44	124	669.61	405.00	47,930.40
5	249.42	825.19	98,773.34	65	410.37	664.24	79,298.07	125	675.19	399.42	47,255.21
6	251.50	823.11	98,521.84	66	413.79	660.82	78,884.28	126	680.82	393.79	46,574.39
7	253.59	821.02	98,268.25	67	417.24	657.37	78,467.04	127	686.49	388.12	45,887.90
8	255.71	818.90	98,012.54	68	420.72	653.89	78,046.32	128	692.21	382.40	45,195.69
9	257.84	816.77	97,754.70	69	424.22	650.39	77,622.10	129	697.98	376.63	44,497.71
10	259.99	814.62	97,494.71	70	427.76	646.85	77,194.34	130	703.80	370.81	43,793.91
11	262.15	812.46	97,232.56	71	431.32	643.29	76,763.02	131	709.66	364.95	43,084.25
12	264.34	810.27	96,968.22	72	434.92	639.69	76,328.10	132	715.57	359.04	42,368.68
13	266.54	808.07	96,701.68	73	438.54	636.07	75,889.56	133	721.54	353.07	41,647.14
14	268.76	805.85	96,432.92	74	442.20	632.41	75,447.36	134	727.55	347.06	40,919.59
15	271.00	803.61	96,161.92	75	445.88	628.73	75,001.48	135	733.61	341.00	40,185.98
16	273.26	801.35	95,888.66	76	449.60	625.01	74,551.88	136	739.73	334.88	39,446.25
17	275.54	799.07	95,613.12	77	453.34	621.27	74,098.54	137	745.89	328.72	38,700.36
18	277.83	796.78	95,335.29	78	457.12	617.49	73,641.42	138	752.11	322.50	37,948.25
19	280.15	794.46	95,055.14	79	460.93	613.68	73,180.49	139	758.37	316.24	37,189.88
20	282.48	792.13	94,772.66	80	464.77	609.84	72,715.72	140	764.69	309.92	36,425.19
21	284.84	789.77	94,487.82	81	468.65	605.96	72,247.07	141	771.07	303.54	35,654.12
22	287.21	787.40	94,200.61	82	472.55	602.06	71,774.52	142	777.49	297.12	34,876.63
23	289.60	785.01	93,911.01	83	476.49	598.12	71,298.03	143	783.97	290.64	34,092.66
24	292.02	782.59	93,618.99	84	480.46	594.15	70,817.57	144	790.50	284.11	33,302.16
25	294.45	780.16	93,324.54	85	484.46	590.15	70,333.11	145	797.09	277.52	32,505.07
26	296.91	777.70	93,027.63	86	488.50	586.11	69,844.61	146	803.73	270.88	31,701.34
27	299.38	775.23	92,728.25	87	492.57	582.04	69,352.04	147	810.43	264.18	30,890.91
28	301.87	772.74	92,426.38	88	496.68	577.93	68,855.36	148	817.19	257.42	30,073.72
29	304.39	770.22	92,121.99	89	500.82	573.79	68,354.54	149	824.00	250.61	29,249.72
30	306.93	767.68	91,815.06	90	504.99	569.62	67,849.55	150	830.86	243.75	28,418.86

Month	Principal	Interest	Balance	Month	Principal	Interest	Balance	Month	Principal	Interest	Balance
31	309.48	765.13	91,505.58	91	509.20	565.41	67,340.35	151	837.79	236.82	27,581.07
32	312.06	762.55	91,193.52	92	513.44	561.17	66,826.91	152	844.77	229.84	26,736.30
33	314.66	759.95	90,878.86	93	517.72	556.89	66,309.19	153	851.81	222.80	25,884.49
34	317.29	757.32	90,561.57	94	522.03	552.58	65,787.16	154	858.91	215.70	25,025.58
35	319.93	754.68	90,241.64	95	526.38	548.23	65,260.78	155	866.06	208.55	24,159.52
36	322.60	752.01	89,919.04	96	530.77	543.84	64,730.01	156	873.28	201.33	23,286.24
37	325.28	749.33	89,593.76	97	535.19	539.42	64,194.82	157	880.56	194.05	22,405.68
38	328.00	746.61	89,265.76	98	539.65	534.96	63,655.17	158	887.90	186.71	21,517.78
39	330.73	743.88	88,935.03	99	544.15	530.46	63,111.02	159	895.30	179.31	20,622.48
40	333.48	741.13	88,601.55	100	548.68	525.93	62,562.34	160	902.76	171.85	19,719.72
41	336.26	738.35	88,265.29	101	553.26	521.35	62,009.08	161	910.28	164.33	18,809.44
42	339.07	735.54	87,926.22	102	557.87	516.74	61,451.21	162	917.86	156.75	17,891.58
43	341.89	732.72	87,584.33	103	562.52	512.09	60,888.69	163	925.51	149.10	16,966.07
44	344.74	729.87	87,239.59	104	567.20	507.41	60,321.49	164	933.23	141.38	16,032.84
45	347.61	727.00	86,891.98	105	571.93	502.68	59,749.56	165	941.00	133.61	15,091.84
46	350.51	724.10	86,541.47	106	576.70	497.91	59,172.86	166	948.84	125.77	14,143.00
47	353.43	721.18	86,188.04	107	581.50	493.11	58,591.36	167	956.75	117.86	13,186.25
48	356.38	718.23	85,831.66	108	586.35	488.26	58,005.01	168	964.72	109.89	12,221.53
49	359.35	715.26	85,472.31	109	591.23	483.38	57,413.78	169	972.76	101.85	11,248.77
50	362.34	712.27	85,109.97	110	596.16	478.45	56,817.62	170	980.87	93.74	10,267.90
51	365.36	709.25	84,744.61	111	601.13	473.48	56,216.49	171	989.04	85.57	9,278.86
52	368.40	706.21	84,376.21	112	606.14	468.47	55,610.35	172	997.29	77.32	8,281.57
53	371.47	703.14	84,004.74	113	611.19	463.42	54,999.16	173	1,005.60	69.01	7,275.97
54	374.57	700.04	83,630.17	114	616.28	458.33	54,382.88	174	1,013.98	60.63	6,261.99
55	377.69	696.92	83,252.48	115	621.42	453.19	53,761.46	175	1,022.43	52.18	5,239.56
56	380.84	693.77	82,871.64	116	626.60	448.01	53,134.86	176	1,030.95	43.66	4,208.61
57	384.01	690.60	82,487.63	117	631.82	442.79	52,503.04	177	1,039.54	35.07	3,169.07
58	387.21	687.40	82,100.42	118	637.08	437.53	51,865.96	178	1,048.20	26.41	2,120.87
59	390.44	684.17	81,709.98	119	642.39	432.22	51,223.57	179	1,056.94	17.67	1,063.93
60	393.69	680.92	81,316.29	120	647.75	426.86	50,575.82	180	1,065.74	8.87	(1.81)

*Amount of loan = $100,000   Interest rate = 10%
Loan period = 15 years   Monthly payment = $1,074.61

16.11

TABLE 16.2 Spreadsheet for Solution of Example 16.4

Interest rate = 10%

Year	Cash flow	Present value a*	bt
1	11,000.00	$11,000.00	$10,000.00
2	8,000.00	7,272.73	6,611.57
3	28,000.00	23,140.50	21,036.81
4	4,000.00	3,005.26	2,732.05
5	45,000.00	30,735.61	27,941.46
6	3,000.00	1,862.76	1,693.42
7	15,000.00	8,467.11	7,697.37
8	0.00	0.00	0.00
9	82,000.00	38,253.61	34,776.00
10	9,000.00	3,816.88	3,469.89
	Total	127,554.44	115,958.59

*Assuming cash flow to be at the beginning of the year

†Assuming cash flow to be at the end of the year

16.3.2 Inflation/Escalation

Inflation is a general word used to describe the reduced purchasing power of currency. Some common examples of inflation-related indices familiar to the general public are the Consumer Price Index (CPI) reported by the U.S. Commerce Department, the Wholesale Index, and the Cost of Living Adjustment (COLA). These indices are a blend of different items. Each of the items such as commodities, industrial products, labor cost, and taxes increase at different rates. The increase in prices of individual items or a group of items is called *escalation,* as compared to *inflation,* which describes the general economy (as denoted by the above-mentioned indices).

Escalation impacts economic analysis in different ways:

1. The expenditure for the same items increases in future years. In common terminology, the same dollar buys less in the future unless there is a price drop. Hence, in computing expenditures in future years, proper escalation should be factored into the calculation. This requires forecasting ability of future prices. We will assume that such capability is available, however imperfect it may be. In practice, different escalation rates for major components, such as direct material and labor, are used and a common escalation rate is used for remaining miscellaneous items.

2. The general inflation rate affects the interest rate of borrowing money. Hence, this has an impact on the time value of money or in the calculation of present worth. The cost of borrowing money is called the *cost of capital.* This has three components:

Real interest rate

Inflation adjustment

Risk premium

Historical analysis of interest rate and inflation shows a base interest rate. This is called the *real interest rate.* Inflation adjustment accounts for changes in the real purchasing power of the currency. When borrowing money for a long term, usually the interest rate is fixed for the entire period. A fixed-rate home mortgage is a common example. In this situation, if the interest rate goes up, then the lender faces the risk of either getting back less than the original amount if the note is sold to a third party, or getting a lower interest (original rate) than the current market rate. Hence, a risk premium is usually demanded by the lenders. In some instances, inflation is not used in calculating future cash flows. In such a case, the inflation adjustment component should not be used in calculating the interest rate. This is also referred to as analysis using *real dollars.*

In many large companies, the financial department, accounting departments, and economists can provide estimates for escalation, inflation, and interest rates. For large projects, properly documented and established rates for these factors, as well as agreement with management, are very important. There are publications which put out indices for construction, etc.[1-8]

16.4 ECONOMIC ANALYSIS

Before presenting different methods of economic analysis, it is instructive to list systematic steps that are used in selecting a most economic alternative. These steps are:

1. Define mutually exclusive alternatives.
2. Perform a technical/engineering analysis so that the alternatives meet the performance requirements.
3. Determine all the costs (year by year) over the period of the study horizon.
4. Compare different alternatives.
5. Perform sensitivity analysis (Sec. 16.7).
6. Select a preferred alternative.

These steps are performed iteratively until a clear picture emerges. They provide a means for quantifying the various aspects of different alternatives. However, there are many other issues, such as risk, investment needs, safety, environment, and the marketplace. The final decision needs to factor in all these nonquantitative issues. However, in this chapter we are dealing only with economic quantification issues so as to make the final selection easier.

The main purpose of engineering economic analysis is to compare different alternatives. This comparison will be valid and meaningful only if:

- The alternatives, technically speaking, perform the same function. (If the alternatives have minor differences in performance, then adjustments are made in the costs to bring alternatives to the same level of performance.)
- The cost assumptions are all consistent.

In addition, all the relevant cash flows occurring at different times throughout the life of the alternatives should be included. A method to convert these different cash flows at different times to one or a few indices is required. This may be called *equivalencing the alternatives. Computing present worth* is one of the methods of equivalencing.

There are several methods of comparing different alternatives. Some of these methods are:

- Present worth
- Benefit-to-cost ratio

- Payback period
- Rate of return

Sometimes more than one method is used so that a better understanding of all the implications is achieved.

16.4.1 Present Worth Method

The present worth procedure, as described in Sec. 16.3.1, may be used to calculate *net present value* (NPV) for each of the alternatives. The lowest NPV will be the most economical alternative. The annual cash flows should include all relevant costs. A life cycle cost approach is usually the most comprehensive way of approaching the annual cost calculations. One of the characteristics of the present worth method is to postpone the capital investment as far into the future as possible. This method will be illustrated through an example.

EXAMPLE 16.5 *An electrical engineer has an application for a 50-hp three-phase induction motor. There are two choices: a standard motor with 92 percent efficiency at an initial cost of $2400 and a low-loss motor with 95 percent efficiency at a cost of $3000. The data for these two motors are given below. Which motor is more economical?*

	Losses	
	Standard motor	Low-loss motor
Capital cost	2400	3000
Interest rate	10.00%	10.00%
Useful life (yrs)	10	10
Annual payment	390.59	488.24
Discount rate	10.00%	10.00%
Loading	80.00%	80.00%
HP output	40	40
kW output	53.6	53.6
Efficiency	0.92	0.95
Total kW loss	4.29	2.68
Run hours/day	16	16
Total kWh loss (per year)	25,053.6	15,651.2
Energy cost ($/kWh)	0.05	0.05

Let us make the following assumptions to solve this problem:

Escalation in cost of electricity	5% per year
Interest rate	10%
(Present worth factor/discount factor)	
Motor loading (constant)	80%
Motor running, 365 days	16 h per day
Useful life	10 years
Ignore demand loss reduction	

There are only two main cost components in this problem, viz., the annual cost due to capital and the cost of losses. Let us compute losses first.

	Losses	
	Standard motor	Low-loss motor
hp output	$0.8 \times 50 = 40$	$0.8 \times 50 = 40$
kW output	$\dfrac{40}{0.746} = 53.6$	$\dfrac{40}{0.746} = 53.6$
kW loss	$53.6 \times 0.08 = 4.29$	$53.6 \times 0.05 = 2.68$
kWh loss per year	$4.29 \times 16 \times 365 = 25{,}053.6$	$2.68 \times 16 \times 365 = 15{,}651.2$

The initial capital has to be recovered in 10 years. Using the capital recovery formula,

$$CR = \left[\frac{(i/100)(1 + i/100)^n}{(1 + i/100)^n - 1} \right]$$

we get annual cost due to capital

Standard motor $390.59
Low-loss motor $488.24

The remaining analysis is shown in a spreadsheet format in Table 16.3. It is easy to do these calculations in a spreadsheet.

16.4.2 Benefit-to-Cost Ratio Method

In this method, as the name implies, the ratio of the present worth of all the benefits to the present worth of all the costs is calculated. If this ratio is greater than one, then the benefits exceed the costs. Its simplicity has a certain appeal, placing a dollar value on

TABLE 16.3 Economic Comparison for Selecting Motors

	Standard motor				Low-loss motor				
Year	Annual payment	Cost of losses	Total cost	Total PW	Annual payment	Cost of losses	Total cost	Total PW	Savings
1	390.59	1,252.68	1,643.27	1,493.88	488.24	782.56	1,270.80	1,155.27	338.61
2	390.59	1,315.31	1,705.90	1,409.84	488.24	821.69	1,309.93	1,082.59	327.25
3	390.59	1,381.08	1,771.67	1,331.08	488.24	862.77	1,351.01	1,015.04	316.05
4	390.59	1,450.13	1,840.72	1,257.24	488.24	905.91	1,394.15	952.22	305.02
5	390.59	1,522.64	1,913.23	1,187.97	488.24	951.21	1,439.45	893.78	294.18
6	390.59	1,598.77	1,989.36	1,122.94	488.24	998.77	1,487.01	839.38	283.57
7	390.59	1,678.71	2,069.30	1,061.88	488.24	1,048.71	1,536.95	788.70	273.18
8	390.59	1,762.65	2,153.24	1,004.50	488.24	1,101.14	1,589.38	741.46	263.04
9	390.59	1,850.78	2,241.37	950.56	488.24	1,156.20	1,644.44	697.40	253.16
10	390.59	1,943.32	2,333.91	899.82	488.24	1,214.01	1,702.25	656.29	243.53
			Total	11,719.71			Total	8,822.12	2,897.59

benefits is not very easy. However, a relative ranking of this benefit-to-cost ratio can be quite useful by providing the incremental value of different alternatives.

EXAMPLE 16.6   *Compute the benefit-to-cost ratio for the problem in Example 16.5.*

$$\text{Benefits or savings} = \$2897.59 \text{ (from last column in Table 16.3)}$$

$$\text{Cost} = 3000 - 2400 = \$600$$

$$\text{Ratio} = \frac{2897.59}{600} = 4.8$$

16.4.3 Payback Method

The *payback period* is defined as the time it takes to recover the initial capital outlay. The preferred alternative is one with the shortest payback period. In this method, a dollar is a dollar, irrespective of the time at which this expenditure (or income) occurs. However, it is possible to compute payback period by using present worth of annual costs as well as escalation.

EXAMPLE 16.7   *Compute the payback period for the problem in Example 16.5.*

The additional capital for the low-loss motor is $600 (3000−2400). The annual savings in energy cost without factoring escalation = $1252.68−782.68 = 470.12

$$\text{Payback period} = \frac{600}{470.12} = 1.3$$

As can be seen, this approach has the appeal of simplicity. There is not much confusion about various factors, such as segregation of capital and operating cost, time value of money, and interest rates. In simple cases, this method is useful. Also, this method can give some indication of risk quantification. For example, let us say that the motor selection has to be made by a small machine shop owner. Whether an additional $600 should be spent depends upon the payback period. The longer the payback, the higher the probability of the business going down and the additional investment being lost. In this example, the breakeven is within 18 months, which is very good for buying a low-loss motor.

16.4.4 Rate-of-Return Method

The *rate of return* is defined as that discount rate which converts the annual cash flows equal to the initial required investment. This is expressed as

$$\text{Initial investment} = \frac{A_1}{(1+i)} + \frac{A_2}{(1+i)^2} + \cdots \frac{A_n}{(1+i)^n} \qquad (16.12)$$

where A_n = cash flow at the nth year
       i = internal rate of return to be computed

When the cash flow is the same throughout the year, the right-hand side of the equation is a geometric series, and the rate of return (sometimes also called the *internal rate of return*, or IRR) can be easily determined. However, when this series is not uniform (it may be different each year, including negative cash flows), then computing the IRR is not easy. However, the spreadsheet programs have ready-made functions to compute the IRR.

This method is not necessarily useful to select alternatives based on simple ranking. In practice, usually a lower-net-present-worth alternative has a higher initial investment. Even though the net present worth is lower, this does not necessarily mean a higher initial investment is justified. However, it is possible to compute the net savings differential between alternatives and compute the rate of return for the additional capital investment. If this rate of return is higher than the acceptable or minimum rate of return, then the higher capital investment may be warranted.

EXAMPLE 16.8 *Compute the rate of return for the problem in Example 16.5. The calculations with and without escalation are shown in Tables 16.4 and 16.5.*

TABLE 16.4 Economic Comparison for Selecting Motors

Internal-rate-of-return method with no escalation in energy cost

| Year | Cost of losses | | Savings | PW with IRR |
	Standard	Low loss		
1	1252.68	782.56	470.12	263.95
2	1252.68	782.56	470.12	148.20
3	1252.68	782.56	470.12	83.20
4	1252.68	782.56	470.12	46.72
5	1252.68	782.56	470.12	26.23
6	1252.68	782.56	470.12	14.73
7	1252.68	782.56	470.12	8.27
8	1252.68	782.56	470.12	4.64
9	1252.68	782.56	470.12	2.61
10	1252.68	782.56	470.12	1.46
		Total	4701.20	600.01

Rate of return = 78.11 percent.

TABLE 16.5 Economic Comparison for Selecting Motors

Internal-rate-of-return method with escalation

| Year | Cost of losses | | Savings | PW with IRR |
	Standard	Low loss		
1	1252.68	782.56	470.12	256.82
2	1315.31	821.69	493.63	147.32
3	1381.08	862.77	518.31	84.50
4	1450.13	905.91	544.22	48.47
5	1522.64	951.21	571.43	27.80
6	1598.77	998.77	600.01	15.95
7	1678.71	1048.71	630.01	9.15
8	1762.65	1101.14	661.51	5.25
9	1850.78	1156.20	694.58	3.01
10	1943.32	1214.01	729.31	1.73
		Total	5913.12	600.00

Rate of return = 83.05 percent.

16.5 OTHER FACTORS

Sometimes in evaluating large projects with long useful lives, such as hydroelectric dams and highways, the planning horizon of 20 to 30 years may be shorter than the plant life. Also, capital investment made in the tail end of the planning horizon will have only a few years included in the present worth calculations as compared to earlier investments. This is called *end effect*. Care should be taken to make sure the end effect does not skew the economic comparison results.

Another issue is the comparison of alternatives with unequal lifetimes. For example, consider equipment from two different manufacturers with lifetimes of ten and fifteen years, respectively. A planning horizon of thirty years will give an integral number of replacements, assuming in-kind replacements. Of course, escalation and technology obsolescence will complicate the issue of these in-kind replacements. Some economists recommend a long planning horizon of 50+ years so that the present worth effect of these later years is very small.

Another approach is to factor in the remaining (residual) value of the alternative at the end of a shorter planning horizon. Another issue is the interest rate for large capital, and also in relation to the existing company size.

The discussion in this section is meant to alert the reader that sophisticated economic analysis and comparison gets more complicated. However, so long as the practicing engineer states all the assumptions used, and understands the limitations of these assumptions and the methodology used, then using simpler methods (depending upon the problem) should give satisfactory and useful results.

16.6 SENSITIVITY ANALYSIS

For a practicing engineer, economic analysis and comparison are not complete without sensitivity analysis. This is also called *parametric analysis*. As the name implies, this step involves studying the effect of changes in the final result for different significant parameter changes. The need for such analysis is almost obvious, because we make many different assumptions during the economic analysis. These assumptions for the capital cost, interest rate, inflation, and market sales picture may be different from the predicted values. It is important to understand the significance of these parameters to the overall economics of different alternatives.

Usually, one parameter at a time is changed by a given percentage, and its effect on economics is evaluated. Even though it is easy to perform many sensitivity analysis cases, it is very important to make sure these cases contribute to the basic understanding of the economics of alternatives rather than clouding the base case results.

In some cases, risk analysis—both in terms of technology as well as in the financial sense—may be required. There are probability-based decision models available. Refer to other textbooks[9-11] for further reading.

EXAMPLE 16.9 *Perform a sensitivity analysis for Example 16.5 using the following factors:*

Interest rate	15% (not 10%), but discount rate at 10%
Motor loading	50% (not 80%)
Motor running	8 h per day (not 16 h)
Useful life	5 yr (not 10 yr)

As discussed earlier, sensitivity cases are performed by changing one parameter at a time. The results are shown in four separate spreadsheets in Tables 16.6 through 16.9. The sensitivity cases are summarized in Table 16.10, along with a base case for comparison purposes. The most sensitive parameter (based on NPV savings) is listed at the top and the least sensitive parameter at the bottom of this table. The sensitivity analysis still shows substantial benefit by purchasing the low-loss motor and, of course, provides greater confidence in making the decision. Further, the analyst can examine the assumptions behind the most sensitive parameters, make necessary changes, and perform a final analysis before the final decision is taken.

As an exercise, the reader should repeat Example 16.9 with all of these four changes at the same time and determine whether a low-loss motor is beneficial.

TABLE 16.6 Sensitivity Analysis for Selecting Motors

Interest rate changed from 10 to 15%

	Standard motor				Low-loss motor				
Year	Annual payment	Cost of losses	Total cost	Total PW	Annual payment	Cost of losses	Total cost	Total PW	Savings
1	478.20	1,252.68	1,730.88	1,573.53	597.76	782.56	1,380.32	1,254.84	318.69
2	478.20	1,315.31	1,793.51	1,482.24	597.76	821.69	1,419.45	1,173.10	309.15
3	478.20	1,381.08	1,859.28	1,396.90	597.76	862.77	1,460.53	1,097.32	299.58
4	478.20	1,450.13	1,928.33	1,317.08	597.76	905.91	1,503.67	1,027.03	290.05
5	478.20	1,522.64	2,000.84	1,242.36	597.76	951.21	1548.97	961.79	280.58
6	478.20	1,598.77	2,076.97	1,172.40	597.76	998.77	1,596.53	901.20	271.20
7	478.20	1,678.71	2,156.91	1,106.84	597.76	1,048.71	1,646.47	844.90	261.94
8	478.20	1,762.65	2,240.85	1,045.37	597.76	1,101.14	1,698.90	792.55	252.82
9	478.20	1,850.78	2,328.98	987.71	597.76	1,156.20	1,753.96	743.85	243.87
10	478.20	1,943.32	2,421.52	933.60	597.76	1,214.01	1,811.77	698.51	235.09
			Total	12,258.04			Total	9,495.08	2,762.96

TABLE 16.7 Sensitivity Analysis for Selecting Motors

Motor loading changed

	Standard motor				Low-loss motor				
Year	Annual payment	Cost of losses	Total cost	Total PW	Annual payment	Cost of losses	Total cost	Total PW	Savings
1	390.59	782.56	1173.15	1066.50	488.24	490.56	978.80	889.82	176.68
2	390.59	821.69	1212.28	1001.88	488.24	515.09	1003.33	829.20	172.69
3	390.59	862.77	1253.36	941.67	488.24	540.84	1029.08	773.16	168.50
4	390.59	905.91	1296.50	885.53	488.24	567.88	1056.12	721.35	164.18
5	390.59	951.21	1341.80	833.15	488.24	596.28	1084.52	673.40	159.75
6	390.59	998.77	1389.36	784.26	488.24	626.09	1114.33	629.01	155.24
7	390.59	1048.71	1439.30	738.59	488.24	657.40	1145.64	587.89	150.69
8	390.59	1101.14	1491.73	695.90	488.24	690.27	1178.51	549.78	146.12
9	390.59	1156.20	1546.79	655.99	488.24	724.78	1213.02	514.44	141.55
10	390.59	1214.01	1604.60	618.64	488.24	761.02	1249.26	481.64	137.00
			Total	8222.11			Total	6649.70	1572.41

TABLE 16.8 Sensitivity Analysis for Selecting Motors

Motor operating only one shift

	Standard motor				Low-loss motor				
Year	Annual payment	Cost of losses	Total cost	Total PW	Annual payment	Cost of losses	Total cost	Total PW	Savings
1	390.59	626.34	1016.93	924.48	488.24	391.28	879.52	799.56	124.92
2	390.59	657.66	1048.25	866.32	488.24	410.84	899.08	743.04	123.28
3	390.59	690.54	1081.13	812.27	488.24	431.39	919.63	690.93	121.34
4	390.59	725.07	1115.66	762.01	488.24	452.96	941.20	642.85	119.16
5	390.59	761.32	1151.91	715.25	488.24	475.60	963.84	598.47	116.77
6	390.59	799.39	1189.98	671.71	488.24	499.38	987.62	557.49	114.22
7	390.59	839.36	1229.95	631.16	488.24	524.35	1012.59	519.62	111.54
8	390.59	881.32	1271.91	593.36	488.24	550.57	1038.81	484.61	108.74
9	390.59	925.39	1315.98	558.10	488.24	578.10	1066.34	452.23	105.87
10	390.59	971.66	1362.25	525.21	488.24	607.00	1095.24	422.26	102.94
			Total	7059.86			Total	5911.07	1148.79

TABLE 16.9 Sensitivity Analysis for Selecting Motors

Useful life changed

	Standard motor				Low-loss motor				
Year	Annual payment	Cost of losses	Total cost	Total PW	Annual payment	Cost of losses	Total cost	Total PW	Savings
1	633.11	1,252.68	1,885.79	1,714.35	791.39	782.56	1,573.95	1,430.86	283.49
2	633.11	1,315.31	1,948.42	1,610.27	791.39	821.69	1,613.08	1,333.12	277.15
3	633.11	1,381.08	2,014.19	1,513.29	791.39	862.77	1,654.16	1,242.80	270.49
4	633.11	1,450.13	2,083.24	1,422.88	791.39	905.91	1,697.30	1,159.28	263.60
5	633.11	1,522.64	2,155.75	1,338.55	791.39	951.21	1,742.60	1,082.02	256.54
			Total	7,599.35			Total	6,248.08	1,351.27

TABLE 16.10 Summary of Sensitivity Analysis Cases

Parameter changed	Total savings NPV	Benefits-cost ratio	Payback period	IRR, %
Base case	2897.59	4.8	1.3	83.05
Sensitivity cases				
Interest rate (15%)	2762.96	4.6	1.3	83.05
Useful life (5 years)	1351.27	2.3	1.3	77.71
Loading (50%)	1572.41	2.6	2.1	52.50
Motor running (8 h/day)	1148.79	1.9	2.6	83.05
All 4 changes	(72.53)	0	4.1	10.38

EXAMPLE 16.10 *An electrical engineer in a plant has collected the following information from four different cable manufacturers.*

Brand	Cost	Probability of failure	O&M cost, %
A	1500	0.1	6.0
B	1600	0.08	5.5
C	1750	0.07	4.5
D	1900	0.06	4

Assume that replacement cost is 60 percent of the original cost upon failure of the cable. There is no salvage value. Assume a plant life of 20 years and an interest rate of 10 percent. We will assume that probability of cable failure is uniformly distributed.

This example will be used to illustrate factoring the risk of failure in the cost evaluation. Hence, we will compute a simple expected annual cost by including risk.

	Annual cost		Expected replacement cost	Expected annual cost total
Brand	Capital *a*	O&M *b*	*c*	*a + b + c*
A	176.90	90.00	90.00	356.19
B	187.94	88.00	76.80	352.74
C	205.55	78.75	73.50	357.80
D	223.17	76.00	68.40	367.57

Expected replacement cost = fraction of replacement cost × capital cost
× probability of failure

The probability of failure is interpreted on an average basis. For cable brand A, one failure for every 10 years of in-service life has been recorded. For cable A, this would be 0.6 × 1500 × 0.1 = 90.

16.7 ESTIMATING CAPITAL COST

As a part of engineering economic analysis, there is always a need to estimate the capital cost of either brand new equipment or a plant. Improvements or additions require estimating capital cost. Capital cost is used in the sense of installed cost. Consequently, this cost involves all items and services until the plant is put into production, including material, equipment, land cost, site preparation, construction, labor, transportation, commissioning, testing, etc. The importance of cost estimating is recognized in many major corporations and projects. Personnel specialized in cost estimating are either in a separate department or part of a project team. Professional associations, such as the American Association of Cost Engineers, exist in the United States. In this section we will briefly present some introductory comments on capital cost estimating. For more information, consult appropriate texts on this subject.

Some of the common steps involved in cost estimating are:

- Prepare or obtain a specification and plans.
- Review the specification and plans.

- Become familiar with the project and/or engineering discipline.
- Assemble pertinent per-unit cost information from published sources.
- Adjust costs to local conditions and requirements.
- Prepare estimates in a methodical manner.
- Check for reasonableness.
- Review with planning personnel.
- Make a sensitivity analysis.

The complexity of cost estimating varies from project to project. The availability of cost data banks and spreadsheets on personal computers make the job of cost estimating less tedious and more productive.

Cost estimates, used in day-to-day engineering and planning activities, may be broadly divided into three different types:

- Order-of-magnitude estimate
- Preliminary estimate
- Definitive estimate

The *order-of-magnitude* estimate, as the name suggests, is a rough estimate. The main purpose of this type of estimate is for investment screening or throwing out alternatives which are too costly as compared to others. The project specification is usually very general in scope and the following items may be definable at this stage of the project:

- Capacity of the plant/equipment
- Site or area to be located
- Technology used
- Utility needs
- Land and building requirements including storage
- Transportation and other services requirements

The applicable cost information is obtained from various sources.[1,5] Cost indices are used to project the base year costs to the expected plant in-service date. A good understanding of these indices is essential before they are used. Some of the indices and cost data used by cost estimators may be found in Refs. 4–10. During the order-of-magnitude estimates, details regarding various major items are not available. In such situations, a cost factor approach may be used. A cost factor approach essentially defines the percentage of cost for each major item of the plant with the total adding to 100 percent. The order-of-magnitude estimate may vary in the range of ±30 percent. Hence, it may be used only for gross comparison and screening purposes.

At the *preliminary estimate* stage, more information regarding the project is available, including:

- Flowchart of process or manufacturing
- Engineering specifications
- Conceptual (preliminary) designs
- Site conditions
- Tentative construction plan
- Utility supply requirements (water, electricity, waste disposal)

- Environmental compliance
- Overhead costs (engineering, supervision, contingency, etc.)

Because detailed process flow sheets, project plans, and equipment details will be available, the confidence factor in this type of estimate will be better than in the order-of-magnitude estimates. The confidence range is −15 to +30 percent. This type of estimate may be used for budgetary and expenditure approval purposes.

As the name implies, the *definitive estimate* is based on a detailed specification for plant, equipment, and construction. Detailed design and drawings would have been completed by this time. Construction plans and schedules would be complete. A list of equipment and material requirements and specifications would be completed. The cost estimating here may be called a "bottoms up" approach. The cost of individual units, groups, and processes are assembled together. Cost estimates for different packages such as equipment, material, instrumentation, control, construction, transportation, and installation may be prepared. The confidence in the estimated cost may be in the range of −5 to +15 percent. This type of cost estimate is used for final project approval, bidding, and construction.

EXAMPLE 16.11 *Given the following cost data for equipment, estimate the cost for the equipment to manufacture 250 widgets per day.*

Output (widget/day)	Cost ($)
50	2500
80	3100
90	3600
140	5000

This cost data may be represented by a straight line (by linear regression), as shown in Fig. 16.5. The equation for this straight line is

$$\text{Cost} = 28.33 \times \text{widgets/day} + 1000$$

Based on this equation, the cost for equipment with a capacity of 250 is $8083.33.

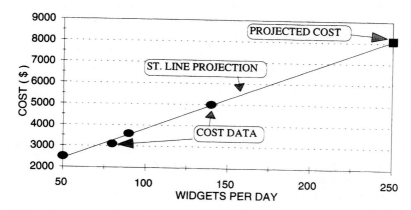

FIGURE 16.5 Cost estimating by projection.

EXAMPLE 16.12   *A bicycle manufacturing plant (we are really moving into high tech here) would have cost $10 million in 1986 when the cost index was 210. The cost index was 265 in 1993. Estimate the cost of this plant for an in-service date of January 1, 1996. Assume the plant cost to escalate at 5 percent per year.*

The estimated plant cost in 1993 is

$$10 \times \left(\frac{265}{210}\right) = 12.62 \text{ million}$$

Assuming the beginning of the year for indexing, the estimated plant cost with an in-service date of 1996 (i.e., three years from 1993):

$$= 12.62 \times (1.05)^3$$

$$= \$14.61 \text{ million}$$

16.8   PROBLEMS

1. An engineer has two alternatives which perform the same function. The costs of these two alternatives are as follows:

	Alternative A	Alternative B
Capital cost (1st cost)	$200,000	$248,000
Labor cost per year	$150,000	$130,000
O&M per year	$ 45,000	$ 42,000

Use an interest rate of 12 percent. Assume taxes and insurance to be 5 percent of the capital. Ignore the income tax consequences. Assume 5 percent per year escalation for labor and 3 percent escalation per year on O&M costs. Assume 10 years useful life and zero salvage value. Calculate (*a*) total present worth for the two alternatives, (*b*) benefit-to-cost ratio, (*c*) payback period for the most economical alternative, and (*d*) the internal rate of return.

2. There are two machines available in the market with the following cost parameters. Make a recommendation to the management on which machine to purchase.

	Machine 1	Machine 2
Installed cost (capital)	$30,000	$50,000
O&M cost	12,500	10,000
Life (years)	5	10
Salvage value	20%	10%
Taxes and insurance	3%	3%

Assume 10 percent escalation in O&M costs. For replacement of machines use 5 percent escalation per year. Salvage value given is a percent of replacement value. Use 12 percent for cost of money.

3. For Prob. 2, find the following breakeven factors for the two machines to be equivalent by performing sensitivity analysis:

Interest rate

O&M cost escalation

Escalation of machine replacement cost

4. A person invests $1000 at 7.2 percent interest, compounded annually. What is the amount at the end of 10 years?

5. Future cash value of a negotiable investment instrument is $100,000 at the end of 10 years from now. If the discount rate is 12 percent, what is its cash value now?

6. What is the rate of return on an investment of $1000 that pays back $1500 in five years?

7. A student deposits $100 every two weeks in a savings account paying 5 percent interest compounded monthly. How much money will the student have at the end of four years?

8. A work automation project involves an investment of $100,000. The benefits over the life is computed to be $140,000 and a net cost of $18,000. Compute the (*a*) net savings, and (*b*) benefit-to-cost ratio.

9. A person buys a car at a cost of $20,000 with a trade-in value of 20 percent at the end of five years. The car is financed for the entire amount on a 60-month loan at 12 percent interest. The owner expects to drive 10,000 miles per year. The car gives 20 mpg and the gasoline costs $1.20 per gallon with escalation of 5 percent per year. Car insurance costs $500 per year. Other maintenance costs average $300 per year. Calculate (*a*) the monthly payment on the loan, (*b*) the annual cost of owning the car, and (*c*) the cost per mile driven.

10. In the preceding problem, the car owner finds out that the car has to be used for business purposes. The business miles are estimated to be an additional 5000 miles. The increase in insurance premium is $250 per year, and additional maintenance is $200 per year. The salvage value of the car is reduced to 5 percent at the end of five years.

   a. The company reimburses the car owner at the rate of $0.30 per mile. Is the car owner making a profit or losing money, and by what amount?

   b. At what reimbursing rate will it be economical to buy a company car for company business only? The cost of the car will be the same, but the salvage, insurance, and maintenance will be 30 percent, $500, and $300 per year.

11. A wizard can make a device at an initial investment of $200,000. This device is expected to bring in $45,000 per year for 15 years. Annual expenses of $10,000 per year have been estimated. What are (*a*) the benefit-to-cost ratio, and (*b*) the rate of return?

16.9 REFERENCES

1. Cost Estimating Standards, published and updated by Richardson Engineering Services, Inc., Mesa, Ariz., annual update.

2. Means Cost Data Services, R. S. Means Company, Inc., 100 Construction Plaza, Kingston, MA 02364-0800.

3. *Electrical Trade Book*, Trade Service Corporation, Mt. Prospect, IL 60056 and San Diego, CA 92121.

4. *Cost Engineering Magazine,* AACE Inc., Box 1557, Morgantown, WV 26507-1557.

5. *The Handy Whitman Index of Public Utility Construction Costs,* Whitman, Requardt and Associates, Baltimore, Md.

6. *Engineering News Record,* The McGraw-Hill Construction Weekly Magazine, McGraw-Hill, New York, NY.

7. Plant Engineering Cost Index, *Plant Engineering,* Cahners Publishing Company, Newton, MA.

8. "Electrical Estimating for the Plant Engineer," *Plant Engineering,* The Cahners Publishing Company, Newton, MA.

9. L. T. Blank and A. J. Tarquin, *Engineering Economy,* 3d ed., McGraw-Hill, New York, 1989.

10. H. M. Steiner, *Basic Engineering Economy,* Books Associates, Glen Echo, Md., 1989.

11. E. P. DeGarmo, W. G. Sullivan, and J. A. Boutadelli, *Engineering Economy,* 8th ed., Macmillan, 1989.

CHAPTER 17
ENERGY CONSERVATION AND MANAGEMENT

17.0 INTRODUCTION

Design and architect professionals as well as plant engineers have to recognize the importance of reducing overall energy consumption, whether it is a building, plant, or process that is under their jurisdiction for design, construction, or operation. The industrial sector in the United States uses the most energy, followed by transportation, residential, and commercial sectors.

The industrial use of energy includes fuel used in boilers, process heating and drying, feedstock for plastics, mechanical drives, space heating and cooling, and lighting. Commercial and institution buildings use most of the energy in heating, cooling, and lighting. The percent cost related to energy varies from one type of industrial product to another. For example, an aluminum reduction plant uses more energy than a metal-grinding workshop.

Reducing the energy cost of any industrial operation or a service organization will not only increase the profit margin, but also enables the enterprise to be competitive in the marketplace. For example, consider an enterprise selling a manufactured product with 20 percent of the price attributed to the energy used during its manufacture. Assume that, based on the current market price, the profit margin is 10 percent. If the energy consumption can be reduced by 25 percent, the profit margin increases to 15 percent; i.e., there is an increase of 50 percent in the profit, assuming, of course, that the product can be sold at the same price. If the competition is severe in the marketplace, this may enable a price cut and, thus, retain or increase the market share.

Hence, energy conservation is important, even though at the outset it looks like we are trying to reduce the cost by only a small percentage (5 percent in the preceding example). Granted, this example is simplistic and nothing in practice is this simple; however, it brings home the point that reduction in the use of energy should be important to businesses and, consequently, to design and plant operation professionals. The plant or facilities engineer needs to keep the energy cost per unit of product or service as low as possible. The per-unit cost itself is a function of many factors, including the number of units produced in a given time.

Reducing the use of energy can take many approaches, including:

- Reducing or reclaiming waste
- Using advanced technologies
- Using substitute materials
- Improving operating techniques and procedures

- Considering alternate energy sources
- Improving maintenance

Heating processes require a certain amount of wasted energy (the second law of thermodynamics). Recovery of part of this wasted heat in some form, either by using existing or new technologies, is possible. For example, the use of steam for different processes is widespread in industry. The same steam may be used to drive turbines and generate electricity. This is called *cogeneration,* which is discussed later in this chapter. Reduction of heat loss in kilns, dryers, etc., is another consideration. Many times, changing the type of technology may be the answer. For example, applying microwave technology for heating and drying, as compared to conventional methods, may be economical and save energy as well. Alternative energy sources to be considered are solar (electrothermal, thermal, photovoltaic), wind farms, and geothermal energy. There may be specific applications for these alternatives. Photovoltaic collectors are being used as energy sources in remote areas.

17.1 ENERGY CONSERVATION

Energy use reduction or conservation[1-5] has four main steps, as illustrated in Fig. 17.1. The planning of energy conservation requires setting goals to meet within the time frame available, prioritizing the use of resources, and determining a schedule for implementation.

The next step is to perform the necessary technical and economic analyses. This step is popularly known as the *energy audit.* The purpose of an energy audit is to identify the locations, items, processes, etc., where the most energy savings are possible at the lowest possible cost. The different phases involved in an energy audit are:

- Survey
- Analysis
- Options
- Cost-effectiveness

Conducting a survey of existing facilities, process, equipment, operation, and maintenance is very important. Some type of measurement may be needed. Data on the operating environment and the type of operation is necessary. A walk through the plant and building may be necessary. Because the survey and data collection may become labor intensive (hence, expensive), it is necessary to plan and prepare survey forms, the type of data to be collected, computer data entry procedures, etc., before starting the survey. In large facilities and complex processes, it may be advisable to break this survey into several steps, such as a reconnaissance, and preliminary, detailed, and final surveys. Whereas the lack of some important data may become a problem later, during the analysis phase, unnecessary, large amounts of data may make the analysis time consuming and, at times, may obscure the problem being solved.

Once the data is collected, further analysis is needed to select where the greatest potential exists for energy conservation. This may require expertise from various engineering disciplines. The type of analysis may range from reviewing simple heat gain and heat loss to involved thermodynamics calculations.

FIGURE 17.1   Energy conservation steps.

Once the places where most of the energy loss is occurring are identified, various options to reduce this energy loss need to be evaluated. The options may be divided into two types: passive and active technologies. Examples of passive technologies include insulation of the building, ventilation, installing flue dampers, providing shading of the building, and reducing heat loss or gain through windows. Active technologies include a range of options, the simplest being changing the lighting system to a more efficient type (see Chap. 12). Options for technologies may also include replacing motors by adjustable-speed drives, using substitute materials, fuel substitution, microwave heating, and cogeneration. Some of these are active and some are passive technologies. The concepts behind some of these options are discussed later in this chapter. Because the selection of a particular energy conservation option may impact production, operation, and maintenance, it is essential that all parties responsible for these functions be part of the decision-making process. Once the options have been evaluated and, possibly, ranked on the basis of the technical merits, the cost of these options needs to be computed. While calculating the cost and computing the savings, a life-cycle costing approach should be used.

The economic evaluation using life-cycle cost analysis considers many different factors, including:

- Capital cost
- Operation and maintenance costs (existing and new)
- Present fuel and electricity bills
- Interest rate
- Plant and equipment life
- Indirect cost
- Additional labor costs
- Savings (O&M, material, labor)
- Other benefits
- Risk

Here, not only the initial cash outlay is important, but the subsequent annual savings and expenditures should be taken into consideration. The engineering economics procedures discussed in Chap. 16 may be used for this purpose. Taxes and depreciation should also be considered as a part of financial analysis.

After performing economic analysis, the most cost-effective options should be selected for implementation. The final selected option may be a simple recommendation to modify the present maintenance frequency and procedure. For example, the energy savings realized through cleaning exhausts, intakes, keeping inlets and outlets free of obstruction, changing filters, using different types of filters, proper frequency of lubrication, etc., may be the most cost-effective option. This does not require any substantial investment.

Once the cost-effective options have been implemented, it is necessary to follow up to make sure that the changes made are effective and functional and that cost savings are being realized. In energy conservation terminology, this is called the *persistence factor*. It has been observed that the amount of energy saved decreases as the number of years elapses. This may be due to a change in maintenance, process, or environment of operation, etc. The success of the energy audit and conservation effort mainly depends upon

- Commitment of all personnel, from management to shop floor worker
- Selection of a suitable technology
- Cost-effectiveness

- Proper implementation
- Periodic monitoring and correcting as needed

Electric energy is the most widely used form of energy due to its availability, convenience, ease of control, etc. Traditionally, electrical utilities built new generation capacity and necessary transmission and distribution facilities to supply customer load requirements. However, due to an increase in the cost of fuel and the high cost of building new power plants, utilities have adopted what is called *integrated resource planning* or *least-cost planning*. One of the main theses of this type of planning is that if it is cheaper to save 1 kW of power than to build to supply the same, the investment should be to save the 1 kW. To meet this objective of utility planning, almost all electrical utilities in industrialized countries have implemented demand-side management and energy conservation. These electrical utilities, through their marketing and customer services departments, offer a wide range of assistance, from simple, ready-made pamphlets on energy conservation to technical assistance or help in locating expertise for the particular process. Electrical utilities deal with many customers and thus contain a reservoir of experience and information. During the 1990s, electrical utilities have entered an era of deregulation and competition. This makes the customer the key point of their business and profit. The plant engineer should make use of the local electrical utility in achieving energy savings.

17.2 ELECTRIC BILL

Reducing the cost of electric energy deals with at least two basic aspects: the electric bill and the electrical energy-saving techniques. The tariff on electricity use and the rate for residential customers is based on a simple cents-per-kWh-used charge and is billed every month. However, for large commercial business buildings and industrial customers, the tariff structure and rates can become more complicated. Almost all of these have a demand charge and an energy use charge. The demand charge is on a $/kW-month basis. The electric power demand kW-month is called *billing demand per month*. This is the highest rate of energy consumption averaged over a given period.

For example, if the integrated energy over 30 min is 30 kWh, then the integrated demand for that half hour is 60 kW (30 kWh divided by 0.5 h). The integrated period used for the billing may be 15 min, 30 min, or 1 h, depending on the particular tariff. Then the highest integrated demand (called *peak demand*) in any one billing period—typically one month—is the billing demand kW-month.

The actual dollars per kW are derived on the basis that, for every peak demand kW, the utility has to install 1 kW of generation and other facilities. The actual dollars per kW are derived from the cost of owning the generation capacity and the coincidence of a class of customers (for example, large industries) to the utility's peak load. Further, there may be a ratchet clause to the billing demand determination. The ratchet clause essentially states that the monthly billing demand is the highest peak demand in any one month, for the previous 11 months. This usually applies to large-peak-demand customers in relation to the utility size.

The concept here is that the electrical utility has to install generation and the necessary equipment to supply the particular peak demand in a given period and in a given month, even though the customer does not use this capacity in the remaining months. Also, the demand charge will be higher if the customers' peak load coincides with the utilities' overall peak load. Thus, one of the incentives to the plant engineer is to reduce the peak demand. This is called *peak clipping* or *shaving*.

Furthermore, integrated demand is measured only during on-peak periods, not during off-peak periods. The on-peak period, for example, may be between 8 A.M. and 5 P.M. on weekdays for a summer-peaking utility. However, if the utility has a substantial number of residential customers with a summer-peaking load, then the peaks may be in the evenings and on weekends. This type of load characteristic has resulted in a rate tariff called *time-of-day metering*. Lower rates are offered for off-peak loads as compared to on-peak loads. This allows the plant engineer to look into load modification or load shifting.

The second important component of an electric bill is the energy charge in cents/kWh. This is a more straightforward computation. The total energy (kWh) consumed in any given billing period is charged at a rate of cents/kWh, which corresponds to the cost of supplying 1 kWh of energy. However, since the early 1980s, the cost of fuel itself has become more volatile when compared to earlier periods. This volatility is factored into the energy charges through a fuel surcharge. If a time-of-day (TOD) rate structure is used, the on-peak energy rate may be substantially higher when compared to off-peak energy rate. Thus, load modification or shifting may save not only demand charges, but also energy charges.

In addition to the demand and energy rates (and associated clauses), there are at least three other clauses which require attention by a plant engineer. These are the block rate structure, the load factor, and the power factor penalty. Until the 1980s, the utility rate structure contained what is called *decreasing block rates*. A certain minimum kW demand was billed at a higher rate, and successive blocks had a declining rate. The reasoning behind this concept was that supplying incremental blocks of power (kW) was cheaper than the base loaded generating plant. However, due to the emphasis on energy conservation and the higher cost of building new power plants, the declining block rate structure has been eliminated. In some cases, an inverted block structure—meaning higher rates for increasing kW consumptions—has been implemented. Thus, the block rate structure has a direct impact on savings due to demand reduction and energy conservation.

The definition of load factor is:

$$\text{Load factor} = \frac{\text{kWh used}}{\text{peak demand (kW)} \times \text{number of hours}}$$

Obviously, the load factor may be measured, in varying periods: daily, monthly, and annually. When plotted graphically, a poor load factor is represented by a peaky-looking load shape as compared to the high load factor waveshapes. In some rate structures, there may be adders for poor load factors.

Another clause in a rate structure is the power factor penalty. This pertains to reactive power. For purposes of maintaining and regulating proper voltage levels of electric power supply, utilities have to maintain reactive power sources. A low power factor means that the utility may have to install either reactive power supply, controlling equipment, or both. Thus, a load with poor power factor may be required to pay additional charges. The topic of power factor correction is discussed in Chap. 6.

EXAMPLE 17.1 *Compute the monthly electric bill for a customer with the following consumption and the declining block rate structure:*

(a) Monthly demand = 3000 kW

(b) Total energy metered = 1,482,000 kWh

(c) Demand charge

Demand block (kW)	Demand charge ($/kW/month)
50	8.76
950	7.70
over 1000	6.70

(*d*) Energy charge

Energy block (kWh)		Energy charge (¢/kWh)
First	72,000	5.8
Second	128,000	5.3
Third	256,000	4.8
Fourth	512,000	4.3
Fifth	excess	3.8

(*e*) Fuel cost adjustment 0.92¢/kWh

The monthly electric bill computation is shown in Table 17.1.

Total demand charge	= $21,870
Total energy charge	= 64,796
Fuel cost adjustment	= 13,338
Total bill	$100,004

TABLE 17.1 Calculation of Monthly Electric Bill for Declining Block Rate

Demand			Energy		
Block (kW)	Rate ($/kW/month	Charges ($)	Block (kWh)	Rate ($/kWh)	Charges ($)
100	8.76	876	72,000	0.058	4,176
900	7.86	7,074	128,000	0.053	6,784
2,000	6.96	13,920	256,000	0.048	12,288
			512,000	0.043	22,016
			514,000	0.038	19,532
Total		21,870	1,482,000		64,796

EXAMPLE 17.2 *Assume an inverted block rate structure for Example 17.1 and compute the monthly electric bill. Here, for simplicity, we will use the same per-unit cost but in increasing order for the inverted block rate structure. The monthly electric bill computation is shown in Table 17.2. As can be seen from this table, demand reduction and energy conservation is much more attractive as compared to the case of declining block rate structure.*

Total demand charge	= $25,290
Total energy charge	= 77,476
Fuel cost adjustment	= 13,338
Total bill	$116,104

TABLE 17.2 Calculation of Monthly Electric Bill for Inverted Block Rate Structure

Demand			Energy		
Block (kW)	Rate ($/kW/month)	Charges ($)	Block (kWh)	Rate ($/kWh)	Charges ($)
100	6.96	696	72,000	0.038	2,736
900	7.86	7,074	128,000	0.043	5,504
2,000	8.76	17,520	256,000	0.048	12,288
			512,000	0.053	27,136
			514,000	0.058	29,812
Total		25,290	1,482,000		77,476

17.3 DEMAND REDUCTION OPTIONS

Some of the demand reduction options considered by utilities are described in this section. Large commercial and industrial premises have specific and individual needs. Additionally, equipment, machinery, and computer systems generate too much heat to allow the absence of air-conditioning for any length of time. Hence, load shape modification and load shifting have to be adopted with careful forethought. Electrical utilities offer incentives to enlist the customers to incorporate internal load management systems to shift the load.

EXAMPLE 17.3 *Consider a conventional chiller system of 160 tons capacity at a cost of $80,000. Installation of a refrigeration and thermal storage system of 1000 ton-hours capacity will cost $150,000. The annual savings on the electric bill are estimated to be $10,000 (savings in demand charges − cost of running refrigeration system). Calculate the incentive needed, assuming a three-year payback.*

The additional cost due to thermal storage system = $150,000−$80,000 = $70,000

Total savings in three years = $30,000

Net excess investment = $40,000

Incentive for breakeven = $40 per ton-hour

This is a simple illustration of economic analysis. Some of the techniques to shift loads in large commercial or industrial companies, based on North American experience, are discussed in the following sections and in Table 17.3.

17.3.1 Efficient Lighting

The purpose is to promote the use of efficient fluorescent lighting and high-intensity discharge exterior lighting (i.e., the replacement of existing mercury vapor fixtures with either high-pressure sodium or metal halide fixtures) in the commercial, office, and retail sectors. This reduces annual energy consumption and weather-sensitive peak load.

Older commercial, retail, and office lighting systems use standard 40- and 75-W fluorescent tubes, with magnetic (core and coil) ballast, incandescent lamps, and mercury vapor fixtures. Experience has shown that standard fluorescent tubes can be replaced with high-efficiency tubes, magnetic (core and coil) ballast can be replaced with electronic ballast (or hybrid ballast), and incandescent lamps can be replaced with self-ballasted fluorescent lamps, all without an appreciable loss of light or color. Existing

TABLE 17.3   Demand Reduction Options

		Options	
Type	Purpose	Industrial	Commercial
Peak clipping	Reduce peak demand	Interruptible rate	Water heater control
Valley filling	Utilize low cost off-peak energy	Shift operation (total or partial)	Thermal storage (heating/cooling)
Load shifting	Peak clipping + valley filling	Thermal storage (heating/cooling)	Heat pumps
Conservation	Reduce demand and energy usage or substitute energy source	Utilize more efficient process/technology Fuel substitution	Lighting Fuel substitution

exterior mercury vapor fixtures can be replaced with either high-pressure sodium or metal halide fixtures without any loss of light. Another option is to add reflectors to existing four-tube fluorescent fixtures and remove two tubes and one ballast. Further discussion on lighting and illumination may be found in Chap. 12. Many utilities will work with customers to determine the most economic options. Incentive payments towards front-end capital expenditures are also available in certain areas.

17.3.2   Cold Storage

Cold storage or thermal energy storage (TES) can be used to provide off-peak air-conditioning or process cooling. Cold storage is primarily a peak-demand reduction or load-shifting mechanism. If on-peak and off-peak energy rates are applicable, then energy savings are also possible. Cold storage involves the use of a thermal mass as a heat sink during on-peak cooling periods. The heat from air-conditioning or process cooling loads is temporarily rejected into the heat sink while simultaneously avoiding or limiting the use of high-energy-consuming chillers. During subsequent off-peak periods, chillers or special ice-making equipment are operated to remove energy from (called *recharging*) the heat sink. Because substantial demand is shifted to off-peak hours, the demand charge is reduced significantly. The off-peak energy charge may also be lower than the peak energy charge. Three types of cold storage techniques are described here.

Chilled Water Storage.   Heat is rejected into a mass of stored, prechilled water. The stored water is subsequently rechilled during off-peak hours. Systems can shift up to 100 percent of peak chiller load; however, economic considerations usually result in partial storage (load-leveling) installation. Where space permits, insulated, above-ground, welded-steel storage tanks provide an economical and efficient means of accomplishing cold storage for large-capacity applications. Suppliers offer such systems on a design/build basis and with a thermal performance guarantee. Chilled water storage can be readily retrofitted to existing systems with any type of water chillers. Their capability to provide a low-capital-cost option has been documented in some new construction cases and retrofit capacity expansions, even in the absence of utility incentives. Typical applications involve thousands or tens of thousands of ton-hours capacity, resulting in hundreds or thousands of kilowatts of peak load shift each, usually with little change in total energy consumption. Due to their large size, chilled water storage installations represent a large portion of all cold storage applications.

Ice Storage. Heat is rejected into a stored mass of ice. The ice is subsequently re-formed, using any of several techniques for ice making, often using positive displacement chillers or modified centrifugal chillers, as appropriate to produce the low evaporator temperatures necessary for ice making. These low temperatures result in higher kW-per-ton consumption than for conventional air-conditioning. However, the high consumption occurs during off-peak periods when there may be no demand charges and energy costs are low. Systems can shift up to 100 percent of peak chiller load. Partial storage (load-leveling) installations may be more economical. Ice storage is often the preferred method of cold storage for small- to medium-capacity applications. Because ice storage systems typically require only 25 to 50 percent of the gross storage volume in comparison to chilled water storage, the ice storage is preferred where space is a consideration. Typical applications involve hundreds or a few thousand ton-hours capacity, resulting in tens or hundreds of kilowatts of peak load shift each, though often with a moderate energy consumption premium. Due to their application in large numbers, ice storage installations also represent a large portion of cold storage applications.

Other (Non-Ice) Phase Change Storage. Instead of using ice as a storage mass, a different phase change material (PCM), which undergoes liquid-to-solid phase change in the 40 to 50°F range, is used. Non-ice PCM storage systems typically require somewhat larger storage volumes than do ice storage systems of similar capacities and cannot shift 100 percent of peak chiller load due to the relatively high phase change temperatures. These types of storage installations currently represent a relatively small portion of all cold storage applications.

Cold Storage Installation Examples. According to recent reports in trade journals, thermal energy storage is making a strong comeback in the commercial sector in the United States.[7] It was first used in the 1930s and 1940s on a commercial basis in theaters, churches, dairies, and other applications where cooling was required for a specific period of the day. Today, thousands of cold storage systems are successfully operated in a variety of businesses. The best candidates for these systems are customers such as schools, hospitals, large office buildings, hotels, convention centers, sports arenas, and places of worship.

One example of thermal energy storage facilities is a system for a high school which utilizes a 28,000-gal, insulated ice storage tank connected to the three existing 100-ton reciprocating chillers. This tank has a capacity of 2132-ton-hours, which provides full storage cooling during the day. Another system for an educational facility was designed to use existing concrete water tanks for chilled-water storage. The tanks were reconditioned and insulated. They hold 400,000 gal of water, which equate to 4015 ton-hours of thermal storage. This system will shift approximately 1000 kW to off-peak time periods. A third example of a chilled-water system uses a million-gallon storage tank to store 6000 ton-hours of cooling. The system is designed to shift 850 kW to off-peak.

These systems have been encouraged by utility incentive programs, which give a certain amount for each kW shifted from on-peak hours. Some have ceilings on total incentive payment, some incentive payments are in terms of total storage capacity, and some are on a declining scale.

17.3.3 Heat Storage

Heat storage or electric thermal storage (ETS) can be used to provide off-peak space heating or process heat. In many ways, it is analogous to cold storage, described earlier. Heat storage is primarily a peak-demand reduction or load-shifting mechanism. Off-peak electric resistance heat (or, alternatively, gas or another heat source) heats a stored mass of water, rock, ceramic, or other material of high heat capacitance. Air or water is subsequently used as a heat transfer medium to discharge the heat during on-peak periods. Applications range from small-capacity (residential) systems using any of various storage media to large-capacity (industrial) systems using hot water storage in a manner quite similar to the chilled-water storage systems.

17.4 COGENERATION

Cogeneration may be defined as the technique of utilizing the steam for both electricity generation and for process, heating, drying, drivers, etc.[8-11] Thus, through judicial use of waste heat, the overall efficiency increases and, thereby, results in cost savings. In the early 1900s, cogeneration technology was widely used and about 50 percent of the electricity was generated by this means in the United States of America. However, the electricity from the electrical utilities became a cheaper and reliable option due to economies of scale in the central generating station cost, the ease with which electricity could be transmitted and distributed, and increasing labor and maintenance costs at the industry-owned cogeneration plants. In fact, the price of electricity continuously declined until the 1970s. Thus, cogenerated electricity declined to a few percent of total electricity generation. However, the two steep price increases for oil during 1973 and 1978, the environmental movement, and the Public Utility Regulatory Policy Act (PURPA) of 1978 passed by the U.S. Congress changed the course of cogeneration. This PURPA legislation created a new class of power generators called *qualifying facilities* (QF) with the idea of reducing oil imports, increasing efficiency, and promoting the use of alternative fuels. Until the passage of 1978 PURPA legislation, any power generator engaged in the sale of electricity would come under the U.S. Federal Power Act (FPA) and Public Utility Holding Company Act (PUHCA) and, therefore, would be subject to regulation just like investor-owned electric utility companies. After the passage of the PURPA legislation, the Federal Electric Regulatory Commission (FERC) made some rules regarding qualifying facilities. Three main items removed the barriers for nonutility generation:

- Electrical utilities are required to buy power from QFs at prices equal to the utilities' avoided cost
- That electrical utilities interconnect with QFs
- That electrical utilities provide backup power to QFs

The avoided-cost rule was a major economic incentive, because, within the electrical utility, the avoided cost corresponds to the energy generated during peak hours, which is the most expensive generated energy. Several requirements had to be satisfied to obtain QF status. These requirements included minimal thermal output, an efficiency standard, alternative fuel use, and ownership. For example, an electrical utility cannot own more than 50 percent of a QF.

Following the PURPA legislation, it is reported that nearly 5000 applications for QF status were filed and about 30,000 MW of cogeneration have been built as QFs in the U.S. Cogeneration may be used in many industries which need steam for some processes, generate waste heat which could be used for electric generation, or produce waste products which could be used as fuel. Such industries include chemical plants, paper mills (waste fuel and process heat), textiles manufacturers, breweries, pharmaceutical companies, and food processing companies.

There are many institutions which need heating and use a substantial amount of electricity, and, therefore, could be prime candidates for cogeneration. Some of these institutions are universities, hospitals, office buildings, prisons, shopping centers, hotels, district heating, local government-owned power plants (refuse-fired or incinerators), and desalinization plants.

There are many examples of cogeneration projects in these industries and institutions. For example, district heating is used in many European cities. Desalinization plants produce potable water from sea water in many arid countries. Refuse-burning plants are located in many cities and towns in the United States, although public opposition to such plants is increasing.

The cogeneration concept using either a steam turbine or a combustion turbine is shown in Fig. 17.2. In the case of a steam turbine driving the generator, the exhaust steam or extracted steam from the turbine is used for the process. In the combustion turbine, the exhaust gas is used to produce steam in a heat-recovery boiler. It is also possible to use waste heat from the process and use it to generate steam for a steam turbine. The increased efficiency and the lower cost come from using part of the heat content in the exhaust steam or gases.

EXAMPLE 17.4 *Consider an industrial plant with 5 MW of electrical generation and 68 Mbtu/h of steam requirement. In general, the efficiency for electrical generation and the steam process boiler are about 33.3 and 85.5 percent, respectively. In some modern*

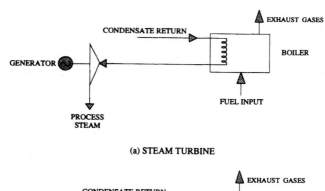

(a) STEAM TURBINE

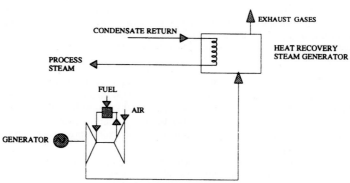

(b) COMBUSTION TURBINE

FIGURE 17.2 Cogeneration concepts.

combined-cycle (combination of gas turbines and steam turbines) plants, this efficiency could be about 50 percent.

The following conversion may be used for changing units to the metric system:

1 btu = 1060 joules = 252 calories

1 kWh = 3412 btu

(*a*) The heat input to stand-alone electric generation is

$$\frac{5000 \times 3412}{0.333 \times 10^6} = 51.2 \text{ Mbtu/h}$$

(*b*) The heat input to a stand-alone process steam boiler is

$$\frac{68}{0.85} = 80 \text{ Mbtu/h}$$

(*c*) Total heat input = 131.2 Mbtu/h. Overall fuel efficiency without cogeneration is

$$\frac{17 + 68}{131.2} = 64.8\%$$

(*d*) In a cogeneration system, assuming that the electrical and steam systems are ideally matched and optimized, the total heat input is

$$\frac{68 + 17}{0.85} = 100 \text{ Mbtu/h}$$

We have ignored miscellaneous losses here. The overall cogeneration efficiency is 85 percent. Assuming that the fuel for process steam production remains the same, the additional fuel for electric generation is = 100−80 = 20 Mbtu/h.

The efficiency assigned to electric power production is 17/20 = 85 percent.

As can be seen, this is much higher than the best efficiency of a power plant producing electricity only. Here, we are assigning all the benefits of cogeneration to the electric generation fuel input (i.e., cost). This treatment is most favorable to the cost of electric generation. In fact, the cost of electricity in cents/kWh, computed in this method as compared to the electric power companies' avoided cost, will directly give the savings per kWh.

In practice, the power and process steam needs of a plant are not exactly matched. If the power requirement is being met, then the process steam through cogeneration may not be sufficient. In such cases, an auxiliary boiler would be used to make up the shortfall in process steam supply. A corollary to this situation is that the process steam needs are met, but excess power is generated. In such cases, the excess power will be sold to the electric utility company.

Whereas a cogeneration project is a complex system involving many disciplines of engineering, in this chapter, we will address only some of the electrical engineering–related items which a plant engineer may deal with during the course of duties at the plant.

17.4.1 Feasibility Study Steps

An overview of a typical feasibility study of a cogeneration plant is shown in Fig. 17.3. Even though this is shown as a sequence of steps, considerable back-and-forth analysis

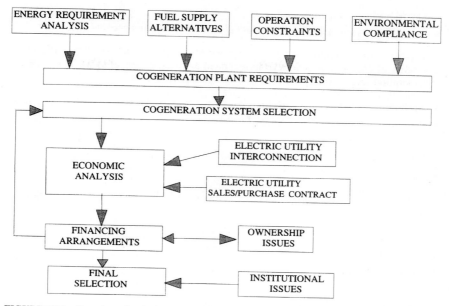

FIGURE 17.3   Overview of typical cogeneration feasibility study steps.

among these steps is not uncommon in practice. The first four boxes at the top in Fig. 17.3 involve some type of plant survey and collection of other information pertaining to the cogeneration feasibility study.

Energy requirement analysis should consider the following:

- Electricity (both demand and energy)
- Electrical rates
- Steam requirements (process, heat)
- Others (such as standby, requirements, and supplementary resources)

The monthly billing demand (kW) and energy (kWh), as well as typical daily and weekly consumption patterns for electricity, are required. Billing data for the previous 12-month period is essential. If it is a new plant or major changes are planned, then projected usage needs to be estimated. A similar requirement for steam should be determined.

The fuel supply alternatives include:

- Type of fuel (coal, oil, gas, plant waste, refuse, etc.)
- Availability
- Cost
- Contract types (take or pay, spot market, etc.)

The operational constraints should be spelled out early in the study. These include:

- Peak electric and steam demands
- Steam quality (pressure, etc.)

- Electricity-to-steam energy ratios
- Reliability requirements for steam and electricity, including process flexibility

Environmental compliance has become one of the key factors which may affect the feasibility of cogeneration, both from technical and cost points of view. This includes consideration of environmental factors and regulation, such as:

- Air quality and allowable emissions
- Water quality and discharge limits
- Land use and siting
- Mitigation to satisfy laws and regulations
- Public acceptance

Based on these factors, one or several cogeneration technologies need to be selected, including:

- Steam turbine
- Combustion turbine
- Combined cycle
- Diesel
- Selection of top or bottom cycle

After selecting one or several technology options and performing a conceptual design for each of these options, the total plant capital cost will be estimated. This capital cost will include cost of equipment, facilities, construction, and installation.

The economic evaluation consists of computing one or several of the following:

- Net present value
- Payback period
- Rate of return
- Benefit-to-cost ratio

In computing these economic performance indices, several factors should be considered, including capital cost, fuel cost, O&M costs, life of equipment or number of years to be considered in this evaluation, and savings in purchase cost of electricity and revenue from sales of electricity. Because the savings are accrued from the sale of electricity to the electrical utility, the negotiation and the type of contract plays an important part in a cogeneration study. The purchasing price from the utility point of view depends upon many variables, including:

- Capacity rate ($/kW)
- Energy rate ($/kWh)
- Firm or nonfirm power
- Peaking or base load
- Utility need for peaking capacity
- Other cogeneration contracts in place
- Reliability of supply
- Economic dispatch

Thus, the cogenerator needs to assume appropriate prices to compute revenue from sales. Several scenarios of electricity sales contracts may need to be evaluated.

The cost of electrical interconnection to the electrical utility and associated communication and control costs should be included.

The financial analysis takes into account the tax consequences, depreciation, ability to raise the capital, ownership options, risk of investment, etc. In the overview depicted, we have purposefully separated the engineering economic analysis and financial analysis so that the technical-related and financial issues influencing the final decision are clearly evident. However, it is not absolutely necessary to make this separation. Before making the final selection, other factors such as government regulations, electric utility response, and industry outlook should be taken into account.

17.4.2 Interconnection with Utility

One of the key factors in the justification of cogeneration is selling electric energy to the electrical utility. This involves establishing a new or strengthening the existing connection to the utility.[12] An electrical utility grid is a complex network with a large range of requirements, such as protection, control, communication, metering, safety, reliability, and system stability. Hence, it is essential to make sure that all requirements for this interconnection are met. A reliable, safe, and lowest-cost interconnection is as important to the cogenerator as to the electrical utility. Any loss of interconnection capability may result in the loss of revenue to the cogenerator. Discussion with the utility engineering department is recommended during the planning, design, construction, and commissioning stages of the project. The main topics for discussion with the utility are:

- Interconnection capacity and transfer limits
- Interconnection facility requirements
- Facility plans
- Normal, faulted, and emergency operation
- Protection and relaying
- Communication
- Metering
- Maintenance coordination
- Safety

The interconnection capacity and transfer limits are dependent, not only on the size of the electrical generators, but also on the connecting substation and electrical transmission lines. Electrical utilities may need to perform power-flow, short-circuit, and stability studies to ensure that the cogeneration units can be integrated into the electric grid. The reactive power requirements for voltage control need to be addressed. The equipment ratings and interconnection voltage level may be finalized after these studies.

For system operation during normal and emergency operation, utilities install supervisory control and data acquisition systems. Plant control and communication should interface properly with the utility system. Scheduling of generation, governor response, type of control, load-following capability, and normal and emergency modes of generator operation are operation and control items on which the cogenerator and utility must agree. Protection, relaying, and systems grounding are other issues that need to be resolved. Sometimes the electric power may pass through a third party (wheeling utility). Under such situations, real power loss and revenue metering are important considerations.

The question of safety is directly related to switching. Both plant and utility personnel safety may be at risk due to switching errors. Thus, plant operators and maintenance personnel should coordinate with their utility counterparts. Wherever possible, written procedures, along with sufficient training, should be provided.

17.4.3   Cogeneration Plant Electrical System Design Factors

The plant electric system (including transformers, switchgear, and cables) must be designed for the safety of personnel and equipment, ease of maintenance, and high availability. Equipment should be selected that has the proper and suitable characteristics for the specific application. When selecting equipment, the following factors should be considered:

- Continuous capability
- Emergency rating
- Interrupting rating
- Voltage insulation level
- Environmental conditions
- Ambient conditions

The selected system configuration should provide the highest possible availability that is consistent with cost and safety. Station service should be provided through the auxiliary transformers. These transformers should be placed such that electric power from the utility is available for startup in case of a cogenerating station blackout.

The auxiliary transformer and associated station service system should be able to carry that assigned part of the station load during both normal and outage conditions. Thus, some reserve capacity is needed. The auxiliary substation should contain the necessary switchgear, motor control centers, uninterruptible power supply (UPS) systems, etc.

The generator and main step-up transformer should have proper protection. The plant grounding should be properly designed and installed. The plant communication, lighting, and electrical service should be properly designed.

17.4.4   Other Factors

Selecting a proper technology option for cogeneration is an important factor. The heat and electricity requirements need to be properly matched. Combustion turbines (also called gas turbines), internal combustion engines, steam turbines, and combined cycle plants are some of the prime mover choices available. The technology to be selected depends upon many nonelectrical factors, including heat-to-power ratio, steam quantity, pressure, temperature, etc. The cogeneration plant itself could be custom designed or bought as prepackaged systems, or it could be the product of a range of options between these. A variety of financing methods is available, including whole ownership, partnerships, leaseback, joint venture, and third-party arrangements. The tax consequences and risks involved influence the final decision. Financing through local government, industrial development bonds, etc., lowers the cost of borrowing. When evaluating risk, several items, such as technology risk, fuel cost escalation, availability, capacity factor, utility rates, public service commission attitudes, and investment, should be considered.

During 1992, the U.S. Congress made changes to the PUHCA and allowed the independent power producers (IPP) (also called nonutility generators, or NUG) to generate and sell wholesale power. These IPP-owned generating stations are called *eligible facilities* (EF). The EF owners are not subject to thermal energy limits and are exempt from rate regulation and reporting requirements. Based on the open transmission access concept, EFs may sell to third parties by requesting wheeling through the interposing utility transmission system. But cogeneration enjoys the preference via avoided cost, as compared to EFs competing with other EFs and utilities based on wholesale price. Ownership of EFs by utilities themselves is not prohibited.

EXAMPLE 17.5 *An economic comparison of two cases—namely, with and without cogeneration—is shown in Tables 17.4 and 17.5.*

17.5 PLANT ENERGY MANAGEMENT SYSTEMS

The purpose of an energy management system (EMS) in an industrial or commercial facility is to enable the plant or facility engineer to operate the entire electric system during normal conditions at the least cost and as efficiently as possible. During emergency conditions, the interruption of supply is minimized to the extent possible without jeopardizing safety. Further, restoration of power supply is done as quickly as possible. This basic purpose of energy management may be accomplished with technology ranging from simple time-controlled systems to complex control and monitoring systems. The development of digital, computer-based data collection and new communication

TABLE 17.4 Problem Data

Steam requirement	100	Mbtu/h
Steam system efficiency	82.5%	
Heat input	121.2	Mbtu/h
Natural gas cost	$3.00	/Mbtu
Annual fuel cost	$3,185,136	(1st year)
Peak demand	15	MW
Demand rate	$18.00	/kW-mo.
Annual electric demand cost	$3,240,000	(1st year)
Average demand	10	MW
Energy rate	$0.04	/kWh
Annual electric energy cost	$3,504,000	(1st year)
Annual O&M costs incl. taxes & ins.	$200,000	
Additional data with cogeneration		
Cogeneration added heat input	77.6	Mbtu/h
Additional annual fuel cost	$2,039,328	
Annual O&M costs incl. taxes & ins.	$600,000	
Installed cost of generation	$18.0	million
Interest rate	10%	
Discount rate	8%	
Plant life	15	years
Annual capital recovery	$2,366,528	
Electric energy sold	21900	MWh
Escalation factors (% per year)		
Fuel cost	5%	
O&M	5%	
Demand charge	4%	
Energy charge	5%	

TABLE 17.5 Cogeneration Economic Analysis Example

	Annual costs without cogeneration					Annual costs with cogeneration						Annual savings NPV
Year	Fuel	Electric bill	O&M	Total	Total NPV	Capital recovery	Fuel	Electric sales	O&M	Total	Total NPV	
1	3,185,136	6,744,000	200,000	10,129,136	9,378,830	2,366,528	5,224,464	(876,000)	600,000	7,314,992	6,773,141	2,605,689
2	3,344,393	7,048,800	210,000	10,603,193	9,090,529	2,366,528	5,485,687	(919,800)	630,000	7,562,415	6,483,552	2,606,977
3	3,511,612	7,367,544	220,500	11,099,656	8,811,265	2,366,528	5,759,972	(965,790)	661,500	7,822,210	6,209,522	2,601,743
4	3,687,193	7,700,877	231,525	11,619,595	8,540,750	2,366,528	6,047,970	(1,014,080)	694,575	8,094,994	5,950,062	2,590,688
5	3,871,553	8,049,476	243,101	12,164,130	8,278,702	2,366,528	6,350,369	(1,064,783)	729,304	8,381,417	5,704,252	2,574,451
6	4,065,130	8,414,046	255,256	12,734,433	8,024,853	2,366,528	6,667,887	(1,118,023)	765,769	8,682,161	5,471,234	2,553,618
7	4,268,387	8,795,329	268,019	13,331,735	7,778,939	2,366,528	7,001,281	(1,173,924)	804,057	8,997,943	5,250,213	2,528,726
8	4,481,806	9,194,099	281,420	13,957,325	7,540,708	2,366,528	7,351,346	(1,232,620)	844,260	9,329,514	5,040,446	2,500,262
9	4,705,897	9,611,168	295,491	14,612,555	7,309,916	2,366,528	7,718,913	(1,294,251)	886,473	9,677,663	4,841,241	2,468,675
10	4,941,191	10,047,384	310,266	15,298,841	7,086,324	2,366,528	8,104,858	(1,358,964)	930,797	10,043,220	4,651,954	2,434,370
11	5,188,251	10,503,638	325,779	16,017,668	6,869,703	2,366,528	8,510,101	(1,426,912)	977,337	10,427,054	4,471,985	2,397,718
12	5,447,663	10,980,860	342,068	16,770,592	6,659,833	2,366,528	8,935,606	(1,498,257)	1,026,204	10,830,081	4,300,774	2,359,059
13	5,720,047	11,480,025	359,171	17,559,243	6,456,497	2,366,528	9,382,387	(1,573,170)	1,077,514	11,253,258	4,137,800	2,318,697
14	6,006,049	12,002,153	377,130	18,385,332	6,259,489	2,366,528	9,851,506	(1,651,829)	1,131,389	11,697,595	3,982,575	2,276,914
15	6,306,351	12,548,312	395,986	19,250,650	6,068,608	2,366,528	10,344,081	(1,734,420)	1,187,959	12,164,148	3,834,647	2,233,961
					114,154,945						77,103,398	37,051,547

Total net present value (NPV) 37,051,547

Summary of results:

Savings with cogeneration	$37,051,547
Benefits/cost ratio	2.1
Payback period (years)	4.0
Internal rate of return (before depreciation & taxes)	18.0%

and instrumentation technologies has opened up a wide range of possibilities. A computer-based EMS with appropriate supervisory control and data acquisition capabilities can perform a wide range of functions. The major functions of an EMS may include:

- Monitoring
- Control
- Display
- Alarm
- Communication
- Data storage and retrieval
- Maintenance coordination and record keeping
- Reports

The monitoring function of an EMS includes currents, voltages, power, energy, reactive power, frequency, temperature, status of breakers and switches, fuel quantity used, and reserve margins. The monitoring capability is limited only by the extent of instrumentation, communication, and cost. Because all items need not be measured with the same accuracy and sampling rate, it is possible to set up a different scanning rate and collect a large amount of data. Suitable error checking and procedures for discarding bad data are also available.

The ability to remotely control different activities from one central location is probably the most convenient and cost-saving aspect of an EMS. The control will typically consist of both automatic and operator-directed actions, including load transfer, closing of normally open breakers and opening of normally closed breakers, automatic change of voltage set points, and demand control by changing local generation or load shedding, among many other possibilities. The actual amount of remote switching and control depends upon the need and the cost.

CRT displays can show the operator the status of the system and equipment. The displays may be obtained in graphical or tabular form. System one-line diagrams with actual system operation quantities may be displayed. Use of light pen, keyboard, or cursor-based switching actions is possible. Hard copies of these displays are another benefit.

Alarms to alert operators to the violation of normal or emergency operating limits can be part of an EMS. Further, some type of alarm processing is also possible so that the operator must focus on the most immediate emergency conditions rather than being swamped with minor conditions and multiple alarms.

Data storage and retrieval for analysis, billing, post-disturbance analysis, loading and life calculations, duty cycle, overloading, undervoltage, equipment efficiencies, maintenance, etc., are also possible. Not only is the paperwork avoided, but the quality and quantity of data are also considerably improved.

An EMS may be based on total central, local, or distributed control. The actual configuration depends upon the particular application and preference. Each has its own advantages and disadvantages.

A typical EMS has four major subsystems:

- Hardware
- Software
- Communication
- Instrumentation

The hardware consists of computer, associated display, hard disk, backup, printers, modems, etc. Computer systems may be specific to a vendor or general-purpose micro-processors. Because of the rapid pace of technological change, any selection of computer hardware should permit replacement or modification without major disruption and cost. Further, once an EMS is put into service, the desire and need to add on many other functions will undoubtedly arise. Hence, provision for expansion should be made.

Software essentially runs the EMS, performing all the monitoring, data logging, reporting, etc. The software may include real-time, on-line, and off-line types. The software involved in switching, etc., may require faster execution than off-line can provide. The user should be able to add functions and other analysis software. Different operating conditions, both steady state and transient, should be addressed. Above all, the software should have fail-safe characteristics.

Communication is the link between the computer and the measuring instrumentation, and it should be designed and implemented to meet various needs. The data transmission rate, ability to address from multiple hosts, interference-free transmission, reliability, etc., are all key factors. Local area network (LAN), wide area network (WAN), and interconnection to other systems are some of the possible choices. The communication type may range from simple twisted-pair copper conductors to fiber optics, depending upon the distance, amount of data, the environment, isolation needs, and cost. Some type of multiplexing may be used for economic reasons.

Instrumentation is probably one of the most expensive needs of a monitoring system. However, multiple measurements and/or calculation of many quantities based on a few basic measurements reduces the cost of instrumentation. The instrumentation may not involve only electrical quantities but also others such as temperature, vibration, fuel reserve, cooling liquid temperature, and ambient conditions. The type of instrumentation should properly match the type of data needed, sampling rates, analog/digital conversion, communication, and the computer system. Sampling voltage, current waveforms, and their relationship to each other give the basic information to calculate electric power, reactive power, etc.

There are many vendors supplying EMSs. These include those who supply prepackaged and custom-designed systems and those who mix and match components to integrate into one system. The type of system selected will depend on the type of application, the experience of the vendor in applying the EMS to the particular application, the buyer's technical capability to oversee the selection, the installation and commissioning of the EMS, and, finally, the total cost of such a system.

17.6 BUILDING ENERGY MANAGEMENT SYSTEMS

The preceding discussion was oriented more toward a plant EMS. A similar approach to a building EMS is possible. A building EMS is much simpler. However, the purposes of a building EMS are similar to those of a plant EMS and are restated here. A building EMS should provide for:

- Time-based program control
- Status and alarm monitoring
- Event-initiated program control
- Recording power and energy consumption
- Security control

Time-based program control employs a sophisticated clock timer type of control. Different on/off times for weekdays, weekends, seasons, ambient conditions, etc., can be set. Status and alarm monitoring alerts the facilities personnel to an unacceptable condition in the building, equipment problems, a service outage, etc. The event-initiated programs could include starting an emergency generator or switching safety-related equipment to backup power. Such items as demand limiting and monthly bill calculation could be included. The building EMS and security system could be integrated.

The major energy consumption in a building is the result of heating or cooling; hence, heating, ventilating, and air-conditioning (HVAC) control is one of the main functions of a building energy management system. This could range from a simple clock timer–based control to a sophisticated computer system. A sophisticated building EMS can perform any of the following:

- Automatic functions
- Operator-initiated controls
- Scheduling of operation and maintenance
- Monitoring
- Data collection
- Analysis
- Alarm and security
- Data storage, retrieval, and reporting

Thus, there are EMSs available for industrial plants or buildings, offices, commercial establishments, and other institutions.

17.7 COMPUTER-BASED FLEXIBLE LOAD-BALANCING SYSTEMS

Installation of a flexible load-balancing system would address many of the specific situations faced by an industrial plant, and would be attractive for the following reasons:

- Limiting electrical demand (kW)
- Maintaining supply to critical loads
- Supplying as many loads as possible during outages
- Acquiring and archiving data to analyze disturbances and to meter in-plant energy usage
- Minimizing power supply interruptions to processes
- Selectively shedding load in response to specific contingencies
- Balancing real and reactive power flows
- Maintaining plant generation operating in the desired frequency range and in a stable mode
- Recognizing "hung" frequency conditions and incrementally reducing load to return system to normal conditions
- Balancing the load faster than what is achieved by underfrequency relays, thereby reducing the stresses imposed upon in-plant generation
- Providing a flexible system to accommodate future plant additions and modifications

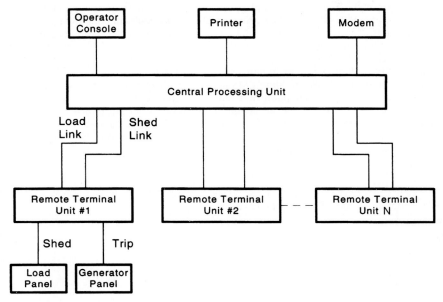

FIGURE 17.4   Example of overall structure and signal flow of the load-balancing system. (*Reproduced with permission from Power Technologies Inc., Schenectady, NY.*)

17.7.1   Hardware Configuration

The load-balancing system is designed in a modular, hierarchical structure, with overall system control being performed using a microprocessor-based computer operating in a multitasking environment. An example of the overall structure and signal flow for a load-balancing system is shown in Fig. 17.4.

From load points to control center, the system consists of the following equipment:

- *Load panels.*   These provide a direct interface to individual load breakers. Their primary purpose is to monitor analog quantities (watts and VARs) and breaker contact status, and to provide load breaker interruption via a trip coil.

- *Source panels.*   These measure the power output of all sources, including in-plant generation and utility ties. They also monitor the source status. Loss of any source is immediately broadcast to the central processing unit.

- *Area remote terminal units (RTUs).*   RTUs receive the analog and contact status information from the load panels and send trip signals to area load breakers based upon load shed table information provided by the central processing unit. All input and output signals are optically isolated. The standard load-balancing system provides for multiplexing of analog and digital signals to the central processing unit. Alternatively, signals can be direct-wired to the central processing unit for increased data transmission speed.

- *Communications.*   Communications between the RTUs and central processing can be either metallic or fiber optic.

- *Central processing unit.*   This serves as the master controller for the load-balancing system. As a multitasking operating system, the central processing unit is responsible for:

interfacing with input/output signals from/to the RTUs

updating load shed tables at periodic intervals

checking critical breaker status and initiating load balancing according to user-specified priorities

providing an interface for the operator

providing local and remote access to enable database modifications by engineering personnel

calculating, storing, and reporting operating history

The exact number of load panels and RTUs is determined from the specific industrial application, with full configuration capability provided in software at the central processing unit.

A flexible load balancing system may be said to perform four primary functions:

1. Intelligent load balancing
2. Alarm and/or logging functions
3. Utility demand prediction and alarming
4. Report preparation

These four primary functions are described in the following sections.

17.7.2 Intelligent Load Balancing

The load-balancing system scans both analog signals and digital contact status at user-selectable rates. Analog signals are scanned at the lower rate (measured in seconds) while the digital contacts are scanned at the higher rate (measured in milliseconds). Analog signals typically consist of load, utility tie, and in-plant generation watts, which are used to predetermine a prioritized shed table for each loss of utility tie or in-plant generation contingency. The priority of each load in the shed table is preset by the operator from the operator console. At each analog scan, the shed tables are updated according to the existing load/supply conditions and the individual load priorities. Loss of multiple sources can also be accommodated.

Digital contact information typically consists of breaker statuses for all utility ties and in-plant generation, and are used to identify a loss of supply contingency. At each digital scan, the status of each supply source is read. If a loss of supply is detected, the appropriate shed table is immediately selected. In those cases where extraordinary speed is required, direct CPU interrupts are generated upon loss of a source. The central processing unit initiates a load balance request to each load in the shed table by sending the appropriate instructions to each of the RTUs. The RTUs in turn send out trip signals to the load panels where the load breaker trip coils are energized through interposing relays.

The load-balancing system can automatically initiate startup of additional in-plant generation, if available, to allow the operator to balance load and supply. Through the console, the operator can manually restore individual loads as the additional in-plant generation comes on-line or the utility tie is reestablished.

Load monitoring. In order to fully specify the conditions under which load balancing is required, it is necessary to monitor the following quantities:

- Contact status of utility tie breakers
- Watt and VAR flow (using watt and VAR transducers) on each utility tie

- Contact status of the breakers associated with the existing in-plant generating units
- Output of each generator
- Reverse power relay contact status on the utility ties

Load balancing.   Load balancing is usually required under the following situations:

1. If load exceeds the preset demand.
2. If utility ties are lost. This is indicated by different possible conditions that can separate the utility ties from the plant.
3. If one line-side utility breaker opens and the bus undervoltage relay contact closes. This situation may occur if one of the ties experiences a fault and primary relaying fails to operate.
4. If systems are self-contained, meaning no separate utility connections such as plants on islands, the load-balancing system may be made to function as an energy management and control system or it may supplement the existing control system.
5. If utility ties are lost and there is a subsequent islanded condition of operation.

A review of the substation configuration is necessary to identify independent groups of feeders and individual loads. A priority list of loads to be dropped/retained is also necessary. Table 17.6 illustrates an example of the relationship between individual load

TABLE 17.6   Example of Load Buses Served by Primary Feeders

(See Fig. 17.5)

| | Feeder identification | | | | | | | | | | | | | | | | | |
| | Sub-A-Feeders | | | | | Sub-B-Feeders | | | | | | | | Sub-C-Feeders | | | | |
Load group	A1	A2	A3	A4	A5	101	102	103	104	105	106	107	108	501	502	511	512	513
16 TO 71	71		16															
3 TO 71	71	3																
10 TO 72				10	72													
24 TO 46							24	46										
35 TO 43							35								43			
35 TO 51							35							51				
28 TO 51								28						51				
51 TO 80						80								51				
41 TO 75											75						41	
21 TO 28								28									21	
41 TO 46									46								41	
28 TO 46								28	46									
73 TO 74												74	73					
56																	56	
58																		58

bus groups and primary feeders. Focusing on the first two load groups (one group fed from buses 16 and 71, the other fed from buses 71 and 3), it is necessary to trip feeders N1, N2, and N3 to ensure that only the desired load group is interrupted.

A simplified one-line diagram of an industrial plant, with monitoring and control points noted on the appropriate feeder loads, is shown in Fig. 17.5.

Load shed control can be at the feeder level, the load substation level, or the individual load level, depending upon the distribution of critical loads and the shed selectivity desired. Due to the interconnected nature of loads on some feeders, some of which may be critical, a combination of feeder and substation load shedding may need to be used. In other cases, loads served by some feeders may be shed as a block. Loads or feeders which are not eligible for shedding do not need to be monitored, because total plant load is determined by the combined power output of the sources.

17.7.3 Alarm and Logging Functions

When a load shed situation occurs, the operator is notified through an audible alarm and a visual alarm condition reported on the operator's console. Alarm messages are also printed on the system printer. The operator must acknowledge the alarm before it can be reset. Alarm messages include:

- RTU off-line
- Remote or control room power down

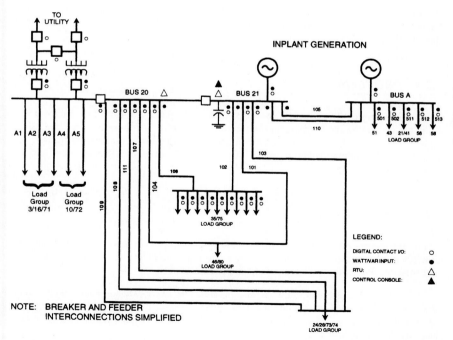

FIGURE 17.5 A simplified one-line diagram with monitoring points. (*Reproduced with permission from Power Technologies Inc., Schenectady, NY.*)

- Communications link bad
- System operating on UPS
- RTU returned to normal operation
- Load balance initiated or recorded—amount and location of load interrupted is identified

In addition to the system failure alarms, the system displays the availability of interruptible load. If sufficient interruptible load is not available for the worst loss-of-supply contingency, the operator is informed so that corrective action can be taken before an emergency occurs.

17.7.4 Utility Demand Prediction and Alarming

The load-balancing system may include a demand prediction and alarming function that permits the recording of utility kW and kWh consumption over a 15- to 30-min interval. From the immediate past loading profile, a demand projection is made for the end of the demand period. In situations where demand limits are set, this logic will alarm the operator and indicate the extent of the load reduction required to remain within the demand limit for the current demand interval. On-screen histograms of kW and kWh utility demand are updated at the analog sampling rate.

17.7.5 Report Preparation

A number of different reports can be prepared, covering any combination of input quantities over any time interval, such as by load point, area, total plant, hour, plant shift, day, week, or month.

17.8 PROBLEM

1. In Example 17.2, (a) if there is a penalty of $1.0 per kW for a load factor below 80 percent, what is the increase in the monthly charge? (b) If there is a reactive power charge of 25¢/kVAR for excess kVAR consumption if the power factor is below 0.9. The maximum kVA demand metered is 3750. Compute excess kVAR consumed and the reactive penalty charges.

17.9 REFERENCES

There are innumerable books written on the subject of energy management, resources, conservation, etc. Books with similar titles and contents offer the reader more detailed information. Presented here are some sample references.

1. L. C. Witte, P. S. Schmidt, and D. R. Brown, *Industrial Energy Management and Utilization,* Hemisphere Publishing Corporation, New York, 1988.

2. T. E. Smith, *Industrial Energy Management for Cost Reduction,* Ann Arbor Science Publishers, Inc., Ann Arbor, Mich., 1979.

3. T. A. Lehr and J. D. Stranahan, *Industrial Energy Management—A Cost Cutting Approach,* Society of Manufacturing Engineers, Dearborn, Mich., 1983.

4. H. B. Zackrison, *Energy Conservation Techniques for Engineers,* Van Nostrand Reinhold Company, Inc., New York, 1984.

5. R. W. Roose, *Handbook of Energy Conservation for Mechanical Systems in Buildings,* Van Nostrand Reinhold Company, New York, 1978.

6. IEEE Std 739-1984, *IEEE Recommended Practice for Energy Conservation and Cost-Effective Planning in Industrial Facilities* (IEEE Bronze Book).

7. J. Katzel, "Ice Thermal Storage Systems," *Plant Engineering,* July 8, 1993, pp. 59–62.

8. G. Polimeros, *Energy Cogeneration Handbook—Criteria for Central Plant Design,* Industrial Press, New York, 1981.

9. R. H. McMahan, *Cogeneration—Why, When and How to Assess and Implement a Project,* Marcel Dekker, Inc., New York, 1987.

10. D. J. Ahner and R. J. Mills, "An Application of Integrated Thermal and Electrical Energy Cogeneration Optimization," presented at *Industrial Energy Technology Conference,* Houston, Texas, April 13–14, 1994.

11. D. J. Ahner and R. B. Priestley. "Combined Cycle Power and Cogeneration Optimization Requirements," presented at *American Society of Mechanical Engineers (ASME-IGTI) Cogeneration Turbo Power 94 Conference,* Portland, Oregon, October 25–27, 1994.

12. *Reliability Considerations for Integrating Non-Utility Generating Facilities with the Bulk Electric Systems,* North American Electric Reliability Council, April 1987.

CHAPTER 18

ELECTRICAL POWER SYSTEM STANDARDS, PUBLICATIONS, AND NAMEPLATES

18.0 INTRODUCTION

The electrical industry thrives on standardization. Standards cover the manufacture, testing, application, installation, and maintenance of electrical devices and equipment. Some standards (UL and NEMA) apply mainly to the manufacturer. Other standards (NEC and NECS) primarily concern the installation of equipment. Some (NEC) are adopted by various governmental authorities and require compliance for any industrial installation. International standards (IEC and ISO) may require compliance for equipment installed outside the United States.

In addition to standards, application guides or recommended practices help users install, maintain, and operate electric systems. Manufacturer instruction books contain information which supplements that contained in standards. Technical societies, universities, manufacturers, and publishers sponsor conferences and seminars to help users keep up to date and maintain technical competence.

18.1 PUBLISHED INDUSTRY STANDARDS

18.1.1 American National Standards Institute (ANSI)

ANSI Sales Department
11 W. 42nd Street
New York, NY 10036
(212) 642-4900
(212) 302-1286 (FAX)

Founded in 1918, The American National Standards Institute (ANSI) is a private, non-profit membership organization that coordinates the U.S. voluntary consensus standards system and approves national standards. ANSI consists of approximately 1300 national and international companies, 30 government agencies, and 250 professional, technical, trade, labor, and consumer organizations including ISO, IEC, IEEE, NEMA, and UL, as well as other national and international standards groups.

In addition to approving standards generated by other organizations, ANSI publishes standards generated by its own committees not necessarily affiliated with other organizations. Two such standards are listed as follows:

ANSI no.	Standard title or description
C34.2	*Practices and Requirements for Semiconductor Power Rectifiers*
C84.1	*Electric Power Systems and Equipment—Voltage Ratings (60 Hz)*

18.1.2 Institute of Electrical and Electronics Engineers (IEEE)

IEEE Customer Service
445 Hoes Lane, P.O. Box 1331
Piscataway, NJ 08855-1331
(800) 678-IEEE in the U.S. and Canada
(908) 981-1393 outside the U.S. and Canada
(908) 981-9667 FAX

The Institute of Electrical and Electronics Engineers (IEEE) was formed in 1962 by a merger of the American Institute of Electrical Engineers (AIEE), founded in 1884, and the Institute of Radio Engineers (IRE), founded in 1912. IEEE is the world's largest technical professional society, consisting of more than 320,000 members who conduct and participate in its activities in 150 countries. Thirty-five societies composed of committees, subcommittees, and working groups write consensus standards, sponsor meetings, conduct seminars, and engage in other professional activities. A Standards Board coordinates the standards activities through more than 20 Standards Coordinating Committees.

Consensus standards written by IEEE entities include three grades: Guides, Recommended Practices, and Standards. Each standards-writing entity is composed of representatives of all parties concerned with the standard: utility, user, consultant, government, plus any involved engineer. These standards are coordinated with ANSI, IEC, and NEMA where such coordination is required. IEEE standards are reviewed every five years and are either revised, reaffirmed, or withdrawn.

In addition to publishing individual standards, IEEE offers collections of standards as well as a color book series of Recommended Practices. The IEEE Press also publishes numerous textbooks and other technical information publications, including the National Electrical Safety Code (NESC), ANSI Standard C2. Standards collections include IEEE, ANSI/IEEE, and ANSI Standards, as well as drafts of proposed standards and, in some cases, other standards pertaining to the collection. Standards collections of interest to the industrial electric power engineer include:

Title	Contents
Circuit Breakers, Switchgear, Substations, and Fuses Stds. Coll. and Supplement (2 vols.).	All C37 series of standards plus Stds. 120, 495, 525, 605, 693, 837, 979, 980, Draft 999, 1030,1031, 1109, 1119.
Distribution, Power, and Regulating Transformers Stds. Coll. (2 vols.)	All C57 series of standards plus Stds. 259, 637, 799.
Electric Machinery Stds. Coll.	Stds. 11, 43, 56, 67,85, 95, 112,113, 115, 115A, 116, 117, 252, 275, 290, 304, 432, 433, 434, 492, 522, 792, 1129.

Title	Contents
Protective Relaying Systems Stds. Coll.	C37 and C57 standards that apply to protective relaying.
Surge Protection Stds. Coll.	C62 standards that apply to surge protection plus Std. 32.
Power Capacitor Stds. Coll.	Stds. 18, 824, 1036, C37.012, C37.99 plus shunt power capacitor bibliography.
Petroleum and Chemical Applications Stds. Coll.	Stds. 303, 463, 515, 576, 841, 844, 1017, 1018, 1019, 1068.

   IEEE Color Books provide guidelines for industrial applications which many industry engineers find helpful in the execution of their responsibilities. The following is a list of the published and in-process Color Books.

IEEE Std no.	Color	Title
141	Red	*Recommended Practice for Electric Power Distribution for Industrial Plants*
142	Green	*Recommended Practice for Grounding of Industrial and Commercial Power Systems*
241	Gray	*Recommended Practice for Electric Power Distribution in Commercial Buildings*
242	Buff	*Recommended Practice for Protection and Coordination of Industrial and Commercial Power Systems*
399	Brown	*Recommended Practice for Industrial and Commercial Power Systems Analysis*
446	Orange	*Recommended Practice for Emergency and Standby Power Systems for Industrial and Commercial Power Systems*
493	Gold	*Recommended Practice for the Design of Reliable Industrial and Commercial Power Systems*
602	White	*Recommended Practice for Electric Systems in Health Care Facilities*
739	Bronze	*Recommended Practice for Energy Conservation and Cost-Effective Planning in Industrial Facilities*
1100	Emerald	*Recommended Practice for Powering and Grounding Sensitive Electronic Equipment*
	Work in progress	
551	Violet	*Guide for Calculating Short-Circuit Currents in Industrial & Commercial Power Systems*
902	Yellow	*Guide for Maintenance, Operation & Safety of Industrial & Commercial Power Systems*
1015	Blue	*Application Guide for Low Voltage Circuit Breakers Used in Industrial & Commercial Power Systems*

Other IEEE standards	
IEEE std. no	Title
100	*IEEE Standard Dictionary of Electrical and Electronic Terms*
499	*Recommended Practice for Cement Plant Electric Drives and Related Electrical Equipment*
625	*Recommended Practices to Improve Maintenance and Safety in the Cement Industry*
428	*Definitions and Requirements for Thyristor AC Power Controllers*
444	*Standard Practices and Requirements for Thyristor Converters and Motor Drives: Part 1—Converters for DC Armature Supplies*
519	*Recommended Practices and Requirements for Harmonic Control in Electric Power Systems*
597	*Practices and Requirements for General Purpose Thyristor DC Drives*
936	*Guide for Self Commutated Converters*

18.1.3 National Electrical Manufacturers Association (NEMA)

NEMA Publication Distribution Center
2101 L Street, NW, Suite 300
Washington, DC 20037
(202) 457-8474
(202) 457-8473 (FAX)

Founded in 1926, NEMA is the largest trade association in the United States, representing the interests of electroindustry manufacturers. More than 600 companies compose the NEMA membership. These companies manufacture products used in the generation, transmission and distribution, control, and end use of electricity. Many of the standards developed by NEMA technical committees are approved by ANSI. NEMA standards are reviewed and updated every five years. Some of the standards applicable to industrial products include the following list.

NEMA no.	Title
	Industrial batteries
1B 1	*Definitions for Lead-Acid Industrial Storage Batteries*
1B 4	*Determination of Amperehour and Watthour Capacity of Lead-Acid Industrial Storage Batteries for Stationary Service*
	Busways
BU 1	*Busways (600 Volts or Less)*
BU 1.1	*Instructions for Handling, Installation, Operation and Maintenance of Busways Rated 600 Volts or Less*
BU 2	*Major Requirements Related to the Application and Testing of Low Voltage Busway*
	Circuit breakers
AB 1	*Molded Case Circuit Breakers*
AB 3	*Molded Case Circuit Breakers and Their Application*
AB 4	*Guidelines for Inspection and Preventive Maintenance of Molded Case Circuit Breakers Used in Commercial and Industrial Applications*

NEMA no.	Title
SG 4	*Alternating-Current High Voltage Circuit Breakers (Above 1000 Volts)*
	Industrial control and systems
ICS 1	*General Standards for Industrial Control Systems*
ICS 1.1	*Safety Guidelines for the Application, Installation and Maintenance of Solid State Control*
ICS 1.3	*Preventive Maintenance of Industrial Control and Systems Equipment*
ICS 2	*Industrial Control Devices, Controllers and Assemblies*
ICS 2.2	*Maintenance of Motor Controllers after a Fault Condition*
ICS 2.3	*Instructions for the Handling, Installation, Operation and Maintenance of Motor Control Centers*
ICS 2.4	*NEMA and IEC Devices for Motor Service—A Guide for Understanding the Differences*
ICS 3	*Industrial Systems*
ICS 3.1	*Safety Standards for Construction and Guide for Selection, Installation and Operation of Adjustable-Speed Drive Systems*
ICS 6	*Enclosures for Industrial Control and Systems*
	Enclosures
250	*Enclosures for Electrical Equipment (1000 Volts Maximum)*
	Fuses
FU 1	*Low Voltage Cartridge Fuses*
FU 2	*High-Voltage Fuses*
	Ground-fault circuit interrupters
280	*Application Guide for Ground Fault Circuit Interrupters*
	Motors and generators
MG 1	*Motors and Generators*
MG 2	*Safety Standard for Construction and Guide for Selection, Installation and Use of Electric Motors and Generators*
MG 3	*Sound Level Prediction for Installed Rotating Electrical Machines*
MG 10	*Energy Management Guide for Selection and Use of Polyphase Motors*
MG 11	*Energy Management Guide for Selection and Use of Single-Phase Motors*
MG 13	*Frame Assignments for Alternating-Current Integral-Horsepower Induction Motors*
RP 1	*Renewal Parts for Motors and Generators (Performance, Selection and Maintenance)*
	Panelboards and distribution switchboards
PB 1	*Panelboards*
PB 1.1	*General Instructions for Proper Installation, Operation and Maintenance of Panelboards Rated 600 Volts or Less*

NEMA no.	Title
PB 2	*Deadfront Distribution Switchboards*
PB 2.1	*General Instructions for Proper Handling, Operation and Maintenance of Deadfront Distribution Switchboards Rated 600 Volts or Less*
PB 2.2	*Application Guide for Ground Fault Protective Devices for Equipment*

Power equipment

SG 6	*Power Switching Equipment (High Voltage)*

Surge arresters

LA 1	*Surge Arresters*

Switches

KS 1	*Enclosed and Miscellaneous Distribution Switches (600 Volts Maximum)*

Switchgear

SG 3	*Low-Voltage Power Circuit Breakers*
SG 5	*Power Switchgear Assemblies (1000 Volts or less including AC network protectors as well as DC low voltage power circuit breakers up to 3200 volts)*

Transformers

ST 1	*Specialty Transformers (Except General Purpose Type)*
ST 20	*Dry Type Transformers for General Applications*

Steam turbines

SM 23	*Steam Turbines for Mechanical Drive Service*
SM 24	*Land Based Steam Turbine Generator Sets 0 to 33,000 kW*

18.1.4 Underwriters Laboratory (UL) and Other Testing Laboratories

Underwriters Laboratory, Inc.
P.O. Box 75330
Chicago, IL 60675-5330
(708) 272-8800, ext. 42612
(708) 272-8129 (FAX)

Factory Mutual Research Corporation, Approvals Division
1151 Boston-Providence Turnpike
Norwood, MA 02062
(617) 762-4300
(617) 762-9375 (FAX)

Canadian Standards Association (CSA)
178 Rexdale Boulevard
Ontario, Canada
M9W 1R3

Met Electrical Testing Laboratory, Inc.
916 W. Patapsco Ave.
Baltimore, MD 21230
(301) 354-2200

The testing laboratories listed here represent some of the testing laboratories authorized
to perform product certification testing. Laboratories may issue their own standards or
test in accordance with UL standards. This section covers UL standards only. Many UL
standards are recognized by ANSI. Some standards of interest to the industrial electric
power engineer are:

UL no.	Title or description
44	*Rubber-Insulated Wires and Cables*
67	*Panelboards*
83	*Thermoplastic-Insulated Wires and Cables*
96	*Lightning Protection Components*
96A	*Installation Requirements for Lightning Protection Systems*
98	*Enclosed and Dead-Front Switches*
198B	*Class H Fuses*
198C	*High-Interrupting-Capacity Fuses, Current-Limiting Types*
198D	*Class K Fuses*
198E	*Class R Fuses*
198F	*Plug Fuses*
198G	*Fuses for Supplementary Overcurrent Protection*
198H	*Class T Fuses*
198L	*D-C Fuses for Industrial Use*
198M	*Mine-Duty Fuses*
347	*High-Voltage Industrial Control Equipment*
363	*Knife Switches*
467	*Grounding and Bonding Equipment*
489	*Molded-Case Circuit Breakers and Circuit Breaker Enclosures*
506	*Specialty Transformers*
508	*Industrial Control Equipment*
512	*Fuseholders*
519	*Impedance-Protected Motors*
547	*Thermal Protectors for Motors*
810	*Capacitors*
840	*Insulation Coordination Including Clearances and Creepage Distances for Electrical Equipment*
845	*Motor Control Centers*
854	*Service-Entrance Cables*
857	*Busways and Associated Fittings*
869	*Service Equipment*

UL no.	Title or description
869A	*Reference Standard for Service Equipment*
877	*Circuit Breakers and Circuit-Breaker Enclosures for Use in Hazardous (Classified) Locations*
891	*Dead-Front Switchboards*
943	*Ground-Fault Circuit Interrupters*
977	*Fused Power-Circuit Devices*
1004	*Electric Motors*
1008	*Automatic Transfer Switches*
1053	*Ground-Fault Sensing and Relaying Equipment*
1054	*Special-Use Switches*
1062	*Unit Substations*
1066	*Low-Voltage AC and DC Power Circuit Breakers Used in Enclosures*
1072	*Medium-Voltage Power Cables*
1077	*Supplementary Protectors for Use in Electrical Equipment*
1087	*Molded-Case Switches*
1236	*Battery Chargers for Charging Engine-Starter Batteries*
1277	*Electrical Power and Control Tray Cables with Optional Optical-Fiber Members*
1363	*Temporary Power Taps*
1429	*Pull-Out Switches*
1449	*Transient Voltage Surge Suppressors*
1558	*Metal-Enclosed Low-Voltage Power Circuit Breaker Switchgear*
1561	*Dry-Type General-Purpose and Power Transformers*
1562	*Transformers, Distribution, Dry-Type—Over 600 Volts*
1564	*Industrial Battery Chargers*
1569	*Metal-Clad Cables*
1581	*Electrical Wires, Cables, and Flexible Cords*
1585	*Class 2 and Class 3 Transformers*
1670	Medium-Voltage Switchgear over 1000 V (tentative title pending publication)
1778	*Uninterruptible Power Supply Equipment*
1917	*Solid-State Fan Speed Controls*

18.1.5 International Electrotechnical Commission (IEC)

IEC Sales Department
P.O. Box 131
3 de Varembé,
1211 Genève 20,
Switzerland
+ 41 22 734 01 50
+ 41 22 733 38 43 (FAX)
(Standards available in U.S. from ANSI Sales Department)

The IEC, founded in 1906, consists of about 200 technical committees and subcommittees responsible for developing international standards within well-defined sectors of electrical technology. IEC composition consists of about 42 national committees representing 80 percent of the world's population that produces and consumes 95 percent of the world's electric energy. The IEC maintains working relationships with some 200 international bodies, both governmental and nongovernmental, as well as the International Organization for Standardization (ISO). National committees represent manufacturers, users, trade associations, government, and the academic and engineering professions.

Some of the IEC standards of interest to the industrial power distribution engineer are the following.

IEC no.	Standard title and description
	Rotating electrical machines
34-1	Part 1: Rating and Performance.
34-2	Part 2: Methods for determining losses and efficiency of rotating electrical machinery from tests (excluding machines for traction vehicles).
34-3	Part 3: Specific requirements for turbine-type synchronous machines. For three-phase machines used as generators, rated 10 MVA and above.
34-4	Part 4: Methods for determining synchronous machine quantities from tests. Machines ≥ 1 kVA, frequency ≤ 400 Hz and ≥ 15 Hz.
34-5	Part 5: Classification of degrees of protection provided by enclosures of rotating electrical machine enclosures.
34-6	Part 6: Methods of cooling.
34-7	Part 7: Symbols for types of construction and mounting arrangements of rotating electrical machinery.
34-8	Part 8 : Terminal markings and direction of rotation of rotating machines.
34-9	Part 9: Noise limits.
34-10	Part 10: Conventions of description of synchronous machines.
34-11	Part 11: Built-in thermal protection.
34-12	Part 12: Starting performance of single-speed three-phase cage induction motors for voltages up to and including 660 V.
34-13	Part 13: Specification for mill auxiliary motors.
34-14	Part 14: Mechanical vibration of certain machines with shaft heights 56 mm and higher—Measurement, evaluation, and limits of vibration severity.
34-15	Part 15: Impulse voltage withstand levels of rotating a.c. machines with form-wound stator coils.
34-16	Part 16: Excitation systems for synchronous machines.
72	Dimensions and output series for rotating electrical machines.
892	Effects of unbalanced voltages on the performance of three-phase induction motors.
	Electrical quantities and miscellaneous standards
38	IEC standard voltages.
59	IEC standard current ratings.
152	Identification by hour numbers of the phase conductors of three-phase electric systems.

IEC no.	Standard title and description
196	IEC Standard Frequencies.
300	Reliability and maintainability management.
364	Electrical installations of buildings. Covers most premises except street-lighting, mines, lightning protection, and radio interference suppression.
781	Application guide for calculation of short-circuit currents in low-voltage radial systems.
865	Calculation of the effects of short-circuit currents.
909	Short-circuit current calculation in three-phase a.c. systems.
949	Calculation of thermally permissible short-circuit currents, taking into account non-adiabatic heating effects.

Steam turbines

45	Specification for steam turbines.
46	Recommendations for steam turbines.
842	Guide for application and operation of turbine-type synchronous machines using hydrogen as a coolant.

Circuit breakers and switchgear

56	High-voltage alternating-current circuit breakers. Voltage > 1000 V, Frequency ≤ 60 Hz.
298	A.C. metal-enclosed switchgear and controlgear for rated voltages above 1 kV and up to and including 52 kV.
466	A.C. insulation-enclosed switchgear and controlgear for rated voltages above 1 kV and up to and including 38 kV.
517	Gas-insulated metal-enclosed switchgear for rated voltages of 72.5 kV and above.
934	Circuit breakers for equipment.
947-1	Low-voltage switchgear and controlgear, Part 1: General rules. Voltages: ≤ 1000 V a.c., 1500 V d.c.
947-2	Part 2: Circuit breakers.

Relays

255	Electrical Relays. Includes several parts.

Industry control

466	A.C. insulation-enclosed switchgear and controlgear for rated voltages above 1 kV and up to and including 38 kV.
470	High-voltage alternating current contactors.
644	Specifications for high-voltage fuse links for motor circuit applications.
947-1	Low-voltage switchgear and controlgear Part 1: General rules. Voltages: ≤ 1000 V a.c., 1500 V d.c.
947-4-1	Low-voltage switchgear and controlgear Part 4: Contactors and motor starters. Voltage ≤ 1000 V.

Fuses

269	Low-voltage fuses. Voltages ≤ 1000 V a.c. or 1500 V d.c. This standard consists of several parts.

IEC no.	Standard title and description
282-1	Part 1: Current limiting fuses. Voltages > 1000 V, frequencies = 50, 60 Hz.
282-2	Part 2: Expulsion and similar fuses. Voltages > 1000 V, frequencies = 50, 60 Hz.
787	Application guide for the selection of fuse-links of high-voltage fuses for transformer circuit applications.

Power transformers

76-1	Part 1: General.
76-2	Part 2: Temperature rise.
76-3	Part 3: Insulation levels and dielectric tests.
76-4	Part 4: Tappings and connections.
76-5	Part 5: Ability to withstand short circuit.
214	On-load tap-changers.
354	Loading guide for oil-immersed power transformers.
542	Application guide for on-load tap-changers.
606	Application guide for power transformers.
726	Dry-type power transformers.
905	Loading guide for dry-type power transformers.

Reactors

289	Reactors. Includes most types except small reactors < 2 kVAR single-phase, 10 kVAR three-phase, motor starting, high frequency line traps or mounted on rolling stock.

Surge arresters and lightning protection

99-1	Non-linear resistor type gapped surge arresters for a.c. systems.
99-2	Expulsion-type lightning arresters.
99-4	Metal-oxide surge arresters without gaps for a.c. systems.
1024	Protection of structures against lightning.

Switches

129	Alternating current disconnectors (isolators) and earthing switches. Voltages > 1000 V, ≤60 Hz.
265-1	Part 1: High-voltage switches for rated voltages above 1 kV and less than 52 kV.
265-2	Part 2: High-voltage switches for rated voltages 52 kV and above.
420	High-voltage alternating current switch-fuse combinations.
947-3	Part 3: Switches, disconnectors and fuse-combination units. Voltages: 1000 V or less.

Semiconductor equipments (drives)

146	Semiconductor convertors. This standard consists of several parts including Part 1-2: Application guide.
147	Essential ratings and characteristics of semiconductor devices and general principles of measuring methods. This standard consists of several parts including Part 1: Essential ratings and characteristics and Part 4: Acceptance and reliability.

IEC no.	Standard title and description
	Cables
227	Polyvinyl chloride insulated cables of rated voltage up to and including 450/750 V. Includes several parts.
228	Conductors of insulated cables. 0.5 mm^2 to 2000 mm^2.
230	Impulse tests on cables and their accessories.
245	Rubber insulated cables of rated voltages up to and including 450/750 V. Includes several parts.
287	Calculation of the continuous current rating of cables (100% load factor). All a.c. voltages, up to 5 kV d.c.
502	Extruded solid dielectric insulated power cables for rated voltages from 1 kV up to 30 kV.
702	Mineral insulated cables with a rated voltage not exceeding 750 V.
986	Guide to the short-circuit temperature limits of electric cables with a rated voltage from 1.8/3 (3.6) kV to 18/30 (36) kV.
	Capacitors and capacitor fuses
70	Power capacitors. All voltages, frequencies ≤100 Hz.
549	High-voltage fuses for the external protection of shunt power capacitors.
593	Internal fuses and internal overpressure disconnectors for shunt capacitors.
831	Shunt power capacitors of the self-healing type for a.c. systems having a rated voltage up to and including 660 V.
871	Shunt capacitors for a.c. power systems having a rated voltage above 660 V.
931	Shunt power capacitors of the non-self-healing type for a.c. systems having a rated voltage up to and including 660 V.

18.1.6   International Organization for Standardization (ISO)

Case postale 56
CH-1211 Genève 20
Switzerland

ISO is a nongovernmental, specialized international agency for standardization (established in 1947), with members consisting of standards bodies of some 90 countries representing 95 percent of the world's industrial production. ISO develops worldwide standards to improve international communication and collaboration, in addition to promoting the smooth and equitable growth of international trade.

The most controversial series of standards from ISO is ISO 9000, a series of quality management standards adopted by ANSI as ANSI/ASQC Q90-94 (available from the ANSI Sales Department). While it is not necessary to be ISO 9000–registered to sell goods in the European Common Market, it is sometimes helpful.

18.1.7   European Country Standards

Many European nations issue standards which apply to their countries. Two organizations that attempt to coordinate these standards are the European Committee on Standardization (CEN) and the European Committee for Electrotechnical Standardization (CENELEC).

18.1.8   National Electrical Code (NEC) ANSI/NFPA 70

National Fire Protection Association
1 Batterymarch Park
P.O. Box 9146
Quincy, MA 02269-9959

ANSI/NFPA Standard 70, the National Electrical Code (NEC), has been sponsored by the National Fire Protection Association since 1911. The original code document was developed in 1897 as a result of the combined efforts of various insurance, electrical, architectural, and allied interests.

Responsibility for revisions to the NEC every three years rests with the NEC Committee, made up of 20 Code Panels plus a Correlating Committee. Each Code Panel consists of representatives from all interested sectors of the electrical industry, including users, electrical utilities, manufacturers, contractors, unions, inspectors, insurance companies, the IEEE, testing laboratories, municipalities, and others.

The NEC is a consensus standard adopted by many governmental entities, including the Occupational, Safety and Health Administration (OSHA) of the U.S. government. The purpose of the NEC is "the practical safeguarding of persons and property from the hazards arising from the use of electricity" [Article 90-1 (a)]. It is "not intended as a design specification nor an instruction manual for untrained persons" [Article 90-1 (c)]. The NEC covers all electric conductor and equipment installations, public and private: domestic, commercial, industrial, mobile homes, floating buildings, and other premises such as yards and carnival, parking, and other lots. It does not cover ships, watercraft other than floating buildings, railway rolling stock, aircraft, automotive vehicles other than mobile homes and recreational vehicles, or mines and mining machinery. The NEC also does not cover installations by railways for generation, transmission, or distribution of power for operation of rolling stock or installations used exclusively for signaling or communication, nor are installations under the control of electrical utilities covered when they are used exclusively for communication, metering, generation, control, transformation, transmission, or distribution. Communication installations by communication utilities are also not covered.

The NFPA publishes other standards and reference books of use to the industrial electric power distribution engineer, listed as follows:

NFPA no.	Title
	Standards
70B	*Electrical Equipment Maintenance*
70E	*Electrical Safety Requirements for Employee Workplaces*
75	*Protection of Electronic/Data Processing Equipment*
77	*Static Electricity*
79	*Electrical Standard for Industrial Machinery*
110	*Emergency and Standby Power Systems*
	Other Publications
70HB96	*1996 NEC Handbook*
HLH-88	*Electrical Installations in Hazardous Locations*
RES-47	*OSHA Electrical Regulations Simplified*
RES-32	*Designing Electrical Systems*
RES-23	*American Electrician's Handbook*
RES-34	*Electrical Wiring: Industrial*

18.1.9   National Electrical Safety Code (NESC) ANSI C2

IEEE Customer Service
445 Hoes Lane, P.O. Box 1331
Piscataway, NJ 08855-1331
(800) 678-IEEE in the U.S. and Canada
(908) 981-1393 outside the U.S. and Canada
(908) 981-9667 (FAX)

ANSI C2, the National Electrical Safety Code (NESC), originated in 1913 at the National Bureau of Standards (NBS). In January 1973, the IEEE was designated as administrative secretariat for C2, assuming functions formerly performed by the NBS.

This standard covers basic provisions for the safeguarding of persons from hazards arising from the installation, operation, and maintenance of (1) conductors and equipment in electric supply stations and (2) overhead and underground electric supply and communication lines. It also contains work rules for the construction, maintenance, and operation of electric supply and communication lines and equipment. The NESC applies to the systems and equipment operated by utilities, or similar systems and equipment, of an industrial establishment or complex under the control of qualified persons.

Standards committee membership includes representatives from the utility industry, insurance industry, municipalities, the IEEE, television industry, NEMA, railroad and transit industry, Edison Electric Institute (EEI), liaisons to the Canadian Standards Association (CSA) and Canadian Electrical Code, labor unions, the U.S. government, and individual members. The Standards Committee consists of eight subcommittees each responsible for certain sections of the code:

Subcommittee	Sections	Subjects
1	1, 2, 3	Purpose, scope, application, definitions, references
2	9	Grounding methods
3	10–19	Electric supply stations
4	23	Overhead lines—clearances
5	24–26	Overhead lines—strength and loading
6	20–21, 27	Overhead lines—general and insulation
7	30–39	Underground lines
8	40–44	Work rules

Procedures have been established by the committee to revise and reissue the code on a three- to four-year cycle. Some editions and parts of the code have been adopted by state and local governmental jurisdictional authorities.

18.1.10   Local Codes

Several jurisdictional authorities throughout the United States require compliance with locally issued codes. Most of these codes incorporate the provisions of the National Electrical Code and add certain provisions determined to be required in the jurisdiction indicated. Examples of jurisdictions where such codes apply are the State of California, City of Chicago, and the City of Los Angeles.

18.2   TRADE PUBLICATIONS

Trade publications provide a good way to stay up to date on the new products and procedures prevalent in the electrical industry through advertisements and technical articles. Some of the magazines of interest to the industrial electric power distribution engineer are:

Plant Engineering
Cahners Publishing Company
1350 East Touhy Ave.
Des Plaines, IL 60018

EC&M (Electrical Construction & Maintenance)
P.O. Box 12960
Overland Park, KS 66282-2960

Consulting-Specifying Engineer
P.O. Box 7525
Highlands Ranch, CO 80126-9325

Power
McGraw-Hill
11 West 19th Street
New York, NY 10011

Power Engineering
P.O. Box 1440
Tulsa, OK 74101-1440

Electrical World
McGraw-Hill
11 West 19th Street
New York, NY 10011

18.3   MANUFACTURER'S INSTRUCTION BOOKS

Manufacturer's instruction books not only contain information required to operate a particular device or equipment but also often contain application tips, maintenance recommendations, renewal parts lists as well as warnings concerning certain operating hazards. Upon receipt of new equipment, early perusal of the instruction book can save considerable time in understanding some of the installation, operating, and maintenance procedures necessary to avoid future troubles.

18.4   TECHNICAL MEETINGS

Technical meetings and seminars sponsored by technical societies, publishers, government agencies, manufacturers, and regional technical groups offer a forum for the industrial electric power engineer to meet and mingle with his or her peers as well as keep up to date on new industry developments. Some of these meetings are held on a regular periodic basis.

Some of the technical meetings sponsored by IEEE are:

Industrial Application Society (IAS)

Sponsor	Usual Time of Year
Applied Power Electronics Conference	Mid-February
Rural Electric Power Technical Conference	Late April
Rubber & Plastics Industry Technical Conference	Late April
Textile, Film & Fiber Technical Conference	Early May
Appliance Industry Technical Conference	Early May
Industrial & Commercial Power Systems Technical Conference	Early May
Cement Industry Technical Conference	Mid-May
Pulp and Paper Industry Conference	Early June
Petroleum and Chemical Industry Conference	Late September
Industry Application Society Annual Meeting	Early October

Power Engineering Society (PES)

Power Engineering Society Winter Meeting	Mid-January
Power Engineering Society Summer Meeting	Mid-July

The meetings listed are held in different parts of the United States (and sometimes abroad) to permit the largest spectrum of attendance.

18.5 EQUIPMENT NAMEPLATE DATA

Nameplates identify a particular piece of electric apparatus and specify its rating. In addition, nameplate data contains information helpful for correct equipment operation and maintenance. This information may be an aid for determining the suitability of apparatus for application under changing power system conditions or if the apparatus is relocated on the same or a different electric power system. Some essential information not presented on nameplates must be assumed or obtained elsewhere.

Table 18.1 lists ANSI/IEEE Standards which contain requirements for nameplates plus the equivalent IEC Standard.

18.5.1 Transformers

The ANSI/IEEE Standard specifies three basic types of nameplates plus additional requirements for each particular type of transformer. Identification information required is the manufacturer's name, a serial number, and identification of the insulating medium. Rating information given is kVA, frequency, number of phases, voltage, temperature rise, and percent impedance. Basic insulation level (BIL) is required for all transformers with a BIL less than 150 kV.

Information given that is helpful for correct operation includes tap voltages, phasor diagram (for polyphase transformers), polarity (for single-phase transformers), a con-

TABLE 18.1 Applicable ANSI/IEEE Standards for Industrial Power Distribution Equipment Containing Nameplate Requirements with Cross-Reference to Equivalent IEC Standards

ANSI/IEEE	Subject	IEC
C37 Series	*Power Switchgear*	
C-37.04-1979	Rating Structure of AC High-Voltage Circuit Breakers Rated on a Symmetrical Current Basis	56
C37.13-1990	Low-Voltage AC Power Circuit Breakers Used in Enclosures	157-1
C37.20.1-1987	Metal-Enclosed Low-Voltage Power Circuit-Breaker Switchgear	947-1
C37.20.2-1987	Metal-Clad and Station Type Cubicle Switchgear	298
C37.20.3-1987	Metal-Enclosed Interrupter Switchgear	947-3
C37.29-1981	Low-Voltage AC Power Circuit Protectors Used in Enclosures	—
C37.30-1992	Definitions and Requirements for High-Voltage Air Switches	265-1
C37.42-1989*	Distribution Cutouts and Fuse Links— Specifications	282-2
C37.44-1981*	Specifications for Distribution Oil Cutouts and Fuse Links	420
C37.45-1981*	Specifications for Distribution Enclosed Single-Pole Air Switches	265-1
C37.46-1981*	Specifications for Power Fuses and Disconnecting Switches	282-2
C37.47-1981*	Specifications for Fuse Disconnecting Switches, Fuse Supports, and Current-Limiting Fuses	282-1
C37.66-1969*(R1982)	Requirements for Oil-Filled Capacitor Switches for Alternating-Current Systems	—
C57 Series	*Transformers, regulators, and reactors*	
C57.12.00-1993	Liquid Immersed Distribution, Power, and Regulating Transformers	76-1
C57.12.01-1988†	Dry-Type Distribution and Power Transformers Including Cast-Coil Transformers	726
C57.12.20-1988*	Overhead-Type Distribution Transformers, 500 kVA and Smaller: HV ≤34.5 kV; LV ≤7.97/13.8 kV	
C57.12.21-1980*	Pad-Mounted, Compartmental-Type, Self-Cooled, Single-Phase Distribution Transformers with HV Bushings; HV ≤34.5 GrdY/19.92 kV; LV, 240/120 V; ≤167 kVA	
C57.12.22-1989*	Pad-Mounted, Compartmental-Type, Self-Cooled, 3-Phase Distribution Transformers ≤2500 kVA: HV ≤34.5 GrdY/19.92 kV; LV ≤480 V	
C57.12.50-1981 (R1989)*	Ventilated Dry-Type Transformers, 1-500 kVA, Single-Phase, and 15-500 kVA, 3-Phase, HV 601-34500 V, LV 120-600 V	

(Continued)

TABLE 18.1 Applicable ANSI/IEEE Standards for Industrial Power Distribution Equipment Containing Nameplate Requirements with Cross-Reference to Equivalent IEC Standards (*Continued*)

ANSI/IEEE	Subject	IEC
C57.12.51-1981 (1989)*	Ventilated Dry-Type Transformers, ≥501 kVA, 3-Phase, and 15-500 kVA, 3-Phase, HV 601-34500 V, LV 208Y/120-4160 V	
C57.12.52-1981 (1989)*	Sealed Dry-Type Transformers, ≥501 kVA, 3-Phase, and 15-500 kVA, 3-Phase, HV ≤601-34500 V, LV 208Y/120-4160 V	
C57.15-1986	Step-Voltage and Induction-Voltage Regulators	
C57.16-1958‡(R1971)	Current Limiting Reactors	289
C57.21-1990	Shunt Reactors Rated Over 500 kVA	289

*ANSI Standard only

†IEEE Standard only

‡Standard withdrawn by ANSI but copies available.

nection diagram, and a reference to an instruction book or sheet. Maintenance information includes total weight for smaller transformers and specified component and total weights, as well as gallons of the required insulating liquid for larger transformers.

Performance information, including efficiency and losses, does not normally appear on transformer nameplates. This information can be obtained from either a specific test report or a representative test report. The X/R ratio can be calculated from test reports or from estimates in ANSI/IEEE Standard C37.010 for use in short-circuit calculations. Maximum rated ambient temperature is given in ANSI/IEEE Standard C57.12.00, which also gives insulation derating factors for altitude. Loading guides are given in ANSI/IEEE Standards C57.91, C57.92, C57.94, and C57.96. For operation at other than rated frequency or other unusual requirements, consult the manufacturer.

18.5.2 Switchgear

ANSI/IEEE Standards C37.20.1, C37.20.2, and C37.20.3 require that each switchgear assembly nameplate give the manufacturer name, address, type designation (optional), identification reference, plus, where applicable, rated frequency and maximum voltage. The standards do not specify the location of this nameplate. On newer equipment, this equipment nameplate usually is mounted on the front of the switchgear assembly. On older equipments, it is sometimes found on the wiring grille inside the left-hand end unit for metal-clad equipment or a bus support for low-voltage equipment. From the information given on the equipment nameplate, the manufacturer can ascertain the applicable drawings: front view and floor plan, elementary and wiring diagrams, summary or bill of material and instruction books.

Circuit-breaker nameplates usually are easy to locate, although some may be oriented horizontally and the inscriptions obscured by dust or other accumulation. Modern high-voltage circuit-breaker nameplates may combine operating mechanism and circuit-breaker rating information on one nameplate. Older oil circuit breakers usually have separate mechanism and circuit-breaker nameplates. Sometimes the circuit-breaker nameplate is located on or near the tank top or arc chute support, horizontal or vertical, accessible or inaccessible. Low-voltage circuit-breaker nameplates contain only circuit-breaker information. Trip devices have separate nameplates. Fuses have labels giving the appropriate information.

All nameplates require that the maximum voltage, interrupting rating, and continuous current rating appear on the nameplate or label. For high-voltage circuit breakers, the interrupting capability must be calculated from the rating for the specific application voltage in accordance with the procedure outlined in Chap. 7.

Year of manufacture is required on all modern circuit-breaker nameplates. Instruction book numbers and weights are required only on high-voltage circuit-breaker nameplates. Most low-voltage power circuit-breaker nameplates list an instruction book number. High-voltage switch nameplates must contain a rated short-time current in addition to the manufacturer's name, address, type, designation number, plus-rated voltage, continuous current, BIL, and frequency. In addition to these items, the ANSI/IEEE Standards require interrupter switches to show the rated interrupting current and some other specified rating information where applicable.

18.5.3 Motors

Motor nameplates are prominently located on the motor frame. In many installations of motors in special equipments, however, reading the nameplates is a problem because of interference from other equipment.

Motor nameplate information is specified by NEMA Standard MG1 and ANSI/NFPA 70 (NEC).

ANSI/IEEE Standards C50.13 and C50.14 cover only cylindrical-rotor synchronous generators. In addition to the manufacturer's name, rated motor volts and full-load amperes, rated frequency, full-load speed, and rated horsepower, the NEC requires a time rating and code letter to appear. The time rating specifies a short-time or a continuous rating. The code letter signifies a range of locked-rotor kVA per horsepower. From this information, a range of full-voltage starting currents can be estimated. Since short-circuit reactance is approximately equal to full-load current divided by full-voltage locked-rotor current, a range of short-circuit reactances can be ascertained.

Article 430-7 of the NEC permits a choice for nameplate information between rated temperature rise or insulation system class and rated ambient temperature. Generally, larger than NEMA frame-size motors show rated temperature rise. By referring to NEMA Standard MG1 (Section 20.40), the rated ambient temperature and insulation rise can be determined. Nameplates for NEMA-size motors usually display both rated ambient temperature and insulation class.

Motor nameplate identification includes manufacturer's name, model or catalog number, and/or serial number. Electrical ratings displayed are output horsepower, input voltage, current in amperes, locked-rotor code letter, frequency, rpm, and phase. Field current and voltage are shown for synchronous motors. Secondary current and voltage are shown for wound-rotor motors.

One important datum not included on the nameplate is inertia information (Wk^2). The motor rotor inertia varies from motor to motor in relation to the load inertia. For any particular motor, the motor inertia must be obtained from the manufacturer because of its absence on the nameplate. Normal load inertia for specific frame motors is given in NEMA Standard MG1. Comparing this to the actual load inertia, a multiple of load inertia can be determined. In general, fans have high load inertia and pumps or M-G sets have low load inertia.

To aid motor operation and maintenance, larger motors usually are equipped with special-purpose or warning nameplates covering space heaters, bearing oil level, phase sequence, rotation direction, and any special starting directions. Special-purpose nameplates usually are grouped near the rating nameplate. Also, rating nameplates for larger motors usually contain an outline drawing number, connection diagram number, and instruction book number.

Ac motors do not show the type of enclosure; dc motors do indicate the enclosure type on the rating nameplate.

18.5.4 Reactors and Voltage Regulators

Reactor and voltage regulator manufacturers usually follow the practices of transformer manufacturers concerning information displayed on nameplates, even though the applicable standards are not as detailed for required nameplate information. This is particularly true for liquid-filled equipments.

One particular rating item which may be confusing is the kVA rating of this equipment. Usually this rating is the physical kVA of the equipment, not the through-kVA. On voltage regulators of 1000 V and over, the rated kVA is equal to the maximum per-unit of the maximum voltage change times the maximum current. For three-phase equipments, this product is multiplied by the square root of three. Voltage regulators below 1000 V usually are rated in circuit through-kVA.

For example:

1. Three-phase step-voltage regulator

   10 percent raise and lower

   107 maximum A

   27,000 V

$$kVA = 0.10 \times 27\ kV \times 107\ A \times \sqrt{3} = 500\ kVA$$

2. Three-phase current-limiting reactor

   1.5 percent V drop (0.2 Ω)

   600 A

   13,800 V

$$kVA = 0.015 \times 13.8\ kV \times 600\ A \times \sqrt{3} = 215\ kVA$$

Reactor designers usually arrive at the three-phase kVA by the formula: $3 \times I^2 X \times 1/1000$. (For the reactor in this example, the kVA $= 3 \times 600^2 \times 0.2 \times 1/1000 = 216$ kVA).

Voltage regulators under 1000 V, regulating transformers, and regulator and transformer combinations are usually rated in continuous through-kVA as indicated on the nameplate.

The *volts drop* shown on a current-limiting reactor is not, except in unique circumstances, the difference between the reactor input and output voltages in the circuit under normal operating conditions. It is the impedance voltage of each phase of the reactor expressed as a percentage of rated system line-to-neutral voltage when carrying rated current. In other words, it is the reactor impedance on the rated through-kVA base.

Motor-starting reactors are not specifically covered by ANSI/IEEE standards. Since they are designed for specific motors, motor-starting reactor nameplates refer to the specific motor for which they are designed.

Other special-purpose reactors such as neutral grounding, smoothing, interface, filter, tuning, or subsynchronous reactors have nameplate ratings consistent with their application. All these reactors show an impedance either in henries or ohms at a specified voltage and frequency.

18.5.5 Drives and Industry Control

This equipment, like switchgear, has a number of nameplates on individual devices plus an equipment nameplate. The equipment nameplate is usually displayed in an accessible location on the front of the enclosure. It usually gives an elementary or schematic diagram number. From this drawing, the bill of material, the physical location on the panel, and individual device identity can be determined. Each device in the equipment has a nameplate which corresponds to the identification on the schematic diagram. Each panel is identified with a panel nameplate showing the function of the panel.

Individual motor controllers are equipped with nameplates describing the controller in some detail. High-voltage fused motor controller nameplates are usually located on the outside of the controller enclosure but sometimes are accessible only by opening the controller door. Low-voltage controller nameplates usually are accessible only by opening the controller door.

Motor controllers are covered by NEMA Standard ICS 2, and motor control centers by NEMA Standard ICS 2.3. Several UL standards cover controllers and the circuit breakers or fuses contained therein (see Sec. 18.1.4 for a listing).

High-voltage fused controllers usually show the following:

1. Manufacturer's name, type of controller, and catalog number
2. Diagram and instruction book number
3. Power fuse type, size, and catalog number
4. Control volts
5. HP, phase, frequency, voltage, and full-load current plus PF (for synchronous motor) or secondary current in amperes (for wound-rotor motor)

Current transformer ratio and overload relay information can be determined from reference diagrams or individual device nameplates.

Low-voltage motor controller nameplates usually give:

1. Manufacturer's name, type, serial number, and size
2. Diagram and instruction book number
3. Control volts
4. Horsepower or current, voltage, and frequency

Current transformer ratio, fuse size, circuit-breaker information, overload relay information, and type of motor controlled can be determined from reference diagrams. Obtaining information from device nameplates may be difficult since devices may be mounted in a configuration which partially obscures its nameplate.

INDEX

ABOUT THE AUTHORS

F. S. Prabhakara is a Senior Engineer at Power Technologies, Inc. He has 30 years of consulting engineering experience in power system planning, designing, and performing system studies. He is a Registered Professional Engineer and a Senior Member of the Institute of Electrical and Electronics Engineers (IEEE).

Robert L. Smith, Jr. is an Associate in the Technology Assessment Group at Power Technologies, Inc., as well as Volts and Vars I (his own consulting firm). He previously worked for more than 40 years at the General Electric Company, and he has extensive industrial power distribution experience with low-, medium-, and high-voltage electrical and electronics systems. He is a fellow of the Institute of Electrical and Electronics Engineers (IEEE).

Ray P. Stratford is an Associate in the Technology Assessment Group and retired as a Senior Consultant at Power Technologies, Inc. He previously worked for more than 40 years at the General Electric Company and is an expert on the problems of industry with both utilization and power supply equipment. He is a Registered Professional Engineer and a Fellow of the Institute of Electrical and Electronics Engineers (IEEE).